CROP IMPROVEMENT IN THE ERA OF CLIMATE CHANGE

CROP IMPROVEMENT IN THE ERA OF CLIMATE CHANGE

Editor

RAJIB ROYCHOWDHURY

Dept. of Biotechnology
Visva-Bharati, Santiniketan
West Bengal, India

I.K. International Publishing House Pvt. Ltd.

NEW DELHI • BANGALORE

Published by

I.K. International Publishing House Pvt. Ltd.
S-25, Green Park Extension
Uphaar Cinema Market
New Delhi–110 016 (India)
E-mail:info@ikinternational.com

ISBN 978-93-82332-61-9

Published by Krishan Makhijani for I.K. International Publishing House Pvt. Ltd., S-25, Green Park Extension, Uphaar Cinema Market, New Delhi–110 016 and Printed by Rekha Printers Pvt. Ltd., Okhla Industrial Area, Phase II, New Delhi–110 020.

Preface

From the ancient time period to recent one, human beings depend on agriculture to cultivate and produce various crop plants (food and non-food based) for their civilization and survivability. But in the present scenario, climate change and global population increase jointly challenge the sustainable agriculture and crop productivity. One of the imposing tasks for the crop researchers globally is to diminish the negative effects of climate change on the performance of crops, especially in relation to yield and quality of harvested products. This is of special significance in view of the impending climate change, with complex consequences for economically profitable and both ecologically and environmentally sound global agriculture. The challenge at the hands of the crop scientist in such a scenario is to promote a competitive and multifunctional agriculture, leading to the overall crop improvement. Hence, crop improvement is an urgent need for us. As a result of continuous climatic changes in the form of environmental hazards, abiotic stresses (water scarcity, salinity, heavy metals, oxidants and others), global warming, etc., natural and cultivated habitats of crop plants are continuously disturbed. Crop productivity and crop biology are seriously affected by such changing pattern. Beyond such kind of hazardous barriers, this book is aimed at gross crop improvement through agricultural processes, soil reclamation, fertilizer and biofertilizer application, organic farming, hybridization and breeding programme and biotechnological approaches including genetic engineering, transgenesis, tissue culture and different 'omic' technologies. Thus, it covers a wide range of topics under present-day environmental challenges, agronomy and agriculture processes, and biotechnological approaches. The book focusses on broad research areas in relation to crop improvemental aspects which are in the forefront in contemporary crop based research in this changing climate. Different diverse chapters in this book are contributed by high dignity and experienced scientists, researchers, and academicians of both national and international reputation, from around the world who has worked on the challenging problems of achieving crop improvement in the variable environments. Every chapter is written in such a way that it deals about the theoretical as well as applied aspects of regarding themes. Hope such hot topics will be helpful to formulate the future keys toward the crop improvement in the present challenging and changing climate.

Words are too inadequate to express my feelings of indebtedness and deep sense of gratitude to respected Dr. N. Dey (Dept. of Biotechnology, Visva Bharati) and Dr. Jagatpati Tah (Botany Dept., The University of Burdwan) for their motivation and continuous support for making this book. Acknowledgement is extended to Prof. Parthadeb Ghosh (Botany Dept., University of Kalyani), Prof. Nirmalya Banerjee (Botany Dept., Visva-Bharati), Prof. Aloke Bhattacharya, Dr. Tushar Kanti Maiti, Dr. Abhijit Bandyopadhyay and Dr. Tinkari Dalal (Botany Dept., The University of Burdwan), Dr. Khalid Rehman Hakeem (Dept. of Forestry, Universiti

Putra Malaysia) and Dr. Mirza Hasanuzzaman (Dept. of Agronomy, Sher-e-Bangla Agricultural University). I express my profound regards to my parents whose heartiest wishes, endless blessing and special interest have led to this glorious exposure. I am not unmindful of my responsibility to acknowledge for constant help and cooperation directly extended by my intimate and nearer colleagues, namely, Upasana Banerjee, Subhankar Chakraborty, Saptarshi Mondal, Suman Kundu, Bibhas Bhunia, Moumita Das, Reha Labar, Subhrajyoti Mondal, Joydip Karmakar, Amrita Paul, Subhajit Galguly, Nibedita Chakraborty, Nisha Patwa, Sonali Bej, Anindya Pattyanayak, Molla Jannatul Ferdosul Alam, Sunanda Bishnu, Anuj Mamgain, Showkat Ahmad Ganie and others. I am grateful to all the contributors who gave their valuable time to make chapters for the possible success of this book. I shall be highly thankful to the readers for pointing out the errors and omissions which in spite of all care might have crept in. All the suggestions from anybody for further improvement of this edited volume will be highly appreciated and accepted. I, along with the contributors, apologize to those researchers for those parts of their work that were not cited in this edited volume/chapters due to the space limitation.

Hope, this book will be useful for the scientists, researchers and students in traditional and interdisciplinary streams related to plant biology, agriculture, biotechnology and environmental science.

Rajib Roychowdhury

E-mail: rajibroychowdhurybu@gmail.com

Contributors

Abdelmoumen Taoutaou. Department of Molecular Biology and Biotechnology, University Babes-Bolyai, Clinicilor 5-7, Cluj-Napoca, Romania. *E-mail: tamoumen@gmail.com*

Anindya Pattanayak. Integrated Science Education and Research Centre (ISERC), Visva-Bharati, Santiniketan-731235, West Bengal, India. *E-mail: pattanayakanindya07@gmail.com*

Arvind K Shukla. Indian Institute of Soil Science, Bhopal, Madhya Pradesh, India. *E-mail: akshukla@iiss.ernet.in*

Ashim Chakravorty. Cytogenetics and Plant Breeding Section, Department of Botany, University of Kalyani, Kalyani-741235, Nadia, West Bengal, India. *E-mail: ashimbot@gmail.com*

Ashish Kumar. Department of Pharmaceutical Sciences, BIT Mesra, Ranchi-835215, Jharkhand, India. *E-mail: ashish.pp17@gmail.com*

Ashok Kumar. Manager-EIA, En-Vision Enviro Engineers Pvt. Ltd., 208, G-Tower, Sagrampura, Surat-395002, Gujarat, India. *E-mail: asokumr@gmail.com*

Athar Ali Khan. Department of Botany, Aligarh Muslim University, Aligarh-202002, India. *E-mail: khan_ath@yahoo.com*

Ayon Alipatra. Mandan Bharti Agriculture College, Agwanpur, Saharsa-852201, Bihar, India. *E-mail: ayon.1987@gmail.com*

Budheswar Dehury. Department of Agricultural Biotechnology, Assam Agricultural University, Jorhat-785013, Assam, India.

Dharman Dhakshanamoorthy. Division of Molecular Biology and Biotechnology, Department of Botany, Annamalai University, Annamalainagar-608002, Tamilnadu, India. *E-mail: biofueldd2009@yahoo.com*

Gh Mohd Rather. Department of Botany, Aligarh Muslim University, Aligarh-202002, India. *E-mail: ratherbio@gmail.com*

Hirak Banerjee. Bidhan Chandra Krishi Viswavidyalaya (BCKV), Mohanpur-741252, West Bengal, India. *E-mail: hirak.bckv@gmail.com*

Jagatpati Tah. Department of Botany (UGC-CAS), The University of Burdwan, Burdwan-713104, West Bengal, India. *E-mail: jt_botbu2012@yahoo.in*

Kannan Pandian. Dryland Agricultural Research Station, Tamilnadu Agricultural University, Chettinad, Tamilnadu, India. *E-mail: pandian.kannan@gmail.com*

Khalid Rehman Hakeem. Faculty of Forestry, Universiti Putra Malaysia, Serdang-43400, Selangor, Malaysia. *E-mail: kur.hakeem@gmail.com*

Krishnendu Ray. Bidhan Chandra Krishi Viswavidyalaya (BCKV), Mohanpur-741252, West Bengal, India. *E-mail: krishnenduray.bckv@gmail.com*

Madhumita Barooah. Department of Agricultural Biotechnology, Assam Agricultural University, Jorhat-785013, Assam, India. *E-mail: m17barooah@yahoo.co.in*

Mahalingam Govindaraj. Centre for Plant Breeding and Genetics, Tamilnadu Agricultural University, Coimbatore, Tamilnadu, India; Millet Breeding, ICRISAT, Patancheru, Andhra Pradesh, India. *E-mail: m.govindaraj@cgiar.org*

Mahendra K Modi. Department of Agricultural Biotechnology, Assam Agricultural University, Jorhat-785013, Assam, India.

Maya Kumari. Defence Institute of Bio-energy Research (DRDO), Goraparao, Haldwani, Nainital-263139, Uttarakhand, India. *E-mail: madhulogin@gmail.com*

Md Mahabubur Rahman. Research Institute for Sustainable Humanosphere, Kyoto University, Uji, Kyoto 611-0011, Japan.

Meghana Agrawal. Department of Botany, The IIS University, S.F.S., Gurukul Marg, Mansarovar, Jaipur, Rajasthan-302020, India. *E-mail: meghana.ktm@gmail.com*

Mohamed Ragab Abdel Gawwad. Natural Sciences Department, Genetics and Bioengineering Program Coordinator, Faculty of Engineering and Natural Sciences, International University of Sarajevo, Bosnia and Herzegovina. *E-mail: mragab@ius.edu.ba*

Mohammad Imran Kozgar. Department of Botany, Aligarh Muslim University, Aligarh-202002, India. *E-mail: m.i.kozgar@gmail.com*

Mohd Mazid. Advanced Plant Physiology Division, Department of Botany, Faculty of Life Sciences, Aligarh Muslim University, Aligarh-202002, India. *E-mail: mazidmohd699@gmail.com*

Mousumi Sahu. Department of Agricultural Biotechnology, Assam Agricultural University, Jorhat-785013, Assam, India.

Mridul Das. Department of Geography, Visva-Bharati, Santiniketan-731235, West Bengal, India. *E-mail: mrisgeovb@gmail.com*

Natarajan Sritharan. Agricultural Research Station, Tamil Nadu Agricultural University, Kovilpatti, Tamil Nadu, India. *E-mail: sritnau@gmail.com*

Papiya Mitra Mazumder. Department of Pharmaceutical Sciences, BIT Mesra, Ranchi-835215, Jharkhand, India. *E-mail: papiyamm@gmail.com*

Paramasivam Arunachalam. Dryland Agricultural Research Station, Tamilnadu Agricultural University, Chettinad, Tamilnadu, India. *E-mail: arunachalp@gmail.com*

Parthadeb Ghosh. Cytogenetics and Plant Breeding Section, Department of Botany, University of Kalyani, Kalyani-741235, Nadia, West Bengal, India. *E-mail: pdggene@rediffmail.com*

Radhakrishnan Selvaraj. Division of Molecular Biology and Biotechnology, Department of Botany, Annamalai University, Annamalainagar-608002, Tamilnadu, India. *E-mail: selvarajphd14@yahoo.co.in*

Rajib Roychowdhury. Department of Biotechnology, Visva-Bharati, Santiniketan-731235, West Bengal, India. *E-mail: rajibroychowdhurybu@gmail.com*

Rajni Shukla. Botany Department, University of Lucknow, Lucknow-226007, India. *E-mail: rajni_shukla1@yahoo.co.in*

Ramesh Namdeo Pudake. Department of Agronomy, School of Agriculture, Lovely Professional University, Jalandhar-Delhi G.T. Road (NH-1), Phagwara-144411, Punjab, India. *E-mail: ramesh.17503@lpu.co.in*

Rubaiyat Sharmin Sultana. Department of Botany, University of Rajshahi, Rajshahi-6205, Bangladesh. *E-mail: sultanaru@yahoo.com*

Saima Quddusi. Amity Institute of Biotechnology, Amity University, Lucknow-227106, India. *E-mail: saimabiotech@gmail.com*

Shilpi Paul. G. B. Pant Institute of Himalayan Environment and Development, Kosi-Katarmal, Almora-263643, Uttarakhand, India. *E-mail: shilpipaul01@gmail.com*

Shrilekha Misra. Banasthali Vidhyapeeth, Rajasthan, India. *E-mail: shreelekha2@rediffmail.com, misraliza@yahoo.co.in*

Smita Purohit. Department of Botany, The IIS University, S.F.S., Gurukul Marg, Mansarovar, Jaipur, Rajasthan-302020, India. *E-mail: smitaarun4@gmail.com*

Subhankar Chakraborty. Department of Geography, Visva-Bharati, Santiniketan-731235, West Bengal, India. *E-mail: subhageographervb@gmail.com*

Subhrajyoti Mandal. Department of Geography, Visva-Bharati, Santiniketan-731235, West Bengal, India. *E-mail: subhrajyotimandal@ibibo.com*

Taqi Ahmed Khan. Department of Biochemistry, Faculty of Life Sciences, Aligarh Muslim University, Aligarh-202002, India; Applied Biotechnology Department, College of Applied Sciences, Sur, Oman. *E-mail: lovingfareed@gmail.com*

Viralkumar Bhikhalal Mandaliya. Plant Biotechnology and Genetic Engineering (UGC-CAS), Department of Biosciences, Saurashtra University, Rajkot-360005, Gujarat, India. *E-mail: viral mandaliya@yahoo.com*

Vrinda Someshwar Thaker. Plant Biotechnology and Genetic Engineering (UGC-CAS), Department of Biosciences, Saurashtra University, Rajkot-360005, Gujarat, India. *E-mail: casprogramme@gmail.com*

Yogesh K Sharma. Botany Department, University of Lucknow, Lucknow-226007, India.
E-mail: yogesh_s26@yahoo.com

Contents

PART C: Biotechnological Approaches to Crop Improvement

PART D: Implications of Advanced Technologies to Improve Demanding Crops in Changing Climate

PART A

Environmental Challenges to Crop Improvement

1

An Overview of Climate Change: Causes, Trends and Implications

Subhankar Chakraborty[1], Anindya Pattanayak[2], Subhrajyoti Mandal[1], Mridul Das[1] and Rajib Roychowdhury[3*]

[1]Department of Geography, Visva-Bharati, Santiniketan-731235, West Bengal, India
[2]Integrated Science Education and Research Centre (ISERC), Visva-Bharati Santiniketan-731235, West Bengal, India
[3]Department of Biotechnology, Visva-Bharati, Santiniketan-731235, West Bengal, India

ABSTRACT

Global climate change is the most troublesome hindrance that the world is currently facing in its struggle for existence. Earth's climatic systems are one of the most dynamic and complex systems which still have not been properly understood. The present climates and their elements are far different from the past, introducing newer complexities within the climatic sciences. It is of no doubt that, climate change is entirely a natural process, the examples of which can be drawn from the ice ages, but the anthropogenic side cannot be ignored. In fact, humans act as moderating factor through their everyday activities which somehow attempt to introduce desired modifications in natural climatic processes. Thus, the concept of Global Warming has evolved. A clear distinction between climate change and global warming is utmost necessary. Lots of evidences of climate change and global warming have already been recorded through modern instrumentation, historical temperature analyses, global precipitation studies, etc. A number of events prior to climate change have already been seen, like the melting of glaciers, coral bleaching, tree line shifting, sea level rising, and erratic precipitation patterns. Climate and their elements no doubt are the most important determinants of all type of life forms on the earth. This chapter marks the train of thoughts followed by man, not only in the path of realization of the climate change, but also its modifications in the form of global warming which caused people to question their environmentally harmful intentions and activities. It gradually moves on to focus on the advent of ideas regarding the safeguard of the environment in order to mitigate climate change, to the current trends being followed all over the world to bring the change under control. The discussion about its implications is inevitable, particularly in case of agriculture and crop production which supports the entire mankind.

Keywords: Adaptive capacity, agriculture, climate change, global warming, greenhouse effect.

**Corresponding author's Email: rajibroychowdhurybu@gmail.com*

Roychowdhury R. (Ed.), Crop Improvement in the Era of Climate Change.

Abbreviations

ADB, Asian Development Bank; **AR4**, IPCC's Fourth Assessment Report; **CGIAR-CCAFS**, Consultative Group on International Agricultural Research- research programme on Climate Change, Agriculture and Food Security; **COP**, Conference of the Parties; **ESRL-NOAA**, Earth System Research Laboratory-National Oceanic & Atmospheric Administration, **FAO**, Food and Agriculture Organization; **GDP**, Gross Domestic Product; **GHGs**, Green House Gases; **ICIMOD**, International Centre for Integrated Mountain Development; **IPCC**, Intergovernmental Panel on Climate Change; **UNDP**, United Nations Development Programme; **UNEP**, United Nations Environmental Programme; **UNFCCC**, United Nations Framework Convention on Climate Change; **UNICEF**, United Nations Children's Fund; **WHO**, World Health Organization.

1. INTRODUCTION

Climate is one of the most complex systems which is merely understood by us as climate is always marked by its spatial and temporal variability. In fact, the atmosphere remains mostly in the states of turmoil and instability due to its elemental variations leading to spatio-temporal heterogeneities. There is no doubt that, the world climates are changing continuously. The earth's climate is vastly different now from what it was 100 million years ago when dinosaurs roamed the planet and tropical plants thrived closer to the poles (Mac *et al.*, 1998). Warming of the world climates is confirmed by the evidence provided by the observations of increases in global average air and ocean temperatures, comprehensive melting of glaciers and ice sheets and rising global average sea level (IPCC, 2007a). Global warming becomes such an issue which has led numbers of summits, conferences around the world, however, the mitigation strategies are yet to be properly introduced. Meanwhile, there is conceptual difference between the terms 'climate change' and 'global warming'. IPCC's AR4 has defined climate change as *"to a change in the state of the climate that can be identified (by using statistical tests) by changes in the mean and/or the variability of its properties, and that persists for an extended period, typically decades or longer. Climate change may be due to natural internal processes or external forcings, or to persistent anthropogenic changes in the composition of the atmosphere or in land use. It can manifest as extreme climatic events or as gradual systemic change leading to a slow deterioration in environmental conditions until customary practices or habitation become non-viable"* (IPCC, 2007a). Singh (2007) considered climate change as the variations and shifts in weather conditions over space and time of different scales and magnitude resulting into change of climatic type, for example, from warm and moist climate to warm and dry climate, from warm and moist climate to cool and moist climate (as happened during carboniferous period in India) etc. In fact, climatic changes refer to drastic or secular changes in heat balance of the earth-atmosphere system, moisture, cloudiness and precipitation caused by either external factors such as variations in orbital characteristics of the earth, solar variability (fluctuations in radiation from the photosphere of the sun), tectonic processes (mainly plate tectonics, displacement of continents and ocean basins), vulcanicity, changes in atmospheric composition in terms of concentration of atmospheric aerosols and carbon dioxide contents etc. or by internal factors such as exchanges of energy between the atmosphere, hydrosphere, lithosphere and cryosphere (ice-covered surfaces of both lithosphere and hydrosphere) or by both, at local, regional and global levels. On the other hand, the present-day global warming is much because of human activities. Actually, global warming is the gradual rise in atmospheric and ground surface air temperatures and consequent changes in global radiation balance caused mainly by anthropogenic processes (although natural processes also cause global warming and cooling) leading to climatic changes at different levels (Singh, 2007). In fact, climate change is not a new phenomenon in the climatic history of the earth but the worrisome concern is that, the climates are changing comprehensively and we the human beings have evolved as the largest important reason behind this. Mac *et al.* (1998) considered that earth's climate system will surely go through certain changes; climatic changes which occurred in the distant past were guided by natural causes, such as variations in the earth's orbit or the carbon dioxide (CO_2) content of the atmosphere. Climatic changes in the future, however, will probably have another source very likely—the human activities. Thus, it is evident that, the conceptual dilemma is revolving around the concerns that, while natural processes cause climate change, global warming is merely due to anthropogenic activities. In fact, climochronology (history of palaeo climates) suggests that the world climates have changed in the geologic past and hence, the real fact is *"the world's climates have changed in the past, are changing now, and there is every reason to expect that they will change in future"* (Hobbs, 1980). The trends in climatic changes are supposed to be quick and rapid processes. Singh (2007) exemplified this by citing the rapid and instantaneous onset of cold climate during the Jurassic period which caused mass extinction of dinosaurs.

Whatever the ongoing debates are, it is a fact that the world climates are changing gradually and human activities have become major concern in this regard far from the natural processes. The processes of global warming is of no doubt has raised the changing trends of climates and is modifying the climatic systems throughout the globe. Meanwhile, the international communities and scientific authorities are now worried as these climatic changes could incur large scale adverse effects in near future. The rising sea levels, melting glaciers, steady depletion of ozone layer, massive coral bleaching etc. are some blazing examples of such adversities of climatic changes. Moreover, the changing atmospheric chemistry and resulted calamities like unusual floods, severe droughts, changing global radiation balance could invoke several damages to ecological processes, plant and animal communities and may cause catastrophic hazards for the human communities throughout the world. Actually, the truth is that, climatic changes are interminable, we cannot cease these immediately or by near future. However, we can try to resolve the problem of global warming which is the effect of our prolonged invasion and interferences into the natural processes to avail a healthy and balanced earth.

2. INDICATORS AND EVIDENCE OF CLIMATE CHANGE

This ongoing discussion needs a complete set of evidences of climate change as well as also for global warming to understand the variability of world climates from the very beginning. Human induced warming of the global climates is thus, a recent phenomenon if we consider the climatic variabilities throughout the geologic history of the earth. Oliver and Hidore (2003) have presented a list of indicators of climate change. These indicators are quietly able to reveal the nature of climate changes since the Precambrian. The biological indicators such as floral and faunal indicators support the climatic changes in the past. The present distribution of vegetation types highly corresponds to the present distribution of world climates. The studies of fossilized plants and animals provided significant proxy data for contemporary climatic conditions particularly the temperature and moisture conditions (Table 1).

Table 1 Summary of climatic conditions during past geologic periods

Eras	*Epochs*	*Periods*	*Duration (million years)*	*Ice Age*	*Summary of climatic conditions*
Neozoic	Quaternary	2. Holocene			Beginning of ice age-4 or the Pleistocene Glaciation which witnessed consequent periods of cooling (Nebraskan, Kansan, Illinoin and Wisconsin in North America and Guntz, Mindel, Riss and Wurm in Europe) and warming (inter-glacial periods i.e. Aftonian, Yarmouth, Sangamon and Holocene)
		1. Pleistocene	0.990	Pleistocene Ice Age- 4	
Cenozoic	Tertiary	4. Pliocene	10.000		Warm and temperate climates during Eocene and Oligocene periods, humid climates during Miocene periods and repetition of warm and cold climates during the Pliocene period.
		3. Miocene	14.000		
		2. Oligocene	15.000		
		1. Eocene	30.000		
Mesozoic	Secondary	3. Cretaceous	65.000		Warm and dry climates during Triassic period and wet later on, evolution of Angiosperms, mammals
		2. Jurassic	45.000		
		1. Triassic	45.000		
Palaeozoic	Primary	6. Permian	45.000	Carboniferous Ice Age-3	Cold phase was replaced by warm climates during Silurian and Devonian periods, evolution of life mainly plants and vertebrates in the seas, beginning of Ice age-2 during middle carboniferous period to early Permian period.
		5. Carboniferous	80.000		
		4. Devonian	50.000		
		3. Silurian	40.000		
		2. Ordovician	60.000	Ice Age-2	
		1. Cambrian	100.000		
Pre-palaeozoic	–	Precambrian or Algonican	–	Precambrian Ice Age	Formation of dense atmosphere, beginning of processes of precipitation, average atmospheric temperature of 37° C during 4250 MYBP which dropped to 25° C about 2500 MYBP, occurrence of a few glacial periods, the first ice age or the Precambrian ice age or the Huronian Glaciation.
Azoic or Achaean	–	Achaean	–		–

*MYBP denotes million years before present (compiled by authors).

3. CAUSES AND THEORIES OF CLIMATE CHANGE

Generally, climatic changes are regulated by the atmospheric circulation and interaction between the major five component parts of the earth, i.e. atmosphere, hydrosphere, lithosphere, biosphere and cryosphere. The causes for such interactions leading to climatic changes are related to (a) outside sources or extraterrestrial sources and (b) inside sources or terrestrial sources. The causes and theories of climate change are viewed in terms of periodicity of climatic changes which are of two types, namely (i) short-term climatic changes and (ii) long-term climatic changes. As the nature and pattern of climatic changes vary spatio-temporally their causes also vary greatly. This is why, no single theory can explain all types of climatic changes and hence, there are a number of causes and theories of climate change (Singh, 2007). Another important dimension of climate change is based on the aggressive evolution of human activities as significant climate moderators. Thus, the sources of climate change are threefold, namely: (a) outside or extraterrestrial sources, (b) inside or terrestrial sources and (c) anthropogenic sources.

The principal causes and theories of climate change are described as follows:

3.1 Theories Related to Change in Solar Radiation

These theories are based on the premise that the output of solar energy is subject to a considerable variation. These theories state that, terrestrial climatic fluctuations occur in accordance with variations in the absolute quantity of radiant energy emitted by the sun (Lal, 2009). The important theories under this category are as follows:

- The ***Variable Sun Theory*** is based on the idea that the sun is a variable star and its radiant energy undergoes changes through time. It suggests that increased output of solar radiation leads to the warming of the atmosphere, while its decline would result in decrease of atmospheric temperature.
- The ***Sunspot Theory*** depicts a correlation between sunspot cycles and climatic variations. Sunspots which appear as dark spots are simply the huge magnetic storms that take place on the sun's surface. The number of sunspots is regulated in a cyclic manner. This cycle averages 11 years in length. As climatic conditions are primarily regulated by the sun and since sunspots are clear indications of increased or decreased solar activity, atmospheric temperature and precipitation are highly correlated with these. In general, increased sunspot activity results in warming of the atmosphere whereas decreased sunspot activity causes reduction in the air temperature.
- Another theory revolves around the ***variable solar constant*** (the rate of radiation from the sun which is of the value of 2 cal/cm^2/min.) which correlates climatic fluctuations with variations in the solar constant.

3.2 Theories of Continental Displacement and Polar Wandering

These theories are based on the premise that the arrangements of the present continental and ocean basins were different in the past geological periods.

- ***Wegner's Continental Drift Theory:*** Wegner believed that all the present continents were united together in the form of a huge landmass known as 'Pangaea' surrounded by a huge oceanic body 'Panthalasa' during the Carboniferous period. The locations of the poles were also different from today. For example, the North Pole was at 14° N and 124° W during the Silurian period which moved to 51°N and 153°W at the tertiary epoch (Singh,

2005). Later on the Pangaea was disrupted and the broken land units drifted away from each other to form the present arrangement of continents. These continental drifts might have influenced the present climatic distributions as the relative locations of continents from the poles have immense control on climatic parameters.

- ***Plate Tectonic Theory:*** The Plate Tectonic theory has removed the uncertain mistakes of Continental Drift theory and has solved many historical riddles regarding the origin of the continental crust and arrangement of ocean basins. Plate tectonics consider that, the earth is formed of rigid lithospheric slabs or rigid and solid crustal layers known as plates floating above the underlying asthenosphere. Except the Pacific Ocean basin all the other plates are made of both continental and oceanic crusts. These plates are in motion in relation to each other and these plate motions influence the arrangement of continental landmasses which in turn influences the climatic shifts. For example, the present-day glacial features found in Africa, Australia, South America and India do not match with the present climatic conditions of these landmasses which clearly indicate that may be these landmasses went through ice ages during the geological pasts (probably about 230 million years ago).

3.3 Theories Involving Tectonism and Topographic Controls

It is an established fact that topographic elements particularly relief largely shapes weather and climatic phenomena at local-regional and global scales. We all know how temperature, air circulation and atmospheric moisture change with elevation. This is why attempts have been made to correlate ice ages with active Tectonism and mountain building, 'as with increasing height of landmasses, the potential for ice formation is greatly increased' (Oliver and Hidore, 2003). The proponents of tectonic theory of ice ages and glaciation have attempted to demonstrate close relationship between Permo-Carboniferous ice age and Pleistocene ice age and glaciation with large scale vertical tectonic movements and orogenesis (mountain building) as these two great periods of ice age and comprehensive glaciation preceded large-scale mountain building activities i.e. Caledonian mountain building was followed by Permo-Carboniferous ice age and late Cenozoic mountain building, leading to the formation of Alpine-Himalayan mountain chains, North American Cordillera, upliftment of Tibetan Plateau, western North America including Great Plains etc. was followed by Pliocene-Pleistocene ice age (Singh, 2007).

3.4 Theories Related to Anthropogenic Activities

These theories are more relevant particularly in case of automobile based civilizations. It has already been mentioned that, the potential effects of increasing human interferences have come into scenario since the industrial revolution from when the climatic parameters have showed abrupt changes.

- ***Carbon dioxide theory and Greenhouse Effect:*** The carbon dioxide theory was postulated by T.C. Chamberlin during the 1890's. It considers that the variations in the carbon dioxide content of the atmosphere would influence the global climatic changes. This gas is transparent to incoming short-wave solar radiation but absorbs the outgoing long-wave terrestrial radiation and thus, leads to warming of the earth's surface. Carbon dioxide, water vapour and aerosols are the most important heat absorbers which result into the greenhouse effect. The gases with the properties of heat absorbing are known as greenhouse gases (GHGs), such as carbon dioxide (CO_2), nitrous oxide (N_2O), methane (CH_4), fluorinated gases or F-gases etc. Out of the total emissions of the GHGs, various gases share their relative proportions (Fig. 1). Furthermore, various anthropogenic sectors contribute different amounts of GHGs by volume. The energy sector by so far has appeared as the largest contributor (26%) in emitting GHGs followed by industrial sector (19%), landuse changes and deforestation activities (17%), agricultural operations (14%), transportation activities (13%) and various commercial and household usages (8%) and wastewater and garbage disposal (3%).

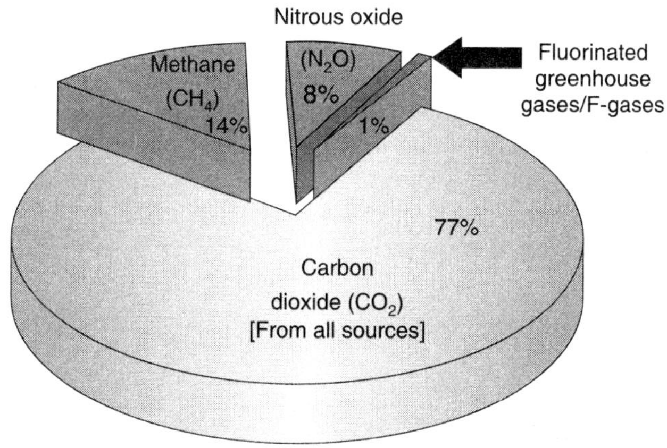

Fig. 1 Global greenhouse gases emission (2004) by relative shares of gases.

The major concern regarding the increase of the GHGs particularly the increasing concentrations of carbon dioxide and F-gases appear in the form of rapid depletion of the ozone layer and creation of ozone hole (ozone less patches) over Antarctica. It is well known that the ozone layer absorbs the ultraviolet radiation from the sun and thus, protects the biosphere. The global average thickness of the ozone layer which is necessary for our survival is about 300 Dibson units (equivalent to 3 mm). Contrastingly, it is about 100 Dibson units which is warning us that we all are in trouble. The ozone level fell to 90 Dibson units, which is historically the second lowest value on September, 1998. Due to ozone depletion, the UV-B radiation is now reaching to the earth surface in a range between 280 and 320 nm (Rao, 2008).

The most important of these gases is the CO_2 which is mainly emitted from combustion of fossil fuels in energy-industrial and transport sectors with additional contributions by activities like deforestation, landuse transformation etc. Methane (CH_4) is emitted through agricultural operations, livestock farming, meat industries, waste management and energy usages. Nitrous oxide (N_2O) primarily originates from farming operations particularly from fertilizer usages. F-gases are derived into the atmosphere from industrial sectors and consumer goods such as refrigerators, air coolers etc. which include hydrofluorocarbons (HFCs), chlorofluorocarbons (CFCs), perfluorocarbons (PFCs) and sulfur hexafluorides (SF_6).

The present industrialization, deforestation, landuse transformation and increasing number of vehicles are largely responsible for adding large amounts of carbon dioxide to the atmosphere. Although some of this increased flux of CO_2 is absorbed by vegetation and dissolved in oceans but still a huge amount of this CO_2 flux is in the atmosphere. It is significant to note that during the pre-industrial period (around 1750 A.D.), the total atmospheric content of CO_2 was within 280-290 ppm which recorded an observed value of 379 ppm at the end of the 20th century (2000 A.D.). The most recent observed value of atmospheric CO_2 concentration reached to 392.13 ppm during October, 2012 (ESRL-NOAA, 2013). Now it is projected that, by 2050 A.D. the CO_2 concentration will reach up to 445-640 ppm and in 2100 A.D. it will go as high as between 540-970 ppm. Consequently, the global temperature may increase by about 2°C by 2040 A.D.

However, there are some scientific debates regarding the increase of GHGs in the atmosphere and the warming of the earth. Schneider (1990) has shown that there are variations in opinions of the scientists regarding the role of GHGs. Some scientists argue that, during the last century much of the warming took placed between 1915 and 1940's which was followed by cooling at such a time when the concentrations of GHGs in the atmosphere were increasing. Based on this fact, they have denied attributing the 20th century warming to the increase in GHGs concentrations. On the other hand, the scientific bodies like the IPCC attributed the global warming (since 1950s) solely to the observed increase in anthropogenic GHG concentrations (IPCC, 2007a).

4. CLIMATE MODELLING STRATEGIES FOR PREDICTING FUTURE CLIMATE CHANGES

Different mathematical models have been employed to assess the interrelationship between various natural-anthropogenic variables and variations in climatic parameters. Climate models are mathematical representations of the climate system, expressed as computer codes and run on powerful computers. Models show significant and increasing skill in representing many important mean climate features, such as the large-scale distributions of atmospheric temperature, precipitation, radiation and wind, and of oceanic temperatures, currents and sea ice cover. Models are routinely and extensively assessed by comparing their simulations with observations of the atmosphere, ocean, cryosphere and land surface. The predicted climate changes for the selected years will depend greatly on the model and scenario selected (White *et. al.*, 2010). Recently, the Fourth Assessment Report of IPCC has concluded that there is considerable confidence that the current fully coupled global models can supply credible quantitative estimates of future climate change, particularly at continental and larger scales (IPCC, 2007b). In general there are three modelling strategies identified by Barry and Chorley (1998). These models are as follows:

(a) **Black box Modelling:** It involves the statistical extrapolation of an historical time series (e.g., past temperatures or rainfall) into the future, without real concern for the mechanisms involved. Generally, it is a primitive technique which yields unsatisfied predictions.

(b) **Grey box Modelling:** It is based on the assumption that the influence of most important determinants can be identified, measured and superimposed to produce a satisfactory simulation of a past record, and the derived mathematical model is capable of being projected meaningfully into the future of predictive purposes. Although these models are scientific improvements over the black box models and provide better satisfactory results, are not free from defects. The problems of these models are that, neither all the important variables can be taken into consideration in these models nor the future prediction of CO_2 concentration can be satisfactorily determined by them.

(c) **White box Modelling:** These models are based on detailed understanding of the structure and operation of the earth-atmosphere-ocean systems, such that its possible future states can be simulated by applying assumed forcing mechanisms, particularly the anthropogenic ones. Thus construction of such numerical models needs bringing together locational, temporal and attribute information into a database that allows hypotheses regarding climatic processes and interactions to be simulated.

5. IMPLICATIONS OF CLIMATE CHANGE

Climate change has now become probably the most highlighted topic as it is now clear to all of us that, this may pose the largest ever threat to the human civilizations in the future. In fact, climate change especially the human induced global warming has a wide spectrum of influence and it can definitely be expressed that, there will be no such sector which will get rid from climate change. The impacts of climate change can be categorized into extreme and non-extreme types. The extreme impacts are those which may be very significant, resilient implications on the biotic landscape (i.e. natural ecosystems), natural environment and human societies caused due to a single extreme event or by consecutive extreme or non-extreme events (IPCC, 2012). The non-extreme events, on the other hand are the temporary impacts may be caused due to either changes in climatic parameters (sudden onset of cyclones due to steady rise in temperature and resultant low pressure) or through changes non-climatic ones (such as landslides due to massive rainfall on the hill-slopes or sudden flash floods due to heavy rains in the upper catchments of a river). However, there is no such evidence that climate change will only have negative impacts rather various scholars and institutional bodies believe that, climate change will ensure some benefits also. These will be clear, if we go through the consequences of climate change which are already evident as well as anticipated for the future. The various impacts of climate change can be presented through the following sub-headings:

5.1 Climate Change and Extreme Weather Conditions

Climate change has largely modified the weather condition throughout the world. It will be really hard to deny the fact that, weather extremes are increasing steadily as all weather parameters are now developing in a different environment than before. While natural variability continues to play a key role in extreme weather, climate change has shifted the odds and changed the natural limits, making certain types of extreme weather more frequent and more intense. While our understanding of how extreme weather conditions are getting affected by climate change is still developing, evidence suggests that extreme weather conditions may be affected even more than anticipated (Cutting *et al.*, 2011). Before going to further discussion, it will be better to know what extreme weather really means. Simply, extreme weather means, the occurrence of a value of a weather or climatic parameter above or below a particular/threshold value near the upper or lower ends of the range of observed values of the variable. In general, both extreme weather events and extreme climate events are collectively referred to as 'climate extremes' (IPCC, 2012). Extreme weather events do not have a single cause rather they possess various possible contributing factors – and human-induced climate change i.e. global warming is now one of those factors (Cutting *et al.*, 2011). Actually, these weather extremes are rare, even though there are sufficient examples which prove that the weather conditions are comprehensively changing which may lead to a rapid increase in future climate extremes.

Various observations on diurnal temperature records (both for past and recent times) have revealed that, the number of cold days and nights are decreasing while there is an overall increase in the number of warm days and nights throughout the world (IPCC, 2012). Hot days have become hotter and more frequent as well. Since the 1950's the number of heat waves has increased and heat waves have become longer. In the United States the record high temperature each year now outnumber the record lows by 2:1 (Meehl *et al.*, 2009). In Europe, global warming is accused for an estimated 29% of the new record highs set each year (Wergen and Krug, 2010). The European heat wave in 2003 took the lives of thousands. In Asia, China has witnessed a remarkable increase of hot days across all over the country (Ding *et al.*, 2009). Globally, extremely warm nights that used to come once in 20 years are now occurring in every 10 years (Zwiers *et al.*, 2010). There has been a reduction in the number of cold extremes, such as very cold nights observed across the whole world along with a widespread reduction in the number of frost days particularly in the mid-latitude regions. Snow cover has decreased in most of the regions, especially in the spring, and mountain snowpack has also decreased in the northern hemisphere. The freeze date for lakes and rivers in the Northern Hemisphere has moved later in the year, at an average rate of six days per 100 years, while the ice break-up date has moved up earlier in the year, at an average rate of a little more than six days over the last 100 years (Cutting *et. al.*, 2011).

Such weather extremes can prove very fatal for majority of the countries of the world particularly for the developing and underdeveloped nations as these changes in weather conditions could culminate in severe damages to agriculture and food production, economy, livelihood, health and natural resources of these countries. The case of India may provide a really good example where 70% of the population is directly or indirectly rely on agriculture, thus, an early or delayed monsoon can hamper not only the cultivation rather it will also affect these huge number of people. For example, the erratic and abnormal monsoon of 2012 caused widespread draughts in most of the parts of the country; about 644 farmers in Maharashtra committed suicide due to repeated crop failure during 2001 to 2004.

5.2 Climate Change and the Geophysical Environment

The climate and weather events also have certain geomorphological and geological impacts. The steady rise in global atmospheric temperatures and resultant loss of glaciers, ice sheets and permafrosts, gradual rise in sea level across the globe along with changing precipitation trends may modify the geophysical components significantly. Here are some instances:

(1) The temperature rise also will result in gradual degradation of mountain permafrost. Observations show that permafrost temperatures have increased since the 1980s. Temperatures in the colder permafrost of northern Alaska, the Canadian Arctic, and Russia have increased up to 3°C near the permafrost table and up to 1 to 2°C at depths of 10 to 20 m since the late 1970s/early 1980s.

(2) Another significant geophysical consequence of climate change has been the retreat of glaciers particularly in the high mountainous environment. Since, the 1950s most of the glaciers have retreated. For example, the Himalayan glaciers are retreating at rates between 10 and 60 metres per year. The terminus of most high valley glaciers in Bhutan, Nepal, and neighbouring parts of China are retreating very fast. Vertical shifts over the last fifty years of 100 metres have been recorded and retreat rates of 30 metres per year have become common (UNEP, 2009). The 'Glacial Lake Outburst Floods' (GLOFs) are solely a result of glacier retreat and formation of lakes behind unstable natural dams. In the 20th century, GLOFs have culminated in several disasters in many high-mountain regions of the world including the Andes, the Caucasus and Central Asia, the Himalayas, and the European Alps.

(3) Transient sea level extremes and extreme coastal high water is caused by severe weather events (also by tectonic disturbances which cause tsunamis) may influence the coastal geomorphological processes. Mean sea level has always varied considerably over glacial time scales because of the fact that, the extent of ice caps and glaciers have fluctuated with global temperatures. Previously, it is mentioned that the global mean sea level is rising at an average rate of 1.7 mm/year over the 20th century, 1.8 mm/year over 1961 to 2003, and at a rate of 3.1 mm/year over 1993 to 2003 (IPCC, 2012) and at a rate of 2.5 mm/year since 2003 (UNEP, 2009). The IPCC AR4 (2007) projected a rise in global sea levels of between 18 and 59 centimetres by 2100. More recent studies indicate an average global sea level rise between 75 and 190 centimetres by 2100 (Lawler *et al.*, 2011). There are many evidences that, this sea level rise is the contribution of massive loss of ice covers from Greenland and Antarctica in recent past. By 2006, glaciers and ice caps contributed about 32% of the total sea-level rise, while the ice sheets on Greenland and Antarctic together contributed about 20% (UNEP, 2009).

(4) The Greenland and Antarctica ice sheets are by far the largest ice fields, containing 12% and 87% of land ice volume, respectively. Glaciers and ice caps contain only about 1 per cent of global land ice volume. The massive loss of these ice-covers has become one of the largest scientific concerns in this era as this wastage of ice covers is undoubtedly exacerbated by global warming since the post 1970's (IPCC, 2007a). In 2007, the sea-ice in the Arctic Ocean shrank to its smallest extent on record, 24 per cent less than the previous record set in 2005 and 34 per cent less than the average minimum extent over 1970-2000. Ice loss from the Antarctic Peninsula, which extends from West Antarctica towards South America, increased by 140 per cent. A large part of the 13,000 square kilometres Wilkins Ice Shelf, between 70° and 74°W along the Antarctic Peninsula, collapsed in February 2008 (UNEP, 2009).

(5) Severe coastal hazards such as marine erosion and inundation of salt water are nourished by climate change through rising sea levels and changes in extreme events. Increasing sea levels will increase the incidents of saltwater intrusion into coastal aquifers. The AR4 state with very high confidence that the impact of climate change on coasts is exacerbated by increased pressures on the physical environment arising from human settlements in the coastal zone. The IPCC (2007a) shows much confidence that the coastal areas will be more exposed to increasing consequences, including coastal erosion and inundation over the coming decades due to both climate change and sea level rise. Such events will prove to be fatal especially for the small island nations.

5.3 Climate Change and Water Resources

Hydrological processes and water resources are other sectors which could be largely modified by climate change. The increasing air temperature will probably increase the amount of evaporation on one hand while the fluctuating precipitations could affect the run-off processes which in turn will influence both surface and groundwater resources in the forthcoming decades. The hydrological cycle is of no doubt, intimately linked with changes in atmospheric temperature and global radiation balance (IPCC, 2008). Thus, in response to global warming, the hydrological cycle is expected to accelerate as rising temperatures increase the rate of evaporation from both land and sea. The distribution of rainfall is anticipated to be of slightly different, i.e. rainfall is predicted to rise in the tropics and higher latitudes but it will decrease in the already dry semi-arid to arid mid-latitudes and in the interior of large continents (Turral *et al.*, 2011). Besides, climate change could result in periodic and chronic shortage of water, particularly in arid and semi-arid areas of the world. The developing countries which are located in such regions are highly vulnerable to climate change because these countries mostly derive their water resources from single-point systems such as bore holes or isolated reservoirs etc. (IPCC, 1997). Although the prediction of run-off behaviour (including distribution, duration, frequency and magnitude of run-off) is a difficult job as run-off behaviours are not always consistent with precipitation only rather some other factors such as terrain slope, surface roughness etc. also play dominant roles in it. However, several evidences have been found in recent years which demonstrate a broad and coherent pattern of change in annual run-off, with some regions experiencing an increase in run-off (e.g., high latitudes and large parts of the USA) while others (such as parts of West Africa, southern Europe and southernmost South America) experiencing a decrease in it (IPCC, 2008). Some studies have shown substantial reductions (up to 40%) in regional run-off in southeastern Australia and in other areas where annual potential evapotranspiration exceeds rainfall. In glacier-fed river systems, the timing of flows are changing because of increasing loss of glaciers in the upstreams, however, the mean annual run-off has not been much affected (Turral *et al.*, 2011).

The events of floods will be a larger problem in many temperate and humid regions, as already evident by several indicators (IPCC, 1997). Moreover, climate change has the potential to change river flood characteristics through changing the volume and timing of precipitation, by altering the proportions of precipitation falling as snow and rain, and to a lesser extent, by changing evaporation and hence, accumulated soil moisture deficits. However, there is considerable uncertainty in the magnitude, frequency, and direction of change in flood characteristics (IPCC, 2012).

Sea-level rise will affect drainage and water levels in coastal areas, particularly in low-lying deltas, and may result in saline intrusion into coastal aquifers and river estuaries (Turral *et al.*, 2011). Further influences in the earth's water resources could be the increasing acidification of the oceans. The ocean's role in absorbing anthropogenic CO_2 released into the atmosphere has been underway for over two centuries. This has altered the chemistry of the global ocean fundamentally, by acidifying the top 2,000 metre layer of the oceans' waters (UNEP, 2009).

5.4 Climate Change and Biotic Resources

The increasing global temperature, rapid changes in rainfall patterns, sea level rise, epidemic loss of ice covers and permafrosts, acidification of oceans, frequent drought events and associated wildfires along with massive deforestation and agricultural expansion into the forest areas could largely hamper the biological processes as well as also the biotic communities and biotic landscapes.

Possible impacts on ecosystems: Climate change could culminate in severe dangers for both marine and terrestrial ecosystems. The possible and most challenging irreversible climate-related changes which can hamper the ecosystems include altered chemical characteristics of the ambient environment, inundation of many small islands and low-lying coastal ecosystems by sea-level rise, loss of wetland quantity and quality, and increased aridity in subtropical areas. These anticipated irreversible changes and their cumulative effects which are proceeding at very fast rates, will transform the ecosystem characteristics (especially ecosystem functioning and species' interaction) resulting in potential species extinction. Increases in temperature over the 20th century were directly responsible for the shifts in species' distributions. In marine ecosystems, climate change will probably result in the shifts from cold-adapted to warm-adapted communities, phenological changes along with modifications in species' interactions while in terrestrial ecosystems, these changes will include shifts to earlier onset of spring events such as leaf unfolding, blooming date, migration, and reproduction timing; change in species distribution; and modification of community structure. Researchers are now assessing how gradual sea-level rise and accelerating intensification of windstorms will influence the structure, function, and capacity of coastal and inland forest ecosystems. Another worrisome concern is the ocean acidification which is solely because of rapid emission of GHGs; the increasing trends in ocean acidification have the potentiality to cause mass species extinction in the future. If CO_2 concentration continues to the current trends, it can be predicted that acidification of the ocean water will deplete carbonate ions in majority of the global oceans by 2100, turning the waters corrosive for many shell-building animals. The global terrestrial ecosystems, on the other hand, have shown significant inertia in their responses to climate change. However, sudden extreme events can provoke the crossing of mortality thresholds for certain organisms and trigger population decline, decrease resilience, and alter community and ecosystem structure over time. Indeed, it is a fact that the future of several major ecosystems now in great threats. For example, ecosystems in sub-polar regions, the tropics, and semi-enclosed seas will face numerous local extinctions by 2050. Contrastingly, the Arctic and Southern Oceans will undergo severe species invasions (UNEP, 2009).

Ecosystems can be committed to long-term change long before any response is observable: For example, the prolonged drought in the Amazon rainforest which started in 2005 as a response to decreasing rainfall (3.2%/year during 1970-1998) and steadily declined water availability in the region (during 1995-2005) caused widespread (270,000 mile^2) and observable damage to the canopy (CBS NEWS, 2013).

Influences on Biodiversity: Global warming and regional variations of biodiversity have very close correspondence. It is well known that, temperature and rainfall-humidity have immense influence on the distribution of both plant and animals. The present distributions of phyto-geographic and zoogeographic regions reveal such intimate linkages. Meanwhile, most of the changes in climatic parameters are predicted to occur at high latitudes and high altitudes, many tropical species are incapable of tolerating anything beyond mild temperature variations. As every single species has its own critical temperature (maximum temperature beyond which it cannot exist) even a slight warming may threaten their survival. Under such circumstances, species migration probably can be seen. Moreover, ecosystem niche gaps caused by such species migrations in tropical lowland ecosystems may endanger those species that are able to adapt to changes within an ecosystem at a particular location, but not to the absence of a key player in that ecosystem (UNEP, 2009). Global warming driven permanent or semi-drought conditions and desertification in areas for example, the Sahel and Southern Africa could lead to severe changes for both the ecosystems and their biotic communities; such changes will obviously incorporate forest fires and degradation of grasslands. It is expected that 25%-40% of animal species in national parks in sub-Saharan Africa will become endangered (UNFCCC, 2007). Besides, there are future potentialities that the ongoing global warming will cause species extinction too. There are certain evidences which demonstrate that, global warming is directly responsible for the extinction of at least one endemic vertebrate species, the golden toad, from the cloud forests of Costa Rica. It is now projected that, by mid 21st century, global warming will cause the extinction of a substantial number of the subset of 1000 endemic species currently being analyzed globally. It is also predicted that, global warming will hit the hardest to those species which have intrinsically low population numbers, those inhabiting restricted or patchy areas, and those circumscribed to limited climatic ranges, such as coral reefs, mangrove forests, cloud forests, inland water ecosystems, and ecosystems overlying permafrost (Guariguata, 2006). It is estimated that each temperature increase of 1°C will force animals to migrate 160 km from their customary habitats as the plant species on which those animals depend for food are unable to subsist under the new, warmer weather conditions (UNDP, 2009). The mass events of coral bleaching (loss of algae from the corals resulting in whitening which cause death of the corals due to starvation) are the blazing examples of how harder global warming can cause catastrophe for the biotic communities. Although, coral bleaching can be caused by several natural processes, there are no doubts that the increasing events of coral bleaching are the major outcome of rising oceanic temperatures driven by global warming. For example, more than 60 coral bleaching events out of 105 mass coral moralities were reported between 1979 and 1990. Nearly all of the world's major coral reef regions (Caribbean/ western Atlantic, eastern Pacific, central and western Pacific, Indian Ocean, Arabian Gulf, Red Sea) have undergone some degree of coral bleaching and mortality during the 1980s (Buchheim, 1998). The coral bleaching of 1997-98 was the most catastrophic events accounted for the death of huge number of corals in the tropical oceans. Another example can be provided by the rapid loss of Antarctic Penguins due to starvation (as there are acute shortages of zooplanktons) since the 1970's. About 40% of these penguins are now wasted since then (Singh, 2007).

Possible influence on Forestry: Climate change will possibly alter the areas covered by forests along with changes in forest types due to changing temperature and precipitation regimes, while in some regions, forest area is assumed to expand (e.g., temperate regions) and in others to contract (e.g., boreal, tropical and mountain forests). Globally, planted forests and natural regeneration have increased the forest areas in the United States, Europe, China, and some countries in Latin America and the Caribbean e.g., Chile, Uruguay, Cuba and Costa Rica while, on the other hand, some countries in Africa, Asia and the Pacific and the tropical countries of Latin America still facing rapid deforestation, mainly due to agricultural conversion; while deforestation in the boreal forests of Siberia is mainly caused by forest fires. Although the boreal forests are expected to move northward, temperate forests are expected to increase their area northward to a greater extent than the boreal forests, thus reducing the total area of boreal forests which will be replaced by the later (FAO, 2012; Burton *et al.,* 2010). It is expected that the combination of climate change, land use conversion and unsustainable land use practices will interact in the forthcoming decades. Changes in water availability are considered to be a key factor for the survival and growth of many forest species, although the response to prolonged droughts will vary among species and also among different varieties of the same species (Lucier *et al.*, 2009). Besides these, climate change will increase the risk of frequent and more intense wildfires, especially where changing climate is accompanied by lower precipitation or longer dry periods as in the boreal, Mediterranean and sub-tropical forests, and traditional deforestation activities in the Amazon. Moreover, climate change may profoundly hit the health and vitality of the world's forests. In some cases, vitality may increase due to a combination of a more favourable climate for growth and CO_2 fertilization. In most cases however, increasing temperatures favour the growth of insect populations that is detrimental to the health of forests (Lucier et al., 2009). This is more likely to occur in forests dominated by few tree species (tropical deciduous or monsoon forests) or where specific temperatures or moisture levels control insect populations. For example, the absence of consistently low temperatures over a long time has enhanced the spread of the mountain pine beetle, *Dendroctonus ponderosae*, in boreal forests, which allowed an existing outbreak to spread across montane areas and into the colder boreal forests (Burton et al., 2010). In the tropics, on the other hand, increased warming reduces the life cycle of many insect pests; while at the same time increased fire damage makes trees more susceptible to insect attacks and vice versa. Species growth and survival depends for a large part on climate variables especially temperature and rainfall-humidity. Most species have a particular survival range within which they grow best, are competitive and are able to adapt to slight environmental changes and respond to insect attacks, diseases and other adverse environmental and human influences. Many of the ecological processes that are needed for tree and other plant and animal species to live together are influenced by climatic conditions. In general, many species have a tendency to move to higher latitudes or higher altitudes. In a study, Lucier *et al.* (2009) in their revision of climate change impacts on forests, found reports of phenological changes in a number of species, with pronounce changes observed in higher latitudes. Common changes which were observed including the changing flowering times and changing time of bud break, affecting productivity and carbon sequestration potential. Moreover, climate change could affect the forest productivity too. However, the impact of climate change on productivity varies according to geographic area, species, stand composition, tree age, soils (in particular water holding capacity), effects of CO_2 and nitrogen fertilization and interactions between any of these factors. Some of the changes may be temporal, reverting once saturation levels have been reached. This is projected to be the case for water availability, where reduction of water generally reduces plant growth but in areas of water surplus may initially increase growth when waterlogging is being reduced. Similar reactions have been noted for CO_2 and nitrogen fertilization as well as temperature increases. In general, productivity was found to increase with rising temperatures in most forest areas, including the Amazon, probably due to CO_2 fertilization. However, in contrast to temperate forests, there will be temporal increase of productivity the tropical forests which will decrease as the CO_2 saturation levels have been reached once. There is an important interaction between carbon storage and sequestration by forests and changing temperatures and precipitation. On the one hand, the more carbon is stored in forests; less will be in the atmosphere. Hence, the increasing storing of carbon into the forests will lead to control the rates of temperature rise. But contrastingly, the increasing temperatures, longer dry seasons and increasing CO_2 concentrations in the atmosphere in the long term, are expected to reduce the capacity of forests to store and sequester carbon, possibly converting forests from carbon sinks to carbon sources. Since carbon sequestration depends on productivity, all factors that affect productivity will also affect carbon sequestration (FAO, 2012).

5.5 Climate Change and Agriculture-Food Security

In this era of global warming, agriculture undoubtedly is standing at the nexus of three hardest challenges — to adapt with global warming, to mitigate global warming and to achieve food security (CGIAR-CCAFS, 2012). Agriculture undoubtedly forms the backbone of current human civilizations throughout the world in spite of rapid industrialization. About 2.6 billion are still engaged in agriculture. Discussion through certain sub-headings may be the best way to describe the potential impacts of climate change on agricultural sectors.

Expected Changes: The changing climatic variables particularly temperature and rainfall will introduce several challenges to agriculture in future. Climate variabilities are projected to bring changes in the agriculture sectors through changes in land and water regimes i.e. the primary mediums of climate change. Changes in the frequency and intensity of droughts, flooding, and storm damage are anticipated. Climate change is expected to cause long-term water and other resource shortages, depleting soil conditions, drought and desertification, disease and pest outbreaks on crops and livestock. Vulnerable areas are anticipated to face losses in agricultural productivity, primarily due to reductions in crop yields. Contrastingly, climate change may have some benefits also, particularly for the temperate regions (Kurukulasuriya and Rosenthal, 2003). The lengthening of growing seasons, carbon fertilization effects, and improved conditions for crop growth are projected to stimulate gains in agricultural productivity in high-latitude regions, such as in northern China and many parts of northern America and Europe. Early global estimates predict (without consideration of CO_2 fertilization effects or adaptation) a 20–30% reduction in grain production (Darwin *et al.*, 1995). More recently, studies that reflect CO_2 fertilization impacts and adaptation suggest that global agricultural supply is likely to be robust in the face of moderate warming (Kurukulasuriya and Rosenthal, 2003). Besides hampering the agricultural productions, climate change may result in shifts in crop calendars and crop health also. Climate change thus represents an additional burden that for farmers translates into production risks associated with crop yields, probabilities of extreme events, timing of field operations, and timing of investments in new technologies (CGIAR-CCAFS, 2012). Moreover, climate change is likely to increase pest outbreaks in agriculture. Another important concern is demand of irrigation which is expected to increase by 5-20% as a result of rising temperature and changing rainfall regimes.

How climate change can hit agriculture: Kurukulasuriya and Rosenthal (2003), in their revision of how climate change can influence agriculture have described the ways by which climate change can affect agriculture in the future. They have followed the work of Hulme (1996) who described that by means of four ways climate change would have a physical effect on crops. *First*, the changes in temperature and precipitation regimes will obviously change the distribution of agro-ecological zones. There will be regional variations in the changes in soil moisture content and the timing and length of growing seasons. Rosenzweig and Hillel (1995) stated that, in middle and higher latitudes, higher temperatures will lengthen growing seasons and expand crop-producing areas pole-ward, thus, benefiting countries in these regions. In contrast, in lower latitudes, it is expected that higher temperatures will have negative impacts on growing conditions, especially in areas where temperatures are close to or at the optimal level for crop growth to begin with. Irrigation availability and demand will also be affected by both changes in temperature and precipitation. An increase in potential evapotranspiration would intensify drought stress, particularly in the semiarid tropics and subtropics. *Second*, carbon dioxide effects are expected to have a positive impact due to, for example, greater water use efficiency and higher rate of photosynthesis. Rising carbon dioxide concentrations in the atmosphere will increase the rate of photosynthesis and water use efficiency. These effects are strongest for crops with the C_3 photosynthetic pathway which include crops such as wheat, rice, and soybean etc. Carbon dioxide enrichment is also positive—though not by as much— for C_4 plants such as maize, millet, and sorghum, and many grasses. IPCC (1996) and Reilly *et al.* (1996) have estimated that a doubling of carbon dioxide concentrations would lead to yield improvements ranging from 10–30%. *Third*, water availability will be another important variable for assessing climate change impacts on agriculture especially for the drier regions of the world, for example, Africa. Precipitation and cropping (particularly the length of growing season) are no doubt intimately linked. Future water availability will largely depend on rainfall distribution. Previously, it has been mentioned that the future rainfall patterns are likely to increase in the high latitudinal regions and will decrease in the lower latitudinal areas which will obviously create acute water shortages in the lower latitudes that in turn will hamper the crop productions particularly in the sub-Saharan Africa. *Fourth*, increasing trends in weather and climate extremes such as frequent events of droughts, floods, absence of proper seasonal variations and shifting rainfall and water regimes could raise the agricultural losses in the upcoming future. Hulme (1996) exemplified the droughts of 1984/85 and 1991/92 in southern Africa which left significant impacts on maize production of the concerned region.

The World Bank (2007) identifies five main factors through which climate change will affect the agricultural productivity: changes in precipitation, temperature, CO_2 fertilization, climate variability, and surface water runoff. Reilly *et al.* (2003) found that higher precipitation leads to a reduction in yield variability. Therefore, higher precipitation will reduce the differences in crop yields between rainfed and irrigated agriculture, but it may also have a negative impact if extreme precipitation causes flooding (Falloon and Betts, 2009). Temperature and soil moisture determine the length of growing season and control the crop's development and water requirements. In general, higher temperatures will shorten the freeze periods, promoting cultivation in cool-climate marginal croplands. However, in arid and semi-arid areas, higher temperatures will shorten the crop cycle and reduce crop yields. A higher atmospheric concentration of carbon dioxide enhances plant growth and increases water use efficiency (CO_2 fertilization) and so affects water availability. Climate variability, especially changes in rainfall patterns, is particularly important for rainfed agriculture. Lesser rainfall and resultant deficiency in soil moisture will reduce crop productivity in case of rainfed farming systems. Although, rainfall variability generated risks are insignificant in case of irrigated farming systems, this can also be exposed to changes in the spatial and temporal distribution of river flow as irrigation systems entirely depend on run-off regimes (Calzadilla *et al.*, 2010).

Climate extremities and variabilities will enhance the fragility of food production systems and the water resources, particularly in regions having acute water scarcity. Cultivators will be forced to clear the historical records of temperature and rainfall from their memories and will be in trouble to decide when to start cultivation of a particular crop. Modest climate change (a global mean temperature rise of around 2°C) will cause a decrease in agricultural production particularly in the hot and dry regions and increase in areas within middle to high latitude regions. Even a 2°C increase could culminate enough risks in the crop productivity (CGIAR-CCAFS-2012). Aggarawal and Sinha (1993) have shown that in North India, a 1°C rise in mean temperature would have very insignificant effect on wheat yields while a 2°C increase would reduce yields in most places. Climate change is likely to change rainfall patterns, resulting in shorter growing seasons in the future, particularly for subsistence farmers in Africa and parts of South Asia who rely on rainfed agriculture (CGIAR-CCAFS-2012). However, climate change effects on crop production will be more complex than those related to changes in average temperature and rainfall: combinations of stress are common and important. It is expected that agriculture (without any further adaptation), especially in the dry and wet tropics, will be more affected by an increased frequency of extreme events, rather than the mean change in climate (Porter and Semenov, 2005). Thus the factors affecting crop productivity are entirely weather based, rather than determined by long-term climate i.e., (1) changes in mean weather (temperature and rainfall); (2) changes in variability or distribution of weather; (3) combinations of both changes in the mean weather and changes in its variability (Turral *et al.*, 2011).

Potential Regional Variations: It is very likely that there will be higher variations in the impacts of climate change both at local and regional scales. However, there is a doubtless agreement among the scholars that, the tropical regions will suffer the worst. Battisti and Naylor (2009) found that, growing season temperatures of the crops probably will exceed the most extreme seasonal temperatures recorded from 1900 to 2006 in the tropical and sub-tropical areas by end of this century. Based on such consensus, scientists are now expecting that, there will be shifts of tropical crops and agricultural practices towards the temperate and polar regions in the forthcoming decades. Moreover, the agricultural losses will largely affect the economy of the developing nations. Early estimates suggested that, developed countries will face about 4 to 24% losses in their agricultural production while this toll will be around 14 to16% in case of developing countries (IPCC 1996). Dryland areas (arid, semi-arid and drought-prone areas) are likely to be among the most vulnerable.

It has been projected that, there will be declines in gross agricultural production in most of Africa and South and East Asia. Murdiyarso (2000) stated that rice production in Asia may decline by 3.8% of production levels of 2000 (estimated at 430 metric tonnes) as a result of future climate variabilities. Rosenzweig and Parry (1994) have found that a 4°C rise in temperature could result in losses in grain yields by about 25 to 40% for India. Another study in Indonesia (Lawler *et al.*, 2011) found that a one-month delay in rainfall would result in 11% decline in the yield of wet season rice in East Java and Bali. Besides, climate change can shift altitudinal locations of the agricultural fields particularly in mountainous areas of the tropical and sub-tropical regions. A perception study conducted by ICIMOD (2011) in the Himalayan states of India along with Nepal and Bhutan has revealed that due to climate change fruit and other crops can be grown at higher altitudes where it was previously not possible. On the other hand, at high latitudes crop yields are expected to rise with temperature increases of 1–3°C, but fall, due to declining crop health, once 3°C is exceeded. Rosenzweig (1985) stated that climate change would increase the winter wheat production in Canada.

Most climate change models suggest future strengthening of the summer monsoon. In Asia this might increase rainfall by 10 to 20%, but more importantly will be accompanied by a dramatic increase in inter-annual variability (Kumar *et al.*, 2006). However, since the beginning of the 21st century the nature of Asian monsoon has changed greatly, both spatio-temporally. For example, large parts of the rice belts of India faced water scarcity due to erratic monsoonal rainfall in recent years. Moreover, the monsoon trough is largely shifting from its previous positions i.e. over the Indo-Gangetic plains in northern India. These tendencies have been accompanied by late onset and early retreat of monsoon rainfall along with greater spatio-temporal variability. Such changes in monsoon can cause massive loss of rice production in India. The climate models have further indicated that, some arid areas will become even drier, including the Near East, parts of China, the Mediterranean Basin (southern Europe and North Africa), northeastern Brazil, and west of the Andes in southern South America, West Africa and southern Africa. Most of the climate models predicted that, the absolute rainfall in Africa will decrease while variability will increase and thus, the crop productions will be largely affected. In Latin America, the yield of annual crops such as wheat, maize, rice and soybean in several regions will be reduced due to higher temperatures and shorter growing seasons. Moreover, the coffee producing regions will move to different locations. Besides this, coffee yields and quality are likely to change with slight temperature increases (1-2°C). Above all, the disappearing glaciers and reduced snow melt is likely to reduce water availability for irrigation (Ludwig *et al.*, 2007).

Threaten Global Food Security: The World Food Summit of 1996 defined food security as existing "*when all people at all times have access to sufficient, safe, nutritious food to maintain a healthy and active life*" (WHO, 2013). Agriculture is the mainstay of about half of the world population. Besides feeding the 7 billion world population, it forms the economic backbone of the developing and under-developed countries. From 1961 to 2003, world food trade has increased from 1500 Gkcal/day to over 7000 Gkcal/day (CGIAR-CCAFS, 2012). Thus, the questions regarding achieving food security and a good economic health in the future world are increasing day by day. Although some regions may experience an increase in crop productivity as a result of more favourable growing conditions, the overall impacts of climate change on agriculture are expected to be negative, threatening the global food security. An estimate by FAO has revealed that, currently 80% of the world's agricultural fields are rainfed which will suffer in the upcoming decades from water scarcity and thus, the agricultural yields and food production will be hampered largely. Conjointly, the total irrigated agricultural land (18%) of the world which is currently producing more than a billion tonnes of grains is projected to decline steadily due to acute water shortages in the upcoming future (Turral *et al.,* 2011). Beyond these, the Millennium Development Goals (MDGs) can hardly meet the demands of food by 2015 (IPCC, 2008). It is now projected that, the subsistence agricultural sectors particularly in Africa and parts of Asia will face hard challenges due to climate change by 2080 and by that time about 75% of the African population will face hunger and malnutrition (Turral *et al.,* 2011). Rainfalls and runoff are anticipated to decline in North Africa and along the Sahel margin, with some dramatic changes in land use (increased desertification or encroachment) and reduced growth potential by 2050. Further south, in West Africa, agriculture GDP is projected to decline between 2 and 4%. In southern Africa, increased water stress is anticipated as a result of lower rainfall, higher potential evapotranspiration and higher temperatures. In Asia, areas which depend upon Himalayan glaciers for their irrigation purposes will face water shortages due to mass retreat and loss of the Himalayan glaciers and ice covers. In addition, irrigation demand is projected to increase by 10% for every 1°C rise in temperature in arid and semi-arid East Asia, bringing water scarcity to between 0.12 and 1.2 billion people by 2050. Crop yields are predicted to rise by up to 20% in eastern and Southeast Asia, but expected to decline by 30% and become more variable in central and southern Asia (Turral *et al.*, 2011).

6. COPING WITH CLIMATE CHANGE

By so far, it is now clear what climate change is and what it can yield for both the geo-physical and socio-economical-political systems of ours. In fact, coping with climate change is in reality, would be the most difficult job for the mankind as from the very beginnings climate systems of the world are in change but worries came to the scene only after injecting the manmade GHGs into atmosphere which have somehow worsened the picture. However, it is quite a bit relaxing that, we have at least recognized what we have done to our climatic systems and no doubt, we are now vigorously in action to bring down the scenarios at least to certain relaxing stages. Many efforts from various scientific platforms are now in operation to assess how we can cope with climate change, speaking more precisely the global warming. The very outcome of these efforts clearly indicates that, it would be better to adapt the ongoing climate changes rather than to mitigate these. Since the initial stages, the international climate effort has predominantly kept its eyes on mitigation, i.e. reducing GHG emissions to prevent dangerous climate change. The next stage of the international effort must deal squarely with adaptation i.e. coping with the unavoidable impacts (Burton *et al.*, 2006). Thus, the two sets of concept i.e. adaptation and mitigation have emerged into the set. Actually, both these concepts are necessary to deal with global warming.

Adaptation to climate actually, is not a new phenomenon. Indeed, throughout the history of human civilizations there are numerous instances of adaptation such as transforming agriculture, lifestyles, reallocating settlements etc. Although, human induced warming has broken the natural trends of climatic changes but considering the long span of human history and pre-history, human adaptation to climate has always been comprehensively successful (Burton *et al.*, 2006). Adaptation, as defined by the UNFCCC (2013) "*refers to adjustments in ecological, social, or economic systems in response to actual or expected climatic stimuli and their effects or impacts. It refers to changes in processes, practices, and structures to moderate potential damages or to benefit from opportunities associated with climate change*". Adaptation will be required in dealing with the climate change risks already induced by historical emissions and those projected in the forthcoming future. However, it should be remembered that though adaptation has the potentiality to reduce the adversities of climate change, but all the damages may not be prevented (IPCC, 2001). Despite this, adaptation is no doubt most urgent step for those risks which have already been faced and for those which are sensitive to even small or little changes in the climatic parameters. Adaptation will be difficult and costly in case of higher level of warming which will require a great deal of accepted loss (Jones *et al.*, 2006). Adaptation takes place at a range of inter-linking scales and can occur either as a response to climatic changes (reactive adaptation) or as an anticipation of those changes (proactive or anticipatory adaptation). For example, artificial snow making in the Alps can be attributed as a reactive adaptation while the construction of confederation bridge in Canada at higher elevations considering the effect of future sea level rise on ship clearance under the bridge is a proactive adaptation (Wikipedia, 2012). The concepts of vulnerability and adaptive capacity are very important in the context of adaptation. Vulnerability is the context in which adaptation occurs and adaptive capacity on the other hand denotes the capability or potentiality of a system to response successfully to the variabilities and changes in climate in order to avoid the negative impacts and to take benefits of the new opportunities. Where vulnerability is high and adaptive capacity is low, it has really become difficult for the system to sustain without transformations (IPCC, 2012). However, this adaptive capacity is partly a function of a society's material and socio-cultural developments i.e. the level of wealth, education, institutional strength, and technological access. Therefore, the nature and extent of a society's development, largely determine both its degree of exposure to risks associated with climate change and its adaptive capacity (Burton *et al.*, 2006). Adaptation covers both spatio-temporal dimensions based on the contexts. Much adaptation which has followed temporary climatic extremes may lead to a false feeling of safety (IPCC, 2012) or maladaptation (Wikipedia, 2013d). For instance, temporary guard walls on hill-slopes constructed after a landslide due to heavy and unprecedented rains may be a temporary solution which can no longer provide safety during massive landslides or block failures, as it can be seen in the hill-slopes of Eastern Himalayas. Thus, during the framing of adaptation policies policymakers should keep in their minds certain important principles (following Scheraga and Grambsch, 1998):

(i) There may be regional variations in climate change effects.

(ii) Besides, the effects may vary across demographic groups too.

(iii) Climate change may have both adversities and opportunities.

(iv) The effects of climate change should be taken into account in the context of multiple stressors and factors which may be important to framing the adaptive responses as the sensitivity of the change.

(v) Adaptation needs a cost.

(vi) Adaptive responses vary in effectiveness as demonstrated by present efforts to cope with climate variability.

(vii) The systematic nature of climate effects complicates the development of adaptation policy.

(viii) Maladaptation can yield adverse effects as serious as the climate change impacts that are being avoided.

(ix) Many opportunities for adaptation make sense whether or not the effects of climate change are realized.

The various ways (both adaptation and mitigation) of reducing climate relate risks and to avoid or manage the climate related influences upon different geophysical, biotic as well as social, economic and human resources are as follows:

- **Ecosystems, Biodiversity and Forestry:** Afforestation or reforestation, prevention of deforestation, sound forest management, monitoring of forests, animals and wildfires, increasing diversity of plants in forests, biodiversity conservation, conservation of wetlands, removal of old and damaged trees and silviculture to reduce diseases and pest attacks (FAO, 2012), enhancing ecosystem resilience, restoring protective ecosystem services, use of Ecosystem-based Adaptation (EbA), and agroforestry (IPCC, 2012).
- **Water Resources:** CO_2 capture and storage, CO_2 sequestration to enhance water efficiency, afforestation to maintain hydrological cycle, use of geothermal energy, biomass electricity and hydropower to reduce the energy sector usages of water, water management for croplands, Integrated Water Resource Management (IWRM), storage plans, waste water treatment, desalinification, joint river basin management, negotiations for water resources allocation (IPCC, 2008 and 2012), rain water harvesting etc.
- **Agriculture:** Cropland and grazing management, soil conservation, reduced use of fertilizers which emit N_2O, changing the timing of farm operations, crop and livestock diversification, improved nutrient and pest control management, seasonal migration of agricultural labour, insurance mechanisms to reduce climate related risks in agriculture, changing crop type and location, modernization of agricultural technology, improving agricultural water management (Kurukulasuriya and Rosenthal, 2003), supply of climate stress tolerance seeds, integrated pest and disease management (IPCC, 2012), conservation agriculture measure such as reduction in tillage, retention of adequate levels of crop residues and soil surface cover, use of crop rotations (Verhulst *et al.*, 2012).
- **Health:** Monitoring and control of vectors, better sanitation, rationing of medical facilities, waste water treatment, water and air quality monitoring, improvement of health care services (IPCC, 2008 and 2012), improving medical and health facilities, formulation of health policies and plans including climate related health risks (ADB, 2011), improving the mechanism of medical and health insurance.
- **Island Nations and Coastal Zones:** Construction of sea walls, conservation of mangrove habitats, integrated Coastal Zone Management (ICZM), formulation and enforcement of laws to prevent growth of settlements along the coastlines, desalinification of ocean water, embankments to avoid inundation of sea water into the residential areas and agricultural fields.

- **Floods and Droughts:** Early warnings, construction of embankments, dikes to avoid penetration of floodwater; dams, reservoirs to store water; canals to channelize the river water into agricultural fields as irrigation water; construction of wells, ponds to store water during rains, rainwater harvesting, arrangements of roofs to use water during droughts.
- **Tourism:** Technical measures (artificial snowmaking, cyclone-proof building design and construction, enable early warning equipment, e.g., radios to tour operators, weather forecasting and early warnings); managerial initiatives (low season closures, product and market diversification, encourage environmental management with firms, business subsidies, e.g., insurance); policy issues (hurricane interruption guarantee, coastal management plans, building design standards etc.); research operations (site and location, monitoring programmes, extreme event risk exposure); educational efforts (water conservation education for guests and employees, public education campaign, campaign on the risks of UV radiation) etc. (UNEP, 2008).

7. CONCLUSION

From the above discussions, it is now quite obvious that, there is a clear-cut difference between climate change and global warming; while climate change is a natural one, global warming is solely the culmination of ours actions. Although, climate change poses enough risks and opportunities in the futures but we cannot ignore neither our roles in global warming nor the uncertain climate related adversities in the future. In fact, there are certain people who still believe that, climate change (speaking more precisely global warming) is not a big problem and humans can easily handle it. Is this really possible? Can we easily handle the aforesaid climate vulnerabilities which seem to threat not only the geophysical and biological environments but also our survival? By so far we have successfully adapted to our climates and have avoided many climatic troubles, but the future climates will be quite uncertain and we will no longer be able to rely on the historical records or averages. Under such anticipated circumstances, our survival will be really difficult. Undoubtedly, the mitigation and adaptation strategies undertaken so far have taught us well about how to adapt to a world where your surrounding climatic parameters are altering. But, we cannot depend on the policymakers and institutions only, as the inhabitants of this world, we have our own responsibilities and we cannot run away from this truth. We have to keep in our minds that, climate change is an eternal truth, i.e. it was in operation since the geologic pasts and it will always be active in the future, and this cannot be denied. At such a condition, we have to drive our every action in the way to keep our atmosphere safe. Indeed, adaptation starts from an individual and ultimately, a whole community follows him or her. Although, it may take time but it will foster a relatively safer climate in the future.

References

Aggarawal, P.K., and Sinha, S.K. (1993). Effect of probable increase in carbon dioxide and temperature on wheat yields in India. *Journal of Agricultural Meteorology,* 48(5): 811-814.

Barry, R.G., and Chorley, R.J. (1998). *Atmosphere, weather, climate.* Routledge, London and New York, USA. pp. 325-349.

Battisti, D., and R. Naylor R. (2009). Historical warnings of future food insecurity with unprecedented seasonal heat. *Science*, 323: 240-244.

Beddington, J., Asaduzzaman, M., Clark, M., Fernández, A., Guillou, M., Jahn, M., Erda, L., Mamo, T., Van Bo, N., Nobre, C.A., Scholes, R., Sharma, R., Wakhungu, J. (2012). *Achieving food security in the face of climate change: Final report from the Commission on Sustainable Agriculture and Climate Change*. CGIAR Research Program on Climate Change, Agriculture and Food Security (CCAFS), Copenhagen, Denmark. Available online at: www.ccafs.cgiar.org/commission.

Buchheim, J. (1998). *Coral reef bleaching*. Odyssey expeditions- Marine Biology Learning Center Publications. Retrieved on 09/10/2012 from www.marinebiology.org/coralbleaching.htm.

Burton, I., Diringer, E., and Smith, J. (2006). *Adaptation to climate change: international policy options*. Pew Center on Global Climate Change, VA, USA.

Burton, P.J., Bergeron, Y., Bogdansky, B.E.C., Juday, G.P., Kuuluvainen, T., McAfee, B.J., Ogden, A., Teplyakov, V.K., Alfaro, R.I., Francis, D.A., Gauthier, S. and Hantula, J. (2010). Sustainability of boreal forests and forestry in a changing environment. In: *Forests and society – responding to global drivers of change*. Mery, G., Katila, P., Galloway, G., Alfaro, R., Kanninen, M., Lobovikov, M., Varjo, J. (Eds.). IUFRO World Series. pp. 249-282.

Calzadilla, A., Rehdanz, K., Betts, R., Falloon, P., Wiltshire, A. and Tol, R. S. J. (2010). *Climate Change Impacts on Global Agriculture* (Working Paper No. 1617), Kiel Institute for the World Economy, Kiel, Germany.

CLIMATE CHANGE. (2007). *Impacts, vulnerabilities and adaptation in developing countries*. United Nations Framework Convention on Climate Change, Bonn, Germany. pp. 68.

Cutting, H., Chung, S. and Hassol, S. (2011). *Current extreme weather and climate change*. Climate Communication. pp. 28.

Darwin, R., Tsigas, M., Lewandrowski, J. and Raneses A. (1995). *World agriculture and climate change: economic adaptations*. Agricultural Economic Report, 703, U.S. Department of Agriculture, Economic Research Service, Washington D.C.

Ding, T., Qian, W. and Yan, Z. (2010). Changes in hot days and heat waves in China during 1961–2007. *International Journal of Climatology*, 30: 1452–1462.

Earth System Research Laboratory- National Oceanic & Atmospheric Administration. (2013). *Trends in atmospheric carbon dioxide*. Retrieved on 02/10/2012 from www.esrl.noaa.gov/gmd/ccgg/trends/global.html.

Falloon, P.D., and Betts, R.A. (2009). Climate impacts on European agriculture and water management in the context of adaptation and mitigation - The importance of an integrated approach. *Science of the Total Environment*, 408(23): 5667-5687.

Food and Agricultuaral Organization of United Nations. (2012). Forest Management and Climate Change: a literature review. *Forests and Climate Change Working Paper 10,* Rome, Italy.

Guariguata, M. (2006). *Interlinkages between Biodiversity and climate change in 'Migratory Species and Climate Change: Impacts of a Changing Environment on Wild Animals'*. UNEP/CMS Secretariat, Bonn, Germany. pp. 8-11.

Harvey, F. (2012). Doha climate change deal clears way for 'damage aid' to poor nations. *theguardian|TheObserver*. Retrieved on 12.11.2012 from http://www.guardian.co.uk/environment/2012/dec/08/doha-climate-change-deal-nations.

Hobbs, J.E. (1983). *Applied climatology*. Butter Worths, London and Boston. pp. 218.

Hulme, M. (1996). *Climate change and southern Africa*. Climatic Research Unit, University of East Anglia, Norwich, United Kingdom.

IPCC. (2001). *Climate change 2001: impacts, adaptation, and vulnerability, summary for policymakers, A Report of Working Group II of the Intergovernmental Panel on Climate Change.* (Based on a draft prepared by: Ahmad, Q.K., Anisimov, O., Arnell, N., Brown, S., Burton, I., *et al.*), Intergovernmental Panel on Climate Change, Geneva, Switzerland. pp. 18.

IPCC. (2007a). *Climate Change 2007: Synthesis Report, Summary for Policymakers, An Assessment of the Intergovernmental Panel on Climate Change.* (draft prepared by Bernstein, L., Bosch, P., Canziani, O., Chen, Z., Christ, R., Davidson, O., *et al.*). Intergovernmental Panel on Climate Change. pp. 22.

IPCC. (2000). *EMISSIONS SCENARIOS, Summary for Policymakers, A Special Report of IPCC Working Group III, Published for the Intergovernmental Panel on Climate Change.* (Draft Prepared by Nakic′enovic′, N., Davidson, O., Davis, G., Grübler, A., Kram, T., *et al.*). pp. 21.

IPCC. (1996). *Impacts, Adaptations, and Mitigation of Climate Change: Scientific-Technical Analyses - Contribution of Working Group II to the IPCC Second Assessment Report.* Cambridge University Press, Cambridge, U.K.

IPCC. (2012). *Managing the Risks of Extreme Events and Disasters to Advance Climate Change Adaptation. A Special Report of Working Groups I and II of the Intergovernmental Panel on Climate Change,* [Field, C.B., Barros, V., Stocker, T.F., Qin, D., Dokken, D.J., *et al.* (Eds.)]. Cambridge University Press, Cambridge, UK, and New York, NY, USA. pp. 582.

IPCC. (1997). *THE REGIONAL IMPACTS OF CLIMATE CHANGE: AN ASSESSMENT OF VULNERABILITY, Summary for Policymakers, A special Report of IPCC Working Group II Published for the Intergovernmental Panel on Climate Change* [Watson, Robert T., Zinyowera, Marufu C., Moss, Richard H. and Dokken, David J. (Eds.)] Intergovernmental Panel on Climate Change. pp. 16.

IPCC. 2008. Technical paper of the Intergovernmental panel on climate change bates. In: *Climate Change and Water*, Kundzewicz, B.C., Wu, S., and Palutikof, J.P. (Eds.). IPCC Secretariat, Geneva. pp. 210.

Jhonson, B. (2013). Severe droughts in Amazon linked to climate change, says study. *CBSNEWS*, retrieved on 19.10.2012 from http://www.cbsnews.com/8301-205_162-57564795/severe-droughts-in-amazon-linked-to-climate-change-says-study/.

Jones, R.N. and Preston, B.L. (2006). *Climate Change Impacts, Risk and the Benefits of Mitigation.* A Report for the Energy Futures Forum, CSIRO, Victoria.

Kurukulasuriya, P. and Rosenthal, S. (2003). *Climate change and agriculture: A Review of Impacts and Adaptations.* Agriculture and Rural development Department, Paper No. 91. World Bank, Washington, DC. Retrieved on 26.12.2012 from www.c-ciarn.uoguelph.ca/documents/World_Bank_Paper.pdf

Kumar, K.R., Sahai, A.K., Krishna Kumar, K., Patwardhan, S.K., Mishra, P.K., Revadekar, J.V., Kamala, K. and Pant, G.K. (2006). High-resolution climate change scenarios for India for the 21st century. *Current Science*, 90(3): 334-345.

Lal, D.S. (2009). *Climatology.* Sharda Pustak Bhawan, Allahabad, India. pp. 379-391.

Laplante, B., Peralta, G., and Predo, C. (2011). *Accounting for health impacts of climate change.* Asian Development Bank, Publication Stock No. RPT113738, Mandaluyong City, Philippines.

Lawler, J., Urbano, M., Deasey, M., Kusumawardani N., Pardosi, J., Burton D., Mustelin, J., Urich, P., Nenova-Knight, P., Perez, R., Loyzaga A. Y., Narisma, G. T., Vicente, C., Olaguer, D. and Patel M. (2011). *Children's Vulnerability to Climate Change and Disaster Impacts in East Asia and the Pacific.* UNICEF East Asia and Pacific Regional Office, Bangkok, Thailand.

Lucier, A., Ayres, M., Karnosky, D., Thompson, I., Loehle, C., Percy, K. and Sohngen, B. (2009). Forest responses and vulnerabilities to recent climate change. In: *Adaptation of forests and people to climate change.* Seppala, R., Buck, A. & Katila, P. (Eds.), IUFRO World Serie. pp. 22.

Ludwig, F., Scheltinga, C.T. van, Verhagen, J., Kruijt, B., Ierland, E. van, Dellink, R., Bruin, Karianne de, Bruin, Kelly de and Kabat, P. (2007). *Climate change impacts on Developing Countries - EU Accountability*. European Parliament, Brussels, Belgium. pp. 45.

Macchi, M., Gurung, A.M., Hoermann, B., Choudhary, D. (2011). *Climate variability and change in the Himalayas: Community perceptions and responses*. Kathmandu: ICIMOD.

Mac, M.J., Opler, P.A., Puckett Haecker, C.E. and Doran, P.D. (1998). *Status and Trends of the Nation's Biological Resources*. U.S. Department of the Interior, U.S. Geological Survey. pp. 89-93.

McMullen, Catherine P. and Jabbour, J. (2009). *Climate change science compendium*, UNEP.

Meehl, G. A., Tebaldi, C., Walton, G., Easterling, D. and McDaniel, L. (2009). Relative increase of record high maximum temperatures compared to record low minimum temperatures in the U.S. *Geophys. Res. Lett.*, 36(23): L23701.

Meesters, H. and Wanjiru, L. (2009). *Resource guide on gender and climate change*. UNDP.

Murdiyarso, D. (2000). Adaptation to climatic variability and change: asian perspectives on agriculture and food security. *Environmental Monitoring and Assessment*, 61(1): 123-131.

Oliver, J.E. and Hidore, J.J. (2003). *Climatology: an atmospheric science*. Pearson Education, Inc. pp. 410.

Parry, M., Arnell, N., Berry, P., Dodman, D., Fankhauser, S., Hope, C., Kovats, S., Nicholls, R., Satterthwaite, D., Tiffin, R., Wheeler, T. (2009). *Assessing the Costs of Adaptation to Climate Change: A Review of the UNFCCC and Other Recent Estimates*. International Institute for Environment and Development and Grantham Institute for Climate Change. London.

Porter, J. and Semenov, M. (2005). *Crop Responses to Climatic Variation*. Philosophical Transactions of The Royal Society B, 360; 2021–2035.

Prasada Rao, G.S.L.H.V. (2008). *Agricultural Meteorology*. Prentice-Hall of India Private Limited, New Delhi, India. pp. 322-324.

Reilly, J., Baethgen, W., Chege, F.E., Geijn van de, S.C., Erda, L., Iglesias, A., Kenny, G., Patterson, D., Rogasik, J., Rötter, R., Rosenzweig, C., Sombroek, W., and Westbrook, J. (1996). Agriculture in a Changing Climate: Impacts and Adaptation. In: Watson, R.T., Zinyowera, M.C., and Moss, R.H., (Eds.). *Climate Change 1995: Impacts, Adaptations, and Mitigation of Climate Change: Scientific-Technical Analyses. Contribution of Working Group II to the Second Assessment Report of the Intergovernmental Panel on Climate Change*. Cambridge University Press, Cambridge, U.K. and New York, N.Y.

Reilly, J., Tubiello, F., McCarl, B., Abler, D., Darwin, R., Fuglie, K., Hollinger, S., Izaurralde, C., Jagtap, S., Jones, J., Mearns, L., Ojima, D., Paul, E., Paustian, K., Riha, S., Rosenberg, N., and Rosenzweig, C. (2003). U. S. Agriculture and Climate Change: New Results. *Climatic Change*, 57: 43–69.

Rosenzweig, C. (1985). Potential CO2-Induced Effects on North American Wheat Producing Regions. *Climatic Change*, 7: 367-389.

Rosenzweig, C. and Hillel, D. (1995). *Climate Change and the Global Harvest: Potential Impacts on the Greenhouse Effect on Agriculture*. Oxford University Press, New York.

Rosenzweig, C. and Parry, M.L. (1994). Potential Impacts of Climate Change on World Food Supply. *Nature*, 367: 133-138.

Scherafga, J.D. and Grambsch, A.E. (1998). Risk, opportunities, and adaptation to climate change. *Climatic Research*, 10: 85-95.

Schneider, S. (1990). The global warming debate heats up: an analysis and perspective. *Bulletin of the American Meteorological Society*, 71(9): 1292- 1304.

Schwarzer, S. (2012). *Growing greenhouse gas emissions due to meat production*. UNEP Global Environmental Alert Service. pp. 9.

Singh, S. (2007). *Climatology.* Prayag Pustak Bhawan, Allahabad, India. pp. 339-394.

Singh, S. (2005). *Geomorphology.* Prayag Pustak Bhawan, Allahabad, India. pp. 118-121

Smith, T.M. and Reynolds, R.W. (2005). A global merged land–air–sea surface temperature reconstruction based on historical observations (1880–1997), *Journal of Climate*, 18: 2021-2036.

Turral, H., Burke, J. and Faurès, J. M. (2011). Climate change, water and food security, *FAO water reports*, 36, Food and Agriculture Organization of the United Nations, Rome

UN Framework Conventation on Climate Change. (2012). *FOCUS: Adaptations.* Retrieved on 25.12.2012 from http://unfccc.int/focus/adaptation/items/6999.php.

United States Environmental Protection Agency. (2012). *Global Emissions by Gas*. Retrieved on 29/12/2012 from http://www.epa.gov/climatechange/ghgemissions/global.html.

Verhulst, N., Govaerts, B., Sayre, K.D., Sonder, K., Romero-Perezgrovas, R., Mezzalama, M., Dendooven, L. (2012). Conservation agriculture as a means to mitigate and adapt to climate change, a case study from Mexico. In: *Climate Change Mitigation and Agriculture.* Wollenberg, E., Nihart, A., Tapio-Biström, M.-L., Grieg-Gran, M. (Eds.). Earthscan, Oxon. pp. 287-300.

Wergen, G. and Krug, J. (2010). Record-breaking temperatures reveal a warming climate. *Europhysics Letters*, 92(3): 30008.

White, P., Hilario, F.D., Guzman, R.G. de, Cinco, T.A., Abery, N.W. and De Silva, Sena S. (2010). A review of climate change model predictions and scenario selection. *Aqua Climate*, 29.

Wikipedia. (2012). *Adaptation to global warming*. Online source. Retrieved on 18.12.2012 from http://en.wikipedia.org/wiki/Adaptation_to_global_warming.

World Bank. (2007). *World development report 2008: Agriculture for development.* World Bank, Washington D.C.

World Health Organization. (2012). *Food Security*. Retrieved on 08/10/2012 from www.who.int/trade/glossary/story028/en/.

Young, I., Zieger, S., Babanin, A. (2011). Global trends in wind speed and wave height. *Science*, DOI: 10.1126/science.1197219.

Zwiers F., Zhang, X., and Feng, Y. (2010). Anthropogenic influence on long return period daily temperature extremes at regional scales. *Journal of Climate.* doi:10.1175/2010JCLI3908.1.

freeze exists when the air remains below freezing (0°C) for a sufficient period of time (at least 1 or 2 days) over a widespread region (Huschke, 1959). Water within plants may or may not freeze during a frost event. But, when water freezes, intracellular or extracellular freezing in a plant takes place. Intracellular freezing in a plant is immediately fatal. Extracellular freezing in a plant may cause injuries. The extent of injury depends on plant resistance to frost. Plant resistance is associated with a critical temperature below which a malfunction or death of the cells takes place. Frosts are usually classified into:

(a) **Advection frosts:** Advection frosts are a result of large-scale cold wind incursions often from polar regions. These types of frosts have the following characteristics:
  - They can develop during day or night.
  - They are associated with windy conditions and a well-mixed atmosphere.
  - Air temperature may be less than 0°C during the day.

(b) **Radiation frosts:** Radiation frosts are a result of intense long the wave radiation cooling of the earth's surface due to energy loss during calm clear nights. These types of frost have the following characteristics:

- There is an associated temperature inversion.
- Nights are clear and calm.
- Air temperature is greater than 0°C during day. (Kalma *et al.*, 1992)

Snyder and Paulo de Melo-Abreu (1995) have indicated cases where the observed conditions are considered as "micro-scale advection frosts". These frosts occur when radiation frost exists in a region, but cold air drainage on a local scale, leads to rapid loss in temperature within the radiation frost area.

The most damaging frosts are the advective frosts followed by radiation frosts.

2.2 Frost Damages

Total frost injuries are mainly contributed by winter frosts. Table 1 shows weather related agricultural damages and the frost contribution to total damages. It is clear that frost damages account to more than half of the total damages due to weather related phenomena.

Table 1 Weather related phenomena and agricultural damages in Greece (in thousands euros)

Weather related phenomenon	*2005*	*2006*	*2007*	*2008*	*TOTAL*
Wind storm	13.900	11.300	14.100	12.600	51.900
Rainfall	57.000	11.500	131.900	39.100	239.500
Heat wave	29.191	261.000	195.750	431.175	917.116
Frost	175.000	315.950	115.370	69.590	675.910
Flood	7.000	5.350	6.590	11.000	29.94
Hail	51.750	36.150	47.950	55.650	191.500
Snow	119	5.900	1.115	3.750	129.765
Total	452.841	647.150	512.775	622.865	2235.631
Frost contribution (%)	38.64	48.82	22.49	11.17	30.23

2.3 Frost Protection Methods

Various protection methods are used around the globe. These methods are classified as either passive or active. Passive methods are utilized prior to the frost event in order to avoid or minimize injuries.

Main passive methods include the following:

- **Proper site selection:** It is probably the most important passive protection method. Hilltops and middle portions of hillsides are most volatile to advection frosts where the temperatures observed are usually higher than down-wind sides and low spots that are sheltered from the wind. Exactly the opposite is observed during a radiation frost event.
- **Cold air drainage management:** Cold air drained downhill can be diverted by using methods such as building a solid wall or using a wooden fence. This diversion can effectively provide protection.
- **Plant covers:** Canopy trees provide protection from radiation frosts since there is an enhanced long-wave radiation downward from the trees. The effect is that temperatures are higher under these trees than in the open space.

Other passive methods include: Proper plant selection for forestation, avoiding cultivation of the forest land. Active methods are used during a frost night. They mainly include application of wind machines, sprinklers, heaters, surface irrigation, foggers. Passive methods of frost protections are used more widely in agricultural regions, because they are cost-effective and highly beneficial. Active protection methods are costly, since most of them are fuel dependent. Wind machines and microsprinklers are used as active protection methods only in a small scale. Horizontal (conventional) blowing machines are used on a large scale (Rosenberg *et al.*, 1983).

3. HAIL

Hail forms inside thunderstorm cumulonimbus (Cb) clouds, mainly those with intense updrafts, high liquid water content, great vertical extent, large water droplets, and where a good portion of the cloud layer is below freezing temperature. Hail grows in the cloud's main updraft, where most of the cloud is in the form of supercooled water. This is water that remains liquid although its temperature is at or below 0°C.The growth rate is maximized at about –13°C, and becomes vanishingly small much below 30°C as supercooled water droplets become rare. As a hailstone starts to form, it becomes too heavy for the top of the cloud. It starts to fall through the cloud gathering more supercooled droplets as it falls. If the hailstone is heavy enough the updraft pushes the hailstone back up into the upper recesses of the cloud. The hailstone once again gathers more droplets and grows even bigger. It starts to fall again. The hailstone is still not heavy enough, so it gets thrown up into the cloud again. This cycle continues until the hailstone is heavy enough to overcome the updraft. At this point the hailstone leaves the cloud and falls onto the ground. Hail can do serious damage, notably to automobiles, skylights, glass-roofed structures, but most commonly agriculture. Although hailstones have been found weighing more than half a kilogram, even a much smaller hailstone can destroy crops and slice plants to ribbons in a few seconds. Thus, although a hailstorm lasts only for a few minutes (2-8 min) and rarely more the damage caused is of great importance. Hail is most common in mid-latitudes during early summer where surface temperatures are warm enough to promote the instability associated with strong thunderstorms, but the upper atmosphere is still cool enough to support ice. Hail is actually less common in the tropics despite a much higher frequency of thunderstorms than in the mid-latitudes, because the atmosphere over the tropics tends to be warmer over a much greater depth. Also, entrainment of dry air into strong thunderstorms over continents can increase the frequency of hail by promoting evaporational cooling which lowers the freezing level of thunderstorm clouds giving hail a larger volume to grow in. Hail is also much more common along mountain ranges because mountains force horizontal winds upwards (known as orographic lifting), thereby intensifying the updrafts within thunderstorms and making hail more likely. In view of the fact that hail occurrence is a relatively small-scale phenomenon, it has been reported irregularly. Consequently, a proper climatological analysis of hail occurrence is very difficult (Dalezios, 2006).

3.1 Hail Damages

Hail causes considerable damage mainly in agriculture and occasionally in properties. The damaging aspects of a hailstorm include the hailstone size (average and maximum), number of hailstones per unit area and hailstone kinetic energy. Hail risk is a combination of these factors plus the frequency of hail at a point over an area. Crop hail losses in recent years are estimated at several millions euro annually. Damages caused from hail represent the second cause of losses in agriculture after frost, in Greece. Hail losses vary considerably temporally and regionally across worldwide. Damages from hail are examined for the period 2000 - 2004 from data collected by the National Agricultural Insurance Organization (ELGA) and covers compensations given to farmers from hail losses. Hail can damage both agricultural products and human structures and property. Most of the damage is made in agricultural plants and fruit trees, especially if they are hit by hail at their early stage of growing. Since hail, during the growing season, is more frequent in central and northern parts of the country, where most agricultural products are grown, most of the damages takes place at these areas (Dalezios *et al.,* 2002).

3.2 Hail Protection

Protection of hail includes all modern techniques provided by scientists in order to reduce losses in agricultural. Protection from hail is divided into two major methods: active protection and passive protection. Active protection includes modern techniques applied before a hailstorm hits a place in order to reduce or suppress hail and passive protection includes actions taken by farmers so that hailstones do not hit the plants. Active protection methods are cloud seeding techniques from aircraft, hail cannons and hail rockets. Hail cannons ignite a charge of acetylene gas in a specially designed blast chamber releasing an explosive pressure wave creating a cavitation effect which disrupts the formation process of the hail stone embryo. Hail rockets are fired from the ground inside cumulonimbus clouds in order to influence hail clouds to prevent hail and to stimulate rainfall. Cloud seeding from aircraft includes seeding of thunderstorm clouds with silver iodide or other material in order to diminish hail production. Passive protection from hail is performed by applying a safety net over the ground and thus protecting crops or fruit trees (Dalezios and Spanos, 1995).

4. DROUGHT

Drought is one of the major international environmental disasters. In recent years, the damage from droughts to the environment and economies of some countries was extensive, and the death toll of livestock and wildlife was unprecedented (Ungani and Kogan, 1998). Droughts are, by nature, regional phenomena and have been referred to as "non events", since their basic cause is the lack of precipitation events in a region over a period of time. There is no universal quantitative definition of drought, since such a definition must be oriented towards the particular problem (Dalezios *et al.*, 1991). The American Meteorological Society (1997) categorizes drought four types: meteorological or climatological, hydrological, agricultural, and socioeconomic (Keyantash and Dracup, 2002). The first three categories could be defined as environmental droughts, whereas the last group could be considered water resources drought (Loukas *et al.*, 2002).

- Meteorological or climatological drought is generally regarded as precipitation being lower than average for some time period; in some cases air temperature and precipitation anomalies may be combined.
- Agricultural drought occurs when plant available water, from precipitation and water stored in the soil, falls below that required by a plant during a critical growth stage. This leads to below average yields in both pastoral and grain-producing regions.
- Hydrological drought is generally defined by one or a combination of factors such as stream flow, reservoir storage and groundwater.
- Socioeconomic drought is defined in terms of loss from an average or expected return. It can be measured by both social and economic indicators, of which profit is only one (McVicar and Jupp, 1998).

The relationship between the different types of drought is complex. For example, streamflow is the key variable for many water supply activities such as hydropower generation, recreation and irrigated agriculture where crop growth and yield are largely dependent on water availability in the stream (Heim, 2002). During drought, severe water-scarcity results in a region due to insufficient precipitation, high evapotranspiration, and over-exploitation of water resources and/or combination of these parameters. Various methods and indices have been developed by many scientists for drought analysis using different drought-causative and drought-responsive parameters such as rainfall, soil moisture, potential evapotranspiration, vegetation condition, groundwater and surface-water levels. The drought measuring parameters are not linearly related to one another, these drought indices often have little correlation amongst themselves (Bhuiyan *et al.*, 2006). The need for proper quantification of drought impacts and monitoring and reporting of drought development is of critical importance. Traditional methods of drought assessment and monitoring rely on rainfall data, which are limited in the region, often inaccurate and, most importantly, difficult to obtain in near-real time (Thenkabail *et al.*, 2004). In contrast, the satellite-sensor data are consistently available and can be used to detect the onset of drought, its duration and magnitude (Thiruvengadachari and Gopalkrishna, 1993). Several modern drought indices have been developed using remote sensing data, over the past few decades. These indices are calculated from the reflectance in different bands and may be obtained for each pixel (the size of a pixel depends upon the spatial resolution of a sensor). These indices have advantages over traditional methods of drought assessment and monitoring, as they "cover" large areas and may show how drought is progressing over space and time. Even crop yields can be predicted 5 to 13 weeks prior to harvests using remote-sensing techniques (Ungani and Kogan, 1998).

4.1 Damages due to Drought in Agriculture

Drought has a serious impact on agriculture, and causes considerable damage to crop production. The crop reduction is mainly explained by dry conditions and hot temperatures which have had an impact on water resources in the affected areas. Irrigation limitations would additionally affect crop production (Mavromatis, 2006). In 2006, the area affected by the drought impact on wheat production made up 14% of the total EU area, compared with 17% in 2005 and 23% in 2003. The geographic area affected by the drought in 2006 was less extended than that of the extreme droughts in 2005 and in 2003. However, some of the affected regions showed a worse situation. The Iberian Peninsula faced the worst conditions of the last 30 years. In the west and south-west of France, the 2005 drought was as bad as in 2003 (source: agrifish.jrc.it/marsstat). The major environmental factor limiting the range of adaptation for wheat (*Triticum Aestivum*) is drought. Wheat yields could be enhanced if wheat genotypes could be grown in areas too dry to support wheat growth during part or entire the growing season. Drought during the early part of the life cycle is a severe problem in many wheat-growing regions. Wheat genotypes need to be identified which are tolerant of drought during early plant growth (Mavromatis, 2006).

4.2 Protection from Drought

Protection from drought, based on damages in plant production, mainly focus on: improved economic effects secured by introducing agricultural practices, better economic effects of irrigation, improved economic effects resulting from the application of programmes of alternative plant production (integrated, biological agriculture), and models of plant production in conditions of average moisture, drought and irrigation. Irrigation is the only method that provides a complete solution to the problem of drought. However, irrigation should not be treated as a method of combating drought as it is essentially a means of intensive and modern agricultural production. It is crucial to determine a rational regime of irrigation and plants' water requirements. By additionally determining the time, method and rate of irrigation, high and stable yields can be acquired regardless of the duration or intensity of drought spells. Efforts toward solving the problem of drought in plant production are based primarily on the selection of tolerant genotypes. At a practical level, the former method includes recombination of genes of different parents, one of which at least is required to have mechanisms and properties of tolerance to drought. Latter methods include an identification of genes playing part in the expression of tolerance to drought, their isolation and transfer into different genotypes by genetic transformation. These methods primarily aim to improve the germ plasma of some most essential crops growing in drought conditions, to reconsider the criteria and develop new procedures for improving the process of breeding of drought-tolerant genotypes. In many regions, pumping of groundwater remains the dominant solution to face drought event. The alternatives examined are the construction of a number of small off-stream reservoirs to store surface water, constructions of sinks for collection of precipitation water in wet periods to use this in dry periods, and uses of desalinisation plants. Apart from developing tolerant genotypes and providing irrigation, rational agricultural practices have been recognized worldwide as yet another basis for planning plant production in arid regions. By developing cropping systems, an emphasis is being placed on soil cultivation for the purpose of better absorption, conservation and rational distribution of available water (crop rotation, soil cultivation method, conservation tillage, mulching, wind-breaks, choice of crops and cultivars, time of sowing, density, etc.). Depending on available soil moisture, fertility and plant requirements, research should also deal with the problem of plant nutrition (Keyantash and Dracup, 2002).

5. EXCESS RAIN

The duration and frequency of excess rain are important climatological features to be considered in meteorology, agriculture, hydrology and civil engineering. Their knowledge is essential for planning hydraulic works, roads and sewage systems, and for designing rainwater drainage networks in large infrastructures. Moreover, the rainfall rate behaviour could be very useful in order to detect the climatic changes and evaluate their possible effects in a given agricultural area. Heavy precipitation events worldwide have caused great damage to crop production. The cost of crop losses due to heavy precipitation and flooding in the coming decades could rise dramatically. Yet current assessments of the impacts of climate change on agriculture have not quantified the negative effects of increased heavy precipitation and flooding on crop production. A damaging flood is the joint product of societal and climatological factors (as well as ecological, geological, etc.). As societal impacts related to climate have grown, climate variability has become of increasing concern to decision makers (Trenberth, 1998). Recent data show that total annual precipitation and extreme precipitation events have increased in the US and in other parts of the world during the last century, especially in the last two decades (Karl, 1998; Milly *et al.*, 2002), often resulting in large crop losses and other flood-related damages (Chagnon *et al.*, 1997; Pielke and Downtown, 2002). While determining a cause and effect relationship between precipitation increases and flood-related damage is difficult due to concurrent changes in population growth, economic development, flood-control infrastructure and early warning efforts, currently observed trends toward increased precipitation and more extreme events are projected to intensify under future climate change, leading to higher flooding probability (Palmer and Raisanen, 2002), and thus increased damage to agricultural production compared to the present one (McCarthy *et al.*, 2001; Reilly *et al.*, 2001).

5.1 Damages due to Extreme Precipitation

Extreme precipitation can cause considerable damage to agricultural production under current climatic conditions. For example, in 1993, US Midwest flood caused damages to farmers, estimated around $6–8 billion, a figure that was roughly 50% of total losses from the flood (FEMA, 1995). Agricultural production was also negatively impacted by the North Dakota Red River floods of 1997, which caused total damage of roughly $1 billion (Spring Flood Fight Information, 2001). Red River and Mississippi floods occurred again in 2001, delaying planting. Extreme precipitation, combined with floods could also cause soil erosion in the agricultural fields. Examples of soil erosion after heavy rain is presented in the Fig. 1.

Fig. 1 Soil erosion after extreme precipitation.

Excess soil moisture, in addition to direct flood damage, is a major component of crop losses due to extreme precipitation events. Excessively wet soils directly damage crops both above and below ground because of anoxic conditions, increased risk of plant diseases and insect infestation (Ashraf and Habib-ur-Rehman, 1999), and delayed planting or harvesting due to inability to operate machinery. During the 1993 Mississippi floods, about 70% of total crop losses occurred in upland areas due to saturated soils from sustained heavy rains. In the last twenty years, excess soil moisture cost Iowa farmers five times more than direct flood damage, according to crop insurance data (Rain and Hail Insurance Service, historic data base). Despite the risk of increased crop losses due to flooding and excess precipitation under climate change, the potential damage to agricultural production has not been quantified in state-of-the-art assessments of agricultural production. In order to quantify the importance of including the effects of excess soil moisture on current and projected future crop production, a simplified model could be used (Rosenzweig *et al.*, 2002).

5.2 Protection from Excess Rain

Weather forecasting using satellites and radar can predict in advance the geographical areas where heavy precipitation is possible. This will help to prevent the human losses as well as damages in agriculture production. Weather forecasting is the application of science and technology to predict the state of the atmosphere for a given location. Human beings have attempted to predict the weather informally for times immemorial, and formally since the nineteenth century. Weather forecasts are made by collecting quantitative data about the current state of the atmosphere and using scientific understanding of atmospheric processes to project how the atmosphere will evolve. Once an all-human endeavour based mainly upon changes in barometric pressure, current weather conditions, and sky condition, weather forecasting now relies on computer-based models that take many atmospheric factors into account. Human input is still required to pick the best possible forecast model to base the forecast upon, which involves pattern recognition skills, teleconnections, knowledge of model performance, and knowledge of model biases. The chaotic nature of the atmosphere, the massive computational power required to solve the equations that describe the atmosphere, error involved in measuring the initial conditions, and incomplete understanding of atmospheric processes mean that forecasts become less accurate as the difference in current time and the time for which the forecast is being made (the range of the forecast) increases. The use of ensembles and model consensus help narrow the error and pick the most likely outcome. There are a variety of end uses to weather forecasts. Weather warnings are important forecasts because they are used to protect life and property. Forecasts based on temperature and precipitation are important to agriculture, and therefore to traders within commodity markets. Temperature forecasts are used by utility companies to estimate demand over coming days. On an everyday basis, people use weather forecasts to determine what to wear on a given day. Since outdoor activities are severely curtailed by heavy rain, snow and the wind chill, forecasts can be used to plan activities around these events, and to plan ahead and survive them (Pielke, 2002).

6. FOREST FIRES

Natural environmental hazards are internationally recognized as phenomena of special importance. There is an increasing interest of the international community in the study of natural hazards, mainly due to the possible climate intensification by human factors (Domenikiotis *et al.*, 2002). Forest fires, also called wildfires, constitute one of the most widespread environmental hazards. This hazard contributes significantly to climate changes and soil degradation (Domenikiotis *et al.*, 2002). The destruction of vegetation by forest fires can affect the land surface and the hydrologic cycle, by increasing the surface albedo, surface run-off, decrease of evapotranspiration, increase of erosion, appearance of floods and deserts. Moreover, the biomass burning can contribute, with gases, to the greenhouse effect and cause destruction of the ozone layer (Matson *et al.*, 1987). Forest fires have been a natural feature of wild lands ecosystems in the bordering of Mediterranean basin. In fact, fire has played an important role in shaping many plant communities that grow in a climatic region characterized by very hot and dry summer conditions, with high evapotranspiration rates. However, during the last three decades, human activities have disturbed the delicate natural balance between fire action and regeneration processes (Barcelo and Cortina, 1996). Forest fires occur in the Mediterranean countries during the dry season and mainly affect pine forests, brush lands and sometimes cultivated fields. On the average about 2000 forest fires are recorded annually in Greece. These fires affect an area of 50000 ha. Although the vast majority of fires are small-scale fires a small number of them, which is considered large scale fires, burns extended areas. In the USA, large-scale fires are considered those fires, which burn more than 400 hectares (Pyne, 1984). In Greece, large scale fires are considered those fires, which burn more than 100 hectares. These fires represent only a 5.4% of the annual number of fires, but the burned area reaches 72% of the annually burned areas (Kailidis 1990). There are three types of forest fires: ground fires, surface fires and crown fires (Domenikiotis *et al.*, 2002). Ground fires, which burn on the ground or below ground vegetation, are often best controlled by digging trenches or "firelines" down into the mineral soil layer, which cannot burn. When the fire reaches the fireline, it is starved of fuel and extinguishes itself.

Surface fires, which burn along the surface and tend to move quicker. Their fighting requires more manpower and equipment. Portable water backpacks and pumps (where a water supply is available), and firebreaks are the preferred methods. Crown fires are most dangerous and spread the fastest. They occur in the tops of the trees where fire can "jump" from crown to crown, often jumping over firebreaks. Crown fires in extremely windy conditions have been known to jump rivers and even lakes. Fighting crown fires usually calls for extreme measures, generally calling for aerial bombing with water and/or fire retardant chemicals. Meteorological parameters such as rainfall, high temperature, low relative humidity and wind velocity have serious effect in the initiation and spread of forest fires. The most significant of all the factors that affect the ignition and spread of forest fires is the fuel moisture content.

Rainfall

Rainfalls of high duration are capable to deposit significant amounts of water and drench the flammable forest matter, increasing its resistance to initiation and spread of fire. On the other hand, small duration rainfalls have very low effect in increasing the fuel moisture. According to the amount of rainfall, the fuel moisture is proportional to it. Increasing amounts lead to increase of the fuel moisture.

Wind

The effects of wind to forest fires depends on:

- The wind speed.
- The wind direction.

Wind speed is a very significant parameter for the initiation and spread of forest fires. Once ignited, the rate of wildfire spread is closely related to the surface wind strength (Smith, 2002). In the initial stages, wind is the parameter that provides the necessary oxygen supply and in the maturity stage affects the propagation velocity through flame tilting. The mechanism which is similar to the slope effect dominates, when wind speed exceeds 2.8 ms^{-1} (a 45% slope equivalent) (Spanos *et al.,* 1996). So, as wind speed increases the oxygen supply increases and also the flames are forced more rapidly to neighbourhood areas, making them more warm and dry. At the same time burning biomass shedding such as bark, creates a special problem of rapid fire spread known as spotting. This occurs when ignited fuel is blown ahead of an advancing fire front by strong winds to create "spot" fires (Smith, 2002).

The wind direction is also important because it determines the moisture content of the wind and the spread direction of the fire. Dry winds are more dangerous in order to initiate and spread a fire.

Large scale fires show a high propagation velocity (2.5 km/h) and usually evolve to canopy fires and cannot be theoretically described by quasi steady-state models (Rothermel, 1972). Atmospheric environment therefore is very significant to the fire initiation and spread as well. In complex terrains where meteorological parameters are extremely variable, meteorological conditions derived from neighbouring stations are usually not representative of fire conditions (Spanos *et al.,* 1996).

Temperature

High temperatures in combination with dry spells are very dangerous for the initiation and spread of wildfires. Warm and dry winds favour the propagation of forest fires. Most fire outbreaks occur near surface fronts, particularly in warm, dry conditions ahead of a well developed cold front with unstable temperature lapse rate and strong winds at low levels (Smith, 2002). Wild fires with spatial extension over 500 hectares take place when temperature is greater than 30°C.

Relative Humidity

Relative humidity is a significant factor according to the frequency and the propagation velocity of forest fires. It affects the number of incidents and the spatial extension as well (Table 5.1). Dry atmospheric air results to dry and highly flammable forest biomass while during nights the increase of relative humidity results to the increase of the fuel moisture content. Among the other meteorological parameters relative humidity determines the intensity of dehydration of dead organic material and therefore the amount of ignition heat (Spanos *et al.,*1996).

Except from the meteorological conditions, other factors affecting the fire danger are the variety and the amount of fuel (vegetation and dead organic material), the fuel moisture and the topography (altitude, slope). The fuel loading of an area depends on the land use (forest or cultivated area), the vegetation species and the vegetation condition. Moisture content depends both on fuel size and atmospheric conditions. Thin dead material (i.e., needles) respond rapidly to moisture changes in the atmosphere. On the other hand brunches of large size and thick leave layers maintain their moisture for several days after rain episodes especially when calm conditions prevail and saturation vapour pressure deficit is low (Spanos *et al.,* 1996).

Topography plays an important role in the initiation and spread of forest fires. Altitude affects the vegetation period through temperature and moisture changes in the atmospheric environment and the groundwater supply of the plants. During the Mediterranean summers green vegetation is restricted to higher altitudes. Green vegetation contains sufficient moisture and therefore prevents fire initiation and retards fire spread (Spanos *et al.,* 1996). In upslope firs the flame tilts over the potential fuel, thereby increasing radiation and more significantly causing direct flame contact. The steeper the slope the closer the fuel is posed to the flames (Spanos *et al.,* 1996). In no wind conditions even a moderate slope of 20% doubles the propagation velocity compared to a similar fire on a plane surface (Luce and McArthur, 1977).

Concluding, weather conditions are crucial for forest fires. Drought periods followed by dry and hot winds blowing from arid continental interiors over a period of days to give a cumulative heating and drying effect on vegetation are particular culprits. These are also the atmospheric conditions which promote dry lightning storms, which are a frequent ignition source (Smith, 2002).

6.1 Damages due to Forest Fires

It is very important to know and study the principal causes of forest fires, for the purpose of understanding the phenomenon and taking action aimed at preventing and contrasting the fires themselves. Causes of forest fires to differ from one area to another due to environmental factors (vegetation type, lightening frequency, climatic condition etc.) and human activities. So, it is necessary to gather and examine reliable data in order to identify and comprehend the possible causes of a forest fire in a given area and choose the identical protection measures.

6.2 Protection from Forest Fires

Protection from forest fires includes prevention measures and suppression measures. Prevention is the key in helping to reduce the number of human caused wildfires. The programmes of prevention are related with all the activities that aim in the minimization of impacts of forest fires. The various techniques that can be used for reduction of the human caused fires fall into two general categories: to reduce danger-risk and to manage it. The danger is related to the breaking of fire. The reduction of the possibility of having a wildfire ignited is highly related with the increase of the level of awareness of public and various teams of persons responsible for the ignition and the impacts of fires. Use of advertisement via national and local mass media (radio, TV, newspapers, magazines) and educational programmes in schools and social or professional organizations can be used in this direction. Also, the application of laws on fires is potentially a useful preventive technique. The education and the application of legislation are useful preventive measures when the fires are results of ignorance, carelessness, or malice. Other measure in order to prevent fire ignition and spread is the creation of fire breaks or fuel breaks or replacement of flammable fuels (e.g., conifers) by not flammable fuels (e.g., broad leaves) around and across regions of high danger (e.g., dumps, camps, road and railway network). Despite of the very high cost of manufacturing and maintaining of fire breaks or fuel breaks, a lot of countries continue to use it for reduction of fire danger. The use of different type of vegetation cover is considered a very important measure in order to prevent very big fires. Forest fire suppression is dependent on two key factors, communications and reaction time. The sooner a fire is reported, the quicker it can be extinguished. The moment a fire report comes in, a number of key questions need to be answered in order to initiate a suppression strategy — location, accessibility, and size and category of fire. In most instances, where initial reports are unclear or when a fire is in a remote area, evaluation is done from the air. The method of transporting crews to a fire site varies depending on terrain and accessibility. Ground transport will usually relay fire crews to a fire within a reasonable driving distance of a fire base. Where no road access is available, the suppression is being held by air (helicopters, planes). The key to controlling and suppressing a forest fire is getting man-power and equipment to the scene in as short a time as possible. Then, during the evolution of the fire, as information from the field, and data from different sources (e.g., weather and satellite maps) becomes available, the fire fighting strategy can be modified from the first response plan. The operational monitoring and mapping of the burned areas are very important aspects in dealing with emergency situations and the quantitative estimation of the affected area. The assessment of the damaged areas can provide valuable information to the authorities and insurance companies. The contribution of new technologies such as remote sensing, photogrammetry and G.I.S in monitoring forest fires and mapping burned areas is very significant (Domenikiotis *et al.,* 2002). The development of fire-weather forecasting systems is an essential tool in the framework of forest fires prevention planning. In this context, the role of the National Meteorological Services as providers of specific services is decisive. A hierarchy of such services has been presented (Reyfsnayder, 1994), ranging from fire-weather warnings produced by central meteorological offices to fire-weather forecasting services provided to personnel at the scene of existing forest fires (Barcelo and Cortina, 1996). Forest fire risk assessment is an important component of every fire suppression organization, since it allows for proper fire prevention planning and is economically judicious since it requires relatively reasonable investments if compared to the costs of non-scientifically based suppression on active wildfires (Dimitrakopoulos and Bemmerzouk, 1996).

7. PESTS AND DISEASES

7.1 Bertisillioseis/*Verticillium mdahliae, Verticilliumalbo-atrum*

Bertisillioseis is considered the most serious mycetology diseases of pyrinokarpon, of vine and are owed in mushrooms of gender. Adromykoseis develop slow and damage the vascular system of trees, causing withering and desiccation of branches. In the pyrinokarpa, symptoms are expressed unilaterally, in a side of damaged organs (imipligia). Later on, it advances in the other side also. The damage is presented with withering of branches and arms and chlorination of leaves, first in inferior and then in the superior leaves. Afterwards the leaves fall and follow desiccation of branch. Characteristic symptom constitutes the brown metachromatismos that is observed in the vascular of wood of damaged branches. In the vine, the disease is expressed with abrupt withering (apoplexy) and desiccation of leafage of certain or all vines-shoot. Moreover, in at length incision observed the characteristic brown metachromatismos of the vascular of wood of vines-shoot or trunk. Bertisillioseis are caused by adilomykites, *Verticillium dahliae* and *Verticillium albo-atrum*. In Greece, the first type has been found that it causes damage in the multiannual cultures. This is owed rather to the fact that *V. dahliae* is encouraged by mean temperatures, while V. albo-atrum is widespread in regions with humid and cold climate. Bertisillioseis are formally soil diseases. Pathogenic fungi/organisms survives mainly with mikrosklirotia, but also as mycelium and seeds (conidia) in the damaged remains of cultivations and is maintained in the soil for many years (8-14). Another way of perpetuation is the various weeds - hosts. Pathogens are propagated via water, the remains of culture, the weeds and with the soil which is transported with the tools or the farming machines. In long distances their transport becomes with polluted multiplicative material. The fungus pollutes the root and installed in the vascular of wood, where with microscopic examination can be distinguished the structure of the mycelium and its seeds (conidia). Prevention methods include installation of trees far from cultivation fields and in soils exempted from pollutant. Also, for cultivation, healthy multiplicative material and durable varieties, should be used. It is important to avoid the cultivation with frail annual plants (e.g., cotton). Irrigation should not be attained with canals because pollutants are transferred with water at healthy trees. Careful use of pesticide is recommended to avoid wounds in the root system of trees. Uprooting of the damaged trees, burning and disinfection of space is recommended to control transfer of pathogenicity (Paschall M.J., *et al.,* 1992).

7.2 Bacillar Necrosis/*Xylophilusampelinus*

This disease is very widely spread in many regions of world. It is considered an old disease and has acquaintance with the name "tsilik marazi". The bacterium is installed in the vascular of wood and from there is expanded and damages the young vegetation, causing progressive draining. *Soyltanina* and *Razaki* varieties are considered more sensitive. Diagnostic symptom of disease is brown metachromatismos of vascular of wood. The heads in the beginning of spring present inflations because of hyperplasia of tissues. Also, is observed delayed and unilateral ekptyxi of eyes or necrosis of heads. The disease is owed in the bacterium *Xylophilus ampelinus* and damage only the vine. The bacterium overwinters in the vascular of wood, from where it is transported to the young shoots. Transmission occurs with the rain accompanied with wind as the bacteria comes out from the vascular of wood in the incisions or other wounds. Also, the lopping should become with dry weather and the tools of lopping should be disinfected continuously. When the weather is rainy, it is recommended preventive sprayings with suitable preparations, for the protection of tender leaves (Orlandini and Rosa, 1997).

7.3 Botrytis (Ash Decay) Vine/*Botryotiniafuckeli and Botrytis*

It damages almost all cultivated plants. It constitutes real threat for the marketable production, because apart from the quantitative losses it downgrades also the quality of products. In the vine, the damage from botrytis has negative repercussions in the vinification and in the quality of wine, because it affects fermentation. Moreover, it causes losses in the production after collection, during transport and storage of grapes. It damages all green parts of vine above the ground, and particularly the grapes before or even afterwards the harvest. In the grapes is presented initial dark spot, which expands and finally rottens. The damaged tissues become brown and shrink. With humid atmosphere, the damaged organs are covered from a gray mildew (fruitfulness of fungus). The tender shoots, the leaves and blossoms are only damaged when conditions of high humidity prevail in spring. The disease is caused by the fungus *Botrytis cinerea*. Pathogens survive on dead plant tissues, as saprophyte. High relative humidity is a basic condition for the growth of disease, while the temperature does not appear to be an important condition, because the fungus can grow even in a wide temperature range. During favourable conditions, the mycelium of pathogens develop very rapidly and it forms conidia. These conidia are propagated mainly via air, rain, and to some extent with hands, clothes and tools, and cause pollution. Preventive measures include spraying of pesticides from the blossoming of the plant. Additional measures like rejecting the wounds at the implementation of farming and suitable pull of leaves for the better airing of grapes, is also in practice (Paschall *et al.,* 1992).

7.4 Necrosis of Arms/*Eutypalata*

This disease is widespread in Greece and causes very serious damage in pyrinokarpa, in the vine, in the citrus fruits and in pistachio tree. The disease is caused by askomykita, *Eutypa lata*. The growth of reproductive bodies of fungus takes place, after some time. After entering a plant, the fungus is installed initially in the wood, later in the bark, where it develops and causes necrosis of the tissues and progressively creates ulcer and necrosis of arms. After 5 years from the pollution in the withering trees or in the dead offended tissues, are formed its fruitfulness (perithikia), from where are going to released the askosporia, which are transported with air in large distances causing new pollutions. Very important factor for the forming of fruitfulness of fungus and its liberation of sporion is the rain. The mean annual rainfall of region should be at least 420 mm. Preventive measures of this disease include destruction of damaged branches, rejection of big incision at the lopping, utilisation of disinfected tools and disinfection of incision of lopping with the suitable means (Dalezios *et al.,* 2000).

7.5 Use of Models in Protection of Diseases

The use of technological advancements for agrometeorological simulation of the disease can produce useful tools for improving crop protection techniques (Rosa *et al.,* 1995). The use of agrometeorological models allows farmers and agronomists to represent the reality of the system and analyze it to improve their knowledge over certain diseases of plants.

PERO model: The PERO model simulates the infection cycle development according to agrometeorological conditions, allowing the best timing for fungicide treatment. The model simulates the beginning and the ending of the infection, the duration of its development and its intensity. In this way every infectious cycle can be simulated and identified by temporal and quantitative characteristics allowing the creation of a timetable for sprayings with fungicides against *Plasmopara viticola.* If sprayings are applied shortly before the expiry time of the incubation of the fungus, farmers will minimize loss of production (Dalezios, 2000).

PLASMO model: The Plasmo model, (Plasmopora Simulation Model), simulates grapevine downy mildew biological cycle on the basis of agrometeorological parameter, allowing the best timing for fungicide treatments. Meteorological data required for the simulation (temperature, relative humidity, rainfall, etc) that can be collected from agrometeorological station with hourly step. In such a way the obtained information is related to the real pathogen development and it allows identifying the most suitable protection techniques (Orlandini and Rosa, 1997).

Traditional ways of protection of diseases combining with technological advancements for agrometeorological simulation of the disease (use of models), can provide the basis for the best confrontation of diseases and for a more efficient production. Additional, pointless chemical applications could be avoided as well as environmental pollution.

8. WHO GUIDELINES ON GOOD AGRICULTURAL AND COLLECTION PRACTICES (GACP) FOR MEDICINAL PLANTS (WORLD HEALTH ORGANIZATION, GENEVA, 2003)

Interest in traditional systems of medicine and, in particular, herbal medicines, has increased substantially in both developed and developing countries over the past two decades. Global and national markets for medicinal herbs have been growing rapidly, and significant economic gains are being realized. According to the Secretariat of the Convention on Biological Diversity, global sales of herbal products totaled an estimated Rs. 2,917,206 in 2000. As a consequence, the safety and quality of herbal medicines have become increasingly important concerns for health authorities and the public alike. The safety and quality of raw medicinal plant materials and finished products depend on factors that may be classified as intrinsic (genetic) or extrinsic (environment, collection methods, cultivation, harvest, post-harvest processing, transport and storage practices). Inadvertent contamination by microbial or chemical agents during any of the production stages can also lead to deterioration in safety and quality. Medicinal plants collected from the wild population may be contaminated by other species or plant parts through misidentification, accidental contamination or intentional adulteration, all of which may have unsafe consequences.

The collection of medicinal plants from wild populations can give rise to additional concerns related to global, regional and/or local over-harvesting, and protection of endangered species. The impact of cultivation and collection on the environment and ecological processes, and the welfare of local communities should be considered. All intellectual property rights with regard to source materials must be respected. WHO has cooperated with other United Nations specialized agencies and international organizations in dealing with the above-mentioned issues. Such cooperation will be further strengthened through the development and the updating of relevant technical guidelines in these areas.

Safety and quality assurance measures are needed to overcome these problems and to ensure a steady, affordable and sustainable supply of medicinal plant materials of good quality. In recent years, good agricultural practices have been recognized as an important tool for ensuring the safety and quality of a variety of food commodities, and many Member States have established national good agricultural practice guidelines for a range of foods. However, quality control for the cultivation and collection of medicinal plants as the raw materials for herbal medicines may be more demanding than that for food production; possibly for this reason, only China, the European Union, and Japan have recently developed guidelines on good agricultural practices for medicinal plants. Since their guidelines were established to meet the requirements of specific regions or countries, they may not be universally applicable or acceptable.

At a WHO informal meeting on Methodologies for Quality Control of Finished Herbal Products, held in Ottawa, Canada from 20 to 21 July 2001, the entire process of production of herbal medicines, from raw materials to finished herbal products, was reviewed. It was recommended that WHO should give high priority to the development of globally applicable guidelines to promote the safety and quality of medicinal plant materials through the formulation of codes for good agricultural practices and good collection practices for medicinal plants. It was envisaged that such guidelines would help to ensure safety and quality at the first and most important stage of the production of herbal medicines.

The main objectives of WHO guidelines on good agricultural and collection practices (GACP) for medicinal plants are:

- contribute to the quality assurance of medicinal plant materials used as the source for herbal medicines, which aims to improve the quality, safety and efficacy of finished herbal products;
- guide the formulation of national and/or regional GACP guidelines and GACP monographs for medicinal plants and related standard operating procedures; and
- encourage and support the sustainable cultivation and collection of medicinal plants of good quality in ways that respect and support the conservation of medicinal plants and the environment in general.

9. CONCLUSION

From above discussion, it is clear that environmental hazards have a significant effect on crop cultivation. Environmental hazards mainly include frost, hail, drought, excess rain, forest fires, pest and diseases. Frost is a meteorological phenomenon that occurs when the air temperature near the earth's surface drops to 0°C or below causing injuries to sensitive crops. Frost forecasts help farmers to reduce any possible injuries to their crops. Frost protection methods (active and passive) help farmers to combat frost.

Hail forms inside thunderstorm cumulonimbus (Cb) clouds, mainly those with intense updrafts, high liquid water content, great vertical extent, large water droplets, and where a good portion of the cloud layer is below freezing temperature. Protection from hail includes active and passive protection methods. Active protection methods are cloud seeding techniques from aircraft, hail cannons and hail rockets. Passive protection from hail is performed by applying a safety net over the ground and thus protecting crops or fruit trees. Droughts are, by nature, regional phenomena and have been referred to as "non events", since their basic cause is the lack of precipitation events in a region over a period of time. Irrigation is the only method that provides a complete solution to the problem of drought. The duration and frequency of excess rain are important climatological features to be considered in meteorology, agriculture, hydrology and civil engineering. Weather forecasting using satellites and radar can predict the geographical areas where heavy precipitation is possible. This will help to prevent the human losses as well as damages in agriculture production. Forest fires, constitute one of the most widespread environmental hazards. This hazard contributes significantly to climate changes and soil degradation. The destruction of vegetation by forest fires can affect the land surface and the hydrologic cycle, by increasing the surface albedo, surface run-off, decrease of evapotranspiration, increase of erosion, appearance of floods and deserts. Protection from forest fires includes prevention measures and suppression measures. Prevention is the key in helping to reduce the number of human caused wildfires. The programmes of prevention are related with all the activities that aim in the minimization of impacts of forest fires. Forest fire suppression is dependent on two key factors, communication and reaction time. The sooner a fire is reported, the quicker it can be extinguished. Various diseases caused by pest, that affect the crop cultivation include Bertisillioseis caused by Verticillium dahliae and Verticillium albo-atrum, Bacillar necrosis caused by Xylophilus ampelinus, Botrytis vine caused by Botryotinia fuckeliana, Necrosis of arms caused by Eutypa lata, etc. Bacterial diseases include Bacillar necrosis caused by Xylophilus ampelinus. Integrated pest management along with technological advancements for agrometeorological simulation of the disease (use of PERO and PLASMO models), can provide the best confrontation of diseases and increase the crop production. Additional, pointless chemical applications could be avoided as well as environmental pollution.

In spite of various protection methods that are available for these hazards, there is still a need of worldwide discussion to protect the agricultural resources and human damages from such calamities. There should be national and international monitoring bodies, which can keep a strict vigilance over such type of natural calamities. These monitoring bodies should have prompt forecasting methods, so that huge agricultural losses and human damages can be prevented with the help of such fool proof measures.

References

Ashraf M. and Habib-ur-Rehman K., 1999. *Interactive effects of nitrate and long-term waterlogging on growth, water relations, and gaseous exchange properties of maize* (Zea mays L.). Plant Science, 144(1): 35-43.

AGRIFISH Unit, Institute of Protection and Security of the Citizen, [http://agrifish.jrc.it/marsstat].

Bagdonas A., Georg J.C., Gerber J.F., 1978. *Techniques of frost prediction and methods of frost and cold protection.* Technical Note 157; 487. World Medical Organization (WMO).

Balatsos P., 2003. *Preventive measures against Forest Fires in Greece.* Greek Ministry of Rural Development and Foods, Forest Fire prevention and suppression department.

Barcelo A.M. and Cortina I.C., 1996. *Meteorological support activities in fighting against the forest fires in Spain.* International Symposium on Applied Agrometeorology and Agroclimatology, Greece. 541-548.

Bhuiyan C., Singh R.P. and Kogan F.N., 2006. *Monitoring drought dynamics in the Aravalli region (India) using different indices based on ground and remote sensing data.* International Journal of Applied Earth Observation and Geoinformation [available online: www.elsevier.com/locate/jag].

Chagnon S.A., Changnon D., Fosse E.R., Hoganson D.C., Roth R.J. and Totsch J.M., 1997. *Effects of recent weather extremes on the insurance industry: major implications for the atmospheric sciences.* Bulletin of the American Meteorological Society; 78 (3): 425–431.

Dalezios N.R., 1988. *The Palmer Drought Severity Index (PDSI) for rain enhancement in northern Greece.* International Symposium/Workshop on Weather Modifications, OGA, Thessaloniki, Greece.

Dalezios N.R., 2006. *Environmental Hazards.* Lecture notes, University of Thessaly, Greece.

Dalezios N.R., Bampzelis D. and Daoularis G., 2000. *Agrometeorological Simulation Using PERO Model for Grape Vine Downy Mildew in Greece.*

Dalezios N.R., Loukas A. and Bampzelis D., 2002. *Assessment of NDVI and Agrometeorological Indices for Major Crops in Central Greece.* Physics and Chemistry of the Earth. 27 (23-24): 1025-1029.

Dalezios N.R. and Spanos S.I., 1995: Operational Efficiency Assessment of Hail Suppression for Agriculture in Greece. The Journal of Weather Modification. 27 (1): 21-35.

Dalezios N.R., Papazafiriou Z.G., Papamichail D.M. and Karacostas T.S., 1991. *Drought assessment for the potential of precipitation enhancement in Northern Greece.* Theoretical Applications in Climatology, 44: 75-88.

Dimitrakopoulos A.P. and Bemmerzouk A.M., 1996. *Evaluation of the Canadian Forest Danger Rating System (CFFDRS) in the Mediterranean-type environment of Greece.* International Symposium on Applied Agrometeorology and Agroclimatology, Volos, Greece. 583-588.

Domenikiotis C., Dalezios N.R., Loukas A. and Karteris M., 2002. *Agreement assessment of NOAA/AVHRR NDVI with Landsat TM NDVI for mapping burned forested areas.* International Journal of Remote Sensing, 23: 4235-4246.

Dracup J.A., Lee K.S. and Paulson E.G., 1980. *On the definition of droughts.* Water Resources Research. 16 (2): 297–302.

Gumbel E.J. Statistics of extremes. New York: Columbia University Press; 1958. Ch. 6, Meteorological examples; p. 241.

Federal Emergency Management Agency (FEMA), 1995. *The 1993 and 1995 midwest floods: flood hazard mitigation through property hazard acquisition and relocation program.*

Hellenic Agricultural Insurance Organization (ELGA), Statistical Bulletin for the year 2005.

Hellenic Agricultural Insurance Organization (ELGA), Statistical Bulletin for the year 2006.

Hellenic Agricultural Insurance Organization (ELGA), Statistical Bulletin for the year 2007.

Hellenic Agricultural Insurance Organization (ELGA), Statistical Bulletin for the year 2008.

Heim R.R. Jr., 2002. *A Review of Twentieth-Century Drought Indices Used in the United States.* Bulletin of the American Meteorological Society, 83(8): 1149-1165.

Huschke R.E., 1959. Glossary of Meteorology.

International organisation for biological and integrated control of noxious animals and plants (IOBC), 1999. *Integrated Production: Principles and Technical Guidelines* 2nd ed., IOBC /WPRS Bulletin 22.

Intergovernmental panel on climate change (IPCC), 1996: *Climate Change, 1995*. The Science of Climate Change. Cambridge University Press, Cambridge, UK.

McCarthy J.J., Canziani O.K., Leary N.A., Dokken D.J. and White K.S., 2001. *Impacts, Adaptation and vulnerability*. Third Assessment Report of the Intergovernmental panel on climate change, working Group 2. Cambridge University Press, Cambridge, UK.

Kailidis D.S. 3rd ed. Thessaloniki: Giahoudi Giapouli Publications; 1990. Forest Fires; p. 510.

Kalma J.D., Laughlin G.P., Caprio J.M, Hamer P.J.C., 1992. *The Bioclimatology of Frost*, 1-12.

Karl T.R., 1998. *Secular trends of precipitation amount, frequency, and intensity in the USA*. Bulletin of the American Meteorological Society, 79: 231–341.

Karl T.R., Easterling D. R., and Quayle R. G., 1995. *Trends in U.S. climate during the twentieth century; Consequences*: 1: 3–12.

Karl T.R. and Knight R.W., 1998. *Secular trend of precipitation amount, frequency, and intensity in the United States*. Bulletin of American Meteorological Society, 79: 231–242.

Keyantash J. and Dracup J. A., 2002. *The Quantification of Drought: An Evaluation of Drought Indices*. Bulletin of American Meteorological Society, p. 1167-1180.

Kozdroj J. and Elsas van J.D., 2000. *Response of the bacterial community to root exudates in soil polluted with heavy metals assessed by molecular and cultural approaches*. Soil Biology and Biochemistry, 32 (10): 1405 - 1417.

Keyantash J. and Dracup J. A., 2002. *The Quantification of Drought: An Evaluation of Drought Indices*. Bulletin of American Meteorological Society. 1167-1180.

Kogan F. N., 2000. *Contribution of remote sensing to drought early warning*. World Meteorological Organization, Geneva. 75–87.

Komuscu, A.U. (1999). *Using the SPI to analyze spatial and temporal patterns of drought in Turkey*. Drought Network News 11 (1): 7–13.

Luce R.C. and McArthur A.G., 1977. *Bushfires in Australia*. CSIRO Division of Forest Research, Campera. p. 359.

Loukas A., Vasiliades L. and Dalezios N.R., 2002. *Hydroclimatic Variability of Regional Droughts in Greece Using the Palmer Moisture Anomaly Index*. Nordic Hydrology, 33 (5): 425-442.

Matson M., Stephens G., and Robinson J.M., 1987. *Fire detection using data from NOAA-N satellites*. International Journal of Remote Sensing, 8: 961-970.

Mavi H.S., 1986. *Introduction to Agrometeorology*. Oxford & IBH Publishing Co. Pvt Ltd, New Delhi, p. 191-193.

Mavromatis T., 2007. *Drought index evaluation for assessing future wheat production in Greece*. International Journal of Climatology, 27: 911–924.

McVicar T. R. and Jupp D.L.B., 1998. *The Current and Potential Operational Uses of Remote Sensing to Aid Decisions on Drought Exceptional Circumstances in Australia: a Review*. Agricultural Systems, 57(3): 399-468.

Milly P.C.D., Wetherald R.T., Dunne K.A. and Delworth T.L.: 2002, *Increasing risk of great floods in a changing climate*. Nature, 415: 514 - 517.

Palmer T. and Raisanen J., 2002. *Quantifying the risk of extreme seasonal precipitation events in a changing climate*. Nature, 415: 512 - 514.

Paschall M.J., Hellingsworth C., Coli W., Cohen N.L., 1992. *Working the bugs out*. Integrated pest management saves crops and the environment. J. Am. Diet Assoc. 47: 931.

Pielke R.A., 2002. *Mesoscale Meteorological Modeling*. Academic Press. p. 48–49.

Pielke R.A. and Downton M.W., 2002. *Precipitation and damaging floods: Trends in the United States,* 1932-1997. Journal of Climate, 13(20): 3625-3637.

Prout N., Dalezios N., Manore M., Tracey J., Brown R. and Walker G., 1986. *Agricultural Drought: precipitation deficiency and NOAA AVHRR Indices Comparisons, Proceedings, Canadian Hydrology Symposium (CHS: 86), Regina, Sask.,* 3-6 June, NSERC, p. 91-100.

Pyne S.J. Fire management in the United States. New York: John Wiley and Sons; 1984. *Introduction to wildlife fire.*

Reilly J., Tubiello F.N., McCarl B. and Melillo J. Washington DC: In, US National Assessment Foundation Document, National Assessment Synthesis Team, US Global Change Research Program; 2001. *Impacts of climate change and variability on agriculture.*

Reyfsnayder W.E., 1978. *Systems for evaluating and predicting the effects of weather and climate on wildland fires.* World Meteorological Organization, Special Environmental Report 11, WMO-496 (Revised in 1994, with additional material on the use of remote sensing techniques in forest fire management).

Rosa M., Gozzini B., Orlandini S., and Seghi L., 1995. *A Computer program to Improve the Control of Grapevine Downy Mildew.* Computer and Electronics in Agriculture. 12, 311-322.

Rosenberg N. J. and Myers R.E., 1962. The nature of growing season frost in and along the plate valley of Nebraska. Monthly Weather Review, 90: 471-476.

Rosenzweig C., Tubiello N., Golberg R., Mills E. and Bloomfield J., 2002. *Increased crop damage in the US from excess precipitation under climate change,* Global Environmental Change, 12 : 197–202.

Rothernel R.C., 1972. *A mathematical model for predicting fire spread in wilgland fuels.* USDA Forest Service Research paper INT-114, p.40.

Snyder R., Paulo de Melo-Abreu J., 2005. *Frost Protection: fundamentals, practice, and economics,* 1(1-11), 8(1-9).

Smith K., 2002. *Environmental hazards: Assessing risk and reducing disaster,* 3rd edition. Routledge (Taylor and Francis Group), London and New York, p.392.

Spanos S.I., Zarpas C.D. and Dalezios N.R., 1996. *Meteorological and satellite indices for the assessment of forest fires.* International Symposium on Applied Agrometeorology and Agroclimatology, Volos, Greece. p.583-588.

Steinemann A., Hayes M.J. and Cavalcanti L., 2005. *Drought Indicators and Triggers.* In Wilhite D. (ed.) Drought and Water Crises, Science Technology and Management Issues, Marcer Dekker Publ., p. 71-90.

Thiruvengadachari S. and Gopalkrishna H. R., 1993. *An Integrated PC Environment for Assessment of Drought.* International Journal of Remote Sensing, 14: 3201-3208.

Trenberth, 1998. *Atmospheric moisture residence times and cycling: Implications for rainfall rates and climate change,* Climatic Change, 39: 667 - 694.

Ungani, L.S. and F. N. Kogan (1998). *Drought Monitoring and Corn Yield Estimation in Southern Africa from AVHRR data.* Remote Sensing of Environment, 63: 219-232.

Wilson W., 1988. *Synoptic typing forecast scheme.* In R.C. Rudolph et al. Program, 1987. Annual Report. Prepared by Intera Technologies Ltd., Greek National Hail Suppression for ELGA. Athens. p. 209.

WHO guidelines on Good Agricultural and Collection Practices (GACP) for medicinal plants. World Health Organization, Geneva, 2003.

3

Response of Plants to Oxidative Stress

Meghana Agrawal* **and Smita Purohit**

Department of Botany, The IIS University, S. F. S., Gurukul Marg, Mansarovar Jaipur-302020, Rajasthan, India

ABSTRACT

Enhanced formation of reactive oxygen species (ROS) causing damage of plant tissues is a common response of plant cells to abiotic stress or on infection by pathogens. Although they are constantly produced, uncontrolled production of ROS causes oxidative stress and cellular damage. Plants are endowed with an array of radical scavengers and antioxidant enzymes that act in concert to alleviate oxidative stress. The extent of the damaging effect of ROS depends on the effectiveness of the antioxidative system which include low molecular mass antioxidants, which may be hydrophilic (glutathione, ascorbate) or lipophylic (α-tocopherol, β-carotene, phenols) as well as serve as antioxidative enzymes (superoxide dismutase (SOD), ascorbateperoxidise (APX), catalase (CAT), glutathione reductase (GR), monodehydroascorbatereductase (MDHAR), dehydroascorbatereductase (DHAR). In nonstressed plant tissues, antioxidants are able to neutralize the effect of ROS. ROS at low concentrations can be signals that switch on developmental programmes or regulate physiological processes as cell wall loosening required for cell elongation, modulation of cytosol Ca^{2+} concentration, senescence, adaptation to abiotic stresses or resistance to pathogens. Signalling mediated by ROS involves G-proteins, MAP kinases and protein Tyr phosphatases. In addition to concentration, the balance between their positive and negative effects is influenced by the physiological and developmental status of tissues.

While different sources like high irradiance, drought, salinity, heavy metals, etc., cause ROS production and promote activities of antioxidative enzymes, the role of ROS in plant growth and development *in vitro* has also been well documented. Addition of plant growth regulators in media is often essential and the role of ROS in plant growth is further substantiated by the interplay of ROS with a number of phytohormones. ROS or ROS-scavenging antioxidant enzyme-stimulated improvement in growth and regeneration further support the relationship between oxidative stress and recalcitrance. Higher activities of antioxidative enzyme systems promote cell division and somatic embryogenesis. Consequently, elevation of antioxidant capacity *via* plant tissue culture could increase tolerance of plants to biotic and abiotic stresses.

Keywords: Antioxidative enzymes, non-enzymatic antioxidants, oxidative stress, reactive oxygen species, ROS-scavenging mechanisms.

Corresponding author's E-mail: meghana.ktm@gmail.com

Roychowdhury R. (Ed.), Crop Improvement in the Era of Climate Change.

1. INTRODUCTION

Atmospheric molecular oxygen and water are required to support all forms of aerobic life. The present oxygen reservoir (37×10^{18} moles) has built up as a result of photosynthesis, a process that liberates dioxygen (O_2) from water. The approximately constant amount of O_2 is maintained by another process of respiration, in which O_2 is used as the ultimate electron acceptor. However, ever since the introduction of O_2 into our atmosphere by O_2-evolving photosynthetic organisms, about two billion years ago, reactive oxygen species (ROS) have been the unwelcome companions of aerobic metabolism.

The ground-state dioxygen is rather unreactive and can peacefully co-exist with organic matter owing to its biradical, or in other words, its two unpaired electrons making it paramagnetic. The monovalent reductions of oxygen, i.e., addition of four electrons one by one, and finally reducing it to water, forms its three reactive intermediates, namely superoxide (O_2^-), hydrogen peroxide (H_2O_2) and hydroxyl radical (OH^-). Activation of oxygen (i.e., the first univalent reduction step) is energy dependent and requires an electron donation. The subsequent one-electron reduction steps are not energy dependent and can occur spontaneously or require e^-/H^+ donors. In biological systems transition metal ions (Fe^{2+}, Cu^+) and semiquinones can act as e^- donors. Four-electron reduction of oxygen in the respiratory electron respiratory chain (ETC) is always accompanied with a partial one- to three-electron reduction, yielding the formation of ROS (Bolkhina*et al.*, 2003).

Reactive oxygen species include free radicals, e.g., superoxide radical (O_2^-) and hydroxyl radical (OH^-) and molecules such as hydrogen peroxide (H_2O_2), singlet oxygen ($^1O_2$) and ozone (O_2). Both O_2^- and hydroperoxyl radical HO_2^- undergoes spontaneous dismutation to produce H_2O_2. Although H_2O_2 is less reactive than O_2^- =, in the presence of reduced transition metals such as Fe^{2+} in a chelated form (which is the case in biological systems), the formation of OH^- can occur in the Fenton reaction.

The Fenton reaction: $$H_2O_2 + Fe^{2+} \longrightarrow OH^- + OH + Fe^{+3}$$

ROS are constantly produced during the course of photosynthesis and respiration, whereas redox homeostasis in the cell is tightly controlled by redundant protective mechanisms. Disruption of these protective mechanisms can cause oxidative stress, leading to oxidative damage and cell death. These reactive species are biologically toxic (Elstner, 1987; Hamilton, 1991; McKersie and Leshem, 1994; Yu, 1994). Their reactions with organic substrates are complex due to the surface properties of membranes, electrical charges, binding properties of macromolecules, and compartmentalization of enzymes, substrates and catalysts. ROS colliding with an organic molecule may extract an electron from it, rendering it a radical capable of propagating a chain reaction. Thus, various sites even within a single cell differ in the nature and extent of reactions with oxygen. These intermediates, to various extents, cause inhibition of sensitive enzymes, degradation of chlorophyll, lipid peroxidation disrupting lipid structure and membrane organization and integrity (Yu, 1994). A variety of oxidatively altered DNA species can be identified following OH^- attack, including base alterations and strand breaks that may be difficult to repair or tolerate (Kasai *et al*, 1986). Proteins expose to OH^- undergo typical modification, including specific amino acid alterations, polypeptide fragmentation, aggregation, denaturation and susceptibility to proteolysis (Wolff *et al.*, 1986).

This toxicity is reflected by their short half-lives before reacting with cellular components, as compared to that of O_2 (>100 sec) as shown below:

GROUND STATE OXYGEN:

Dioxygenbiradical (•O-O•)... [*Half life, 37°C*: >100 seconds]

PHYSICAL ACTIVATION:

Chlorophyll + •O-O• ⟶ chlorophyll + singlet oxygen (O-O:)........................ [1×10^{-6} sec]

CHEMICAL ACTIVATION:

O-O: + e^- ⟶ superoxide radical (•O-O:)................................... [1×10^{-6} sec]

•O-O: + e^- ⟶ hydrogen peroxide (H:O-:H)................................. [1×10^{-6} sec]

H:O-:H + e^- ⟶ hydroxyl radical (H:O)...................................... [1×10^{-9} sec]

Now it has been well established that an inevitable result of membrane-linked electron transport (chloroplastic, mitochondrial and plasma membrane) is the spilling of electrons onto molecular oxygen in plant cells, with the resultant generation of ROS (Krieger-Liszkay, 2005; Fridovich, 1995; Alscher *et al.*, 1997). The imposition of abiotic and biotic stresses can further aggravate the production of ROS (Arora *et al.*, 2002; Alscher and Hess, 1993; Hammond-Kosack and Jones, 1996). ROS are also generated during normal metabolic processes (Alscher *et al.*, 1997; Arora *et al.*, 2002). It has been estimated that 1% of O_2 consumed by plants is diverted to produce ROS in various subcellular loci (Asada and Takahasi, 1987).

2. ORGANELLES IN PLANT CELLS ACT AS SITES FOR ROS PRODUCTION

The steady state level of ROS in the different cellular compartments is determined by interplay between multiple ROS-producing pathways and ROS-scavenging mechanisms. These are controlled by the ROS-signal transduction pathway. During normal growth and development this pathway monitors the level of ROS, produced by aerobic metabolism, and controls the expression and activity of ROS-scavenging pathways (Fig. 1). The basic ROS-scavenging mechanisms may also perform fine metabolic tuning, e.g., suppression of photosynthesis, to reduce the production rate of ROS. There are many potential sources of ROS in plants (Table 1). Some are reactions of normal aerobic metabolism, such as photosynthesis and respiration, while others belong to pathways enhanced during abiotic stresses, such as photorespiration. In recent years new sources of ROS were identified in plants, including NADPH oxidases, amine oxidases, and cell wall-bound peroxidases. These are tightly regulated and participate in the control of processes such as programmed cell death, stress response, and pathogen defense.

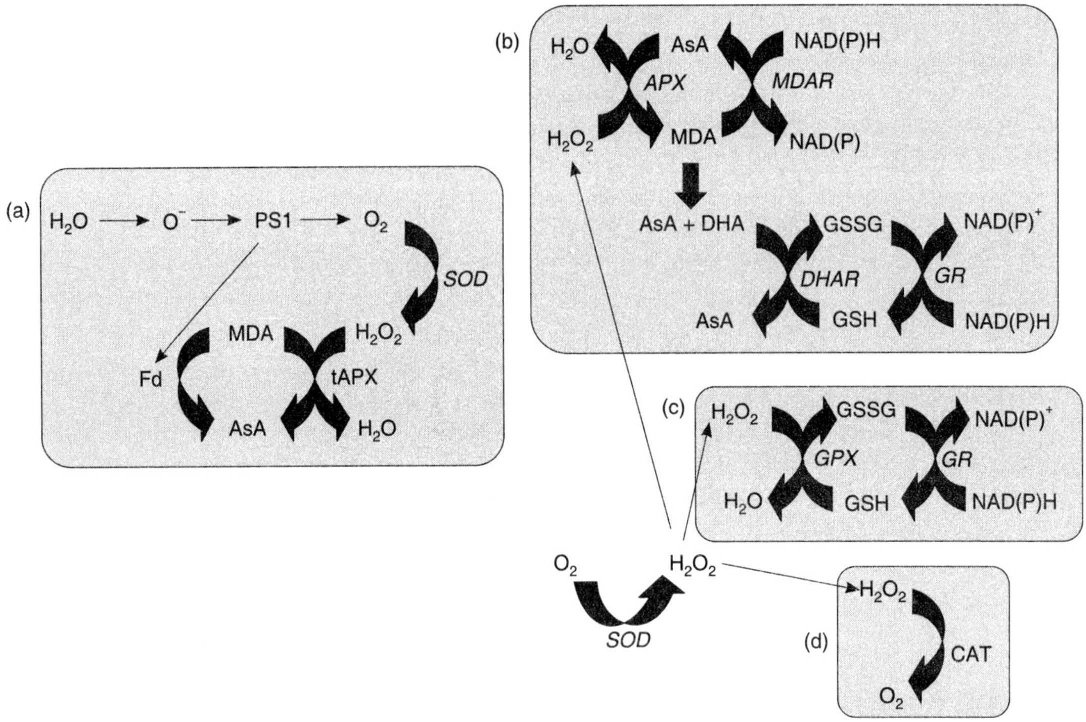

Fig. 1 ROS-scavenging mechanisms in various cellular compartments of a plant cell (a) SOD and thylakoid-bound APX (tAPX) in the chloroplast as part of the water-water cycle, (b) SOD and APX in the stroma, cytosol, mitochondria and apoplast of plants, as part of the ascorbate-glutathine pathway, (c) Glutathione peroxidase (GPX) and its glutathione regenerating cycle, (d) Catalase in peroxisomes. AsA, ascorbate; GSH and GSSG, reduced and oxidized glutathione; DHA, dehydroascorbatereuctase; MDA, monodehudroascorbatereductase; GR, glutathione reductase; PSI, photosystem I; Fd, ferredoxin; e^-, electron.

Table 1 ROS-producing mechanisms and its localization in the cell

Production of ROS	*Localization*	*Primary ROS*
Photosynthesis ET and PSI or II	Chloroplast	O_2^-
Respiration ET	Mitochondrion	O_2^-
Glycolate oxidase	Peroxisome	H_2O_2
Excited chlorophyll	Chloroplast	O_2^1
NADPH oxidase	Plasma Membrane	O_2^-
Fatty acid β-oxidation	Peroxisome	H_2O_2
Oxalate oxidase	Apoplast	H_2O_2
Xanthine oxidase	Peroxisome	O_2^-
Peroxidases, Mn^{2+} and NADH	Cell Wall	H_2O_2, O_2^-
Amine oxidase	Apoplast	H_2O_2

2.1 Chloroplasts

Chloroplasts are the major source of reactive oxygen forms in plants: they harvest light energy at high efficiency, produce reducing equivalents, such as NADPH, and generate fluxes of dioxygen, the most 'radical-prone' conditions. Several independent pathways, or sites of oxygen activation, have been described in chloroplasts, leading to the production of all kinds of ROS (Elstner, 1982; Asada, 1994; Foyer and Harbinson, 1994). The most important is the reducing side of photosystem I (PSI), where an electron may be passed from a membrane-bound carrier to O_2 (the 'Mehler Reaction'), instead of flowing to downstream carriers that finally reduce $NADP^+$ (Figure 1). Under conditions that limit the availability of electron acceptors from PSI, i.e., when the Calvin cycle does not consume NADPH rapidly enough, superoxide will form within the membrane. At low pH, superoxide may spontaneously dismutate to the more diffusible H_2O_2; otherwise it may interact with plastocyanine or cytochrome f and reduce them, resulting in a superoxide-mediated cyclic electron flow around PSI (Hormann *et al.*, 1993). This mechanism actually regulates the superoxide production, namely to divert, or cycle, excessive flow of electrons and, at the same time, prevent the diffusion of radicals away from the membrane.

2.2 Mitochondria

Mitochondria consume oxygen during respiratory electron transport. Different sites of electron leakage and release of superoxide and hydrogen peroxide in respiration have been proposed. One of the sites is specific to plant mitochondria, namely cyanide-insensitive alternate oxidase (Rich and Bonner, 1978; Elstner, 1991).

2.3 Endoplasmic Reticulum, Peroxisomes and Glyoxysomes

The smooth endoplasmic reticulum and the microsomes derived from it harbour various oxidative processes. Mixed-function oxygenases, such as cytochrome P450, perform important hydroxylation reactions in the mevalonic acid pathway, adding oxygen atoms to substrate molecules. NAD(P)H is the electron donor and superoxide may be released by such reactions. Peroxisomes and glyoxysomes are single membrane organelles that compartmentalize enzymes involved in the β-oxidation of fatty acids, and the C_2photorespiratory cycle, where glycolate oxidase transfers electrons from glycolate to oxygen and produces H_2O_2 (Lindquist *et al.*, 1991). Xanthine oxidase, urate oxidase and NADH oxidase generate superoxide (McKersie and Leshem, 1994; Elstner, 1991).

2.4 Plasma Membrane and the Apoplast Compartment

NAD(P)H oxidases are ubiquitous components of plasma membrane and may produce superoxide and H_2O_2 (Vianello and Macri, 1991). Oxygen activation occurs also in the apoplast: being the first site of pathogen invasion, it contains a first line of defense reactions and these involve reactive oxygen. The most common biosynthetic pathway in the apoplast is lignin biosynthesis, where phenylpropanoid precursors of lignin are cross-linked by H_2O_2 in reactions initiated by peroxidases (Gross, 1980). The required NADH s generated by a cell wall malate dehydrogenase, and then used to form H_2O_2, possibly by an NADH oxidase. Amine oxidases produce activated oxygen in the cell wall by using diamines or polyamines to reduce quinine with concomitant formation of peroxides (McKersie and Leshem, 1994; Elstner, 1991; Bolwell and Woftastek, 1997).

2.5 Cell Wall

Cell wall peroxidase is able to oxidize NADH and in the process catalyse the formation of O_2^-. This enzyme utilizes H_2O_2 to catalyse the oxidation of NADH to NAD^+, which in turn reduces O_2 to $O2^-$. Superoxide consequently dismutates to produce H_2O_2 and O_2 (Del-Rio *et al,* 1992; Bolwell and Woftastek, 1997).

3. ANTIOXIDANT SYSTEM IN PLANTS SCAVENGE SUPERFLUOUS ROS

Under optimal growth conditions the production of ROS in cells is estimated at a constant rate of 240μM s^{-1} O_2^-, and a steady state level of 0.5μM H_2O_2. However, biotic and abiotic stresses disrupt the cellular homeostasis of cells resulting in an enhanced production of ROS (up to 720μM s^{-1} O_2^- and a steady state level of 5-15μM H_2O_2). All ROS are extremely harmful to organisms at high concentrations. When the level of ROS exceeds the defense mechanisms, a cell is said to be in a state of "oxidative stress." The enhanced production of ROS during environmental stresses can pose a threat to cells by causing peroxidation of lipids, oxidation of proteins, damage to nucleic acids, enzyme inhibition, activation of programmed cell death pathway and ultimately leading to death of the cells (Shah *et al.*, 2001; Sharma and Dubey, 2005; Maheshwari and Dubey, 2009; Mishra *et al.*, 2011; Srivastava and Dubey, 2011; Verma and Dubey, 2003; Meriga *et al.*, 2004). Nevertheless, we know that a trickle of active oxygen species is constantly produced in all cellular compartments as a by-product of normal cellular metabolism and that cell survival will depend upon adequate protection. All aerobic forms of life have evolved multiple defense lines, which include both scavenging enzymes and non-enzymatic antioxidants (Table 2). Such multiplicity is required because ROS are produced in different cellular and extracellular compartments, and because reactive species differ in properties such as diffusibility, solubility and propensity to react with various biological molecules. We thus need a correspondingly diverse set of defense molecules to act in both aqueous and membranal phases, in all cellular compartments, to promptly inactivate radicals as soon as they are formed. A sequence of detoxification steps is often required to avoid the conversion of one reactive species into a second, more harmful one. This also implies that multistep defense systems, if tilted out of balance, may collapse and get out of control, e.g., under extreme stress conditions. The coordination of the multiple defense components into one integrated and efficient network requires a deep understanding and research. The following text outlines the defense systems found in plants.

Table 2 ROS-scavenging mechanisms and its localization in the cell.Apo, apoplast; Chl, chloroplast; CW, cell wall; Cyt, cytosol; Mit, mitochondrion; O_2^1, singlet oxygen; Per, peroxisome; Vac, vacuole

Production of ROS	*Localization*	*Primary ROS*
Superoxide dismutase	Chl, Cyt, Mit, Per, Apo	O_2^-
Ascorbate peroxidase	Chl, Cyt, Mit, Per, Apo	H_2O_2
Catalase	Per	H_2O_2

Contd...

Contd...

Glutathione peroxidase	Cyt	H_2O_2, ROOH
Peroxidases	CW, Cyt, Vac	H_2O_2
Thioredoxin peroxidase	Chl, Cyt, Mit	H_2O_2
Ascorbic acid	Chl, Cyt, Mit, Per, Apo	H_2O_2, O_2^-
Glutathione	Chl, Cyt, Mit, Per, Apo	H_2O_2
α-tocopherol	Membranes	ROOH, O_2^1
Carotenoids	Chl	O_2^1
Alternate oxidase	Chl, Mit	O_2^-

3.1 Enzymatic Antioxidants

3.1.1 Superoxide Dismutase (SOD)

SOD was first isolated by Mann and Keilis (1938) and thought to be a copper storage protein. Subsequently, the enzyme was identified by a number of names, erythrocuprein, indophenol oxidase, and tetrazolium oxidase until its catalytic function was discovered by McCord and Fridovitch (1969). SOD is now known to catalyze the dismutation of superoxide to hydrogen peroxide and oxygen in all aerobic organisms as well as some anaerobes (Hassan, 1989). Since SOD is present in all aerobic organisms and most subcellular compartments that generate activated oxygen, and also being the first product of univalent reduction of oxygen, it has been assumed that SOD has a central role in the defence against oxidative stress (Beyer *et al.*, 1991; Bowler *et al.*, 1991; Scandalias (1993). SOD is the fastest enzyme known and its three-dimensional structure has been intensively studied (Kitagawa *et al.*, 1991; Getzoff *et al.*, 1992). The dismutation is catalyzed by the metal ion (Cu, Mn or Fe) at the active site. Superoxide is attracted to such a site by appropriately positioned, positively charged amino acid residues. There are three distinct types of SOD classified on the basis of the metal cofactor (Bannister *et al.*, 1987):

(i) copper/zinc (Cu/Zn - SOD) isozymes
(ii) manganese (Mn-SOD) isozymes
(iii) iron (Fe-SOD) isozymes

These isozymes can be separated by native polyacrylamide gel electrophoresis, their activity detected by negative staining and identified on the basis of their sensitivity to KCN and H_2O_2. The Mn-SOD is resistant to both inhibitors, whereas the Cu/Zn SOD is sensitive to both inhibitors; Fe-SOD is resistant to KCN, and sensitive to H_2O_2. The subcellular distribution of these isozymes is also distinctive. The Mn-SOD is found in the mitochondria of eukaryotic cells; some Cu/Zn-SOD isozymes are found in the cytosol, others in the chloroplasts of higher plants. The Fe-SOD isozymes are often not detected in plants, but when detected, Fe-SOD is usually associated with the chloroplast compartment (Bowler *et al.*, 1991). The prokaryotic Mn-SOD and Fe-SOD, and the eukaryotic Cu/Zn-SOD enzymes are dimers, whereas the Mn-SOD of mitochondria are tetramers (Scandalis, 1993). Peroxisomes and glyoxysomes of watermelons (*Citrillus vulgaris*) have been shown to contain both Cu/Zn- and Mn-SOD activity (Sandalio, 1993) but there are no reports of extracellular SOD enzymes in plants. All forms of the SOD are nuclear-encoded and are targeted to their respective subcellular compartments by an amino terminal targeting sequence. Several forms of SOD have been cloned from a variety of plants (Bowler *et al.*, 1991; Scandalis, 1993; Perl-Treves *et al.*, 1988; Kliebenstein *et al.*, 1998). Prokaryotic cells and many eukaryotic algae contain only the Mn-SOD and Fe-SOD isozymes which are believed to be more ancient forms. To date it has been shown that SOD activity is increased in cells in response to diverse environmental and xenobiotic stresses including paraquat, high light, waterlogging and drought. Apparently, each of the SOD isozymes are independently regulated according to the degree of oxidative stress experienced in the respective subcellular compartments, but how this is communicated at the molecular level is unknown.

3.1.2 Catalase (CAT)

CAT is a heme-containing enzyme that catalyses the dismutation of hydrogen peroxide into water and oxygen. The enzyme is found in all aerobic eukaryotes and is important in the removal of hydrogen peroxide generated in peroxisomes (microbodies) by oxidases involved in β-oxidation of fatty acids, the glyoxylate cycle (photorespiration) and purine catabolism. Catalase was one of the first enzymes to be isolated in a highly purified state. All forms of the enzyme are tetramers in excess of 2,20,000 molecular weight. Multiple forms of catalase have been described in many plants. Little or no catalase was found in chloroplasts (Asada, 1994) but mitochondrial isozyme was found in maize. These forms have been cloned (Redinbaugh *et* al., 1988; Scandalias, 1990) and homologous genes have been cloned from several other plants. Maize has three isoforms termed cat-1, cat-2 and cat-3 that are on separate chromosomes and are differentially expressed and independently regulated (Scandalias, 1990). Cat-1 and cat-2 are localized in peroxisomes and the cytosol, whereas cat-3 is mitochondrial. The peroxisomes are linked to the photosynthetic metabolism via the photorespiratory process, and, according to recent findings, may take part in oxidative stress tolerance. CAT isozymes differ in biochemical properties, as well as in developmental specificity: some related to germination, their principal role involving probably fatty acids conversion, whereas others are related to lignifications, photorespiration, or ageing processes.

It is interesting to note that CAT undergoes photoinactivation, i.e., it is very sensitive to light but has a rapid turnover rate similar to that of the D1 protein of PSII (Hertwig and Feierabend, 1992). It requires continuous *de novo* synthesis. This may be a result of light absorption by the heme group or perhaps hydrogen peroxide inactivation. Regardless, stress conditions which reduce the rate of protein turnover, such as salinity, heat shock or cold, cause the depletion of catalase activity. This may have significance in the plant's ability to tolerate the oxidative components of these environmental stresses. This effect may be compensated for by decreased H_2O_2 production through PSII photoinhibition, as well as compensatory increase in ascorbate, glutathione, and ascorbate peroxidases (Volk and Feierabend, 1992). Acclimation to chilling by pre-exposure treatment probably includes induction of CAT (Prasad *et al.*, 1994). Interplay of H_2O_2 and CAT in the hypersensitive reaction (HR) and systemic-acquired resistance (SAR) responses to plant-pathogen infection has been studied. Salicylic acid inhibits catalase and the resulting H_2O_2 burst is part of the pathogenesis signal transduction chain (Durner and Klessig, 1995).

3.1.3 Glutathione Reductase (GR)

GR is a flavoprotein and most of its activity is found in chloroplasts. GR completes the 'Asada-Halliwell pathway' by regenerating the glutathione pool with NADPH as electron (Foyer and Halliwell, 1976). It is worth mentioning how the repair or prevention of oxidative damage finally consumes reducing equivalents from the light reaction.

$$GSSG + NADPH \longrightarrow 2GSH + NADP^+$$

Mitochondrial and cytosolic isozymes have also been described with pea GR even being cloned (Creissen *et al.*, 1994). GR and APX activities increase in response to ethylene, ozone, SO_2 and NO_2. One of the main functions of GSH is to protect –SH (sulphur) groups in enzymes and structural proteins against oxidation, either by acting as a scavenger for oxidizing substances or by repairing the -SH groups via the GSH-disulfide exchange reaction. The GSSG formed in both cases is reduced rapidly by the action of GR. According to the -SH hypothesis, freezing tolerance in plants involves an increase in the resistance toward oxidation of -SH groups in proteins. The theory, however, supposes that an effective system is operating in the leaves of frost-resistant plants which maintains the -SH containing proteins in the reduced state during the frost period (Esterbauer and Dieter, 1977).

3.1.4 Ascorbate Peroxidase (APX)

A large pool of ascorbate, approximately, 10 mM, has been found in the chloroplast, cytosol, vacuole and extra-cellular compartments of the cell. About 20 to 40% of the ascorbate is in chloroplast of leaf mesophyll cell. The chloroplast contains all the enzymes to regenerate reduced ascorbate from its oxidised products. Some researchers proposed that hydrogen peroxide was dissipated in the chloroplast by the coupling of ascorbate and glutathione redox cycling (Anonymous, 2010). APX is a very important enzyme in the ascorbate-glutathione cycle. The enzyme transforms H_2O_2 to H_2O and O_2 (Antmen, 2005). This sequence of reactions is also named as the 'Halliwell-Asada pathway'. It was illuminated that chloroplasts produce superoxide and hydrogen peroxide on the thylakoid. Superoxide is converted into hydrogen peroxide by either spontaneous dismutation or with the action of the SOD enzyme (Koc and Ustun, 2008).

H_2O_2 is reduced to water by this heme-containing enzyme, APX and the ascorbate present in the chloroplast is oxidized to monodehydroascorbate (MDA) in the following reaction:

$$2\ \text{ascorbate} + H_2O_2 \longrightarrow 2\ \text{MDA} + H_2O_2$$

In the chloroplasts, both stromal and thylakoid-bound forms of APX were found (Miyake and Asada, 1992) and cytosolic isozymes were described and cloned (Mittler and Zilinskas, 1992).

3.1.5 Mono-hydroascorbatereductase (MDAR) and Dehydroascorbatereductase (DHAR)

The product of SOD, hydrogen peroxide requires further detoxification. This is achieved by other enzymes and non-enzymatic antioxidants that may differ among the various cellular compartments. The MDA has two routes of regeneration, one via MDAR, the other via DHAR and glutathione. The terminal electron donor is NADPH (Miyake and Asada, 1992). This pathway serves two functions. One is the detoxification of hydrogen peroxide; otherwise it participates in Fenton reactions, and another function is it oxidizes NADPH. The latter function is an apparently energy-consuming, wasteful process analogous to photorespiration. The chloroplast contains catalase and dissipates hydrogen peroxide without "wasting" NADPH. However, it should be realized that favouring conditions for electron transfer from PSI to oxygen, generally causes a high redox potential like high NADPH/NADP ratio. By reducing this redox potential through the Halliwell-Asada pathway, the tendency of PSI to reduce oxygen is minimized (Anonymous, 2010).

In the plastids, MDA may give rise to dehydroxyascorbate (DHA). Both must be reduced to regenerate the ascorbate pool, which can be achieved by several reactions:

non-enzymatic reduction by ferrodoxin:

$$\text{MDA} + \text{FD}_{red} \longrightarrow \text{ascorbate} + \text{FD}_{ox}$$

reduction of MDA by MDA reductase (MDAR) in the stroma, using NADPH:

$$2\ \text{MDA} + \text{NADPH} \longrightarrow 2\ \text{ascorbate} + \text{NADP}^{+}$$

reduction of DHA to ascorbate by DHA reductase (DHAR) with GSH as the reducing substrate:

$$\text{DHA} + 2\ \text{GSH} \longrightarrow \text{ascorbate} + \text{GSSH}$$

Scavenging is thought to occur at the thylakoid surface, near PSI, minimizing the risk of escape and reaction of ROS with each other. A second line of defense operates in the stroma, to protect the sensitive enzymes of the Calvin cycle.

3.1.6 Thioredoxin and Glutathione-S-transferases (GST)

Thioredoxin is a small ubiquitous protein that plays a redox-regulatory role in plants. It has been suggested to protect organisms by scavenging reactive oxygen, as well as regenerating oxidized proteins (Fernando *et al.*, 1992; Takemoto *et al.*, 1998). Some additional cellular processes are activated after severe oxidative damage and repair biological macromolecules. Such mechanisms allow the cell to regain its homeostasis and hence regarded as secondary defenses. The large and variable class of plant GST are regarded as 'secondary defense enzymes' (Keser, 2005). These enzymes catalyze the conjugation of a GSH molecule to a variety of compounds like in the detoxification of herbicides. The conjugate is marked for secretion to the apoplast or vacuole through glutathione pumps. GST conjugations are also important in the synthesis of secondary metabolites. GSTs detoxify toxic breakdown products of lipid peroxidation or oxidative DNA degradation; they may also function as peroxidases and scavenge radicals. Interestingly, GSTs are induced by ROS, ozone, wounding, ethylene, heavy metals and also pathogen attack.

3.2 Non-enzymatic Antioxidants

3.2.1 Ascorbate

Ascorbate, also known as vitamin C, is an important antioxidant (Fig. 2). It reacts with H_2O_2, O_2^-, OH^- and lipid hydroperoxides (Yu, 1994). Its role as the APX substrate has already been discussed (section 4.1.4). Ascorbate is water-soluble and protects or regenerates oxidized carotenes and tocopherol from tocopheroxyl radicals on the thylakoid surface, thus, providing membrane protection. It also acts as cofactor of violaxanthin de-epoxidase, thus sustaining dissipation of excess excitation energy (Ahmad *et al.*, 2008; Smimoff, 2000). It can react indirectly by regenerating c-tocopherol or in the synthesis of zeaxanthin in the xanthophylls cycle (Sade *et al.*, 2011). Although ascorbate metabolism has been studied in most detail in the chloroplast, it is likely that all enzymes for its regeneration also exist in the cytosol of both photosynthetic and non-photosynthetic cells. The cell wall is also an important site of ascorbate metabolism because it contains mM concentrations of ascorbate. Here, ascorbate may have a role in cell wall biosynthesis (Polle *et al.*, 1990). The cell wall does not contain ascorbate peroxidase, but contains ascorbate oxidase (Chichiricco *et al.*, 1989). Since the enzymes to recycle oxidized forms of ascorbate are not present in the cell wall, it has been proposed that the plasmalemma may have an ascorbatetranslocator to shuttle oxidized and reduced forms between the cytosol and cell wall (Foyer, 1993).

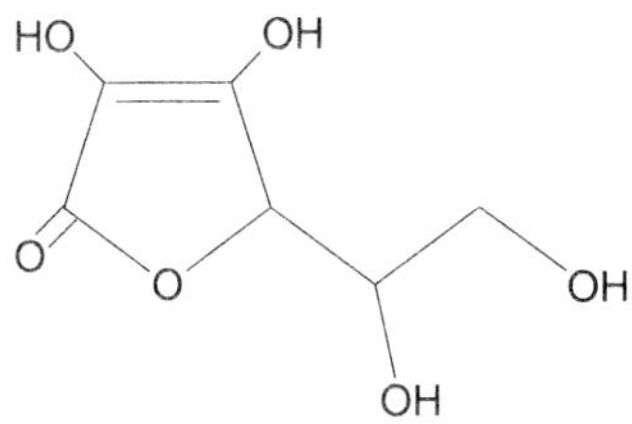

Fig. 2 Structure of ascorbic acid.

3.2.2 Tocopherol

Tocopherol is actually a family of antioxidants of which α-Tocopherol is generally considered to be the most active form (Hess, 1993). The major isomer of vitamin E is α-tocopherol, a phenolic antioxidant present in both plants and animals (Fig. 3). Its hydrophobic nature dictates that α-tocopherol is located exclusively in cell membranes and is oriented with the benzoquinone ring in close association with the carbonyl of the glycerol component of the phospholipid and with the phytyl chain associated with the fatty acids in the hydrophobic inner regions of the membrane bilayer. α-tocopherols prevent lipid auto-oxidation and this makes it an effective free radical trap. The high degree of lipid unsaturation in chloroplast membrane requires large amounts of α-tocopherol. Plants synthesize tocopherol by enzymes localized in the inner chloroplast membrane (Soll *et al.*, 1984).The increase in tocopherols in conjunction with ascorbate has been implicated as one of the primary responses of water deficit conditions in rice (Bhattacharjee, 2005). Oxidative stress activates the expression of genes responsible for the synthesis of tocopherols in higher plants. Antioxidants including α-tocopherol and ascorbic acid have been reported to increase following triazole treatment in tomato and these may have a role in protecting membranes from oxidative damage, thus contributing to chilling tolerance (Ahmad *et al.*, 2008; Smimoff, 2000; Polle *et al.*, 1990; Chichiricco *et al.*, 1989; Foyer, 1993; Sade *et al.*, 2011; Hess, 1993, Boo and Jung, 1999; Wu *et al.*, 2007; Shao *et al.*, 2007). The antioxidant properties of tocopherol are the result of its ability to quench both singlet oxygen and peroxides (Fryer, 1992). Tocopherol is a less efficient scavenger of singlet oxygen than β-carotene and therefore in the thylakoid membrane it may function to break carbon radical chain reactions by trapping peroxyl radicals.

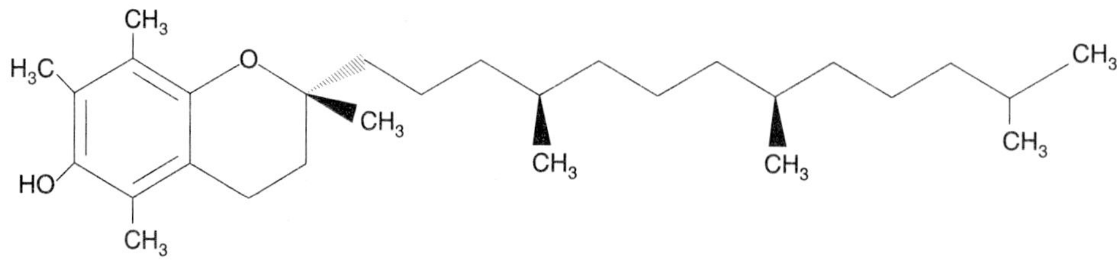

Fig. 3 Structure of tocopherol.

3.2.3 *Carotenoids*

Carotenoids are lipid soluble, C_{40}isoprenoids and tetraterpenes that are located in the plastids of both photosynthetic and non-photosynthetic plant tissues. In chloroplasts, the carotenoids function as accessory pigments in light harvesting, but perhaps a more important role is their ability to detoxify various forms of activated oxygen and triplet chlorophyll that are produced as a result of excitation of the photosynthetic complexes by light. It quenches both excited triplet-state chlorophyll and singlet oxygen, preventing them from initiating lipid peroxidation (Parker and Joyce, 1967). β-carotene is the main precursor of vitamin A (Fig. 4).

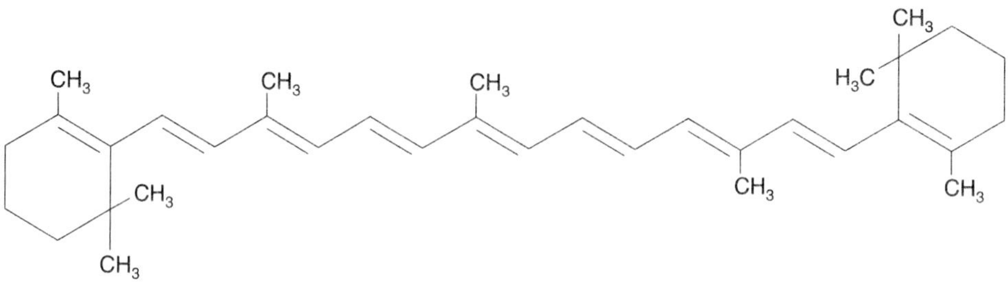

Fig. 4 Structure of β-carotene.

In terms of its antioxidant properties carotenoids can protect the photosystems in one of four ways:

(i) by reacting with lipid peroxidation products to terminate chain reactions (Burton and Ingland, 1984);

(ii) by scavenging singlet oxygen and dissipating the energy as heat (Mathis and Kleo, 1973);

(iii) by reacting with triplet or excited chlorophyll molecules to prevent formation of singlet oxygen; or

(iv) by the dissipation of excess excitation energy through the xanthophyll cycle (Demming-Adams and Adams, 1996).

3.2.4 Glutathione

Glutathione (GSH) is a ubiquitous γ-tripeptide (Glu-Cys-Gly) whose antioxidant function is facilitated by the sulphydryl group of cysteine (Rennenberg, 1982). On oxidation, the sulphur forms a thiyl radical that reacts with a second oxidized glutathione forming a disulphide bond (GSSG) (Fig. 5). GSH is found in most tissues, cells and subcellular compartments of higher plants. Its levels declines with tissue age and vary among growth environments; for example, glutathione levels are higher in the light than in the dark. At the subcellular level, GSH concentration is highest in the chloroplast, averaging between 1 and 4 mM, but significant quantities also accumulate in the cytosol. GSH can function as an antioxidant in many ways. It can react chemically with singlet oxygen, superoxide and hydroxyl radicals and therefore functions directly as a free radical scavenger. GSH may stabilize membrane structure by removing acyl peroxides formed by lipid peroxidation reactions (Price *et al.*, 1990). As mentioned in section 4.1.5, GSH is the reducing agent that recycles ascorbic acid from its oxidized to its reduced form by the enzyme DHAR (Loewus, 1988). GSH can also reduce DHA by a non-enzymatic mechanism at pH >7 and GSH concentrations greater than 1mM. This may be a significant pathway in chloroplasts whose stromal pH in the light is about 8 and GSH concentrations may be as high as 5 mM (Foyer and Halliwell, 1976). GSH is important for recycling all the above vitamins, while its own regeneration depends on NAD(P)H consumption. A decrease in reduced GSH pools accompanies many stress and disease conditions, but it is not always clear whether such a decrease is the cause of the disease or stress situation, or rather reflects a greater demand for GSH (Yu, 1994).

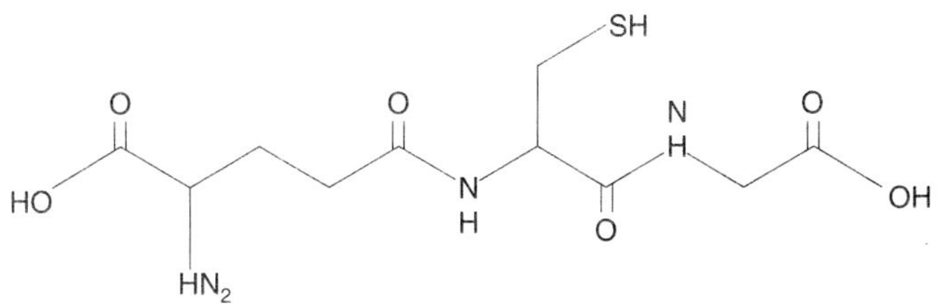

Fig. 5 Structure of glutathione.

3.2.5 Phenolic Compounds

Phenolics are diverse secondary metabolites (flavonoids, tannins, hydroxycinnamate esters and lignin) abundant in plant tissues (Grace and Logan, 2000). Polyphenols possess ideal structural chemistry for free radical scavenging activity, and they have been shown to be more effective antioxidants *in vitro* than tocopherols and ascorbate (Bolkhina*et al*, 2003). Antioxidative properties of polyphenols arise from their high reactivity as hydrogen or electron donors, and from the ability of the polyphenol-derived radical to stabilize and delocalize the unpaired electron (chain-breaking function), and from their ability to chelate transition metal ions (termination of the Fenton reaction) (Grace and Logan, 2000). Another mechanism underlying the antioxidative properties of phenolics is the ability of flavonoids to alter peroxidation kinetics by modification of the lipid packing order and to decrease fluidity of the membranes (Arora *et al.*, 2000). These changes could sterically hinder diffusion of free radicals and restrict peroxidative reactions. Moreover, it has been shown recently that phenolic compounds can be involved in the hydrogen peroxide scavenging cascade in plant cells (Takahama and Oniki, 1997). Their synthesis in plants increases under strong or UV-enriched light and they may protect the cell against lipid peroxidation (Torel *et al.*, 1986).

3.2.6 Additional Antioxidative Compounds

Sugar alcohols, such as mannitol, are produced by many plants undergoing osmotic stress. They may serve as compatible solutes and osmoprotectants, but may also function as antioxidants (Smirnoff and Cumbes, 1989; Stoop *et al.*, 1996). Mannitol is a more efficient carbon sink for light reaction products (such as NADPH), and may therefore alleviate photooxidative stress under some circumstances.

4. ROS ARE A PART OF SIGNALLING NETWORK IN A PLANT CELL

In recent years a new role for ROS was identified: the control and regulation of biological processes such as programmed cell death, hormonal signalling, stress responses and development. In order to exert such a profound effect on plant metabolism, ROS either utilize and/or interfere with other signalling molecules. H_2O_2 was found to induce accumulation of stress hormones such as C_2H_4 and salicylic acid (Leon *et al.*, 1995; Chamnongpol, 1998). Plant hormones are not only downstream of the ROS signal; ROS themselves are secondary messengers in many hormonemediated signalling pathways (Chen *et al.*, 1993; Orozco-Cardenas and Ryan, 1999). Therefore, it is conceivable that feedback/feed forward interactions between ROS and hormones occur. At low/moderate concentration, ROS have been implicated as second messengers in intracellular signalling cascades that mediate several plant responses in plant cells, including stomatal closure (Neill *et al.*, 2002; Yan *et al.*, 2007; Kwak *et al.*, 2003), programmed cell death (Mittler, 2002; Bethke and Jones, 2001), gravitropism (Bethke and Jones, 2001; Joo *et al.*, 2001), and acquisition of tolerance to both biotic and abiotic stresses (Torres *et al.*, 2002; Miller *et al.*, 2008). Plants can sense, transduce and translate ROS signal into appropriate cellular responses with the help of some redox-sensitive proteins, calcium mobilization, protein phosphorylation, and gene expression. ROS can be sensed directly also by key signalling proteins such as a tyrosine phosphatase through oxidation of conserved cysteine residues (Xiong *et al.*, 2002). ROS can also modulate the activities of many components in signalling, such as protein phosphatases, protein kinases and transcription factors (Cheng and Song, 2006) and communicate with other signal molecules and the pathway forming part of the signalling network that controls response downstream of ROS. The strength, lifetime and size of the ROS signalling pool depends on balance between oxidant production and removal by the antioxidant (Sharma *et al.*, 2012).

5. ANTIOXIDATIVE SYSTEMS PLAY ROLE DURING *IN VITRO* GROWTH

Since the past half century, plant tissue culture has transformed from an experimental curiosity to a largely commercialized industry. Nevertheless, its widespread use is restricted by the formation of plantlets of abnormal morphology, anatomy and physiology induced by special conditions during *in vitro* culture e.g., high air humidity, decreased air turbulence, low irradiance, low CO_2 concentration during light period, cultivation media supplemented with sugars and growth regulators. The *ex vitro* transfer or change in the environmental condition further impairs the plant system. Upon imposition of such stresses, production of ROS from cellular organelles like mitochondria, chloroplasts, peroxisomes and cytosol is unequivocally a regular feature in plants. This is a consequence of the fact that the photoperiod, temperature and relative humidity in laboratory conditions, beyond influencing the stomata aperture, regulates the, process of photosynthesis and respiration on chloroplasts and mitochondria and thus the natural production of ROS in cells.While different sources like high irradiance, drought, salinity, heavy metals, etc., cause ROS production and promote activities of antioxidative enzymes, the role of ROS in plant growth and development *in vitro* has also been well documented. Addition of plant growth regulators in media is often essential and the role of ROS in plant growth is further substantiated by the interplay of ROS with a number of phytohormones.

There are reports that provide evidence that antioxidant enzymes are correlated with the morphogenic process in plants. There are differences in antioxidant metabolism between the calli showing different morphogenic potential. These differences might be linked to different metabolic pathways accompanying the formation of roots or shoots. Also, role of antioxidative enzymes during different stages of organogenesis is substantiated in many plants. Further, ROS induce accumulation of stress hormones, such as salicylic acid and ethylene. On the other hand, abscissic acid (ABA) can affect ROS production and ROS are secondary messengers in many ABA signalling pathways including antioxidative defence induction. Cytokinins too have been reported to influence the antioxidative system in *in vitro* grown plants. ROS or ROS-scavenging antioxidant enzyme-stimulated improvement in growth and regeneration further support the relationship between oxidative stress and recalcitrance. Higher activities of antioxidative enzyme systems promote cell division and somatic embryogenesis. Proline, another important stress hormone, involved in releasing stress, protection of enzymes, organizing cellular structure and act as a free radical scavenger. Peroxidase (POD), a multifunctional enzyme, takes part in growth control, development, differentiation and morphogenesis. The tissue, organ and ontogenic specifications of POD (iso-peroxidases) could also be used as markers for the plant physiological status. The role of POD in vitrification is specific and important for *in vitro* cultures. Along with plant stress tolerance, in some cases, the activity of POD and SOD go parallel. The time course of H_2O_2 generation in microshoots of plants is an important factor confirming narrow connection between hyperhydricity and oxidative stress. Hyperhyricity is also a consequence of increased Malinodialdehyde (MDA) content and increased basic isoform of POD.

Further, when the *in vitro* grown plantlets are transferred to *ex vitro* conditions, they are usually accompanied by water-stress or photo inhibition. These physiological stresses contribute to the production of reactive oxygen species and in consequence oxidative stress. The extremely short life-time of ROS makes the study of their production in *planta* very difficult. In some plants antioxidative enzyme activities were affected by CO_2 concentration during *in vitro* cultivation and after *ex vitro* transplantation as well as ABA application. Therefore, to acclimatize the micro shoots efficiently *ex vitro*, study of the non-enzymatic antioxidants as well as the activities of antioxidative enzymes formed during and after *in vitro* growth are very important.

6. SUMMARY AND FUTURE PERSPECTIVES

Unlike animals, plants synthesise a complete range of molecular defences against oxidative attack, the chloroplasts being the richest source of antioxidants including vitamins C and E. It has been demonstrated that under certain stress conditions there is an increase in the cellular levels of ROS. Under normal physiological conditions there is a balance between the formation of ROS and the protective antioxidant mechanisms of cells. But when the generation of these reactive oxygen intermediates overcomes the cellular antioxidant capacity, cells undergo oxidative stress situations, which can eventually produce the cell death. In plants, the main cellular loci where reactive oxygen species are generated are chloroplasts, mitochondria, peroxisomes, plasma membranes, and cell walls. In recent years, it has become increasingly apparent that reactive oxygen species are implicated in oxidative stress situations induced by many environmental stresses including drought, extremes of temperature, flooding and post-anoxia stress, a range of gaseous pollutants (ozone, nitrogen oxides, volatile organic compounds, etc), heavy metals, herbicides through to UV-B radiation, plant diseases and senescence. A substantial progress has been made in understanding the molecular defence responses following perturbations of temperature, water stress, UV-B irradiance and nutritional deficiencies and toxicities. In the past three years, important achievements have been made in the area of plant pathology, particularly on the molecular responses to disease. By using molecular biology methods, a better understanding has been achieved of the complex responses of elicitor challenge (bacterial, viral and fungal) and the remarkable similarity with responses induced by exposure to ozone. Apparently, there could be a homology between the hypersensitive response in plants and the neutrophil killing mechanisms in mammalian studies, with a direct role played by ROS in killing the plant pathogen. There is general agreement in that the oxygen burst induced in plants by elicitors is probably directly involved in molecular signalling and in triggering the orchestrated cascade of events associated with physical isolation of the pathogen. It is becoming increasingly evident that oxygen radicals and antioxidants have roles in modulating gene expression and early growth response genes. These effects can be important both in normal growth as well as in pathological conditions. Evidence accumulated in recent years has demonstrated the importance of natural plant antioxidants in medicine, human nutrition and food industry. An appropriate dose of antioxidants in the human diet can help to avoid the risk of contracting diseases where ROS are involved in its pathogenesis. In this sense, special attention is being paid to the role of the non-enzymatic antioxidants (glutathione, ascorbic acid, vitamin E, vitamin A, etc) and enzymatic antioxidants (superoxide dismutase, ascorbate peroxidase, glutathione reductase, glutathione peroxidase, etc) in the cell response to stress situations via plant tissue culture. It has been demonstrated in model plants that it is possible to enhance the levels of key components of ROS metabolism and antioxidants by genetic manipulation. This way, transgenic plants could be obtained having an elevation of oxidative stress tolerance.

References

Ahmad, P., Sarwat. M. and Sharma, S. (2008) Reactive oxygen species, antioxidants and signaling in plants. *J. Plant Biol.*3: 167-173.

Alscher, R. G. and Hess, J. L. (1993) In *Antioxidant in Higher Plants*, CRC Press, Boca Raton, Florida.

Alscher, R. G., Donahue, J. L. and Cramer, C. L. (1997) Reactive oxygen species and antioxidants: Relationship in green cells. *Physiol. Plant.*100:224-233.

Anonymous (2010). Natural disaster center. http://afetmerkezi.com (Accessed 08.11.2012).

Antmen, E. (2005)Oxidative stress at *Beta talasamide. Cukurova Uni. Pub.*

Arora, A., Byrem, T. M., Nair, M. G. and Strasburg, G. M. (2000) Modulation of liposomal membrane fluidity by flavonoids and isoflavonoids. *Arch. Biochem. Biophys.* 373:102-109.

Arora, A., Sairam, R. K. and Srivastava, G. C. (2002) Oxidative stress and antioxidative system in plants. *Curr. Sci.*82:1221-1235.

Asada, K. (1994) Production and action of active oxygen species in photosynthetic tissues. In C. H. Foyer and P. M. Mullineaux, (eds.) *Causes of Photooxidative Stress and Amelioration of Defense Systems in Plants,* CRC Press, Boca Raton, pp. 77-104.

Asada, K. and Takahashi, M. (1987) Production and scavenging of active oxygen in photosynthesis, In Kyle, D. J., Osmund, C. B. and Arntzen, C. J(eds.) *Photoinhibition*Elsevier, Amsterdam, pp. 227-287.

Bannister, J.V., Bannister, W.H. and Rotils, G. (1987) Aspects of the structure, function and applications of superoxide dismutase.*CRC Crit. Rev. Biochem.* 22:110-180.

Bethke, P. C. and Jones, R. L. (2001) Cell death of barley aleuroneprotoplasts is mediated by reactive oxygen species. *Plant Journal* 25(1): 19-29.

Beyer, W., Imlay, J., Fridovich, I. (1991) Superoxide Dismutases. *Prog. Nucl. Acid Res.* 40: 221-253.

Bhattacharjee, S. (2005) Reactive oxygen species and oxidative burst: Roles in stress, senescence and signal transduction in plants. *Curr. Sci.* 89 (7): 1113-1121.

Bolkhina, O; Virolainen, E. and Fagerstedt, K. V. (2003) Antioxidants, Oxidative Damage and Oxygen Deprivation Stress: a Review.*Annals of Botany*. 91: 179-194.

Bolwell, G. P. and Woftastek, P. (1997) Mechanism for the generation of reactive oxygen species in plant defense – Broad perspective. *Physiol. Mol. Plant Pathol.* 51: 347-349.

Boo, Y. C. and Jung, J. (1999)Water deficit induced oxidative stress and antioxidative defense in rice plants. *J. Plant Physiol.* 51: 255-261.

Bowler, C., Slooten, L., Vandenbranden, S., De Rycke, R., Botterman, J., Sybesma, C., Van Montagu, M. and Inzé, D. (1991) Manganese superoxide dismutase can reduce cellular damage mediated by oxygen radicals in transgenic plants. *EMBO. J.* 10: 1723-1732.

Burton, G.W. and Ingold, K.U. (1984) β-carotene: an unusual type of lipid antioxidant. *Science*. 224: 569-573.

Chamnongpol, S. (1998) Defense activation and enhanced pathogen tolerance induced by H_2O_2 in transgenic plants. *Proc. Natl. Acad. Sci. USA*. 95: 5818-5823.

Chen, Z.,Shilva, H. and Klessig, R. F. (1993) Active oxygen species in the induction of plant systemic acquired resistance by SA. *Science*. 262: 1883-1886.

Cheng, Y. and Song, C. (2006) Hydrogen peroxide homeostasis and signaling in plant cells. *Science in China. Series C, Life sciences* 49(1): 1-11.

Chichiricco, G., Ceru, M.P., D'Alessandro, A., Oratore, A. and Avigliano, L. (1989) Immunohistochemicallocalisation of ascorbate oxidase in *Cucurbitapepo*medullosa, *Plant Sci.* 64: 61-66.

Creissen, G., Edwards, A. and Mullineaux, P. (1994) Glutathione reductase and ascorbate peroxidase. In C. H. Foyer and P. M. Mullineaux, (eds.) *Causes of Photooxidative Stress and Ameloration of Defense Systems in Plants*, CRC Press, Boca Raton, pp. 343-364.

Del Rio, L. A., Sandalio, L. M., Palma, J. M., Bueno, P. and Cospus, F. J. (1992) Metabolism of oxygen radicals in peroxysomes and cellular implications. *J. Free Radicals Biol. Med.*3: 557-580.

Demming-Adams, B. and Adams, W. W. I. (1996) The role of xanthophyll cycle carotenoids in the protection of photosynthesis. *Trends Plant Sci.* 1: 21-26.

Durner, J. and Klessig, D. F. (1995) Inhibition of ascorbate peroxidase by salicylic acid and 2, 6-dichloroisonicotinic acid, two inducers of plant defense responses. *Proc. Natl. Acad.Sci. USA*. 92: 11312-11316.

Elstner, E. F. (1982) Oxygen activation and oxygen toxicity. *Annu. Rev. Plant Physiol.* 33: 73-96.

Elstner, E. F. (1987) Metabolism of activated oxygen species. In D. D. Davies, (ed.), *Biochemistry of Metabolism* (The Biochemistry of Plants: a Comprehensive Treatise, Vol. 11), Academic Press, San Diego, pp. 253-315.

Elstner, E. F. (1991) Mechanisms of oxygen activation in different compartments of plant cells. In E. J. Pell and K. L. Steffen (eds.) *Active Oxygen/Oxidative Stress and Plant Metabolism*. American Society of Plant Physiologists, Rockville, pp. 13-25.

Esterbauer, H. and Dieter, G. (1977). Seasonal variation of glutathione and glutathione reductase in needles of *Piceaabies*. *Plant Physiol.* 61: 119-121.

Fernando, M. R., Nanri, H., Yoshitake, S., Nagata-Kuno, K. and Minakami, S. (1992) Thioredoxin proteins inactivated by oxidative stress in endothelial cells. *Eur. J. Biochem.* 209: 917-922.

Foyer, C. (1993) Ascorbicacid.*In*R.G. Alscher and J.L. Hess (eds.)*Antioxidants in Higher Plants.* CRC Press, Boca Raton, pp. 31-58.

Foyer, C. H. and Halliwell, B. (1976) The presence of glutathione and glutathione reductase in chloroplasts: a proposed role in ascorbic acid metabolism. *Planta.* 133: 21-25.

Foyer, C. H. and Harbinson, J. (1994) Oxygen metabolism and the regulation of photosynthetic electron flow. In C. H. Foyer and P. M. Mullineaux, (eds.) *Causes of Photooxidative Stress and Amelioration of Defense Systems in Plants,* CRC Press, Boca Raton, pp. 1-42.

Fridovich, I. (1995) Superoxide radical and superoxide dismutase. *Annu. Rev. Biochem.*64: 97-112.

Fryer, M.J. (1992) The antioxidant effects of thylakoid vitamin E (a-tocopherol). *Plant Cell Environ.* 15: 381-392.

Getzoff, E. D., Cabelli, D. E., Fisher, C. L., Parge, H. E., Viezzoli, M. S., Banci, L. and Halliwell, R. A. (1992) Faster superoxide dismutase mutants designed by enhancing guidance. *Nature*. 358: 347-351.

Grace, S. and Logan, B. A. (2000) Energy dissipation and radical scavenging by the plant phenylpropanoid pathway. *Phil. Trans. Royal Soc. London B.* 355: 1499-1510.

Gross, G. G. (1980) The biochemistry of lignifications. *Adv. Bot, Res.* 8: 25-63.

Hamilton, G. A. (1991) Chemical and biochemical reactivity of oxygen. In E. J. Pell and S. K. Steffen, (eds.), *Active Oxygen/ Oxidative Stress and Plant Metabolism,* American Society of Plant Physiologists, Rockville, pp. 6-12.

Hammond-Kosack, K. E. and Jones, J. D. G. (1996) Resistance gene dependant plant defense responses. *Plant Cell.* 8: 1773-1791.

Hassan, H. M. (1989) Microbial superoxide dismutase. *Adv. Genet.* 26: 65-97.

Hertwig, B., Steb, P. and Feierabend, J. (1992) Light dependence of catalase synthesis and degradation in leaves and the influence of interferring stress conditions. *Plant Physiol.* 100: 1547-1553.

Hess, J.L. (1993) Vitamin E, a-tocopherol.*In:* R.G. Alscher and J.L. Hess (eds.) *Antioxidants in Higher Plants.* CRC Press. *Boca Raton*, pp. 111-134.

Hormann, H, Neubauer, C., Asada, K. and Schreiber, U. (1993) Intact chloroplast display pH 5 optimum of O_2 reduction in the absence of emthylviologen: indirect evidence for a regulatory role of superoxide protonation. *Photosynthesis Res.* 37: 69-80.

Joo, J. H., Bae, Y. S. and Lee, J. S. (2001) Role ofauxin-induced reactive oxygen species in root gravitropism. *Plant Physiol.*126(3): 1055-1060.

Kasai, H., Crain, P. F., Kuchino, Y., Nishimura, S., Ootsuyama, A. and Tanooka, H. (1986) Formation of 8-hydroxyguanine moiety in cellular DNA by agents producing oxygen radicals and evidence for its repair. *Carcinogenesis.* 7: 1849-1851.

Keser, G. (2005) Effects of lead on the activities of stress related enzymes, contents of mineral and chlorophyll, and growth in *Nasturtium officinale* R. Br. PhD Thesis. Cukurova Univ. Sci. Instit. pp. 12-23.

Kitagawa, Y., Tanaka, N., Hata, Y., Kusunoki, M., Lee, G. P., Katsube, Y., Asada, K., Aibara, S. and Morito, Y. (1991) Three-dimensional structure of Cu, Zn-superoxide dismutase from spinach at 2.0 A resolution. *J. Biochem.* 109: 477-485.

Kliebenstein, D. J., Monde, R. A. and Last, R. L. (1998) Superoxide dismutase in Arabidopsis: An eclectic enzyme family with dispartate regulation and protein localization. *Plant Physiol.* 118: 637-650.

Koc, E. and Ustun, A. S. (2008). Defense against pathogen in plants and antioxidants. *Erciyes Uni. Sci. Instit. J.* 24: 82-100.

Krieger-Liszkay, A. (2005) Singlet oxygen production in photosynthesis. *J. Exp. Bot.* 56: 337-346.

Kwak, J. M., Mori, I. C. and Pei, Z. M. (2003) NADPH oxidaseAtrbohD and AtrbohF genes function in ROS-dependentABA signaling in *Arabidopsis*. *EMBO J.* 22(11): 2623-2633.

Leon, J., Lawton, M. A. and Raskin, I. (1995) H_2O_2 stimulates salicylic acidbiosynthesis in tobacco. *Plant Physiol.* 108: 1673-1678.

Lindquist, Y., Branden, C. I., Mathews, F. S. and Lederer, F. (1991) Spinach glycolate oxidase and yeast flavocytochrome b2 are structurally homologous and evolutionarily related with distinctly different function and flavin mononucleotide binding. *J. Biol. Chem.* 266: 3198-3207.

Loewus, F.A. (1988) Ascorbic acid and its metabolic products.*In:* Preiss, J. (ed) *The Biochemistry of Plants.* Academic Press, New York, pp. 85-107.

Maheshwari, R. and Dubey, R. S. (2009) Nickel-induced oxidativestress and the role of antioxidant defence in rice seedlings. *Plant Growth Regul.* 59(1): 37-49.

Mann, T. and Kleilin, D. (1938)Homocuprein and heptacuprein, copper-protein compounds of blood and liver in mammals. *Proc. R. Soc. London B.* 126: 303-315.

Mathis, P. and Kleo, J. (1973) The triplet state of ß-carotene and of analogpolyenes of different length. *Photochem.Photobiol.* 18: 343-346.

McCord, J.M. and Fridovich, I. (1969) Superoxide dismutase, an enzymatic function for erythrocuprein. *J. Biol. Chem.* 244: 6049-6055.

McKersie, B. D. and Leshem, Y. Y. (1994) *Stress and Stress Coping in Cultivated Plants*, Kluwer Academic Publishers, Dordrecht.

Meriga, B., Reddy, B. K., Rao, K. R., Reddy, L. A.andKishor, P. B. K. (2004) Aluminium-induced introduction of oxygen radicals,lipid peroxidation and DNA damage in seedlings of rice (*Oryza sativa*). *J. Plant Physiol.* 161(1): 63-68.

Miller, G., Shulaev, V. and Mittler, R. (2008) Reactive oxygensignaling and abiotic stress. *Physiol. Planta.* 133(3): 481-489.

Mishra, S., Jha, A. B. andDubey, R. S. (2011) Arsenitetreatmentinduces oxidative stress, upregulates antioxidant system,and causes phytochelatin synthesis in rice seedlings. *Protoplasma.* 248(3): 565-577.

Mittler, R. (2002) Oxidative stress, antioxidants and stress tolerance. *Trends Plant Sci.* 7(9): 405-410.

Mittler, R. and Zilinskas, B. A. (1992) Molecular cloning and characterization of a gene encoding pea cytosolic ascorbate peroxidase. *J. Biol. Chem.* 267: 21802-21807.

Miyake, R. and Asada, K. (1992) Thylakoid-bound ascorbate peroxidase in spinach chloroplasts and photoreduction of its primay oxidation product, monodehydroascorbate radicals, in thylakoids. *Plant Cell Physiol.* 33: 541-533.

Neill, S., Desikan, R. and Hancock, J. (2002) Hydrogen peroxide signaling. *Curr. Opinion Plant Biol.* 5(5): 388-395.

Orozco-Cardenas, M. and Ryan, C. A. (1999) H_2O_2 is generated systematicallyin plant leaves by wounding systemin via the octadecanoidpathway. *Proc. Natl. Acad. Sci. USA.* 96: 6553-6561.

Parker, C. A. and Joyce, T. A. (1967) Delayed fluorescence and some properties of the chlorophyll triplets. *Photochem. Photobiol.* 6: 395.

Perl-Treves, R., Nacmias, B., Aviv, D., Zeelon, E. P. and Galun, E. (1988) Isolation of two cDNA cloned from tomato containing two different superoxide dismutase sequences. *Plant Mol. Biol.* 11: 609-623.

Polle, A.,Chakrabarti, K., Schürmann, W. and Rennenberg, H. (1990) Composition and properties of hydrogen peroxide decomposing systems in extracellular and total extracts from needles of Norway spruce (*Piceaabies* L., karst). *Plant Physiol.* 94: 312-319.

Prasad, T. K., Anderson, M. D., Martin, B. A. and Stewart, C. R. (1994) Evidence for chilling induced oxidative stress in maize seedlings and a regulatory role for hydrogen peroxide. *Plant Cell.* 6: 65-74.

Price, A., Lucas, P.W. and Lea, P.J. (1990) Age dependent damage and glutathione metabolism in ozone fumigated barley: a leaf section approach. *J. Exptl. Bot.* 41: 1309-1317.

Redinbaugh, M.G., Wadsworth, G.J. and Scandalias, J.G. (1988) Characterization of catalase transcripts and their differential expression in maize. *Biochim. Biophys. Acta.* 951: 104-116.

Rennenberg, H. (1982) Glutathione metabolism and possible biological roles in higher plants.*Phytochem.* 21: 2771-2781.

Rice-Evans, C. A., Miller, N. J. and Paganga, G. (1997) Antioxidant properties of phenolic compounds. *Trends. Plant Sci.* 2: 152-159.

Rich, P. R. and Bonner, W. D. (1978) The sites of superoxide anion generation in higher plant mitochondria. *Arch. Biochem. Biophys.* 188: 206-213.

Sade, B., Soylu, S. and Yetim, E. (2011) Drought and oxidative stress. *Afr. J. Biotechnol.* 10(54): 11102-11109.

Sandalio, M. and Del Rio, L.A. (1988) Intraorganellar distribution of superoxide dismutase in plant peroxisomes (glyoxysomes and leaf peroxisomes). *Plant Physiol.* 88: 1215-1218.

Scandalias, J.G. (1990) Response of plant antioxidant defense genes to environmental stress.*Adv. Genet.* 28: 1-41.

Scandalias, J.G. (1993) Oxygen stress and superoxide dismutase. *Plant Physiol.* 101: 7-12.

Shah, K., Kumar, R. G.,Verma, S.andDubey, R. S. (2001) Effectof cadmium on lipid peroxidation, superoxide anion generationand activities of antioxidant enzymes in growing riceseedlings. *Plant Sci.* 161(6): 1135-1144.

Shao, H. B., Chu, L. Y., Wu, G., Zhang, J. H., Lu, Z. H. and Hu, Y. C. (2007) Changes of some anti-oxidative physiological indices under soil water deficits among 10 wheat *(Triticumaestivum* L.) genotypes at tillering stage. *Biointerface.* 59: 113-119.

Sharma, P. and Dubey, R. S. (2005) Drought induces oxidativestress and enhances the activities of antioxidant enzymes ingrowing rice seedlings. *Plant Growth Regul.* 46(3): 209-221, 2005.

Sharma, P., Jha, A. B., Dubey, R. S. and Pessarakli, M. (2012) Reactive oxygen species, oxidative damage and antioxidative defense mechanism in plants under stressful conditions. *Journal of Botany*.doi: 10.1155/2012/217037.

Smimoff, N. (2000)The role of active oxygen in the response of plants towater deficit and desiccation. *New Phytol.* 1125: 27-58.

Smirnoff, N. and Cumbes, Q. J. (1989) Hydroxyl radical scavenging activity of compatible solutes. *Phytochemistry.* 28: 1057-1060.

Soll, J., Schultz, G., Joyard, J., Doucce, R. and Block, M. A. (1984) Localization and synthesis of prenylquinones in isolated outer and inner envelope membranes from spinach chloroplasts. *Arch. Biochem. Biophys.* 238: 290-299.

Srivastava, S. and Dubey, R. S. (2011) Manganese-excess inducesoxidative stress, lowers the pool of antioxidants and elevatesactivities of key antioxidative enzymes in rice seedlings. *Plant Growth Regul.* 53: 1-16.

Stoop, J. M. H., Williamson, J. D. and Mason Pharr, D. (1996) Mannitol metabolism in plants: a method for coping with stress. *Trends Plant Sci.* 1: 139-144.

Takahama, U. and Oniki, T. (1997) A peroxide/phenolics/ascorbate system can scavenge hydrogen peroxide in plant cells. *Physiol. Planta.* 101: 845-852.

Takemoto, T., Zhang, Q. M. and Yonei, S. (1998) Different mechanisms of thioredoxin in its reduced and oxidized forms in defense against hydrogen peroxide in *Escherichia coli. Free. Rad. Biol.Med.* 24: 556-562.

Torel, J., Cillard, J. and Cillard, P. (1986) Antioxidant activity of flavonoids and reactivity with peroxy radicals. *Phytochemistry.* 25: 383-385.

Torres, M. A., Dangl, J. L. and Jones, J. D. G. (2002) *Arabidopsis* gp91*phox* homologues Atrbohd and Atrbohf are required foraccumulation of reactive oxygen intermediates in the plantdefense response. *Proc. Nat. Acad. Sci. USA*. 99(1): 517-522.

Verma, S. and Dubey, R. S. (2003) Lead toxicity induces lipidperoxidation and alters the activities of antioxidant enzymesin growing rice plants. *Plant Sci.* 164(4): 645–655.

Vianello, A. and Macri, F. (1991) Generation of superoxide anion and hydrogen peroxide at the surface of plant cells. *J. Bioenerg. Biomembr.* 23: 409-423.

Volk, S. and Feierabend, J. (1989) Photoinactivation of catalase at low temperature and its relevance to photosynthetic and peroxide metabolism in leaves. *Plant Cell Environ.* 12: 701-712.

Wolff, S. P., Garner, A. and Dean, R. T. (1986) Free radicals, lipids and protein breakdown. *Trends Biochem. Sci.* 11: 27-31.

Wu, G., Wei, Z. K. and Shao, H. B. (2007) The mutual responses of higher plantsto environment: physiological and microbiological aspects. *Biointerface,* 59: 113-119.

Xiong,L., Schumaker,K. S.andZhu,J. K. (2002) Cell signalingduring cold, drought, and salt stress. *Plant Cell.* 14: S165-S183.

Yan, J., Tsuichihara, N. Etoh, T. and Iwai, S. (2007) Reactive oxygen species and nitric oxide are involved in ABA inhibition of stomatal opening. *Plant, Cell and Env.* 30(10): 1320-1325.

Yu, B. P. (1994) Cellular defenses against damage from reactive oxygen species. *Physiol. Rev.* 74: 139-162.

4

Metal Toxicity and Tolerance in Plants

Gh Mohd Rather*, **Athar Ali Khan and Mohammad Imran Kozgar**

Department of Botany, Aligarh Muslim University, Aligarh, India

ABSTRACT

Majority of heavy metals are toxic to all biological entities at all the concentration except some trace elements having an irreplaceable importance in the plant growth and development at lower concentration. The intensity of metal toxicity varies with the nature of the element as well as the plant species. These metals find their way into the living habitats mainly from growing number of anthropogenic sources. Plants being sessile are unable to escape the environmental hazards like heavy metals and hence have developed a broad spectrum of response types from toxicity to tolerance in response to these toxic metals. The toxic responses vary from replacement of essential metals in pigments or enzymes disrupting their function (photosynthesis, respiration, cellular transport etc.) and causing various oxidative damages inside the cellular organelles as well as to membranes. Some of the plant species have developed a unique property of exclusion of these metals from the sensitive areas inside the body like retention at root level, stored in the vacuolar compartments to minimize their toxic effects and enhance the chances of survival in the metal polluted areas. In this chapter the toxic effects of heavy metals on plant growth and survival as well as the special detoxification mechanisms developed by plants to tolerate high levels of metals present in the soil environment will be discussed.

Keywords: Abiotic stress, heavy metals, metal toxicity, phytoaccumulation, phytoremediation, phytotoxicity.

1. INTRODUCTION

Plants are composed largely of water and complex organic molecules. These complex organic compounds are synthesized from water and nutrients from the soil/aqueous media, CO_2 from air and energy from sunlight. The nutrients absorbed from soil/water and needed by plants include a variety of metal ions for the optimum integration of growth and metabolic progress (Hewitt and Smith, 1975; Taiz and Zeiger, 2006; Marschner, 2012). These elements can be grouped as macronutrients and micronutrients and elements having beneficial or restricted essentiality regarding their amount essential for optimal growth and development of plants. Metals and metalloids are found associated with a wide variety of cellular activities in plants either as

*Corresponding author's E-mail ID: ratherbio@gmail.com

Roychowdhury R. (Ed.), Crop Improvement in the Era of Climate Change.

structural components of key molecules. (e.g., chlorophyll, cytochrome) or they are involved in a wide range of metabolic processes such as enzymes (e.g., peroxidises, polyphenol oxidases, nitrate reductase) (Kabata-Pendias and Pendias, 2001; Taiz and Zeiger, 2006; Marschner, 2012). These elements being one of the main building blocks of plant growth and development may become toxic and detrimental at higher concentrations and may prove fatal as well. Soil-root interface, in addition to these essential elements, transports some heavy metals like Pb, Cd, Hg which show toxic effects in plants at very low concentrations (Hewitt and Smith, 1975; Zengin and Munzuroglu, 2006). The effects of heavy metal deficiencies and excesses on plant growth and composition have been discussed by several authors (Hewitt and Smith. 1975; Lepp, 1981; Zengin and Munzuroglu, 2006; Elloumi *et al.*, 2007; Chary *et al.*, 2008; Anjum *et al.*, 2008; John *et al.*, 2009; Ghani, 2010; Marschner, 2012).

1.1 Definition

All metals having the specific gravity greater than four (Nieboer and Richardson, 1980), or five or more are included in this category. The term heavy metal includes all the lower members (metals) of the periodic table and is always used in context with environmental pollution. Only those members of periodic table which are (1) relatively abundant in the earth's crust, (2) extracted and used in a reasonable amount, (3) used in places where the public may come in contact with them and (4) toxic to human beings are generally referred to as heavy metals (Martin and Coughtrey, 1982). The applicability of the term heavy metal in relation to biological entities has come under severe criticism (Duffs, 2002) on the basis that biological organisms are lacking any sort of specificity with regard to the specific gravity of heavy metals. Even though the review paper of Duffs (2002) was commissioned by IUPAC (International Union of Pure and Applied Chemistry) but has remained ineffective in biological literature (including this paper).

1.2 Classification

The definition based on the specific gravity of heavy metals being least related with the biochemical properties of these elements has paved a way for the other ways to discuss and examine categorization of heavy metals with respect to biological organisms. Nieboer and Richardson (1980) classified heavy metals, based on their chemical properties with regard to their affinity towards different ligands, into three classes: Class 'A', Class 'B' and Borderline.

Class A: Metal ions of this class are characterized by small size and low polarizability (sometimes referred to as hard acids). This class includes: Li, Be, Na, Mg, Al, K, Ca, Fe (III), Rb, Zr, etc.

Class B: The metal ions of this class are characterized by large size and high polarizability (sometimes referred to as soft acids). This class includes: Cu (I), Pb, Ag, Cd, Au, Hg, Pb (II), etc.

Class borderline: The metal ions of this class are clearly distinct from those of class A but show increasing degrees of class B characteristics. This class includes: V, Cr, Mn, Fe (II), Co, Ni, Cu (II), Pb (IV), Sn, etc.

The above classification of metals by their Lewis acidity indicates the form of bonding in their complexes. Class A metal ions preferentially form complexes with similar nonpolarizable ligands, particularly oxygen donors and the bonding in these complexes is mainly **ionic**. Class B ions preferentially form complexes with polarizable, soft ligands to give a **covalent** bond formation. Hence the **hard-hard** or **soft-soft** combinations are preferred wherever possible.

2. SOURCES OF HEAVY METALS

2.1 Natural

Heavy metals are natural components of lithosphere and hydrosphere (Forstner and Wittmann, 9179; Temmerman *et al.*, 1984). Soil or water, which contain heavy metals in excess of natural concentrations, are said to be heavy metal contaminated. Metal concentration in different soils varies considerably like Ni, Cr and Cu are abundant in serpentine soils whereas calamine soils are rich in Zn, Pb and Cd. The concentration of heavy metals present in different natural media is presented in **Table 1**.

Table 1 Heavy metal levels in natural water sources and non-polluted soils

Heavy metal	*Heavy metal levels in different media*			
	Natural water $\mu g l^{-1}$		*Soil $\mu g g^{-1}$*	
	Sea water	*Fresh water*	*Sandy soil*	*Loam*
Cadmium	0.01-0.08	0.07	1.00	1.00
Chromium	0.08-0.15	0.50	15.00	30.00
Copper	0.04-0.10	1.81	15.00	25.00
Manganese	0.20	4.00-5.00	500.00	750.00
Molybdenum	10.00	1.00-2.00	5.00	5.00
Nickel	0.20-0.70	0.30	0.50-1.00	1.00
Zinc	0.001-0.62	10.00	100.00	150.00
Cobalt	0.04	0.05	5.00	15.00
Lead	0.001-0.015	0.20	40.00-50.00	40.00-50.00
Mercury	0.01	0.01	0.15	0.15

2.2 Anthropogenic

While metals are ubiquitous in nature, human activities have caused metals to accumulate in soil. Metal pollution results when human activity disrupts normal biogeochemical activities (Nriagu, 1988; Cheng, 2003; Abou-Shanab *et al.*, 2007; AnnCuyperus, 2009). Such contaminated soils provide a metal source from which surface waters, groundwaters etc., can become contaminated. Soil usually exhibit higher concentrations of metals than waters because metals are more likely to accumulate in soil versus being diluted or carried elsewhere in water and soils are composed of minerals which can naturally contain high concentrations of metals. The cation exchange capacity of soils allows metals to attach to soil particles in response to ionic attractions and accumulate. These pollutants enter the environment in a variety of ways, which include mining, metal smelting, ore smelting, electroplating, gas exhaust, energy and fuel production, down wash from power lines, intensive agriculture, industrial manufacturing of batteries, metal alloys, electric components, paints, preservatives and insecticides, sludge dumping and military operations. (Kumar *et al.*, 1995; Nedelkoska and Doran, 2000; Padmavathiamma and Li, 2007). These anthropogenic pollutants in addition to metals may also include a variety of chemicals like phenolics, dust, oxides of sulfur, nitrogen and carbon, pesticides and organic explosives (Muller *et al.*, 1989; Van Asche and Clijsters, 1990). Sometimes a single metal is involved but more often mixtures of metals are present. Examples of specific metal contaminants include copper and zinc salts that are used extensively as pesticides in agricultural settings, silver salts that are used to treat skin burns, lead which is utilized in the production of batteries, cable sheathing and alloys. The extent of metal pollution becomes even more obvious when one considers the reluctant character of metals for degradation. Metals remain persistent in the environment and cannot be degraded through biological, chemical or physical means to an innocuous by-product. The chemical nature and, thus bioavailability of a metal can be changed through oxidation or reduction; however elemental nature remains same because metals are neither thermally decomposable nor microbiologically degradable. Consequently metals are difficult to remove from environment.

3. PLANT-HEAVY METAL INTERACTIONS

3.1 Heavy Metal Uptake by Plants

Living plants have the ability to accumulate heavy metals from soil and water (Miller, 1996; Boyajian and Carrecira, 1997; Hussain *et al.*, 2004; Courbot *et al.*, 2007; Hanikenne *et al.*, 2008; Ramamurthy and Memarian, 2012). The metal uptake from soil by plants, through their roots, to both their aboveground and underground parts depend on amount of metals present in the soil in exchangeable form. Moreover, the metal uptake from soil by plants is regulated by ability of plants to transfer the metals across the soil-root interface.

The transfer of heavy metals from soil media to plant body is very complex and depends on various factors like **quality factor** (total amount of potentially available elements)**, intensity factor** (activity as well as the ionic ratios of elements in the soil solution), **reaction kinetics** (rate of element transfer from solid to liquid phase and plant roots) (Brummer *et al.*, 1986).

The other aspects which show profound impact on the transfer of metals across the soil root interface include a variety of biotic and abiotic factors including soil dwelling organisms, sorption to clay minerals, pH, soil solution concentrations, the cation exchange capacity of the soil, redox potential, soil texture, clay content, organic matter content and temperature of the soil, the type and concentration of metal and the species of plants (Verloo and Eeckhout, 1990; Antosiewicz, 1992; Salim *et al.*, 1993). These metals enter the soil in the form of dissolved ions and move with the inflow of water apoplastically through the root hairs and into the cortex and are then translocated to other parts of plants (Punz and Sieghardt, 1993). Some metals may become firmly bound to the soil particles by higher values of clay and/or organic matter and soil pH which eventually hinder the bioavailability of essential/toxic metals for plant uptake.

Plants distribute metals internally in many different ways. They may localize metals mostly in roots and stems or they may accumulate and store them in nontoxic forms for later distribution and use. A mechanism of tolerance or accumulation in some plants apparently involves binding of potentially toxic metals at cell walls of roots and leaves, away from sensitive sites within the cell or storing them in vacuolar compartment.

Metal uptake efficiency of plants is highest at their lower concentration in soil media. Uptake of metals from soil media by roots increases with increase in metal concentration in soil media up to a certain limit beyond which the linear relation between uptake and concentration becomes non-linear due to the saturation of the uptake sites at higher concentrations of metals in soil media (Greger *et al.*, 1991).

Metals are taken up via root symplasm to enter into the xylem (Tester and Leigh, 2001). In xylem metal ions are transported by membrane transport proteins. Metal elements which are not needed by the plants effectively compete the essential heavy metals for their transport using the same transmembrane carriers (Shah *et al.*, 2010). Majority of the plants absorbing metals from soil attain a higher root/shoot concentration ratios with some exceptions which transfer a substantial amount of metal absorbed to aboveground harvestable part (Zayed and Terry, 2003). A higher concentration of metals in the edible part of the plants (aboveground) may prove detrimental for humans and animals which is being prevented by the plant regulatory mechanisms like prevention of uptake of metal element(s) due to its insolubility in soil, prevention of metal transfer from soil to above ground parts by making them immobile (Chaney, 1980). Metals like B, Cd, Mn, Mo, Se, Zn are easily absorbed and translocated inside the plant tissues while other metals like Al, Ag, Cr, Fe, Hg, Pb are less mobile due to their strong binding to soil components or root cell walls (Chaney, 1983).

Amongst the factors influencing the metal accumulation in plants, soil pH is usually the most important parameter (Ramos *et al.*, 1995; Piechalak *et al.*, 2003; Kirkham, 2006; Deng *et al.*, 2006). At higher soil pH, metal elements in soil solution form low soluble compounds and decrease their bioavailability, while metal bioavailability to plants increases at lower soil pH (Seregin and Ivanov, 2001). Metal accumulation in the plants is also effected by other metals as well. Like Cr is reported to enhance Cd accumulation in plants such as *Hydrilla verticillata* and *Chara corallina* (Rai and Chandra, 1992; Rai *et al.*, 1995), but the accumulation of Cr is found to be greater in comparison to Cd when applied separately (Shukla *et al.*, 2005).

The levels of metals found in plants are often correlated to the levels present in the environment (Greger etal., 1991; Vesk and Allaway, 1997; Biro and Takacs, 2007; Zacchini *et al.*, 2009; Jun and Ling, 2012). Salim *et al.* (1993) showed that the concentrations of different heavy metals like lead, cadmium, copper increased in radish plants (*Raphanus sativus*) when treated with increasing concentrations of these metals. Imran *et al.* (2007) also showed similar results for aluminum and cadmium in soybean (*Glycine max*). Similar results of cadmium accumulation in five poplar species were also discussed by Jun and Ling (2012).

3.2 Biological Significance of Heavy Metals

The whole plant life on this planet is sustained by the mineral nutrients (Aerts and Chapin, 2000; Epstein and Bloom, 2004). Thirty-five out of fifty-three metals are of biological significance and twenty-three of them are known as heavy metals. These metals include arsenic (As), antimony (Sb), bismuth (Bi), cadmium (Cd), cerium (Ce), copper (Cu), chromium (Cr), cobalt (Co), gallium (Ga), iron (Fe), gold (Au), lead (Pb), nickel (Ni), manganese (Mn), mercury (Hg), platinum (Pt), silver (Ag), thallium (Tl), tellurium (Te), and zinc (Zn) (Kabata-Pendias and Pendias, 2001). Majority of these metals are easily absorbed and accumulated by plant species (Wang *et al.*, 2003). Based on their solubility under physiological conditions seventeen heavy metals may be available for living cells and are of importance for organisms and ecosystem (Weast, 1985; Taiz and Zeiger, 2006; Timbrell, 2005; Pechova and Pavlata, 2007). Among these metals Fe, Mo and Mn are important as micronutrients, Zn, Ni, Cu, V, W and Cr are toxic elements with high or low importance as trace elements (Nriagu 1988; Graham and Stangoulis, 2003). Due to their availability in normal soil, their adequate amounts are naturally found in our foodstuffs, fruits and vegetables (Ghafoor *et al.*, 1996; Islam *et al.*, 2007). Heavy metals which cause most detrimental effects on plant growth and development are Ar, Pb, Hg, Cd, Ni, Fe and Al (Hutton and Symon, 1986; Chaney and Ryan, 1994). The essential metals may act as a number of cofactors of metabolic reactions, as enzyme activators or as integral components of a number of metalo-enzymes, taking place in the body (Seiler *et al.*, 1994; Taiz and Zeiger, 2006). Due to this biological importance of metals, their deficiency may lead to suppression of growth and development with observable symptoms like chlorosis and necrosis of plant tissues (Ghani and Wahid, 2007).

Due to their biological importance the deficiency of essential nutrients can seriously disturb normal development, while as the excess of metals can also adversely affect the physiological and biochemical functions in the organism (Leonard *et al.*, 2004; Zengin and Munzuroglu, 2006; Ghani and Wahid, 2007; Elloumi *et al.*, 2007; Chary *et al.*, 2008; Anjum *et al.*, 2008; John *et al.*, 2009; Marschner, 2012). There is a thin line of demarcation in the physiological range of essential metals between deficiency and toxicity and is controlled by metal homeostasis network (Nelson, 1999; Clements, 2006).

Some heavy metals at low doses are essential micronutrients for plants, but in higher doses they may cause metabolic disorders and growth inhibition for most of plant species while other non-essential metals may show toxic effects even at low concentrations as well (Verkleij and Prast, 1990; Islam *et al.*, 2007). Heavy metals are generally toxic above a threshold level and can be transferred to and concentrated in plant tissues from the soil media. Their toxicity to plants manifests itself in various ways and may become a health hazard to man and animals. The threshold levels up to which animals and plants can tolerate heavy metals vary with type and oxidation state of heavy metals as well as the organisms under influence. Beyond the threshold level and even over a narrow range, heavy metals become toxic (Babich and Stotzky, 1980; Babich *et al.*, 1982; Hermle *et al.*, 2006; Walter *et al.*, 2006). Moreover, these toxic metals adversely affect natural microbial populations leading to the disruption of vital ecological balance (Sterritt and Lester, 1980; Nriagu and Nieboer, 1988; Brynhildsen and Rosswall, 1997). The toxicity of metals in agricultural and horticultural fields may lead to the overall decline of yield of the economically important crops (Balaguer *et al.*, 1998; Ahmad *et al.*, 2007, Gupta *et al.*, 2009). Heavy metals which cause most detrimental effects on plant growth and development are arsenic, lead, mercury, cadmium, nickel, iron and aluminum (Hutton and Symon, 1986; Chaney and Ryan, 1994; Gupta *et al.*, 2009). The most common primary characteristics associated with plants affected with heavy metal toxicity include reduced shoot and root growth, poor root development, abnormal and low flower formation, low biomass production, leaf spotting, chlorosis, decline in fruit and seed formation (Goldbold and Hutterman, 1985; Breckle, 1991; Claire *et al.*, 1991; Das *et al.*, 1997; Ewais, 1997; Nies, 1999; Henry, 2000; Madhava Rao and Sresty, 2000; Pandey and Sharma, 2002; Rahman *et al.*, 2005; Gajewska *et al.*, 2006; Prasad, 2004a, 2004b and 2007). Metal concentrations beyond a threshold show detrimental effect on the plant metabolism which may be due to the high affinity of several metals for sulfur and nitrogen donors and which eventually bind to the functional groups of macromolecules and cause metabolic disruption (Patra *et al.*, 2004; Nouairi *et al.*, 2006). These metal species may also lead to substitution of essential micronutrients in functional and structural proteins which effect on the mechanism of various physiological processes and cellular metabolism (Kupper *et al.*, 1996; Suzuki *et al.*, 2001; Vitorello *et al.*, 2005, Rakhshaee *et al.*, 2009).

3.3 Heavy Metal Toxicity

The mechanism of effects of metals on plants has been summarised in the three main pathways: (a) reactive oxygen species (ROS) production; (b) binding of functional groups of essential biomolecules by metals and (c) replacement of functional elements in various important biomolecules. At cellular level heavy metals induce oxidative stress by generation of reactive oxygen species (ROS) like superoxide radical (O2*–), hydroxyl radical (OH*–), singlet oxygen (O*–) and hydrogen peroxide (H_2O_2). Metals like Fe, Cu, Cr, Co and Cd, Zn, Ni, Al are called redox active and redox inactive depending on the way of production of ROS. Physiologically active organelles like mitochondria, chloroplast, peroxisomes are the seat of majority of production of ROS inside the cell (Hossain *et al.*, 2011; Apel and Hirt, 2004). The extremely reactive *OH in the cell causes extensive oxidative damage of membranes and other macromolecules, including membrane transporters, photosynthetic pigments, protein, enzymes, DNA, and lipids (Hossain *et al.*, 2011; Gechev *et al.*, 2006). During normal conditions the ROS produced is scavenged at their production sites by antioxidant machinery of the biological system. However, during the stress conditions ROS production exceeds the antioxidant capacity of the biological system and executes its effects in the form of abnormalities like cell membrane alteration, DNA damage, lipid peroxidation, enzyme deformation, protein degradation and during extreme cases even cell death as well (Hossain *et al.*, 2010; Barconi *et al.*, 2011; Dalcorso *et al.*, 2010). Heavy metals have the inherent ability to bond with many elements like oxygen, nitrogen, sulphur present in active and essential biological compounds like enzymes. These elements may bind with cysteine residue of enzymes and result in the enzyme malfunction (Hall, 2002). Heavy metals can replace a substantial number of essential elements from their respective places and result in the inhibition or loss of the compound activity like replacement of Mg by Zn, Co like divalent atoms from chlorophyll molecule, Ca by Cd in calmodulin signalling molecule (Rivetta *et al.*, 1997, Boddi *et al.*, 1995; Shakya *et al.*, 2008).

These effects manifest in the form of visible symptoms like decline in germination, stunted growth, reduced length, and decline in productivity etc. Roots, even though being primary organs affected by metal toxicity are not the primary toxicity indicators because they remain mostly invisible (Seregin and Ivanov, 2001; Johnson *et al.*, 2011). Hence above-ground affected parts after germination are the foremost symptoms observable due to metal toxicity. The effects include reduced germination, stunted roots, brown root tips, reduced root elongation, chlorosis (progressive, marginal, interveinal), necrosis, reduced stomata activity, limited photosynthesis, increased respiration, loss of apical dominance, reduced shoot growth as well as girth, limited floral organ development, reduced biomass, decline in fruit and seed production (Lee *et al.*, 1976; McKenna *et al.*, 1993; Sharma *et al.*, 2003; Sirkar and Amin, 2007; Lopez *et al.*, 2009). These effects get manifested as a result of the disturbances due to metal contamination at cellular, organelle and bimolecular level. Heavy metals damage membrane homeostasis importantly of plasma mebrane and effect cellular transport machinery (Foyer *et al.*, 1997; Janicka-Russak *et al.*, 2008). Metals may either increase lipid peroxidation, bind with sulfhydryl groups and active sites of membrane proteins, change in activity of membrane H^+-ATPase (Luna *et al.*, 1994; Panda *et al.*, 2003). Moreover, HMs influence various homeostatic events, including water uptake, transport, transpiration, and nutrient metabolism (Fodor, 2002; Poschenrieder and Barcelo, 2004) and interfering with the uptake of Ca, Mg, K, and P (Benavides *et al.*, 2005). Elevated levels of HMs usually decrease photosynthesis due to their direct effect on the photosynthetic apparatus, including thylakoids. Several studies report that HMs can determine the release of proteins, lipids, and element components of thylakoid membranes, causing damage to light-harvesting complexes and Photosystem II (PS II) (Hsu and Kao, 2004; Backor *et al.*, 2007). Additionally, some HMs can replace Mg in the chlorophyll (Chl). Chl synthesis reduction may be a consequence of enzyme inhibition involved in its synthetic pathway (Boddi *et al.*, 1995; Shakya *et al.*, 2008). Enzymes and enzymatic processes involved in the photosynthesis are also reported to be inhibited by heavy metals. Hence, plants exposed to higher heavy metal concentrations show decline in chlorophyll content, reduction in chl a/b ratio, distortion of chloroplast ultra structure, efficiency decline of various enzymes, inhibition in both photosystems I and II activity, electron transport and overall process on whole, as reported in the literature (Bertrand and Poirier, 2005; Mishra and Dubey, 2006).

3.4 Heavy Metal Accumulation and Detoxification

A considerable and significant number of plant species are endemic to metalliferous soils and can tolerate greater than normal levels of heavy metals or other toxic compounds (Banuelos *et al.*, 1997; Blaylock *et al.*, 1997; Raskin and Ensley, 2000; Dahmani Muller *et al.*, 2000, Hussain *et al.*, 2004; Courbot *et al.*, 2007; Hanikenne *et al.*, 2008; Ramamurthy and Memarian, 2012). This tolerance involves immobilization of metals in roots and cell walls (Garbisu and Alkorta 2001). Plant species based on the mechanism adapted regarding the metal tolerance can be classified into:

1. **Metal excluders:** These plant species maintain low and constant metal concentration over a broad range of metal concentration in soil by holding metals in their roots (Cunningham 1995).
2. **Metal indicators:** These plants actively accumulate metals in their aerial tissues due to the production of metal binding compounds (chelators) or alter metal compartmentalisation pattern by storing metals in non-sensitive parts.
3. **Metal hyper-accumulators:** These plant species can concentrate metals in their aerial parts to levels far exceeding than soil (Baker *et al.* 1994a & b; Raskin, 1996; Dahmani Muller *et al.*, 2000; Dahmani Muller *et al.*, 2001, Pilson-Smith, 2005; Hussain *et al.*, 2004; Courbot *et al.*, 2007; Safwan*et al.*, 2008; Ramamurthy and Memarian, 2012).

The plant species recognized as hyper-accumulating have got a substantial scope and importance in the remediation of polluted soils and have become one of the major areas of interest for pollution ecologists throughout the globe. Hyper-accumulating plants are characterized by their capacity to accumulate either As, Cu, Cr, Ni, Pb, or Co >1000 mg/kg or zinc >10000 mg/kg or cadmium > 100 mg/kg in their shoot dry matter (Baker *et al.*, 1994a; Brown *et al.*, 1995; Brooks, 1998; Reeves and Baker, 2000) or Mo>1500 mg/kg (Wei *et al.*, 2002). Very few higher plant taxa have adaptations that enable them to survive and to reproduce in soils heavily contaminated with Zn, Cu, Pb, Cd, Ni, and As (Dahmani-Muller *et al.* 2000; Pulford and Watson 2003). Tree roots of these plants can actively forage towards less contaminated zones of soil (Turner and Dickinson, 1993) and, even with highly reduced growth they can tolerate unfavourable conditions (Watmough 1994).

Plants have developed special mechanisms like to detoxify the metals at cellular levels. The mechanisms include upgradation of antioxidative system, synthesis of stress related amino acids, glutathione, proteins, signalling molecules like salicylic acid and nitric oxide, production of pro and polyamines, chelation byphytochelatins and metallotheoneins, vacuolar sequestration (Mejare and Bulow, 2001). Phytochelatins are a family of Cys-rich polypeptides synthesised from glutathione. These bind metals by their thiolic group (-SH) of Cys and metal gets transported to vacuole to minimize their presence near physiologically active areas inside cell like membranes, membrane bound organells, seats of different enzymatic processes etc. (Yang *et al.*, 2005; Sanita *et al.*, 1995). Similarly, metallothioneins are low molecular Cys-rich polypeptides which bind heavy metals to their thiol groups of their Cys residue and heavy metal gets sequestered from sensitive and physiologically areas of the cell (Hamer, 1986; Gasic and Korban, 2006). The amount and concentration of all these detoxifying agents increases with increase of the stress factor inside plant body (Nouairi *et al.*, 2009).

4. CONCLUSION AND FUTURE PROSPECTS

Heavy metals being one of major sources of agricultural soil contamination have shown deleterious effects on the agricultural production. Heavy metals execute their effects on all the aspects of plant growth and development from germination to seed production. Even though a substantial number of plants have managed different ways of metal detoxifying and accumulation mechanisms to bypass such effects. It is imperative to understand, enhance and utilize these mechanisms for the remediation of heavy metal contaminated soils as it has great consequences for human health and well-being.

References

Abou-Shanab, R., Ghanem, N., Ghanem, K., and Al-Kolaibe, A. (2007). Phytoremediation potential of crop and wild plants for multi-metal contaminated soils. *Res. J. Agric. Biol. Sci.* 3(5): 370-376.

Aerts, R., and Chapin, F. S. (2000). The mineral nutrition of wild plants revisited: a re-evaluation of processes and patterns. *Adv. Ecol. Res.* 30: 1-67.

Ahmad, M.S.A., Hussain, M., Saddiq, R., and Alvi, A.K. (2007). Mungbean: A nickel indicator, accumulator or excluder? *Bull. Environ. Contam. Toxicol.* 78:319-324.

Anjum, N. A., Umar, S., Ahmad, A., and Iqbal, M. (2008). Responses of components of antioxidant system in moongbean genotypes to cadmium stress. *Commun. Soil Sci. Plant Analy.* 39: 2469-2483.

Ann Cuypers, Smeets, K., and Vangronsveld, J. (2009). Heavy metal stress in plants. In: *Plant stress biology*, Hirt, H. (ed.). Wiley, VCH Verlag GmbH & Co. KGaA, Weinheim, pp. 161-178.

Antosiewicz, D.M. (1992). Adaptation of plants to an environment polluted with heavy metals. *Acta Societatis Botanicorum Polaniae.* 16: 2811-2919

Apel, K., and Hirt, H. (2004). Reactive oxygen species: metabolism, oxidative stress, and signal transduction. *Ann. Review Plant Biol.* 55: 373-399.

Babich, H., and Stotzky, G. (1980). Environmental factors that influence the toxicity of heavy metal and gaseous pollutants to microorganisms. *CRC Crit. Rev. Microbiol.* 8: 99-145.

Babich, H., Schiffenbauer, M., and Stotzky, G. (1982). Comparative toxicity of trivalent and hexavalent chromium to fungi. *Bull. Environ. Contam. Toxicol.* 28: 193-202.

Backor, M., Vaczi, P., Bartak, M., Budova, J., and Dzubaj, A. 2007. Uptake, photosynthetic characteristics and membrane lipid peroxidation levels in the lichen photobiont *Trebouxiaerici* exposed to copper and cadmium. *Bryologist.* 110(1): 100-107.

Baker, A.J.M., Reeves, R.D., and Hajar, A.S.M. (1994b). Heavy metal accumulation and tolerance in British populations of the metallophyte *Thlaspi caerulescens* J. & C. Presl (Brassicaceae). *New Phytol.* 127: 61-68

Baker, A.J.M., McGrath, S.P., Sidoli, C.M.D., and Reeves, R.D. (1994a). The possibility of in situ heavy metal decontamination of polluted soils using crops of metal-accumulating plants. *Resour. Conser. Recycl.* 11: 41-49

Balaguer, J., Almendo, M.B., Gomez, I., Navarro-Pedreno, J., and Mataix, J. (1998). Tomato growth and yield affected by nickel presented in the nutrient solution. *Acta. Hort.* 458: 269-272.

Banuelos, G.S., Ajwa, H.A., Mackey, B., Wu, L., Cook, C., Akohoue, S., and Zambruzuski, S. (1997). Evaluation of different plant species used for phytoremediation of high soil selenium. *J. Envir. Qual.* 26(3): 639-646.

Barconi, D., Bernardini, G., and Santucci, A. 2011. Linking protein oxidation to environmental pollutants: redox proteome approaches. *J. Proteomics.* 74(11): 2324-2337.

Benavides, M.P., Gallego, S.M., and Tomaro, M.L. 2005. Cadmium toxicity in plants. *Braz. J. Plant Physiol.*.17(1): 21-34.

Bertrand, M., and Poirier, I. (2005). Photosynthetic organisms and excess of metals, *Photosynthetica.* 43(3): 345-353.

Biro, I., and Takacs, T. (2007). Effects of *Glomus mosseae* strains of different origin on plant macro-micronutrient uptake in Cd polluted and unpolluted soils. *Acta. Agronomica Hungarica.* 55(2): 1-10.

Blaylock, M.J., Salt, D.E., Dushenkov, S., Zakharova, O., Gussman, C., Kapulnik, Y., Ensley, B.D., and Raskin, I. (1997). Enhanced accumulation of Pb in Indian mustard by soil-applied chelating agents. *Environ. Sci. Technol.* 31:860-865.

Boddi, B., Oravecz, A.R., and Lehoczki, E. (1995). Effect of cadmium on organization and photoreduction of protochlorophyllide in dark-grown leaves and etioplast inner membrane preparations of wheat. *Photosynthetica.* 31(3): 411-420.

Boddi, B., Oravecz, A.R., and Lehoczki, E. 1995. Effect of cadmium on organization and photoreduction of protochloro-phyllide in dark-grown leaves and etioplast inner membrane preparations of wheat. *Photosynthetica.* 31:411-420.

Boyajian, G.E., and Carreira, L.H. (1997). Phytoremediation: a clean transition from laboratory to marketplace. *Nat. Biotech.* 15: 127-128.

Breckle, S.W. (1991). Growth under stress: heavy metals. In:*Plant Root. The Hidden Half.* Waisel, Y., Eshel, A., Kafkafi, U. (ed.). NY, USA: Marcel Dekker. 351-373

Brooks, R.R. (1998). General Introduction. In: *Plants that hyperaccumulate heavy metals: their role in phytoremediation, microbiology, archaeology, mineral exploration and phytomining.*Brooks, R.R. (ed.). New York, CAB International. 1-14.

Brown, S.L., Chaney, R. L., Angle, J.S., and Baker, A.J. (1995). Zinc and cadmium uptake by hyperaccumulator *Thlaspi caerulescens* grown in nutrient solution. *Soil Sci. Soc. Am. J.* 59: 125-133

Brummer, G., Gerth, J., and Herms, U. 1986. Heavy metal species, mobility and availability in soils. *Z. Pflanzenernaehr. Bodenkd.* 149: 382-398

Brynhildsen, L., and Rosswall, T. (1997). Effects of metals on the microbial mineralization of organic acids. *Water Air Soil Pollut.* 94: 45-57.

Chaney, R.L. (1983). Plant uptake of inorganic waste constituents. In *Land treatment of hazardous waste*, Parr J.F., Marsh P.B., and Kla J.M.(eds.), *Noyes Data Crop*, Park Ridge, pp. 50-76.

Chaney, R.L., and Ryan, J.A. (1994). *Risk based standards for arsenic, lead and cadmium in urban soils: Summary of information and methods developed to estimate standards for Cd, Pb and As in urban soils.* DECHEMA, *Frankfurt.*

Chary, S.N., Hicks, G.R., Choi, Y.G., Carter, D., and Raikhel, N.V. (2008). Trehalose-6-phosphate synthase/ phosphatase regulates cell shape and plant architecture in *Arabidopsis. Plant Physiol.* 146, 97-107.

Cheng S. (2003). Heavy metals in plants and phytoremediation. *Environ, Sci. Pollut. Res. Int.* 10(5):335-40.

Claire, L.C., Adriano, D.C., Sajwan, K.S., Abel, S.L., Thoma, D.P., and Driver, J.T. (1991). Effects of selected trace metals on germinating seeds of six plant species. *Water Air Soil pollut.* 59: 231-240.

Clements, S. (2006). Toxic metal accumulation, response to exposure and mechanism oftolerance in plants. *Biochemie.* 88: 1707-1719.

Courbot, M., Willems, G., Motte, P., Arvidsson, S., Roosens, N., Saumitou-Laprade, P., and Verbruggen, N. (2007). A major quantitative trait locus for cadmium tolerance in *Arabidopsis halleri* colocalizes with HMA4, a gene encoding a heavy metal ATPase. *Plant Physiol.* 144:1052-1065.

Dahmani-Muller, H., Van Oort, F., Gelie, B., and Blabane, M. (2000). Strategies of heavy metal uptake by three plant species growing near a metal smelter. *Environ. Pollut.* 109: 231-238.

Dahmani-Muller, H., Van-Oort, F., and Balabane, M. (2001). Metal extraction by *Arabidopsis halleri* grown on an unpolluted soil amended with various metal-bearing solids: a pot experiment. *Environ. Pollut.* 114: 77-84.

DalCorso, G., Farinati, S., Maistri,S., and Furini,A. 2008. How plants cope with cadmium: staking all on metabolism and gene expression. *J. Integ. Plant Biol.* 50(10): 1268-1280.

Das, P., Samantaray, S., and Rout, G.R. (1997). Studies on cadmium toxicity in plants: A review. *Environ Pollut.* 98(1): 29-36.

Deng, H.X., Shi, Y., Furukawa, Y., Zhai, H., Fu, R., Liu, E., Gorrie, G.H., Khan, M.S., Hung, W.Y., Bigio, E.H., Lukas, T., Dal Canto, M.C., O'Halloran, T.V.,and Siddique, T. (2006). Conversion to the amyotrophic lateral sclerosis phenotype is associated with intermolecular linked insoluble aggregates of SOD1 in mitochondria. *Proc. Natl. Acad. Sci. USA.* 103(18): 7142-7147

Duffus, J.H. (2002) "Heavy metals" a meaningless term? (IUPAC Technical Report) *Pure and Applied Chemistry.*74: 793-807

Elloumi, N., Ben, F., Rhouma, A., Ben, B., Mezghani, I., and Boukhris, M. (2007). Cadmium induced growth inhibition and alteration of biochemical parameters in almond seedlings grown in solution culture. *Acta Physiol. Plant.* 29: 57-62.

Epstein, E., and Bloom, A.J. (2004). *Mineral nutrition of plants: Principles and perspectives*. 2nd Edn Sinauer Associates, Sunderland, MA.

Ewais, E. A. (1997). Effects of cadmium, nickel and lead on growth, chlorophyll content and proteins of weeds. *Biol. Plant.* 39(3): 403-410.

Fodor, F. 2002. Physiological responses of vascular plants to heavy metals. In *Physiology and Biochemistry of Metal Toxicity and Tolerance in Plants*, Prasad, M.N.V., and Strzalka, K. (eds.), Kluwer Academic, Dortrech, UK. pp. 149-177.

Forstner, U., and Wittmann, G.T.W. (1979). *Metal pollution in aquatic environment*. Springer, Berlin Heidelberg New York.

Foyer, C.H., Lopez-Delgado, H., Dat, J.F., and Scott, I.M. (1997). Hydrogen peroxide- and glutathione-associated mechanisms of acclimatory stress tolerance and signalling. *Physiol Plantarum.* 100:241-254.

Gajewska, E., Sklodowska, M., Slaba M., and Mazur, J. (2006). Effect of nickel on antioxidative enzyme activities, proline and chlorophyll contents in wheat shoots. *Biol. Plant.* 50(4): 653-659.

Garbisu, C., and Alkorta, I. (2001). Phytoextraction: a cost-effective plant-based technology for the removal of metals from the environment. *Bioresource Technol.* 77: 229-236.

Gasic, K., and Korban,S. 2006. Heavy metal stress. In *Physiology and Molecular Biology of Stress Tolerance in Plants*, Rao, K.V.M., Raghavendra, A.S., and Reddy, K.J. (eds.) Springer, Dordrecht, the Netherlands. pp. 219-254.

Gechev, S., Van Breusegem, F., Stone, J.M., Denev, I., and Laloi, C. (2006). Reactive oxygen species as signals that modulate plant stress responses and programmed cell death. *BioEssays.* 28(11): 1091-1101.

Ghafoor, A., Rauf, A. and Arif, M. (1996). Soil and plant health irrigated with Paharang drain sewage effluents at Faisalabad. *Pak. J. Agric. Sci.* 33: 73-76

Ghani, A. (2010). Effect of lead toxicity on growth, chlorophyll and lead (Pb) contents of two varieties of maize (*Zea mays* L.) *Pak. J. Nut.* 9: 887-891.

Ghani, A., and Wahid, A. (2007). Varietal differences for cadmium-induced seedling mortality and foliar toxicity symptoms in mungbean (*Vigna radiata*). *Int. J. Agri. Biol.* 9(4): 555-558.

Goldbold, D.L. and Hutterman, A. (1985). Effect of zinc, cadmium and mercury on root elongation of *Picea abies* (Karst.) seedlings, and the significance of these metals to forest die-back. *Environ. Pollut.* 38:375-381

Graham, R.D., and Stangoulis, J.C.R. (2003). Trace element uptake and distribution in plants. *J. Nutr.* 133: 1502-1505.

Greger, M., Brammer, E., Lindberg, S., Larsson, G., and Idestam A.J. (1991). Uptake and physiological effects of cadmium in sugar beet *Beta vulgaris* related to mineral provision. *J. Exp. Bot.* 42: 729-738

Greger, M., Brammer, E., Lindberg, S., Larsson, G., and Idestam A.J. (1991). Uptake and physiological effects of cadmium in sugar beet *Beta vulgaris* related to mineral provision. *J. Exp. Bot.* 42: 729-738

Gupta, M., Sharma, P., Sarin, N. B., and Sinha, A. K. (2009). Differential response of arsenic stress in two varieties of *Brassica juncea* L. *Chemosphere.* 74 (9): 1201-1208.

Hall, J.L. 2002. Cellular mechanisms for heavy metal detoxification and tolerance. *J. Experi. Bot.* 53(366): 1-11.

Hamer, D.H. (1986). Metallothionein. *Ann. Review Biochem.* 55: 913-951.

Hanikenne, M., Talke, I.N., Haydon, M.J., Lanz, C., Nolte, A., Motte, P., Kroymann, J., Weigel, D., and Kramer, U. (2008). Evolution of metal hyperaccumulation required cis-regulatory changes and triplication of HMA4. *Nature.* 453: 391-395

Henry, J.R. (2000). *In an overview of phytoremediation of lead and mercury*. NNEMS Report Washington. 3-9.

Hermle, S., Gunthardt-Goerg, M.S., and Schulin, R. (2006). Effects of metal-contaminated soil on the performance of young trees growing in model ecosystems under field conditions. *Environ Pollut.* 144:703-714.

Hewitt, E.J., and Smith T.A. (1975). Experimental methods for the investigation of plant nutrient requirements. In *Plant Mineral Nutrition*, Hewitt, E.J. and Smith, T.A. (eds.). The English Universities Press Ltd., London. pp. 30-52.

Hossain, M.A., Hasanuzzaman, M., and Fujita, M. (2010). Up-regulation of antioxidant and glyoxalase systems by exogenous glycinebetaine and proline in mung bean confer tolerance to cadmium stress. *Physio. Mol. Biol. Plants.* 16(3): 259-272.

Hsu, Y.T., and Kao, C.H. 2004. Cadmium toxicity is reduced by nitric oxide in rice leaves. *Plant Growth Regul.* 42(3): 227-238.

Hussain, D., Haydon, M.J., Wang, Y., Wong, E., Sherson, S.M., Young, J., Camakaris, J., Harper, J.F., and Cobbett, C.S. (2004). P-Type ATPase heavy metal transporters with roles in essential zinc homeostasis in *Arabidopsis*. *The Plant Cell.* 16: 1327-1339.

Hussain,S.S., Ali, M., Ahmad, M., and Siddique, K.H.M. (2011). Polyamines: natural and engineered abiotic and biotic stress tolerance in plants. *Biotech. Advances.* 29(3): 300-311.

Hutton, M. and Symon, C. (1986). The quantities of cadmium, lead, mercury and arsenic entering the u.k. environment from human activities. *Sci. Total Environ.* 57: 129-150.

Imran, H.S., Wei, K., Jilani, G., and Zh, G. (2007). Interactions of cadmium and aluminum toxicity in their effect on growth and physiological parameters in soybean. *J. Zhejiang Univ. Sci.* 8: 181-188.

Islam, E., Yang, X., He, Z.,and Qaisar, M. (2007). Assessing potential dietary toxicity of heavy metals in selected vegetables and food crops. *J. Zhejiang Univ. Sci. B.* 8(1): 1-13.

Janicka-Russak, M., Kabala, K., Burzynski, M., and Klobus, G. (2008). Response of plasma membrane H-ATPase to heavy metal stress in *Cucumis sativus* roots. *J. Exp. Bot.* 59: 3721-3728.

John, R., Ahmad, P.K., Gadgil, K., and Sharma, S. (2009). Heavy metal toxicity: Effect on plant growth, biochemical parameters and metal accumulation by *Brassica juncea* L. *Int. J. Plant Prod.* 3(3): 65-76.

Johnson, A., Singhal, N., and Hashmatt, M. (2011). Metal-Plant interactions: toxicity and tolerance. In *Biomanagement of contaminated soils*, Khan, M. S., Goel, R., and Mussarat, J. (ed.). Springer, Dordrecht Heidelberg London New York. pp. 29-64.

Jun, R., and Ling, T. (2012). Increase of Cd accumulation in five poplar (*Populus* L.) with different supply levels of Cd. *Int. J. Phytorem.* 14(2): 101-113.

Kabata-Pendias, A., and Pendias, H. (2001). *Trace Elements in Soils and Plants*. Boca Raton, FL: CRC Press.

Kirkham, M.B. (2006). Cadmium in plants on polluted soils: Effects of soil factors, hyperaccumulation and amendments. *Geoderma.* 137 (1-2): 19-32.

Kumar, N.P.B.A., Dushenkov, V., Motto, H., and Raskin, I. (1995). Phytoextraction: the use of plants to remove heavy metals from soils. *Environ. Sci. Technol.* 29: 1232-1238.

Küpper, H., Küpper, F., and Spiller, M. (1996). Environmental relevance of heavy metal substituted chlorophylls using the example of water plants. *J. Exp. Bot.* 47: 259-236.

Lee, K.C., Cunningham B.A., Paulson, G.M., Liang, G.H., and Moore, R.B. (1976). Effects of cadmium on respiration rate and activities of several enzymes in soybean seedlings. *Physiol. Plant.* 36: 4-6.

Leonard, S.S., Bower, J.J., and Shi, X. (2004). Metal-induced toxicity, carcinogenesis, mechanisms and cellular responses. *Mol. Cell Biochem.* 255: 3-10.

Lepp, N.W. (1981). *Effects of Heavy Metals on Plants.* Vol. I, Applied Science, London.

Lopez, M.L., Peralta-Videa, J.R., Parsons, J.G., Gardea-Torresdey, J.L., and Duarte-Gardea, M. (2009). Effect of indole-3-acetic acid, kinetin, and ethylenediaminetetraacetic acid on plant growth and uptake and translocation of lead, micronutrients, and macronutrients in alfalfa plants. *Int. J. Phytorem.* 11: 131-149.

Luna, C.M., Gonzalez, C.A., and Trippi, V.S. (1994). Oxidative damage caused by an excess of copper in oat leaves. *Plant Cell Physiol.* 35: 11-15.

Madhava-Rao, K.V. and Sresty, T.V. (2000). Antioxidative parameters in the seedlings of pigeonpea (*Cajanus cajan* (L.) Millspaugh) in response to Zn and Ni stresses. *Plant Sci.* 157: 113-128.

Marschner, H. (2012). *Mineral Nutrition of Higher Plants.* Academic Press, London.

Martin, M.H., and Coughtrey, P.J. (1982). *Biological monitoring of heavy metal pollution.* Applied Science Publisher, London.

McKenna, I.M., Chaney, R.L., and Williams, F.M. (1993). The effects of cadmium and zinc interactions on the accumulation and tissue distribution of zinc and cadmium in lettuce and spinach. *Environ. Pollut.* 79: 113-120.

Mejare, M., and Bulow, L. (2001). Metal-binding proteins and peptides in bioremediation and phytoremediation of heavy metals. *Trends Biotechnol.* 19: 67-73.

Memarian, R., and Ramamurthy, A.S. (2012). Modeling of lead and cadmium uptake by plants in the presence of surfactants. *Environ. Monit. Assess.*

Miller, R.R. (1996). *Phytoremediation. Ground-Water Remediation Technologies Analysis Center (GWRTAC)*-Technology Overview Report TO-96-03, GWRTAC-O-Series, pp. 11-12.

Mishra, S. and Dubey, R.S. (2006). Heavy metal uptake and detoxification mechanisms in plants. *Int J.Agri. Res.* 1(2): 122-141.

Muller, J.G., Chapman, P.J., and Pritchard, P.H. (1989). Creosote contaminated sites. *Environ. Sci. Technol.* 23: 1197-1201.

Nedelkoska, T.V., and Doran, P.M. (2000). Characteristics of heavy metal uptake by plant species with potential for phytoremediation and phytomining. *Mineral engi.* 13: 549-561.

Nelson, N. (1999). Metal ion transporters and homeostasis. *EMBO J.* 18: 4361-4371.

Nieboer, E., and Richardson, D.H.S. (1980). The replacement of the non descriptive term heavy metals by a biologically and chemically significant classification of metal ions. *Environ. Pollut. Ser.* B. 1: 3-26.

Nies, D.H. (1999). Microbial heavy metal resistance. *Appl. Microbio. Bitech.* 51: 730-750.

Nouairi, I., Ammar, W.B., Yousef, N.B., Daoud, D.B.M., Ghorbal, N.H., and Zarrouk, M. (2006). Comparative study of cadmium effects on membrane lipid composition of *Brassica juncea* and *Brassica napus* leaves. *Plant Sci.* 170: 511-519.

Nouairi, I., Ben Ammar, W., Ben Youssef, N., Ben-Miled, D.D., Ghorbal, M.H., and Zarrouk, M. (2009). Antioxidant defense system in leaves of Indian mustard (*Brassica juncea*) and rape (*Brassica napus*) under cadmium stress. *Acta Physiologiae Plantarum.* 31(2): 237-247.

Nriagu, J.O. (1988). *Production and uses of chromium: Chromium in natural and human environment.* New York, USA, John Wiley and Sons. pp. 81-105.

Nriagu, J.O., and Nieboer, E. (1988). *Chromium in the natural and human environment.* John Wiley & Sons, New York.

Padmavathiamma, P.K., and Li, L.Y. (2007). Phytoremediation technology: hyper-accumulation metals in plants. *Water Air Soil Poll.* 184: 105-126.

Panda, S., Chaudhury, I., and Khan, M. (2003). Heavy metals induce lipid peroxidation and affect antioxidants in wheat leaves. *Biol. Plant.* 46:289-294.

Pandey, N. and Sharma, C. P. (2002). Effect of heavy metals Co^{2+}, Ni^{2+} and Cd^{2+} on growth and metabolism of cabbage. *Plant Sci.* 163(4): 753-758.

Patra, M., Bhowmik, N., Bandopadhyay, B., and Sharma, A. (2004) Comparison of mercury, lead and arsenic with resect to genotoxic effects on plant systems and the development of genetic tolerance. *Environ. Exp. Bot.* 52, 199-223.

Pechova, A., and Pavlata, L. (2007). Chromium as an essential nutrient: A Review. *Veterinarni Medicina.* 52(1): 1-18.

Piechalak, A., Tomaszewska, B., and Baralkiewicz, D. (2003). Enhancing phytoremediative ability of *Pisum sativum* by EDTA application. *Phytochemistry*. 64: 1239-1251.

Poschenrieder, C. and Barcelo, J. 2004. Water relations in heavy metal stressed plants. In *Heavy Metal Stress in Plants*. Prasad, M.N. V. (ed.). Springer, Berlin, Germany, 3rd edition, 249-270.

Prasad M.N.V. (2007). Aquatic plants for phytotechnology. In *Environmental Bioremediation Technologies,* Singh S. N. and Tripathi, R. D. (ed.). Springer. Verlag. Heidelberg. 259-274.

Prasad, M.N.V. (2004a). Phytoremediation of metals in the environment for sustainable development. *Proc. Indian Nanl. Sci. Acad.* 70(1): 71-98.

Prasad, M.N.V. (2004b). Metallothioneins, metal binding complexes and metal sequestration in plants. In *HeavyMetal Stress in plants: From biomolecules to ecosystems.* Prasad, M.N.V.2nd Ed (ed.). Springer-Verlag. Heidelberg. Narosa New Delhi. 47-83.

Pulford, I.D., and Watson, C. (2003). Phytoremediation of heavy metal-contaminated land by trees - a review. *Environ. Int.* 29: 529-540.

Punz, W.F., and Sieghardt, H. (1993). The response of roots of herbaceous plant species to heavy metals. *Environ. Exp. Bot.* 33: 85-98.

Rahman, H., Sabreen, S., Alam, S., and Kawai, S. (2005). Effects of nickel on growth and composition of metal micronutrients in barley plants grown in nutrient solution. *J. Plant Nutr.* 28: 393-404.

Rai, U.N., Sinha, S., Tripathi, R.D., and Chandra, P. (1995). Wastewater treatability potential of some aquatic macrophytes: Removal of heavy metals. *Ecol. Eng.* 5: 5-12

Rai, U.N., and Chandra, P. (1992). Accumulation of copper, lead, manganese and iron by field population of *Hydrodictyon reticulatum* Lagerheim. *Sci. Total Environ.* 116: 203-211.

Rakhshaee, R. Giahi, M., and Pourahmad, A. (2009). Studying effect of cell wall's carboxyl-carboxylate ratio change of *Lemna minor* to remove heavy metals from aqueous solution. *J. Hazard. Mat.* 163(1): 165-173.

Ramamurthy, A.S., and Memarian, R. (2012). Phytoremediation of mixed soil contaminants. *Water Air Soil Poll.* 223: 511-518.

Ramos, R.L., Rubio, L.F., Coronado, R.M., and Barron, M. (1995). Adsorption of trivalent chromium from aqueous solutions on activated carbon. *J. Chem. Technol. Biotechnol.* 62: 64-72.

Raskin, I. (1996). Plant genetic engineering may help with environmental cleanup (commentary). *Proc. Natl. Acad. Sci. USA*. 93: 3164-3166.

Raskin, I., and Ensley, B.D. (2000). *Phytoremediation of toxic metals: using plants to clean up the environment.* John Wiley & Sons, Inc., New York.

Reeves, R.D., and Baker, A.J.M. (2000). Metal accumulation in plants. In *Phytoremediation of toxic metals: Using plants to clean up the environment.* Ensley, B. D., and Raskin, I. (ed.). New York (NY): John Wiley & Sons, Inc. pp. 193-229.

Rivetta, A., Negrini, N., and Cocucci, M. 1997. Involvement of Ca^{2+}-calmodulininCd^{2+} toxicity during the early phases of radish (*Raphanus sativus* L.) seed germination. *Plant Cell Environ.* 20(5): 600-608.

Safwan, S., Jian, C., and Fengxiang, X. (2008). Phytotoxicity of mercury in *Brassica juncea* L. *Ecotoxicol. Environ. Saf.* 72: 619-625.

Salim, R., Al-Subu, M.M., and Attallah, A. (1993). Effects of root and foliar treatments with lead, cadmium and copper on the uptake, distribution and growth of radish plants. *Environ. Internat.* 19: 393-404.

Sanitadi-Toppi, L., and Gabbrielli, R. (1999). Response to cadmium in higher plants. *Environ Exp Bot* 41: 105-130.

Seiler, H.G., Sigel, A., and Sigel, H. (1994). *Handbook on metals in clinical and analytical chemistry.* Marcel Dekker, Inc., New-York, USA.

Seregin, I.V., and Ivanov, V.B. (2001). Physiological aspects of cadmium and lead toxic effects on higher plants. *Russ. J. Plant Physiol.* 48: 523-544.

Shah, F.R., Ahmad, N., Masood, K.R., Peralta-Videa, J.R. and Ahmad, F.D. (2010). Heavy metal toxicity in plants. In *Plant Adaptation and Phytoremediation,* Ashraf, M., Ozturk, M., and Ahmad, M.S.A. (ed.). Springer Dordrecht Heidelberg London New York. pp. 71-98.

Shakya, K., Chettri, M. K., and Sawidis, T. 2008. Impact of heavy metals (copper, zinc, and lead) on the chlorophyll content of some mosses. *Arch. Environ. Contam. Toxicol.* 54(3): 412-421.

Sharma, D.C., Sharma, C.P., and Tripathi, R.D. (2003). Phytotoxic lesions of chromium in maize. *Chemosphere.* 51: 63-68.

Shukla, S.R., and Pai, R.S. (2005). Adsorption of Cu(II), Ni(II) and Zn(II) on dye loaded groundnut shells and sawdust. *Separation and Purification Technology.* 43: 1-8.

Sirkar, S., and Amin, J.V. (1974). Influence of auxins on respiration of manganese toxic cotton plants. *Indian J. Exp. Biol.* 17: 618-619.

Sterritt, R.M., and Lester, J.N. (1980). Interactions of heavy metals with bacteria. *Sci. Tot. Environ.* 14: 5-17.

Suzuki, N., Koizumi, N., and Sano, H. (2001). Screening of cadmium responsive genes in Arabidopsis thaliana. *Plant Cell Environ.* 24: 1177-1188.

Taiz, L., and Zeiger, E. (2006). *Plant Physiology*, (4th ed.). Sinauer Associates, Inc. Publishers, Massachusetts.

Temmerman, L.O.D., Hoenig, M., and Scokart, P.O. (1984). Determination of "normal" levels and upper limit values of trace elements in soils. *J. Plant Nutr. Soil Sci.* 147(6): 687-694.

Tester, M., and Leigh, R.A. (2001). Partitioning of nutrient transport processes in roots. *J. Exp. Biol.* 52: 445-457.

Timbrell, J.A. (2005). *The Poison Paradox: Chemicals as Friends and Foes.* Oxford University, London.

Turner, A.P., and Dickinson, N.M. (1993). Survival of *Acer pseudoplatanus* L. (sycamore) seedlings on metalliferous soils. *New Phytol.*123: 509-21.

Van Assche, F., and Clijsters, H. 1990. Effects of metals on enzyme activity in plants. *Plant Cell Environ.* 13: 195-206.

Verkleij, J.A.C., and Prast, J.E. (1990). Cadmium tolerance and co-tolerance in *Silene vulgaris. New Phytol.* 111: 637-645.

Verloo, M., and Eeckhout, M. (1990). Metal species transformation in soils: an analytical approach. *Int. J. Environ. Anal. Chem.* 39:179-186.

Vesk, P.A., and Allaway, W.G. (1997). Spatial variation of copper and lead concentrations of water hyacinth plants in a wetland receiving urban run-off. *Aquat. Bot.* 59: 33-44.

Vitorello, V.A., Capaldi, F.R., and Stefanuto, V.A. 2005. Recent advances in aluminum toxicity and resistance in higher plants. *Braz. J. Plant Physiol.* 17: 129-143.

Walter, I., Martinez, F. and Cala, V. 2006. Heavy metal speciation and phytotoxic effects of three representative sewage sludges for agricultural uses. *Environ. Pollut.* 139: 507-514.

Wang, W., Hall, A.E., O'Malley, R., and Bleecker, A.B. (2003). Canonical histidine kinase activity of the transmitter domain of the ETR1ethylene receptor from Arabidopsis is not required for signal transmission. *Proc. Natl. Acad. Sci. USA.* 100, 352-357.

Watmough, S.A., and Dickinson, N.M. (1995). Dispersal and mobility of heavy metals in relation to tree survival in an aerially contaminated woodland soil. *Environ. Pollut.* 90: 135-42.

Weast, R.C. (1984).*CRC handbook of chemistry and physics*, 64 Edition.CRC, Boca Raton, Fla.

Wei, S.H., Zhou, Q.X., Wang, X., Zhang, K.S., Guo, G.L., and MaLena, Q.Y. (2005). A newly discovered Cd hyperaccumulator *Solanum nigrum* L. *Chin. Sci. Bull.* 50: 33-38.

Yang, X.E., Jin, X.F., Feng, Y., and Islam, E. (2005). Molecular mechanisms and genetic basis of heavy metal tolerance/ hyperaccumulation in plants. *J. Integ. Plant Biol.* 47(9): 1025-1035.

Zacchini, M., Pietrini, F., Mugnozza, G.S., Pietrosanti, V.L. and Massacci, A. (2009). Metal tolerance, accumulation and translocation in poplar and willow clones treated with cadmium in hydroponics. *Water Air Soil Pollut.* 197 (1-4): 23-34.

Zayed, A.M., and Terry, N. (2003). Chromium in the environment: factors affecting biological remediation. *Plant Soil.* 249: 139-156.

Zengin, F.K., and Munzuroglu, O. (2006). Toxic effects of cadmium (Cd^{2+}) on metabolism of sunflower (*Helianthus annuus* L.) seedlings. *Acta Agricul. Scand., Sect. B, Plant Soil Sci.* 56: 224-229.

PART B

Agronomy and Agricultural Processes for Improving Crops

5

Soil Salinity: Mitigation Strategies for Crop Improvements Under Global Climatic Conditions

Mohd Mazid[1], Taqi Ahmed Khan[2a, 2b] and Saima Quddusi[3*]

[1]*Department of Botany, Faculty of Life Sciences, Aligarh Muslim University Aligarh-202002, India*

[2a]*Department of Biochemistry, Faculty of Life Sciences, Aligarh Muslim University Aligarh-202002, India*

[2b]*Applied Biotechnology Department, College of Applied Sciences, Sur, Oman*

[3]*Amity Institute of Biotechnology, Amity University, Lucknow-227106, India*

ABSTRACT

The sustainability of crop production is primarily a function of various abiotic environmental stress factors including salinity, drought, high temperature and cold and is associated with the fertility status of soil. Among them salinity is a major constraint raised through the present climatic challenges as well as major abiotic stress factor(s) affecting plant growth and productivity worldwide. Soils with high have amounts of soluble salts are called saline soils. They often exhibit a whitish surface crust when dry. Prolonged dry spells often lead to significant crop losses, a situation that is expected to be exacerbated by climatic changes. Indian farmers are seriously concerned with salinity level high enough to affect production. This salinity-induced reduction in photosynthesis, growth and development of plants is associated with ionic/osmotic effects, nutritional imbalance or oxidative stress. In this chapter, authors attempted to describe the impact of climate change on attainable yields of Indian staple crops in reference of Indian climatic scenario. But, however, it is also focussed on the direct fertilization effect of the expected CO_2 alteration and the indirect effects of the expected temperature and precipitation change. As improved agricultural management practices in Indian agricultural systems are crucial to increase agricultural productivity. So therefore, in this chapter also authors have attempted to expose the role of strategies necessary to overcome the effects of salinity from the cropland including the management of mineral nutrients, status, role of signaling molecules, carbohydrates like substances (e.g., fructan), osmolytes (e.g., mannitol), amino acids (proline), vitamins, plant hormones etc.

Keywords: Agriculture, climate change, global warming, greenhouse gases, productivity and salinity.

Corresponding author's E-mail: saimabiotech@gmail.com

Roychowdhury R. (Ed.), Crop Improvement in the Era of Climate Change.

Abbreviations

GHGs, Greenhouse gases; ROS, Reactive oxygen species; GDP, Gross domestic product; CBF, CRT/DRE binding protein.

1. INTRODUCTION

UN Conference on Climate Change in Copenhagen on 18th December 2009 has not yielded the expected results. It is therefore, necessary that each individual country should formulate its own action plan to minimize the incidence of carbon emission to the maximum extent possible within its own resources and capabilities and monitor meticulously on an annual basis in the interest of human survival and crop productivity. The most imminent climatic changes in recent times is the increase in the atmospheric temperatures due to increased levels of greenhouse gases such as carbon dioxide (CO_2), methane (CH_4), ozone (O_3), nitrous oxide (N_2O) and chlorofluorocarbons (CFCs). Because of the increasing concentrations of those radiative or GHGs, there is much concern about future changes in our climate and direct or indirect effect on agriculture (Houghton *et al.*, 2001). In recent years, with the growing recognition of the possibility of global climate change, an increasing emphasis on world food security in general and its regional impacts in particular have come to forefront of the scientific community. Crop growth, development, water use and yield under normal conditions are largely determined by weather during the growing season. Even with minor deviations from the normal weather, the efficiency of extremely applied inputs and food production is seriously impaired. CO_2 concentration was in the steady state at 280 ppm till the pre-industrial period (1850). It has been rising since then at the rate of 1.5 to 1.8 ppm per year. The concentration of CO_2 is likely to be doubled by the end of 21st century. The increased net photosynthetic rate and greater accumulation of sugar contributed significantly to the accelerated development of leaves and tillers and finally grain yield (Bancal *et al.*, 2007).

The increasing CO_2 concentration in the atmosphere and the anticipated climate change due to global warming are likely to affect future global agricultural production through changes in rate of plant growth. Indian agriculture depends on monsoon and as a result, there have been fluctuations in food production over the years. Despite agriculture being vulnerable to the vagaries of weather, it still remains the oldest and the most crowded occupation of the country as over three-fourths of its population of 1.14 billion is dependent on agriculture for their livelihood. It has about 186 million ha. of arable land of which 57% is rain fed. The average size of holding is only 1.84 ha and constitutes nearly 74.5% of the total holdings in the country posing a great challenge of enhancing productivity, production with sustainability. Indian agriculture will need to produce 265 million tonnes of foodgrains to feed an estimated population of 1.4 billion in 2025 (Byrnes and Bumb, 1998).

Agriculture plays a key role in overall economic and social well-being of India. Though the share of agriculture in both GDP and employment has declined over time, the pace of decline in its share in employment has been much slower than that of GDP. The problem has become acute because urbanization and industrialization have rapidly dwindled the per capita availability of arable land from 0.48 ha in 1950 to 0.15 ha by 2000 and is likely to further reduce to 0.08 ha by 2020. As against the national average of 40% of the total cropped area being irrigated,

60% of the total cropped area is still rain fed and dependent on uncertainties of monsoon. This shows the dependency of Indian agriculture on climate. According to World Meteorological Organization, climate change can adversely impact global environment, agricultural productivity and the quality of human life. More importantly in developing countries, it will be difficult for farmers to carry on farming in the increased temperatures, recognizing this, it is neceaary that India should address the issue of climate change and focus on providing better environment to improve quality of agriculture and gain in productivity.

Moreover, several reports are available that show different strategies adopted to counteract the salinity effects. But, however, the information on mineral nutrient status of plants and salinity tolerance is scarce. Among the mineral nutrients, sulfur (S) is increasingly being recognized as the fourth major essential nutrient element after nitrogen N, P and K that plays an important role not only in growth and development of higher plants but also is associated with salinity stress tolerance in plants. Studies investigating the effect offructan on liposomes indicated that a direct interaction between membranes and fructan was possible during salinity stress tolerance. This new area of research began to move fructan and its association with stress beyond mere correlation by confirming that fructan has the capacity to stabilize membranes during drying by inserting at least part of the polysaccharide into the lipid head group region of the membrane. This helps prevent leakage when water is removed from the system either during freezing or drought (Livingston *et al.*, 2009).

When plants were transformed with the ability to synthesize fructan, a concomitant increase in drought and/or salinity tolerance was confirmed. On the other hand, phytohormones are chemical messengers produced in one part of plant and translocated to the other parts, where they play critical roles in regulating plant responses to stress at extremely low concentration. Phytohormones are natural products and they called plant growth regulators, when they are synthesized chemically. Plants are usually subjected to environmental factors such as drought or high soil and water salinity. The reduction in plant growth exposed to saline environments could be due to either the effects of specific ions on metabolism or adverse water relations. Different strategies are being employed to maximize plant growth under saline conditions. One of them is to produce salt tolerant genotypes of different crops. Attempts to improve tolerance to salinity through conventional plant breeding methods are time consuming, laborious and depend on existing genetic variability. In addition, many attempts have been made to overcome this disorder, including proper management and exogenous application of plant growth regulators (Ashraf *et al.*, 2002).

In addition, glycine betaine (GB) and proline are two major organic osmolytes that accumulate in a variety of plant species in response to environmental stresses including salinity. Although their actual roles in plant osmotic tolerance remain controversial, but however, both compounds are thought to have positive effects on enzyme and membrane integrity along with adaptive roles in mediating osmotic adjustment in plants grown under stress conditions. While many studies have indicated a positive relationship between accumulation of GB and proline and plant stress tolerance, some have argued that the increase in their concentrations under stress is a product of, and not an adaptive response to stress. Since, as not all plant species are capable of natural production or accumulation of these compounds in response to stress, extensive research has been conducted examining various approaches to introduce them into plants. Genetically-engineered (GE) plants containing transgenes for production of GB or proline have thus far faced with the limitation of being unable to produce sufficient amounts of these compounds to ameliorate stress effects. An alternative "shot-gun" approach of exogenous application of GB or proline to plants under stress conditions, however, has gained some attention.

Furthermore, a review of the literature indicates that in many, but not all, plant species such applications lead to significant increases in growth and final crop yield under salinity stress. Moreover, if certain climate change scenarios come to pass, agricultural production in some areas may decrease. Changes in temperature and in precipitation patterns and amount will influence soil water content, run-off and erosion, salinisation, biodiversity, and organic carbon and nitrogen content (Aggarwal *et al.*, 2004). The increase in temperature would also lead to increased evapotranspiration. There is need to quantify the specific regional soil-related problems and that affect the global environmental change will have on soil fertility and its functioning for crop growth and production. Global warming may also threaten food security in India if there is a negative effect on agriculture. Although, the effect of increasing CO_2 concentrations will increase the net primary productivity of plants, but climate changes, and the changes in disturbance regimes associated with them, may lead either to increased or decreased net ecosystem productivity.

In many tropical and subtropical regions, potential yields are projected to decrease for most projected increases in temperature. The impacts of elevated CO_2 should be considered among others, in the context of: (1) changes in air temperature, particularly nocturnal temperature due to increase in CO_2 and other trace gases and changes in moisture availability and their effect on vegetative versus reproductive growth; (2) need for more farm resources (e.g., fertilizers); and (3) survival and distribution of pest populations, thus developing a new equilibrium between crops and pests. The simplest approach is to assume a fixed climate change and to modify the data with a constant number, such as an increase or decrease of 1, 2, 3°C etc. for temperature. Similarly, rainfall and solar radiation can be changed with a certain percentage, such as an increase or decrease of 10, 20, 30% etc. These changes are then applied to the daily water data and the crop simulation models are run with these modified inputs. A more realistic approach is to use the outputs from the General Circulation Models (GCMs) to modify the historical weather data (Mitchell *et al.*, 2004). Eighty-two percent of the cropland worldwide is cultivated under rain-fed conditions. The importance of rain-fed agriculture varies regionally, but it is of utmost significance for India. There, agriculture accounts for 15% of GDP and employs 57% of the population. This sector has been providing employment to 60% of the work force and responsible for feeding more than a billion people. Approximately 95% of the total cropland is managed under rain-fed conditions. The spatial and temporal variations of crop yield may have a profound impact on the national economy which is primarily dependent on the agricultural sector. The high spatial and temporal variability of rainfall, reflected by dry spells and recurrent salinity, may be considered the most important factor affecting agricultural productivity in India. There has been a significantly slower growth in this sector compared to the other two sectors viz, industry and services. This gap has been widening since 1981-82 and is significantly more pronounced since 1996-97, due to more attention paid to the other two sectors and increased demand of foodgrains.

Planting too early might lead to crop failure and, in turn, planting too late might reduce valuable growing time and crop yield (Sacks *et al.*, 2010). However, there is still no consensus in literature about the question of how much rain over which period defines the ORS for agroclimatological impact studies. The vision 2020 document of the National Planning Commission has also reiterated "India needs to sustain an agricultural growth rate of 4.0 to 4.5% in order to reduce food insecurity and poverty. At this growth rate, agriculture development could more rapidly diversify into horticulture, fishery, dairying, animal husbandry and other areas."

2. IMPACT OF PRESENT CLIMATIC CIRCUMSTANCES ON GLOBAL AGRICULTURAL PRODUCTIVITY

In many developing countries, agriculture is the cornerstone of their economy, the basis of economic growth and the main source of livelihood. But agriculture in the developing world is often cited as being one of the sectors most vulnerable to climate change. In India, for example, the majority of available fresh water is used for agriculture; farming techniques are relatively simple; and much of the continent is hot and dry. The frequent climate change, particularly low rainfall and warming has posed serious threat to sustainable agriculture (Kurukulasuriya and Rosenthal, 2003). Low rainfall has resulted in groundwater depletion because of draining of water for irrigation and warming has affected the yield of rice and wheat. Indian agriculture has witnessed momentous growth during the last 60 years. The foodgrain production has increased by 4 times, horticultural crops and milk by 6 times, fish by 9 times and eggs by 27 times since 1950-51.

Any changes in precipitation and temperature patterns will thus have major impacts on the viability and yields in crop production. To exacerbate the situation, several studies warn of an unprecedented confluence of pressures on agriculture with population growth and development driving up global demand for food and competition for land, water and energy intensifying as the impacts of climate change starts to take effect. In this reference, any strategy to enhance agricultural productivity in Asia including India needs to ensure that natural resources are managed sustainably and adapted to climate change. Although agricultural production is sufficient to meet current food demands.

Many of the poorest producers farm in locations where the climate is already marginal for production and farmers with limited access to agricultural knowledge and technology will also be less able to adapt their farming practices to climate change. For these reasons, the poorest farmers are most vulnerable to the potential impacts of climate change. Despite international negotiations to reduce GHG, a 20-30 year lag in our global climate system means we are already committed to a world that will be 0.6 °C warmer, with associated changes in rainfall patterns, by the end of the century (Malhi *et al.*, 2008). Future crop production will thus have to adapt to changes in climate to which we are already committed. Many studies in the research literature describe how agriculture in Asia will be one of the sectors most vulnerable to climate change and variability. This is because a significant proportion of the Asian economy is dependent on agriculture, farming techniques are relatively primitive and the majority of the continent is already hot and dry. Spatial and temporal changes in precipitation and temperature patterns will shift agro-ecological zones and thus have major impacts on the viability of dry landand irrigated farming (Lioubimtseva and Henebry, 2009).

Similarly, agriculture is critical to North Asia's development. More than 75% of the region's poor live in rural areas and are dependent on rain-fed agriculture, livestock, and fragile forests for their livelihoods. The Green Revolution increased foodgrain productivity, improved food security and rural wages bringing a significant reduction in rural poverty. But the challenge now is to replicate and sustain these achievements in the future with a more variable and unpredictable climate. The constraints on food crop production and distribution differ between regions and, in particular, between industrialised and developing countries. The country harvested a record 230 million tonnes of foodgrains, produced 6.87 million tonnes of fish and more than 100 million tonnes of milk during 2008-2009. It could not have been possible but for the pioneering role of Indian Council of Agriculture Research (ICAR) in ushering green revolution and other major developments in agriculture and allied sectors through its research and technological excellence (Sharma, 2002).

Climate change has the potential to exacerbate the stresses on crop plants, potentially leading to catastrophic yield reductions. It is likely to affect hydrological water balances, the availability of fresh water supplies for irrigation and soil moisture balances, with consequent impacts on agricultural productivity. Soils are another essential but non-renewable resource for food crop production so maintaining soil fertility, health and nutrient availability is vital. Significant losses in crop yields also occur through pests, diseases and weed competition, accounting for major inefficiencies in resource use. Reducing these losses represents one of the most accessible means of increasing food supplies. Climate change will aggravate the effects on crops of stresses such as salinity. The adaptation strategies for these crops must be carried out in the face of other constraints such as labour shortages and rising energy costs. Several environmental factors adversely affect plant growth and development and final yield performance of a crop (Ashraf and Foolad, 2005).

Salinity is among the major environmental constraints to crop productivity worldwide. Development of crop plants with stress tolerance, however, requires, among others, knowledge of the physiological mechanisms and genetic controls of the contributing traits at different plant developmental stages. In the past two decades, biotechnology research has provided considerable insights into the mechanism of abiotic stress (salinity) tolerance in plants at the molecular level. Different abiotic stress factors may provoke osmotic stress, oxidative stress and protein denaturation in plants, which lead to similar cellular adaptive responses such as accumulation of compatible solutes, induction of stress proteins, and acceleration of ROS scavenging systems. As climate is a primary determinant of agricultural productivity, any significant changes in climate in the future will influence crop and livestock productivity, hydrologic balances, input supplies and other components of managing agricultural systems. However, the nature of these biophysical effects and human responses are complex and uncertain (Parry *et al.*, 2004). Estimates of impact of climate change on crop production could be biased depending upon the uncertainties in climate change scenarios, region of study, crop models used for impact assessment and the level of management. So it is very important to give these uncertainties due importance while assessing the impacts of possible climate change on crop productivity for formulating response strategies.

Nevertheless, the environmental factors such as cloudiness and solar radiation at the earth's surface will also change but the GCMs are less consistent in their predictions, particularly on a regional basis. Climate models, in general, are subject to several uncertainties, which are especially pronounced at regional scales. The terms 'adaptation' and 'agriculture' were omitted from the primary question as these would excessively broaden the scope of the SR; the adaptation of agriculture to climate change is itself a separate discipline and 'agriculture' could be interpreted to include aspects such as livestock production and forestry (Knox *et al.*, 2011). This SR will focus specifically on the biophysical aspects of crops and the impact that climate change might have on crop productivity. Similarly, the chapter will not consider 'food production', as this is dependent on non-biophysical factors, such as investment in irrigation, international trade policy and world market prices. Nor will it consider the impact of climate related 'shocks' including salinity on food production. Because there is still considerable uncertainty in our understanding of how the climate system varies naturally and reacts to emissions of GHGs and aerosols, current estimates of the magnitude of future warming are subject to future. It is interesting to note that despite considerations of all uncertainties, the simulated rice crop always responded positively to climate change.

Thus, the impact of climate change scenarios of 2010 was nominal and varied between 1–7% depending upon the scenario of climate change, level of N management and impact assessment model used for the analysis and the regions of study. In this context, Agarwal and Mall (2002) concluded that several studies are being done to develop an integrated assessment of climate change impact on regional and global food supplies and demand. Some of these could also find use in directing food policy. These studies use specific GCM based scenarios and a specific crop model. While integrated assessments of climate change on food availability are very much needed, conclusions based on the average simulated grain yields and mean changes in climatic parameters should be used with caution. In spite of the uncertainties about the precise magnitude of climate change on regional scales due to scenarios and crop models on impact assessment, an assessment of the possible impacts of climate change on India's agricultural production under varying socio-economic conditions is important for formulating response strategies, which should be practical, affordable and acceptable to farmers. The identification of suitable response strategies is key to sustainable agriculture. Attri and Rathore (2003) suggested the adaptation strategies for sustainable production of wheat and ensuring food security. Adaptation measures to mitigate the potential impact of climate change included possible changes in sowing dates and genotype selection. Migratory option for reducing the negative impacts of salinity increases indicate that delaying the sowing dates would be favourable for increased soybean yields at all the locations in India. Sowing in the second season would also be able to mitigate the detrimental effects of future increases in surface temperature due to global warming at some locations. Changes in the cropping sequence, irrigation and agriculture land use can be additional alternative options for adaptation in agriculture (Smit and Skinner, 2002).

The loss in farm-level net revenue is estimated to range between 9% and 25% for a temperature rise of 2°C–3.5°C. These adaptation strategies may include altered crops and cropping systems to maintain soil fertility in sustainable manner and improved management practices. Modern technologies in agriculture could also be beneficial with or without climate change; government should encourage farmers to shift towards newer technologies. The government

influence cell proliferation or cell death and trigger specific gene expression essential for plant recovery against various types of environmental stresses including salinity and oxidative stress. Indeed, there are several traits whose corrective association with resistance has been tested in transgenic plants. Several osmotic stresses cause detrimental changes in cellular components. It is believed that osmoregulation would be the best strategy for abiotic stress tolerance, especially if osmoregulatory genes could be triggered in response to salinity. Therefore, a widely adopted strategy has been to engineer certain osmolytes or by over-expressing such osmolytes in plants as a potential route to breed stress-tolerant crops. Moreover, exogenous application of proline can play a significant role in enhancing plant salinity stress tolerance. This role can be in the form of either osmoprotection or cryoprotection. Exogenous application of proline markedly inhibits shoot elongation and root growth of rice seedlings as well as growth of *Brassica napus* callus (Hare and Cress, 1997). If under normal conditions, continuous cycling occurs between proline and its ultimate precursor glutamate and this is blocked by high level of proline, exogenously applied proline may inhibit C-flux through the OPPP and ADP phosphorylation, both of which are likely to be of importance in sustaining a rapid growth rate.

It accumulates in many plant species and can act in association with other components of the antioxidant system, protects plants against oxidative damage resulting from aerobic metabolism, photosynthesis and a range of pollutants. Moreover, compartmentalization of proline biosynthesis and degradation in the cytosol, chloroplast and mitochondria add to the complexity of functional diversification of proline metabolism. Therefore, it was hypothesized that improvement of drought or salt stress tolerance of crop plants via proline metabolism engineering would be an existing possibility in future context. Results from the metabolic engineering experiments emphasize the need for a greater understanding of primary proline metabolic pathways from which proline biosynthesis pathways branch off. Further studies required to identification and exploitation of proline in naturally abiotic stress-tolerant crop plants and the use of multiple genes and reiterative metabolic research to enhance proline flux that could lead to new opportunities to improve plant tolerance of salinity stress. There are many reports about proteins change level in salinity stress. Salinity is the most limiting factor for crop production and it is becoming an increasingly severe problem in many regions of the world. Plants behavioural response to salinity is complex and different mechanisms are adopted by plants when they encounter salinity. The plants under salinity conditions change their metabolism to overcome the changed environmental conditions. One mechanism utilized by the plants for overcoming the salt stress effects might be via accumulation of compatible osmolytes, such as proline and soluble sugar. Production and accumulation of free amino acids, especially proline by plants tissue during salt is an adoptive response (Shahba *et al.*, 2010).

Moreover there are many other reports about increasing and decreasing of proline level in salty stress. The soluble protein and free amino acid in barley organs increased with NaCl increasing. The increasing of amino acid in the plant tissue under stress is related to protein fraction. The cause of protein reduction at salinity conditions is the prevention of nitrate reductase activity (NRA). The salt stress induced some changes in the protein of rice leaf shoots and roots, but not effective on leaf blade. The levels of some protein decreases because of reduction in proline synthesis.

3.4 Osmolytes

While many studies have indicated a positive relationship between accumulation of GB and proline and plant stress tolerance, some have argued that the increase in their concentrations under stress is a product of, and not an adaptive response to stress. Genetically-engineered (GE) plants containing transgenes for production of GB or proline have thus far faced with the limitation of being unable to produce sufficient amounts of these compounds to ameliorate stress effects. An alternative "shot-gun" approach of exogenous application of GB or proline to plants under stress conditions, however, has gained. In many crop plants the natural accumulation of GB is lower than sufficient to ameliorate the adverse effects of dehydration caused by various environmental stresses (Parvaiz and Satyawati, 2008).

Exogenous application of GB to low-accumulating or non-accumulating plants may help reduce adverse effects of environmental stress. Externally applied GB can rapidly penetrate through leaves and be transported to other organs, where it would contribute to improved stress tolerance. This may make extraction and exogenous application of GB an economically feasible approach to counteract adverse effects of salinity on crop productivity. In addition, to its direct protective roles, either through positive effects on enzyme and membrane integrity or as an osmoprotectant, GB may also protect cells from salinity stress indirectly via its role in signal transduction.

High salinity imposes osmotic stress on plants. Plants respond to the stress in part by modulating gene expression, which eventually leads to the restoration of cellular homeostasis, detoxification of toxins and recovery of growth. Water is required as a medium for biochemical activities by all known life forms. For plant cells, water-generated turgor pressure is also a driving force of cell expansion. However, vegetative growth of plants can only occur at a certain range of water status, which can be measured by the free energy state of water molecules.

Under natural conditions, high salinity is the major cause of osmotic stress to plants. The obvious metabolic changes characterized by increased synthesis of osmolytes have attracted researchers to study the underlying mechanisms of hyper osmolarity signalling. The widely reported accumulation of proline in stressed plants makes proline a good candidate marker to study the signalling pathways leading to its accumulation. However, current work on this pathway has mostly been limited to the transcriptional or post-transcriptional regulation of several biosynthesis genes. The power of genetic analysis is being increasingly applied to stress signalling in plants.The osmotic stress signalling pathways that have been more actively studied in plants are those that lead to the synthesis of a group of proteins with no clear biochemical function, and no significant sequence similarity to other well-characterized proteins (Xiong and Zhu, 2002).

3.5 Plant Hormones

Phytohormones such as IAA, gibberellic acid (GA3), and kinetin are known to be involved in the regulation of plant response to salinity stress and counteract the adverse effect of stress conditions. Gifford and Evans (1981) show that plant hormones can be defined as organic substances that are produced in one part of plant and translocated to another parts, where at very low concentration, they stimulate physiological response. Plants are usually subjected to environmental factors such as drought or high soil and water salinity. The reductionin plant growth exposed to saline environments could be due to either the effects of specific ions on metabolism, or adverse water relations. Different strategies are being employed to maximize plant growth under saline conditions. One of them is to produce salt tolerant genotypes of different crops. Attempts to improve tolerance to salinity through conventional plant breeding methods are time consuming and laborious, and rely on existing genetic variability. In addition, many attempts have been made to overcome this disorder, including proper management and exogenous application of plant growth regulators and that produces some benefit in alleviating the adverse effects of salt stress and also improves germination, growth, development and seed yields and yield quality.

It has been reported that exogenous application of ABA reduces the release of ethylene and leaf abscission under salt stress in plants, probably by decreasing the accumulation of toxic Cl^- ions in leaves. In wheat, seed germination decreased with increasing levels of salinity, while the adverse effect of salinity was alleviated by soaking the seed with IAA. In addition, exogenous IAA showed high stimulatory effect on the root and shoot growth of wheat seedling in saline condition (Egamberdieva, 2009). Growth and yield parameters of rice were significantly increased in response to application of cytokinin under salinity stress. ABA is also the major internal signal enabling plants to survive adverse environmental conditions such as salt stress. Exposure of plants to salinity is known to induce a proportional increase in ABA concentration that is in most cases correlated with leaf or soil water potential, suggesting that salt-induced endogenous ABA is due to water deficit rather than specific salt effects. This may not resemble the prolonged increasing of endogenous ABA levels that can occur in association with slowly increasing salinity stresses in nature or field situations. Increases of the endogenous ABA concentration in leaf tissue for salt stressed Brassic, *Phaseolusvalgaris* and *Zea mays* strongly correlated with growth inhibition. There is significant evidence that ABA acts as the root-to-shoot stress signal. A study has demonstrated that ABA contributes to the increase of xylem water potential as well as water uptake to the plant in the presence of salt. Increase of ABA concentration in the xylem is correlated with reduced leaf conductance and general inhibition of leaf growth. Salt stress stimulated ABA synthesis in roots and its xylem transport and well correlated to the stomatal reactions. This may be explained by the fact, when roots are directly exposed to the salt, ABA in roots stimulates ion accumulation in vacuoles of barley roots which may be necessary for adaptation to saline conditions.

Seed enhancement (seed priming) with cytokinins is reported to increase plant salt tolerance. It was hypothesized that cytokinins could increase salt tolerance in wheat plants by interacting with other plant hormones, especially auxins and ABA. Kinetin was effective in increasing germination rate in the salt-intolerant and early seedling growth in the salt-tolerant cultivar when compared with hydropriming under salt stress. Thus, during germination and early seedling growth, the cytokinin-priming induced effects were cultivar specific. In contrast, kinetin-priming showed a consistent promoting effect in the field and improved growth and grain yield in both cultivars under salt stress. The BAP-priming did not alleviate the inhibitory effects of salinity stress on the germination and early seedling growth in both cultivars (Hamdia and Shaddad, 2010).

IAA also responds to salinity in crop plants. However, little information seems to be available on the relationship between salinity stress and auxin levels in plants and the role of auxin in alleviating salt stress. The variations in IAA content under stress conditions appeared to be similar to those of abscisic acid. The increased level of IAA has reportedly been correlated with reduced growth. Therefore, the reduction in plant growth under stress conditions could be an outcome of altered hormonal balance. Hence, their exogenous application provides an attractive approach to counter the stress conditions. GA3 application during the salinization period partly overcomes the effect of salinity on reducing IAA levels and this shows that salinity may influence hormone balances by affecting plant growth and development. There was also a significant reduction in IAA levels in rice five days after NaCl treatment. Cytokinin receptor mutants in stress-response assays showed that all three cytokinin receptors of Arabidopsis act as negative regulators in ABA signaling and in the osmotic stress responses. GAs is generally involved in growth and development. They control seed germination, leaf expansion, stem elongation and flowering.

In order to alleviate deleterious effects of salinity, different types of phytohormones have been used. Among them, gibberellins have been the main focus of some plant scientists. GA3-priming-induced increase in wheat grain yield was attributed to the GA3-priming-induced modulation of ions uptake and partitioning (within shoots and roots) and hormones homeostasis under saline conditions. Under saline conditions, seed germination has been improved by application of GA3 and in this experiment, growth and grain yield of wheat were decreased with increasing salinity levels, but increased relatively by seed treatment with GA3 (Aldesuquy and Ibrahim, 2008). Wheat seeds, after treatment with various growth regulators including GA3, showed highest percent germination when treated with 20 mg/l GA3. Soluble protein content varied not only between the two wheat cultivars but also between the different plant organs. Free radicals induced lipid peroxidation inhibition by GA. These results show that GA3 application could improve salinity tolerance in crop plants grown under saline condition. Brassinosteroids are a new group of phytohormones with significant growth promoting properties with pleiotropic effects, as they influence diverse physiological processes such as growth, seed germination, rhizogenesis, senescence and leaf abscission.

In addition, BRs are implicated in plant responses to salinity stress. The CPD gene of Arabidopsis encodes a cytochrome P450 protein that shares sequence similarity with mammalian steroid hydroxylases. Feeding experiments with different BRs indicate that CYP90 acts at the C-23 hydroxylation step in the BR biosynthetic pathway. BR regulated stress response because of a complex sequence of biochemical reactions such as activation or suppression of key enzymatic reactions, induction of protein synthesis, and the production of various chemical defence compounds. Exogenous applications of BRs under salinity have long been known to increase growth and yield in many economically useful plant species. In cereals, BRs promote the number of ears and in some cases also their length, and the number and weight of kernels per ear (Holá *et al.*, 2010). The findings clearly demonstrate the ability of brassinosteroids to alleviate the inhibitory effects of salinity on germination, seedling growth and plant yield. However, further studies are required to unravel the tolerance mechanism conferred by brassinosteroids in the case of salinity stress treatment may change the balance of endogenous hormones, such as ABA, which provides an important clue for understanding the protection mechanisms against salt stress. Exogenous JA may be involved in the defense not only during wounding and pathogen stress, but also during salt stress. SA induces acidic pathogen-related (PR) genes and inhibits basic PR genes, whereas JA does the opposite. SA also reduced the synthesis of tomato proteinase inhibitors. Antagonistic interactions between SA and jasmonic acid affect the expression of PR protein genes in tomato.

Several methods of application have been shown to protect various plant species against salinity stress factors by inducing a wide range of processes involved in stress tolerance mechanisms. The role of SA in defense mechanism to alleviate salt stress in plants was studied. The ameliorative effects of SA have SA added to the soil also had an ameliorating effect on the survival of maize plants during salt stress and decreased the Na^+ and Cl^- accumulation. Lipid peroxidation and membrane permeability, which were increased by salt stress, were lower in SA treated plants. SA treatment was accompanied by a transient increase in the H_2O_2 level. As seed treatment with H_2O_2 it had an alleviating effect on the oxidative damage caused by salt stress in wheat plants, it seems possible that SA may exert its protective effect partially through the transiently increased level of H_2O_2 (Horváth *et al.*, 2007).

It was found that the SA treatment caused accumulation of both ABA and IAA in wheat seedlings. However, the SA treatment did not influence cytokinin content. The SA treatment reduced the damaging action of salinity and water deficit on seedling growth and accelerated a restoration of growth processes. Treatment with SA essentially diminished the alteration of phytohormones levels in wheat seedlings under salinity. The SA treatment prevented the decrease in IAA and cytokinin content completely which reduced stress-induced inhibition of plant growth. Also, high ABA levels were maintained in SA treated wheat seedlings which provided the development of anti-stress reactions, for example, maintenance of proline accumulation. Thus, protective SA action includes the development of anti-stress programmes and acceleration of normalization of growth processes after removal stress factors.

3.6 Role of Vitamins

Vitamin C is a universal reductant and antioxidant of plants. It is found at concentration 1-2 Mm in legume nodules, and is positively correlated with metabolic for the operation of the ASG-GSH pathways, but it also has beneficial effects that do not require the presence of APX, ASC can directly scavenge ROS. It is also involved in hydroxylation of proline, regulation of the cell cycle and numerous fundamental processes of plant growth and development. Exogenous application of vitamin C to plant could lead to growth stimulation through the activation of some enzymatic reaction (Oertli, 1987).

3.7 Role of Biofertilizers

In present time, the new challenges of the new millennium are to produce more and more agricultural food production from shrinking per capita arable land. Keeping the environmental conditions safe to human population first. In agriculture there are one most important limiting factors is providing plant nutrients, particularly nitrogenous chemical fertilizers are recommended and used by farmers sometimes injudiciously expecting to harvest high yields leading to a high degree of profitability. However, wrong and untimely applications of artificial fertilizers adversely affect the natural balance of soil crop ecosystems microbial ecology and concern environment, resulting in widespread decline in the crop yield.

Soil infertility in arid zones is often due to the presence of large quantities of salt, and the introduction of plants capable of surviving under these conditions (salt-tolerant plants) is worth investigating. Under agricultural condition, the amount of N removal is greater than the N input. To maintain soil fertility, N can be returned to the soil through biological N fixation (BNF) or by N fertilizer applications. Under agricultural condition, the amount of N removal is greater than the N input. To maintain soil fertility, N can be returned to the soil through biological N fixation (BNF) or by N fertilizer applications. A symbiotic N-fixing bacteria, including the genus Azospirillum are considered to be among the most important microorganisms that play a significant role in soil fertility in tropical and temperate regions of the world. Azospirillum strains are commonly isolated worldwide from the rhizosphere of a diverse range of cereals, including corn, millet, sorghum, and wheat. Although more than a decade has passed since the beginning of intensive worldwide Azospirilluminoculation experiments to improve crop productivity of non-leguminous plants, there is still considerable and continued interest in testing the efficiency of Azospirillum strains for their ability to enhance and/or improve yields under salt stress (Tilak *et al.*, 2005).

The Rhizobium-legume symbiosis in arid ecosystems is particularly important for locations where the area of saline soils is increasing and becoming a threat to plant productivity. Legumes, which are usually present in arid ecosystems, may adapt to fix more N under saline conditions than legumes grown in other habitats. Application of mycorrhiza improves plant growth and nutrient acquisition and thus protects plants against salt stress in arid and semi-arid areas. Therefore, the importance of inoculation of soil and plants with the beneficial microorganisms that enhance plant nutrition has been increased showed that plant bacterization had stimulatory effect on amino acids accumulation compared with control uninoculated plants up to –1.2 MPa salinity level. Thus Azospirillum inoculation plays an important role in protein biosynthesis either directly through fixation of N or indirectly by enhancing soil N uptake. However, plant inoculation with *Azospirillum* resulted in significant increases in shoot and root total N-yield up to –0.9 and –1.2 MPa salinity levels, respectively. Increased total N-yield at such high salinity levels could be attributed to factors other than N-fixation, since N fixation was completely inhibited at such NaCl levels.

4. SIGNALING MOLECULES: ASCORBIC ACID, HYDROGEN PEROXIDE, NITRIC OXIDE

There are evidence that H_2O_2 and NO function as stress signalling molecules in plants, mediating a range of defensive mechanisms in plants under stress conditions. Abiotic environmental stresses result in H_2O_2 generation from a variety of sources and have many essential roles in plant metabolism but at the same time, accumulation related to virtually any environmental stress is potentially damaging. NO, another stress responsive molecule, is gaining high attention as a regulator of diverse physiological and biochemical processes in plant science due to its role in regulation in a number of plant cellular mechanisms against responses induced by abiotic stresses like salinity stress and drought. Along with various other abiotic stresses including salinity stress can also induce NO synthesis, but its origin, mode of action and metabolism in plants have not yet been completely resolved to a greater extent. But regarding NO-H_2O_2 signalling and functions, new findings have given an impression of the intricacy of NO-H_2O_2 related signalling network against abiotic stresses. Cellular responses to these molecules are complex, but considerable crosstalk between responses induced to salinity stress is reported by several recent research works. In last few years, the role of NO in H_2O_2 mediating tolerance in plants to above stresses has established much greater consideration in reference to present scenario of agriculture rethinking (Neil *et al.*, 2002).

Plants appear to possess wide arrays of defense strategies to protect themselves from oxidative damage. However, less is known about NO involvement in tolerance of plants to salt stress. The exogenous sodium nitripruside (SNP), a NO donor significantly alleviated the oxidative damage of salinity to seedlings of rice, lupin and cucumber enhanced seedlings growth and increased the dry weight of maize and *Kodtetzky avirginica* seedlings under salt stress. Pretreatment of NO effectively contributed to better balance between C and N metabolism by increasing total soluble protein and by enhancing the activities of endopeptidase and carboxypeptidase in plant under salt stress. Furthermore, NO participates in enhancement of photosynthesis by inducing the photosynthetic pigments under salt stress and also in ATP synthesis and two respiratory electron transport pathways in mitochondria where, it mediates the modulation of ROS and enhances the antioxidant defense system in plants subjected to salinity. It has been initially hypothesized that H_2O_2 might be downstream signal molecules to regulate the activityof PM H+-ATPase.

Ascorbic acid (AA) is found in all eukaryotes including animals and plants and lacks completely in prokaryotes except cyanobacteria, have been reported to have a small amount of AA. It is an antioxidant and, in association with other components of the antioxidant system, protects plants from oxidative damage resulting from aerobic metabolism, photosynthesis and a range of pollutants like ozone, heavy metal and saline stress.

Indeed, in addition to acting simply as an antioxidant and cellular reductant, AA influences transition from the vegetative to the reproductive phase and the final stage of development, senescence. Therefore, there is a need to increase our understanding of this enigmatic molecule since, it could be involved in a wide range of developmental phenomena as well as works against actual salinity stress in order to regulate better growth and development. Plants vary considerably in their physiological response to various kinds of environmental stress including salinity. AA has now gained significant place in plant science, mainly due to its properties (antioxidant and cellular reductant etc.), and multifunctional roles in plant growth, development, and regulation of remarkable spectrum of plant cellular mechanisms against environmental stresses. Indeed, AA plays an important role in resistance to pathogenesis. Foliar application of AA at 200 mg counteracted the adverse effect of salinity that was accompanied by a significant increase in plant growth of flux cultivars (Khan *et al.*, 2011). Also, AA affects many physiological processes including the regulation of growth, differentiations and metabolism of plants under saline conditions and the increasing physiological availability of water and nutrient. In addition, AA protects metabolic processes against H_2O_2 and other toxic derivatives of oxygen affecting many enzyme activities, minimizes the damage caused by oxidative processes through synergistic function with other antioxidants, and stabilizes membranes.

One possibility is that additional AA would inhibit a stress-induced increase in the leakage of essential electrolytes following peroxidative damage to plasma membranes. However, additional AA did not inhibit an increase in the leakage of electrolytes from roots of salt-stressed tomato seedlings. Moreover, the increase in yield and its compounds might be due to the effect of the antioxidant role on enhancing protein synthesis and delaying senescence. Possibly, the protective effect of AA is more related to a reduction in ROS damage to essential proteins and/or nucleic acids. Finally, the fact that new roots and leaves are produced by the seedlings which have recovered from 9 h of salt treatment with AA, suggests that additional AA may have affected meristematic cells in the salt stressed root and shoot tissues. Genetic engineering of the ascorbate pathway enzyme in potato enhanced its ascorbic acid content and subsequently plants suffered minimal oxidative stress-induced damage under salinity. The transgenic plants under salinity maintained a higher reduction in the oxidized GSH:GSSG ratio together with an enhanced activity of GSH dependent antioxidative and glyoxalase enzymes. These results suggest the engineering of ascorbate pathway enzymes as a major step towards the development of salinity tolerant crop plants.

5. FOOD SECURITY AND SALINITY CONSTRAINT IN FUTURE SCENARIO

India needs to sustain an agricultural growth rate of 4.0 to 4.5% in order to reduce food insecurity and poverty. At this growth rate, agricultural development could more rapidly diversify into horticulture, fishery, dairying, animal husbandry and other areas. Sustainable development emerged as an important theme of development in 1987. Sustainable agriculture refers to the ability of a farming system to produce food, without causing severe and irreversible damage to the ecosystem. Generally, sustainable agriculture integrates three things: environmental health, economic profitability and social and economic equity (Alkon, 2008). It desires that all the actors including the farmers, farm labour, credit provider's policy makers, scientists, researchers, and goal. Although it is difficult in the current scenario yet is desirable for the betterment of the future. The government has initiated several direct and indirect measures to ensure food security to its population. These measures can be categorized as: (a) direct measures (i) food subsidy measures and (ii) entitlement feeding programmes; (b) indirect measures (i) employment programme and (ii) social safety net schemes.

After 20 years of low food commodity prices, the price shock of 2007/2010 brought agriculture, food production and food security sharply back into the limelight. Wheat and maize prices peaked at around triple their early 2005 levels, with an even higher peak in rice prices (IMF 2008). High commodity prices quickly fed into increased costs to consumers in developed and developing countries alike (FAO 2008b), escalating into civil unrest in some, ranging from strikes in Italy to riots in Haiti. The FAO estimated that the number of undernourished people in the world increased by 75 million in 2007, mainly attributed to high food prices. This brought the proportion of people in the world without access to sufficient food back to the levels of a decade ago. Commentators from the World Bank, FAO, USDA and the World Food Programme predicted further increase in the number of undernourished people in 2008, though the true figures may not become clear for some time. Even before the price spike, the FAO estimated that more than 850 million people globally were undernourished – the great majority in developing countries (Schmidhuber and Shetty, 2005).

There is an intrinsic link between the challenge we face to ensure food security through the 21st century and other global issues, most notably climate change, population growth and the need to sustainably manage the world's rapidly growing demand for energy and water. Our progress in reducing global poverty and achieving the millennium development goals will be determined to a great extent by how coherently these long-term challenges are tackled. A key question is whether we can feed a future nine billion people equitably, healthily and sustainably. Science and technology can make a major contribution, by providing practical solutions. Securing this contribution requires that high priority be attached both to research and to facilitating the real world deployment of existing and emergent technologies. Put simply, we need a new, 'greener revolution'. Important areas for focus include: crop improvement; smarter use of water and fertilizers; new pesticides and their effective management to avoid resistance problems; introduction of novel non-chemical approaches to crop protection; reduction of post-harvest losses; and more sustainable livestock and marine production. Techniques and technologies from many disciplines, ranging from biotechnology and engineering to newer fields such as nanotechnology.

Poor harvests due to adverse weather conditions in regions such as Australia and the EU reduced the supply of cereals, particularly for wheat, and contributed to low world stocks. At the same time, the increased use of crops for biofuel production increased demand for maize. Moreover, government policies in many countries have been successful in reducing stocks from arguably excessive 1980s levels, but have reduced the capacity to respond to speculation. By 2008, global stocks had declined to a level not seen for more than 30 years. At the same time, policy responses such as export bans and other trade restrictions from some countries facing food security crises exacerbated the problems, especially in the case of rice (Beddington and Beddington, 2010).

A relatively new phenomenon is the extent to which food prices have been influenced by developments in non-food markets. Price transmission across commodities markets is illustrated by the correlation between energy and food prices. Linked to this is the role of large-scale index fund investment in commodities markets and speculation. The depreciation of the US dollar in early 2008 has also been highlighted as an important factor by some. Global population is set to increase to around nine billion by mid-century, rising at a rate of six million people per month, with Africa's population alone projected to double from one billion to two billion. This population increase combines with other transformational change, particularly in the developing world as people move from rural livelihoods to cities, cities that will need to be serviced with food, water and energy. Half the world's population now lives in cities, a figure projected to rise to 60% by 2030.

It is estimated that there will be 26 cities with greater than 10 million inhabitants in 2025, up from 19 today. Five of these new 'megacities' will be in Asia. Population increase will be coupled to increasing prosperity. Economic advances projected for the developing world will help lift millions from poverty, but in other respects will add to the challenges. As incomes rise in developing and middle-income countries, people eat more meat and dairy products, causing a rapid growth in demand for agricultural commodities to feed livestock. Strong growth in demand over the past few decades has been driven particularly by rising consumption in China and Brazil, and the future trend is likely to be strongly influenced also by the extent of income growth in India and sub-Saharan Africa, where per capita meat consumption is still low. The FAO projects total crop and livestock demand and production will rise by around 40% between 2009 and 2030, i.e., a yearly increase of 1.5%. However, this overall figure conceals the larger increase in meat demand. The overall projected rate of demand growth is lower than in previous decades, but must be met within the greater constraints on land, water and energy use outlined below. Agriculture is by far the largest user of water worldwide, at around 70% of total supplies (Beddington and Beddington, 2010).

However, as economies develop and diversify, agriculture declines in significance as a contributor to GDP and as one reflection of this in many developed countries it has ceased to be the dominant user of water. As a consequence, agriculture's share of total water demand is projected to drop to under 60% by 2050. For those developing economies still centred on agriculture, future transitions will see industry and housing put increasing pressure on agriculture's share of water, increasing competition between sectors and heightening the risk of over-extraction of groundwater. Around 1600 million hectares of land are currently cultivated for crops. The FAO estimates that, ignoring impacts on biodiversity and the C-cycle, some 2400 million hectares of land globally would be at least moderately suitable for wheat, rice and grain maize cultivation, around 18% of the total world land area. Other studies have variously suggested between 50 and 1600 million hectares of land to be suitable for agricultural expansion.

The fact that estimates range so widely reflects the major uncertainties involved in assessing the potential of land for agricultural production on a global scale. Indeed, estimates even of current land usage are diverse. It seems inevitable, however, that demand for land will progressively increase, both for food production and linked to the urbanization and energy trends. This growing competition can be illustrated by increased purchases and leases of land in India by a range of investors such as multinational companies and countries with limited domestic production potential. Perhaps the most important competitor for land is the natural environment itself. Sustainable development and agriculture requires safeguarding ecosystems and biodiversity. Forests are also among the areas particularly important from a climate change perspective, given their key role in storing C. The substantial increases in food production required will thus need to fit with efforts to reduce salinization. The need both to mitigate climate change and to adapt to that which it is too late already to avoid is clear. Global GHGs emissions must be reduced by at least 50–60% by 2050 compared with current levels. The UK set an important political lead by committing to an 80 per cent domestic reduction by 2050. The substantial increase in food production that will be required to meet global demand must therefore be achieved in parallel to delivering a steep reduction in GHGs emissions (Ludi and Stevens, 2002).

Agriculture is a major contributor to climate change, responsible for around 10–12% of emissions, even excluding the impact of salinity. Yet, the scope to reduce emissions from agriculture and specific mechanisms and technologies for this are relatively unexplored, in comparison, with energy generation and transport, and are important areas for future research and innovation. One assessment has suggested a realistic abatement potential in the UK of up to 11 Mt CO_2 by 2020, or about a quarter of current emissions from agriculture. These gains could be delivered through a combination of changing farming practices and employing new technologies, in particular focussing on better use of fertilizers, breeding programmes for crops and livestock, and improvements in drainage. Similar analysis at a global level suggests technical abatement potential of up to 60% of sectorial emissions, with C price up to E60/tonne. The realization of such potential will depend on the policies adopted by governments to incentivize change. There is some evidence that global average yields could rise with a small temperature increase (1–38C). However, this conceals large regional variations. Rice production in Asia is also likely to be hard hit by climate change, as rice is particularly vulnerable to high temperatures. Also impacting on food production in Asia is the potential loss of dry-season Himalayan glacial melt water, on which hundreds of millions of people in the Indian subcontinent are dependent. These glaciers are expected to lose 80% of their volume by 2035. Even in temperate regions, farmers will need to adapt to changing temperature and rainfall patterns, and the increased likelihood of extreme events such as floods and salinity. The impact of changing temperatures on the range of pests and diseases is uncertain, but this too cannot be ignored (Beddington and Beddington, 2010).

6. METABOLITE PROFILING AND GENE ANNOTATION: RESISTANCE CREATING TOOL IN CROPS

In recent years metabolite profiling has become routinely applicable across biology. In stark contrast to transcriptomics and proteomics, which rely to a great extent on genome information, metabolomics is mainly species independent, which means that it can be applied to widely diverse species with relatively little time required for re-optimizing protocols for a new species. Metabolite profiling has now been performed on a diverse array of plant species, including but not limited to Arabidopsis, tomato, potato, rice, wheat, strawberry, Medicago, cucumber, lettuce, tobacco, poplar and eucalyptus. Metabolite profiling used metabolite composition as a diagnostic tool to ascertain, for example, the metabolic response to herbicide, the equivalence of GE and conventional crops and the classification of plant genotypes (23–25). More recently, it has been much used in descriptions of the response of plants to a wide range of abiotic stresses like salinity. At the same time as important work is being carried out in these areas, metabolite profiling is also being used increasingly to decipher gene function, investigate metabolic regulation and as part of integrative analyses of the systemic response to saline perturbations. The metabolic response of plants to various stresses of abiotic origin is currently receiving increasing attention (Schauer and Fernie, 2006). This is driven partly by the increasing losses that these processes impose on agriculture and partly by the development of tools to measure such responses. Even so, the majority of such studies are still dominated by analysis at the level of the transcript, with only a minority including measurements of steady-state metabolite levels. However, a huge amount of targeted research has been carried out in the past, studying metabolites at the level of single metabolic pathways that are thought by the researcher to be important within a given process; this work sets the foundation for the broader analyses that have been carried out in years. Alterations in the metabolome of a wide number of species in response to a broad range of stresses have been analysed. The stress that has been best characterized using metabolite profiling is the response and acclimation to salinity stress, which has been the subject of several investigations. Various approaches have been used, for example, the kinetic saline-tolerance response and the induction of salinity tolerance were determined by ecotype of Arabidopsis, and comparative metabolite profiling using the same technology was carried out in ecotypes which are known to be differentially salinity tolerant. These reports have subsequently been complemented by a non-biased analysis of the low salinity metabolome performed by Fourier transform ion cyclotron resonance mass spectrometry. However, even though, in principle, far more metabolites can be detected using this technology, the number that was documented to change in saline stress.

Metabolite profiling provides direct functional information on metabolic phenotypes and indirect functional information on a range of phenotypes that are determined by small molecules, for example, the discovery of functional threonine aldoses. Given this, metabolite profiling has great potential as a tool for functional genomics (Fernie *et al.*, 2004). This research aim has probably received more attention than any other in the metabolomics field in recent years. Although early studies attempted gene annotation by biochemical assay, this approach has not been commonly applied, probably because of the time and labour intensity needed and the difficulty in establishing sensitive and specific assays on a broad scale. Many studies have found that analysis of the metabolite composition of mutants can aid the assignment of functions to genes. In plants, several genes have been annotated on the basis of correlations between transcript and metabolite levels or simply of altered metabolite profiles. So far these approaches have proved more informative in secondary metabolism. Altered metabolite profiles could be correlated to specific genes and heterologous overexpression could then be used to prove gene function.

7. CONCLUSION

Agriculture development needs to focus on reducing GHGs emissions through measures, such as significant reduction of deforestation; improving forest conservation and management; effective control of wildfires; high salinity; promotion of agro-forestry for food or energy; soil carbon sequestration; restoring land through controlled grazing; improving nutrition for ruminant livestock; efficient management of livestock waste; and developing stratcgics thc conscrvc soil and water resources by improving their quality, availability and efficiency of use. Sustainable agriculture and food security are two important aspects of ensuring food for all and food guarantee for all. The future of agriculture will be in jeopardy, if the natural resources like soil and water over-exploited or become saline. The depletion of groundwater will not only raise cost of production, but also crease severe environmental crisis. The excessive use of fertilizers and pesticides will retard soil fertility due to excessive salinity.

The following general conclusions emerge: (a) an important issue to address is how the tolerance to specific salinity-prone area is assessed, and whether the achieved tolerance compares to existing tolerance. The biological cost of production of different metabolites to cope with stress and their effect on yield should be properly evaluated; (b) a well-focused approach combining the molecular, physiological and metabolic aspects of salinity stress tolerance is required for bridging the knowledge gaps between short- and long-term effects of the genes and their products, and between the molecular or cellular expression of the genes and the whole plant phenotype under stress; and (c) through understanding of the underlying physiological processes in response to different abiotic stresses can efficiently/successfully drive the choice of a given promoter or transcription factor to be used for transformation.

Acknowledgements

We thank eminent authorities whose works have been consulted and whose ideas and insights have richly contributed to this work, and my research partners who have shared productively my interest in the study. Financial support from the Indian Council of Medical Research, New Delhi to Taqi Ahmed Khan in the form of S.R.F (45/14/2011-BIO/BMS) and Mohd Mazid in the form of J.R.F (0354-40/2008.E.U-39) is gratefully acknowledged.

References

Aggarwal, P. K., and Mall, R. K. (2002). Climate change and rice yields in diverse agro-environments of India. II. Effect of uncertainties in scenarios and crop models on impact assessment. *Climatic Change*, 52(3): 331-343.

Aggarwal, P. K., Joshi, P. K., Ingram, J. S. I., and Gupta, R. K. (2004). Adapting food systems of the Indo-Gangetic plains to global environmental change: key information needs to improve policy formulation. *Environmental Science & Policy*, 7(6): 487-498.

Aldesuquy, H. S., and Ibrahim, A. H. (2008). Water relations, abscisic acid and yield of wheat plants in relation to the interactive effect of seawater and growth bioregulators. *Journal of Agronomy and Crop Science*, 187(2): 97-104.

Alkon, A. (2008). Paradise or pavement: the social constructions of the environment in two urban farmers' markets and their implications for environmental justice and sustainability. *Local Environment*, 13(3): 271-289.

Ashraf, M., and Foolad, M. R. (2005). Pre-Sowing Seed Treatment—A Shotgun Approach to Improve Germination, Plant Growth, and Crop Yield Under Saline and Non□Saline Conditions. *Advances in Agronomy*, 88: 223-271.

Ashraf, M., Karim, F., and Rasul, E. (2002). Interactive effects of gibberellic acid (GA3) and salt stress on growth, ion accumulation and photosynthetic capacity of two spring wheat (*Triticum aestivum* L.) cultivars differing in salt tolerance. *Plant Growth Regulation*, 36(1): 49-59.

Attri, S. D., and Rathore, L. S. (2003). Simulation of impact of projected climate change on wheat in India. *International Journal of Climatology*, 23(6): 693-705.

Balconi, C., Hartings, H., Lauria, M., Pirona, R., Rossi, V., and Motto, M. (2007). Gene discovery to improve maize grain quality traits. *Maydica*, 52(3): 357-360.

Bancal, M. O., Robert, C., and Ney, B. (2007). Modelling wheat growth and yield losses from late epidemics of foliar diseases using loss of green leaf area per layer and pre-anthesis reserves. *Annals of Botany*, 100(4): 777-789.

Beddington, J., and Beddington, J. (2010). Food security: contributions from science to a new and greener revolution. *Philosophical Transactions of the Royal Society B: Biological Sciences*, 365(1537): 61-71.

Byrnes, B. H., and Bumb, B. L. (1998). Population growth, food production and nutrient requirements. *Journal of Crop Production*, 1(2): 1-27.

Egamberdieva, D. (2009). Alleviation of salt stress by plant growth regulators and IAA producing bacteria in wheat. *Acta Physiologiae Plantarum*, 31(4): 861-864.

Fernie, A. R., Trethewey, R. N., Krotzky, A. J., and Willmitzer, L. (2004). Metabolite profiling: from diagnostics to systems biology. *Nature Reviews Molecular Cell Biology*, 5(9): 763-769.

Gifford, R. M., and Evans, L. T. (1981). Photosynthesis, carbon partitioning, and yield. *Annual Review of Plant Physiology*, 32(1): 485-509.

Hamdia, M. A., and Shaddad, M. A. K. (2010). Salt tolerance of crop plants. *J Stress Physiol Biochem*, 6: 64-90.

Hare, P. D., and Cress, W. A. (1997). Metabolic implications of stress-induced proline accumulation in plants. *Plant Growth Regulation*, 21(2): 79-102.

Holá, D., Rothová, O., Kočová, M., Kohout, L., and Kvasnica, M. (2010). The effect of brassinosteroids on the morphology, development and yield of field-grown maize. *Plant Growth Regulation*, 61(1): 29-43.

Horváth, E., Szalai, G., and Janda, T. (2007). Induction of abiotic stress tolerance by salicylic acid signaling. *Journal of Plant Growth Regulation*, 26(3): 290-300.

Houghton, J. T., Ding, Y. D. J. G., Griggs, D. J., Noguer, M., van der Linden, P. J., Dai, X., and Johnson, C. A. (2001). *Climate change 2001: the scientific basis* (Vol. 881). Cambridge: Cambridge University Press.

Kerepesi, I., and Galiba, G. (2000). Osmotic and salt stress-induced alteration in soluble carbohydrate content in wheat seedlings. *Crop Science*, 40(2): 482-487.

Khan, T. A., Mazid, M., & Mohammad, F. (2011). A review of ascorbic acid potentialities against oxidative stress induced in plants. *Journal of Agrobiology*, 28(2): 97-111.

Knox, J. W., Hess, T. M., Daccache, A., and Ortola, M. P. (2011). What are the projected impacts of climate change on food crop productivity in Africa and S Asia. *DFID Systematic Review Final Report.*

Kurukulasuriya, P., and Rosenthal, S. (2003). Climate change and agriculture. *A Review of Impacts and Adaptations.*

Lioubimtseva, E., and Henebry, G. M. (2009). Climate and environmental change in arid Central Asia: Impacts, vulnerability, and adaptations. *Journal of Arid Environments*, 73(11): 963-977.

Livingston, D. P., Hincha, D. K., and Heyer, A. G. (2009). Fructan and its relationship to abiotic stress tolerance in plants. *Cellular and Molecular Life Sciences*, 66(13): 2007-2023.

Ludi, E., and Stevens, C. (2002). Leo Peskett and Lídia Cabral Overseas Development Institute.

Malhi, Y., Roberts, J. T., Betts, R. A., Killeen, T. J., Li, W., and Nobre, C. A. (2008). Climate change, deforestation, and the fate of the Amazon. *Science*, 319(5860): 169-172.

Mitchell, T. D., Carter, T. R., Jones, P. D., Hulme, M., and New, M. (2004). *A comprehensive set of high-resolution grids of monthly climate for Europe and the globe: the observed record (1901–2000) and 16 scenarios (2001–2100)* (Vol. 55, p. 30). Tyndall Centre Working Paper.

Neill, S. J., Desikan, R., Clarke, A., Hurst, R. D., and Hancock, J. T. (2002). Hydrogen peroxide and nitric oxide as signalling molecules in plants. *Journal of experimental botany*, 53(372): 1237-1247.

Oertli, J. J. (1987). Exogenous application of vitamins as regulators for growth and development of plants—a review. *Zeitschrift für Pflanzenernährung und Bodenkunde*, 150(6): 375-391.

Parry, M. L., Rosenzweig, C., Iglesias, A., Livermore, M., and Fischer, G. (2004). Effects of climate change on global food production under SRES emissions and socio-economic scenarios. *Global Environmental Change*, 14(1): 53-67.

Parvaiz, A., and Satyawati, S. (2008). Salt stress and phyto-biochemical responses of plants-a review. *Plant Soil and Environment*, 54(3): 89-93.

Sacks, W. J., Deryng, D., Foley, J. A., and Ramankutty, N. (2010). Crop planting dates: an analysis of global patterns. *Global Ecology and Biogeography*, 19(5): 607-620.

Schauer, N., and Fernie, A. R. (2006). Plant metabolomics: towards biological function and mechanism. *Trends in Plant Science*, 11(10): 508-516.

Schmidhuber, J., and Shetty, P. (2005). The nutrition transition to 2030. Why developing countries are likely to bear the major burden. *Acta Agriculturae Scand Section C*, 2(3-4): 150-166.

Shahba, Z., Baghizadeh, A., Vakili, S. M., Yazdanpanah, A., and Yosefi, M. (2010). The salicylic acid effect on the tomato (*Lycopersicum esculentum* Mill.) sugar, protein and proline contents under salinity stress (NaCl). *J. Biophys. Struct. Biol*, 2: 35-41.

Sharma, R. (2002). Reforms in Agricultural extension: new policy framework. *Economic and Political weekly*, 9: 3124-3131.

Smit, B., and Skinner, M. W. (2002). Adaptation options in agriculture to climate change: a typology. *Mitigation and Adaptation Strategies for Global Change*, 7(1): 85-114.

Tilak, K. V. B. R., Ranganayaki, N., Pal, K. K., De, R., Saxena, A. K., Nautiyal, C. S., and Johri, B. N. (2005). Diversity of plant growth and soil health supporting bacteria. *Current science*, 89(1): 136-150.

Tuna, A. L., Kaya, C., Ashraf, M., Altunlu, H., Yokas, I., and Yagmur, B. (2007). The effects of calcium sulphate on growth, membrane stability and nutrient uptake of tomato plants grown under salt stress. *Environmental and Experimental Botany*, 59(2): 173-178.

Xiong, L., and Zhu, J. K. (2002). Molecular and genetic aspects of plant responses to osmotic stress. *Plant, cell & environment*, 25(2): 131-139.

6

Implications and Ways to Enhance Nutrient Use Efficiency Under Changing Climate

Kannan Pandian[1*], Paramasivam Arunachalam[1] and Mahalingam Govindaraj[2,3]

[1]*Dryland Agricultural Research Station, Tamilnadu Agricultural University Chettinad, Tamilnadu, India*

[2]*Centre for Plant Breeding and Genetics, Tamilnadu Agricultural University Coimbatore, Tamilnadu, India*

[3]*Millet Breeding, ICRISAT, Patancheru, Andhra Pradesh, India*

ABSTRACT

Introduction of fertilizer and fertilizer responsive crop varieties paved way for green revolution and it was responsible for doubling the crop production to meet the food grain needs of ever growing global population. Over the four decades reduced use of organic manure and injudicious use of inorganic fertilizer led to declining trend of organic carbon and pose serious global warming threats to the environment as well as mankind.

Among the mineral fertilizer, nitrogen fertilizer occupies a major share in the global scenario in terms of production and application, it is the important crop responsive fertilizer at the same time overdose and improper application may cause N_2O emission and NO_3 contamination in groundwater. Phosphorus and potassium fertilizers are applied in large quantity next to nitrogen. Continuous use of primary nutrients and negligence of secondary (Ca, Mg and S) and micronutrients (Fe, Zn, B, Cu, Mn and Mo) application exposed severe multi-nutrient deficiency in the soil. Changing climate highly influence the crop phenology, physiology, root morphology and uptake mechanism. Therefore it is inevitable to understand the climate change mediated changes in soil nutrient availability and plant nutrient uptake in different ecosystem.

In the past two decades decreasing trend of nutrients utilization efficiency and deficiency of micronutrients rendered the land barren under changing climate. But direct impact of climate change on nutrient availability and uptake mechanism has not been understood. In this context, developing or modifying the management strategies to enhance the use efficiency of applied nutrients is of paramount importance. Climate smart management techniques such as appropriate timing of nutrient application, using efficient irrigation practices, converting to nitrogen-fixing plants as cover crops and crop rotation. Employing advanced fertilization techniques such as controlled-release fertilizers and nitrification inhibitors enhances crop productivity in sustainable manner. All these management techniques need to be tested through different quantitative measures of nutrient use efficiency of

*Corresponding author E-mail: pandian.kannan@gmail.com

Roychowdhury R. (Ed.), Crop Improvement in the Era of Climate Change.

crops under different soil conditions. Development of smart fertilizer delivery system for all the crops in different cropping system needs to be explored. Tailoring of higher nutrient use efficiency cultivars of different crops should to be accelerated using advanced breeding techniques to mitigate adverse influence of climate change on crop productivity improvement.

Keywords: Biofertilizer, climate change, fertilizer, nutrient use efficiency, soil management.

1. IMPORTANCE AND PROSPECTS OF FERTILIZER AND BIOFERTILIZER IN AGRICULTURE

Global food production now faces greater challenges than ever before due to changing climate, increasing land degradation and decreasing nutrient use efficiency. There is no single solution to delivering increased crop productivity while improving resource use efficiency and protecting environmental quality. United Nations convention on climate change defined climate change as "change of climate which is attributed directly or indirectly to human activity that alters the composition of the global atmosphere and which is in addition to natural climate variability observed over comparable time periods (IPCC, 2007).

Agricultural crops are exposed to many stresses that are induced by both biotic and abiotic factors. These stresses decrease yields of crops and represent barriers to the introduction of crop plants into areas that are not suitable for crop cultivation. The occurrence and activity of soil microorganisms are affected by a variety of environmental factors as well as plant-related factors. Abiotic stresses include high and low temperature, salinity, acidity, drought, flooding, ultraviolet light and heavy metals. The yield losses associated with abiotic stresses can reach 50 to 82 per cent depending on the crop.

Atmospheric rise of CO_2, N_2O and CH_4 affect agriculture and food production especially during the last decade by increasing global temperature, which led to uncertainty in monsoonal rainfall and increasing the frequency of drought, storms, floods, forest fires and insect pest and diseases besides deteriorating plants and animals habitats. In total climate change severely affect the crop productivity and livelihood security of the farmers in semi-arid tropics of the globe. Among the agriculture practices, fertilizer application plays a major role in green house gases (GHGs) emission. Under changing climate carbon sequestration and reducing N_2O emission through suitable soil and agronomic management techniques are the need of the hour to maximize the nutrient use efficiency and crop yield.

Nutrient mining is a major cause for low crop yields in parts of the developing world. Especially nitrogen and phosphorus move beyond the bounds of the agricultural field because the management practices fail to achieve good congruence between nutrient supply and crop nutrient demand. If it is unchecked, it severely affects the food production. Hence, increasing nutrient use efficiency continues to be a major challenge for agriculture.

Changing climate highly influences the plant growth and nutrients must be available in sufficient and balanced quantities. Soils contain natural reserves of plant nutrients, but these reserves are largely in unavailable forms to plants and only a small portion is released each year through biological activity or chemical processes. This release is too slow to compensate for the removal of nutrients by agricultural production and to meet crop requirements. Therefore, fertilizers are designed to supplement the nutrients already present in the soil. The use of chemical fertilizer, organic fertilizer or biofertilizer has its own advantages under changing climate and its advantages are to be integrated in order to make optimum use of each type of fertilizer and achieve balanced nutrient management for different crops under this starving situation.

2. SCENARIO OF FERTILIZER AND BIOFERTILIZER IN THE GLOBAL CONTEXT

2.1 Nitrogen

World nitrogen fertilizer demand was to increase at an annual rate of about 2.6 per cent upto 2012, for an overall increase of 11 million tonnes N. North America will remain the world's largest nitrogen importer. Consumption in the sub-region is projected to grow at about 1 per cent per year and at almost 3 per cent per year in Latin America. West Europe is projected to show zero growth, while West Asia is forecasted to record the fastest growth in the world at almost 5 per cent per year. Nitrogen fertilizer consumption is also projected to grow at a high annual rate of over 4 per cent in Africa and West Asia. South Asia's nitrogen fertilizer consumption is projected to increase by over 3 per cent per year. The share of urea in nitrogen fertilizer consumption is expected to continue to increase.

2.2 Phosphate

The expected annual growth rate in world demand for phosphate fertilizers was about 2.8 per cent until 2012 (IFA, 2011), for an increase of 5 million tonnes P_2O_5 compared with 2006. About 58 per cent of this growth will take place in Asia; consumption growth in South Asia is projected to surpass growth in East Asia at almost 5 per cent per year. Rapid growth will also occur in East Europe and Central Asia and in Latin America. Phosphate fertilizer consumption will continue to decline marginally in West Europe.

2.3 Potash

World demand for potash fertilizers is projected to increase at an annual average rate of about 2.7 per cent, equivalent to an increment of 3.5 million tonnes. Major fertilizers and raw materials producing countries and its capacity are given in the Fig. 1.

Global use of N and P fertilizer increased 7- and 3.5-fold between 1960 and 1995 and we experience the consequences of this increase in greatly magnified ecological problems. Clearly, we cannot just project the demand for N and P fertilizer to triple by 2050, but have to consider all means at our disposal for the intensification of agricultural production and opt for soil and ecology adapted versions. Globally, crop systems have low nutrient use efficiency with 30–50 per cent and 45 per cent of applied N and P fertilizers respectively utilized by crops (Tilman *et al.*, 2002).

2.4 Biofertilizer

The green revolution saved us from famine; now environmental sustainability is at the heart of innovation in crop production. Next-generation crop systems will increasingly rely on nutrients that supplement or replace synthetic N fertilizers and natural deposit resources. Nutrient efficient

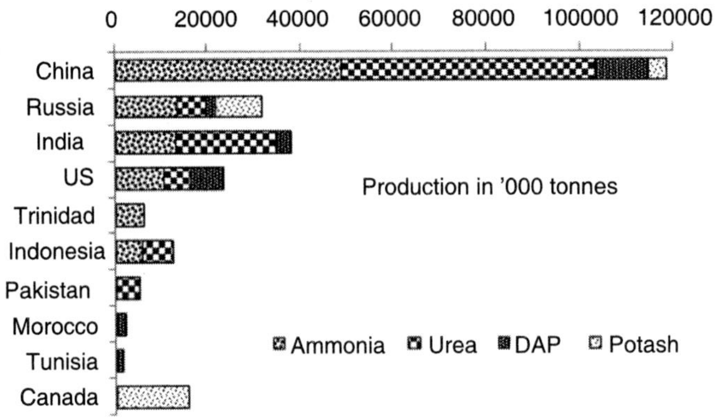

Fig. 1 Global fertilizer production trend in 2010.

crop systems have to integrate microbial symbionts, appropriate soil biota and diverse nutrient sources to combat effect of climate change on soil health and crop production.

Soil microorganisms play a significant role in regulating the dynamics of organic matter decomposition and the availability of plant nutrients such as N, P and S. It is well recognized that microbial inoculants constitute an important component of integrated nutrient management that leads to sustainable agriculture. Biofertilizers represent a promising alternative to synthetic fertilizers. It includes microorganisms, such as bacteria, fungi, cyanobacteria, algae and their metabolites that are capable of enhancing soil fertility, crop growth and yield. Application of biofertilizers to agricultural land could increase the amount of carbon stored in the soils. It contributes significantly to the reduction of greenhouse gas emissions by eliminating the requirement of fossil fuels for production of N: P: K fertilizers. Further, its application through designer seed mechanism and soil application enhance the plant growth promoting bacteria and fungi which are directly responsible for applied fertilizer transformation and its use efficiency. Nowadays farmers and agriculturists are showing more interest in biofertilizers application than the agrochemicals as they are found to be harmful to the soil. The importance of biofertilizers has increased in an increasingly eco-conscious world. Besides, soil quality is improved through the application of these environmental friendly fertilizers and also it contributes in reducing the negative impact of global warming.

3. IMPLICATIONS OF FERTILIZERS AND BIOFERTILIZERS UNDER CHANGING CLIMATE

3.1 Response of Fertilizer Under Changing Climate

Soil degradation is occurring due to inadequate and imbalanced fertilization leading to nutrient mining and development of second generation problems in nutrient management. The estimated supply-demand gap was about 1.8 million tonnes of N and 1.9 million tonnes of phosphorus by 2011-12 and it will be increased in the coming years ahead. Globally it is of concern that the partial factor productivity of fertilizers has been continuously declining. The efficiency of fertilizer N is only 30-40 per cent in rice and 50-60 per cent in other cereal crops. The efficiency of fertilizer P is 15-20 per cent in most of the crops and K use efficiency is 60-80 per cent. Low nutrient recovery efficiency not only increases the cost of crop production but also causes environmental pollution.

Inadequate application of potassium combined with over-excessive application of nitrogen is a serious problem in modern intensive agricultural production systems. It leads to large N losses, environmental pollution and low nitrogen use efficiency. Although fertilizer consumption is increasing quantitatively, the corresponding yield increase per unit of nutrient has diminished over the years (Sharma and Sharma, 2011). The response ratio in foodgrain crops in irrigated areas in India (Fig. 2) substantially declined between 1960 and 2008 (Biswas and Sharma, 2008). Improvement of the NUE is of paramount importance both for economical as well as environmental reasons. The overall fertilizer response in irrigated areas of the country decreased nearly three times from 13.4 kg grain per kg NPK in 1970 to 3.7 kg grain per kg NPK in 2005.

Efficient use of nutrient inputs on a farm is important for improving crop productivity and mitigating the climate change. Enhancing nitrogen efficiency directly responsible for nitrous oxide emission reduction and reducing agriculture's contribution to climate change. Nutrient use efficiency of N, P, K and Mg in all plant parts was significantly increased at elevated CO_2 (SamanSeneweera, 2011).

According to Pregitzer and King (2005), changes in plant and microbial function will be very tightly coupled to changes in soil temperature and moisture in the future and these ecological interactions will play critical roles in nutrient availability and net ecosystem productivity. Impact of climate change on nutrient availability and its use efficiency is illustrated in Fig. 3.

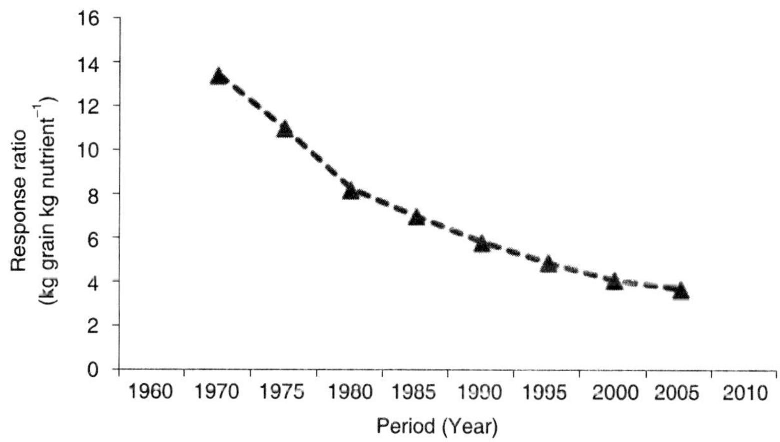

Fig. 2 Fertilizer response of foodgrain crops in irrigated areas in India.

Soil degradation associated with inorganic N fertilizer includes loss of soil organic matter (SOM), declining soil pH and declining crop yields (Guo *et al.*, 2010). Loss of SOM in degraded agricultural soils affects aeration, structure, nutrient availability and microbial diversity. Preventing SOM depletion has been a challenge since the beginning of intensive agriculture. Recycling of organic residues to soil contributes to replenishing of SOM. Adverse effects of continued

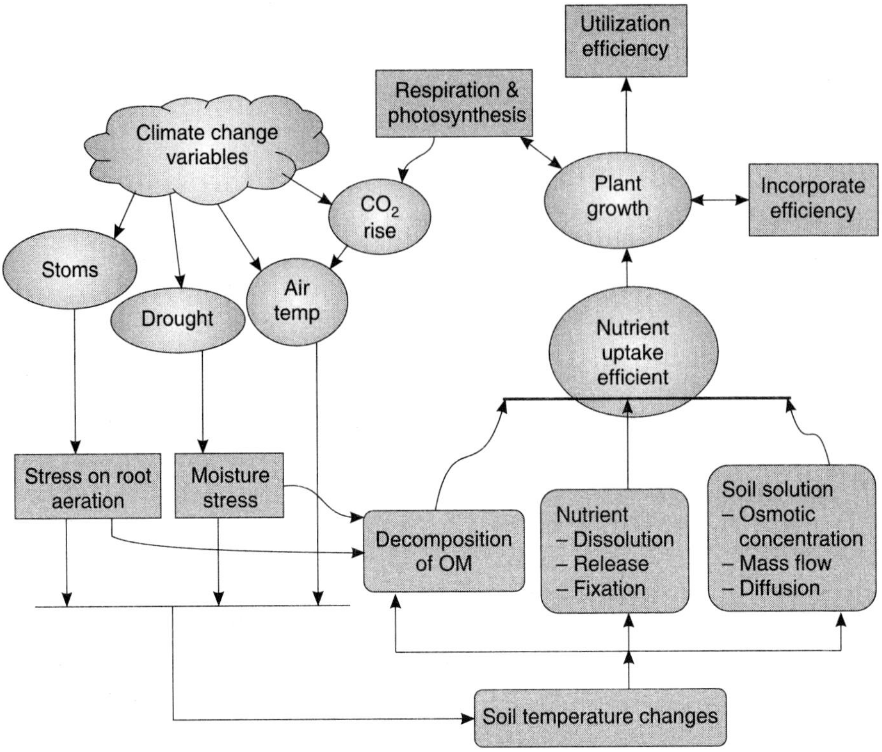

Fig. 3 Impact of climate change on nutrient availability, uptake and its use efficiency.

application of synthetic fertilizers was aptly put by proverb: fertilizer makes the father rich and the son poor". But nitrogen has been reported to play an important role in soil C storage, both by promoting crop dry matter production and by chemically stabilizing C in the soil. Many experiments have shown that fertilizing crops with N results in higher levels of soil C over time. In 25-year study, Wilts *et al.* (2004) reported that total soil organic C declined for all treatments, but at a slower rate in the fertilized treatments than in the unfertilized control and it is depicted in Fig. 4.

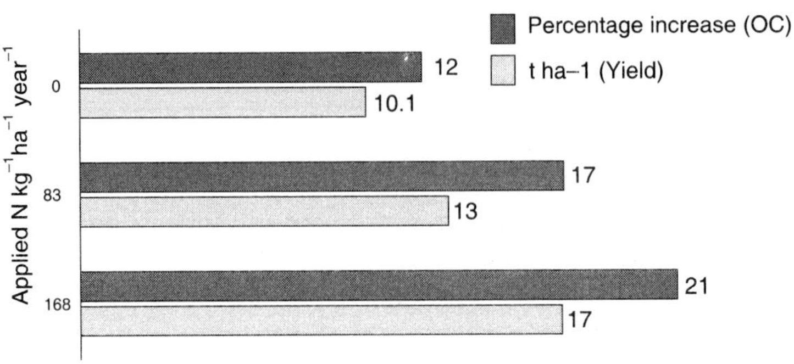

Fig. 4 Soil organic carbon builds up in 30 years corn field and yield performance.

SOM is a key factor for soil fertility due to its role for soil structure, biological processes and nutrient cycling, providing sink and source for nutrients and energy for soil organisms and plants. Long-term trials have shown that yields are often higher in soils with higher SOM (Johnston *et al.*, 2009). Organic cropping systems that supply predominantly organic materials and legumes as nutrient sources have been compared with conventional systems resulted that average yields of 5 to 34 per cent lower in organic compared with conventional systems (Seufert *et al.*, 2012). Compared to conventional agriculture, organic systems are more likely to be N limited, which may reduce off-site N losses.

Improving the efficiency and effectiveness of crop N use can potentially reduce N_2O emissions, by reducing the potential for elevated residual NO_3- N in the soil profile (Dobermann, 2007; Dobermann and Cassman, 2005). Better nutrient management, including fertilizer best management practices, can potentially increase crop N recovery and minimize the cascade to air and water resources. Improvements in N use efficiency can also lead to greater C sequestration and reductions in CO_2 emissions.

Both N fixation and denitrification are affected by temperature, N availability and soil moisture. In developing regions, N fertilizer use was lesser in the early 1960s and increased exponentially during the course of the green revolution. The large increase in N use since the 1960s resulted in a steep decrease in Partial Factor of Productivity of Nitrogen (PFPN) in all developing regions. Regional N rates on cereals range from less than 10 kg N/ha in Africa to more than 150 kg N/ha in East Asia and with the exception of Africa, PFPN continues to decline in all developing regions at rates of –1 to –2 per cent per year (Dobermann and Cassman, 2005). The very high PFPN in Africa (122 kg/kg N applied) and Eastern Europe and Central Asia (84 kg/kg N applied) are indicative of unsustainable soil N mining due to low N rates used at present. In India, PFPN seems to have levelled off in recent years, but in many other developing countries it continues to decline because public and private sector investments in better technologies, services and extension education are far below those made in developed countries. Except in research and limited on-farm demonstrations, there are no documented cases for country-scale increase in N use efficiency. Global fertilizer production trend was explained in Table 1.

Controlled release or enhanced efficiency of fertilizers generally work by controlling the speed at which fertilizer or a coating applied to it, dissolves in soil water. By affecting the timing of nitrogen release from fertilizer, these compounds have the potential to reduce the loss of nitrogen and therefore improve nitrogen use efficiency. Similarly, soluble fertilizers formulated with inhibitors reduce or block the conversion of nitrogen species by affecting specific types of microbes involved. This helps to keep nitrogen in the form of ammonium longer, encouraging uptake by crops and helping to prevent N_2O emissions from either nitrification or denitrification.

The different nitrification inhibitors such as dicyandiamide, nitrapyrin and wax coated CaC_2 in combination with urea were tested to quantify the nitrous oxide emission in corn, rice, wheat and barley crops. Among the N inhibitors urea in combination with wax coated CaC_2 reported low N_2O emission over the urea alone application (Mosier *et al.*, 1994).

Exploitation of specially designed mineral fertilizer such as nitrification inhibitors, coated fertilizer and urease inhibitors certainly curtail the global green house gases (GHGs) emissions under variable climate.

Table 1 Global N fertilizer production and N use efficiency in cereals

Countries	*Total fertilizer production (Mt)*	*N application rate (kg N/ha)*	*N use by cereals (Mt)*	*Total share of N consumption by cereals (%)*	*Partial productivity of N (kg/kg)*
North America	12.5	112	8.3	66	45
North East Asia	0.9	89	0.3	32	68
West Europe	9.5	113	4.3	45	49
East Europe Central Asia	4.9	25	2.5	21	84
Oceania	1.3	48	0.9	67	40
Africa	1.4	9	0.8	56	122
West Asia North East Africa	4.2	68	2.4	56	34
South Asia	14.6	58	7.3	50	41
South East Asia	4	65	2.8	71	49
East Asia	24.9	155	14.5	58	31
Latin America	5.1	55	2.7	53	53
World average	83.2	10	46.7	57	44

Combined application of organic and inorganic nutrients enhanced the yield of groundnut and maize crops in India. Integrated nutrient management practices in the System of Rice Intensification enhanced the crop growth and yield by maintaining favourable soil physio-chemical properties and high water productivity, which in turn lowering cost of production. Application of farmyard manure along with inorganic fertilizers maintained SOM and long-term productivity with high microbial biomass in groundnut based rain-fed cropping system (Kannan *et al.*, 2011). The integrated use of organic and synthetic fertilizers with bio-inoculants may be the best option for improving the soil fertility and crop production under changing climate. Some of the report explained that organic sources application not able to synchronize nitrogen release with crop demands and may increase N_2O emissions in comparison with synthetic fertilizers under certain circumstances (Flynn, 2009).

Practice of integrating green manures, planting nitrogen-fixing crops and incorporating livestock manures into the soil, decreases the amount of nitrogen lost through run-off and emissions of nitrous oxide. Applying these management practices can serve adaptation needs by improving soil quality and also decreasing farmers' costs and dependence on outside inputs.

Phosphorous requirement of plants may increase under the rising atmospheric CO_2 and temperature as indicated by the significantly higher uptake of P despite the reduced yield of wheat under elevated CO_2 and temperature. Although certain compensatory responses by plants, in terms of better root growth and higher recovery efficiency of applied P, may be elicited in response to higher P requirement. The benefits of such responses may get offset by the reduction in P utilization efficiency and overall P use efficiency under elevated carbon dioxide and temperature. Therefore, it underlines the need of P-efficient crop cultivars and suitable nutrient management practices for sustaining crop productivity in a changing climate (Manojkumar *et al.*, 2011).

Plant nutrient acquisition in an elevated CO_2 environment will be regulated by the interaction of compensatory adjustments in root growth, lifespan, physiology and symbiotic association. P concentration in rice grown under elevated CO_2 was found to decline by 5 per cent compared to 28 per cent decline for Zn and 17 per cent for Fe (Seneweera and Conroy, 1997). In his recent study, found no change in P concentration in rice grown under elevated CO_2. Grain P concentration decreased by 3.7 per cent compared to 30 per cent decline for Fe and 15.1 per cent for Zn (Fangmeier *et al.*, 1997). Hence higher doses of P fertilization are considered inevitable to sustain higher crop productivity under elevated CO_2.

Some of the study results confirming that there is no reduction or very small reduction in grain P content in comparison to severe reduction in Zn and Fe concentration. Grain elemental composition of five wheat cultivars grown under elevated CO_2 showed that 2.5 per cent decline in P concentration as against 15 per cent for Fe and 21 per cent for Zn (Loladze, 2002).

Optimized fertilizer management and its significance in climate change

- Fertilizers are used to maximize the increase crop yields, but it consumes more fossil fuel energy for their production. Since fertilizer use itself involves energy consumption, so as to reduce GHGs emission optimum fertilizer application need to be popularized.
- Application of fertilizer enhances the net primary productivity of cropland, which in turn, increases the return of crop residue C to the soil.
- Demand of biofuels increases the need for higher biomass production per unit of land area, it also increases fertilizer use. Efficient fertilizer management is to offset the net energy production or fossil fuel use under prevailing climatic variability.

3.2 Response of Biofetilizer Under Changing Climate

Biofertilizers are important components of integrated nutrients management. These potential biological fertilizers would play a key role in productivity and sustainability of soil and also protect the environment and cost effective inputs for the farming community under changing climate. Biofertilizers are products containing living cells of different types of microorganisms and it is applied through seed, soil and plants, which colonize the rhizosphere, interior of the plant and promotes growth by converting nutritionally important elements such as nitrogen, phosphorus from unavailable to available form through biological process such as nitrogen fixation and solubilization of fixed soil reserved phosphate. Many studies have shown that biofertilizers maintain the soil health and enhances crop yield through the following means:

- increased nutrient transfer
- better crop yield
- enhanced population of beneficial microorganisms
- stabilization of soil aggregates

Nitrogen fixing organisms are used in biofertilizer as a living fertilizer composed of microbial inoculants or groups of microorganisms which are able to fix atmospheric nitrogen. They are grouped into free-living bacteria (azotobacter and azospirillium), the blue green algae (BGA) and symbionts such as rhizobium, frankia and azolla (Gupta, 2004). Many experimental reports revealed that rice yields are increased by 0.5-2 t ha^{-1} due to azolla application. Azobacter and Azospirillum also fix atmospheric nitrogen in cereal crops without any symbiosis. El-Komy

(2005) demonstrated the beneficial influence of co-inoculation of *Azospirillum lipoferum* and *Bacillus megaterium* for providing balanced nitrogen and phosphorus nutrition to wheat plants. Combined application of organics and blue green algae not only recorded higher yield, but found to emit less methane in paddy cultivation than the application of organics alone. Application of BGA and *azolla* reduced methane flux without reducing rice yields and can be used as a practical mitigation option for minimizing the global warming potential of rice ecosystem and increasing N use efficiency.

Phosphate solubilizing bacteria (PSB) are capable of transforming soil P to the forms available to plant. Microbial biomass assimilates soluble P and prevents it from adsorption or fixation (Khan and Joergesen, 2009). Among the soil bacterial communities, ectorhizospheric strains from pseudomonas and bacilli and endosymbiotic rhizobia have been identified as effective phosphate solubilizers. Strains from bacterial genera pseudomonas, bacillus, rhizobium and enterobacteria along with penicillium and aspergillus fungi are the most powerful P solubilizers (Whitelaw, 2000).

Soil application of bacteria with mycorrhizae increased the biological yield and miximized the grain weight of barley (Mehrvarz *et al.*, 2008). Single and dual inoculation of biofertilizer with P fertilizer was 30-40 per cent better than P fertilizer alone for improving grain yield of wheat and dual inoculation without P fertilizer improved grain yield up to 20 per cent against sole P fertilization (Afzal and Bano, 2008). Rhizospheric microorganisms can interact positively in promoting plant growth, as well as N and P uptake. Triple inoculation of *Bradyrhizobium,Glomus fasciculatum* and *Bacillus subtilis* enhanced the seed yield to the tune of 24 per cent in greengram (Zaidi and Khan, 2006).

Mycorrhizal symbioses are nearly ubiquitous in plants and play important roles in mobilization of nutrients from organic matter and acquisition of diffusion limited nutrients, mainly P and Zn. Numerous studies reported increased resource allocation to mycorrhizal roots in plants exposed to elevated CO_2 and this may reflect the relative importance of edaphic constraints to plant growth in most ecosystems. Arbuscular mycorrhizal fungi enhance the decomposition of organic material and retrieve organic N and P for plants. Root exudation of protons, organic acids and extracellular enzymes responsible for improved use of organic nutrients over the inorganic fertilizers. Atmospheric CO_2 enrichment increases the production of a protein called glomalin by common root-dwelling soil fungi and this protein provides a wide range of benefits, including enhancing the stability of soil aggregates, which improves the capacity of soils to store more carbon and better preserve it from further decomposition and decreasing the risk of potentially toxic elements to soil microorganisms and plants.

Biochar application change soil nutrient availability and bio-available nutrients such as N, P and metal ions by altering the soil physicochemical properties. Increases in soil nutrient availability may result in enhanced host plant performance and elevated tissue nutrient concentrations in addition to higher colonization rates of the host plant roots by arbuscular mycorrhizal fungi (DeLuca *et al.*, 2006).

Different mechanisms of rhizosphere microbes positively responded for plant growth and crop yield. Plants can utilize benefits provided by the microbes; agricultural productivity to be enhanced by fully exploiting beneficial microbial benefits such as production of phytohormones, induced systemic resistance and alteration of plant functional traits.

3.3 Nutrient Interaction Under Changing Climate

Nutrient availability may provide one of the most critical controls on the net balance between plant and soil processes. In systems with low soil nitrogen availability, additional nitrogen from deposition is thought to be one of the major causes of increases in crop growth and soil carbon storage in the semi-arid environment. Nutrients such as nitrogen and phosphorus are critical in controlling ecosystem carbon balance as most natural systems are nutrient limited. Thus, the limited availability of nutrients can hamper the dry matter production due to higher temperatures or CO_2 atmospheric concentrations under changing climate. Availability of nitrogen in many ecosystems is now significantly enhanced due to atmospheric nitrogen deposition and this has major consequences for the ecosystem carbon balance.

Unfortunately, the effect of N deposition and nutrient availability in general, on soil organic matter turnover remains largely overlooked by existing models. Assumptions that increased N availability will reduce organic matter decomposition rates. Implications of this greatest effect of nitrogen on carbon storage may be expected in carbon-rich, nutrient-poor systems due to both an increase in production and a decrease in the decay rate of an enlarged recalcitrant organic matter pool. The effects are likely to be considerably smaller in agriculturally managed systems, where nitrogen inputs are higher and where regular soil tillage stimulates soil organic matter turnover. Improving N use efficiency and reducing N fertilizer inputs in crop production is an important goal given the energy and greenhouse gas costs of N fertilizer manufacture and the potency of N_2O as a greenhouse gas 310 times that of CO_2. The basic management goal is to reduce N losses from the soil plant system, especially those by denitrification which is a major source of N_2O emissions from soil. Key parts of improving N efficiency are to avoid excessive N applications and to synchronize N supply with crop demand. The latter is more easily achieved when nitrogen is supplied from fertilizer than from organic N sources, where release is controlled by biological mineralization processes. In general, release of N from organic N sources continues beyond the period of crop production and can contribute to leaching losses and off-site pollution problems, including additional generation of N_2O. Research has also shown that emissions of N_2O from cropland are higher when manure is used as the N source (Duxbury *et al.*, 1982)

K cycling and recycling play an important part in NO_3^- translocation from root to shoot as counter ion and assimilate loading in the phloem. Enhancement in uptake of N through K application ultimately helps in increasing the NUE. Yield response to K application depends to a great extent on the level of N nutrition and the interaction is normally positive (Brennan and Bolland, 2009). A gain of 20 per cent in nitrogen use efficiency can easily be achieved by balanced fertilization with K. A positive relationship between N and K exists for the uptake and utilization of N by plants to form protein and amino acids which ultimately affect the quality and yield of crops. Balanced use of N, P, K and micronutrient fertilizers in cereals and other crops will not only prove more profitable for farmers but also lead to reduced environmental degradation and combat the effect of climate change.

Indirect effects of climate drivers on nutrient availability:

(i) In experimental warming, changes in nitrogen availability have been shown to increase the risk of soil carbon loss. In tundra systems warming results in an increase in the abundance of shrubs (Sturm *et al.*, 2005). This change in vegetation structure causes higher winter soil temperatures and resulting increase in microbial activity and plant-available nitrogen further promotes shrub abundance and a positive feedback loop.

(ii) Drought has been shown to reduce uptake of phosphorus and other nutrients by trees in a Mediterranean system thus increasing P-limitation of growth (Sardans *et al.*, 2008).

(iii) Extreme weather conditions affect the dynamics of nutrient release from litter may also be responsible for oscillations in annual net primary production indifferent ecosystem.

(iv) Patterns across rainfall gradients indicate that concentration of extractable nutrients generally decrease with precipitation with a widening of carbon and nutrient ratios (Austin and Vitousek, 1998). This suggests an asynchrony of carbon and nutrient dynamics driven by the different sensitivity of photosynthesis and decomposition to temperature and water availability but also the effect of rainfall and temperature on other abiotic and biotic processes specific to individual elements.

4. FACTORS GOVERNING THE FERTILIZER AND BIOFERTILIZER USE EFFICIENCY

4.1 Abiotic Factors Affecting Nutrient use Efficiency

4.1.1 Effect CO_2 Concentration on Nutrient Uptake

Uptake of NH_4 and NO_3^- seems to be differentially sensitive to high CO_2, which could influence ecosystem path towards N saturation. Increased soil temperature might increase N and P uptake capacity to a greater extent in species from warm and fluctuating soil habitats than in species from cold and stable soil environments. The few available data also indicate that increased soil temperature elicits a differential effect on uptake of NH_4^+ versus NO_3 (Bassirirad, 2000). Root is the main organ involved in water and nutrient uptake and as one of the major sinks for assimilated C, roots play a critical role in determining plant and ecosystem responses to various aspect of global change ranging from N deposition and elevated CO_2 (Van Noordwijk *et al.*, 1998).

Elevated CO_2 accelerates the crop growth through increased photosynthetic activity. Higher growth rate increase the plant nutrient demand and uptake capacity. A large number of CO_2 studies have demonstrated both a short-term stimulation of growth and increased supply of root respiratory substrates (Cruz *et al.*, 1993). Elevated CO_2 decreased foliar N and P concentration in groundnut and blackgram under rainfed alfisol condition due to the higher biomass production and dilution effect. Crop specific responses of N uptake kinetics is one of the potential mechanisms by which elevated CO_2 might affect competitive balance among co-existing crops, thereby affecting ecosystem productivity and composition.

4.1.2 Moisture Stress

- Water dependent diffusion and mass flow of nutrients to the roots slows with increasing soil moisture deficit as depicted in Fig. 5.
- Impaired root growth under deficient moisture condition decreases the capture of immobile or less mobile nutrients such as phosphorus and zinc.

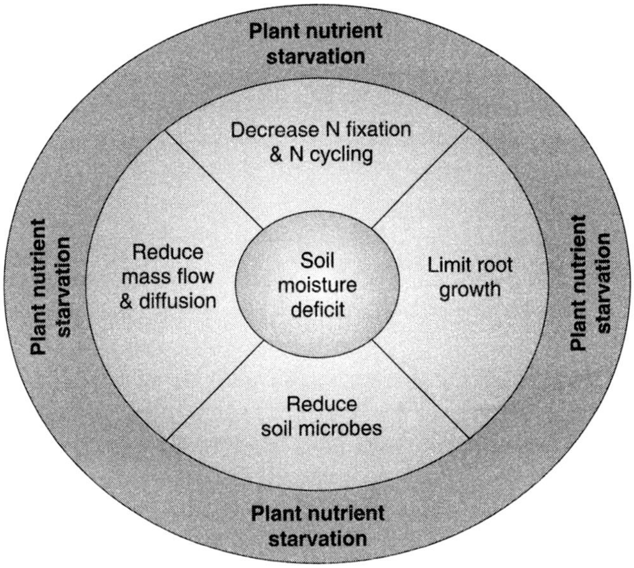

Fig. 5 Impact of soil moisture stress on plant nutrition.

- Drought inhibits N-fixation in legume crops and interrupts N cycling by hindering the soil bacterial activity.
- Root-mycorrhizae symbiosis is not overly sensitive to moderate soil moisture deficits.
- Part of the benefit provided by mycorrhizae under drought conditions is associated with increase in nutrient transfer to the roots.

4.1.3 Excessive Moisture

Impact of flooding on nutrient availability and its use efficiency was illustrated in Fig. 6.

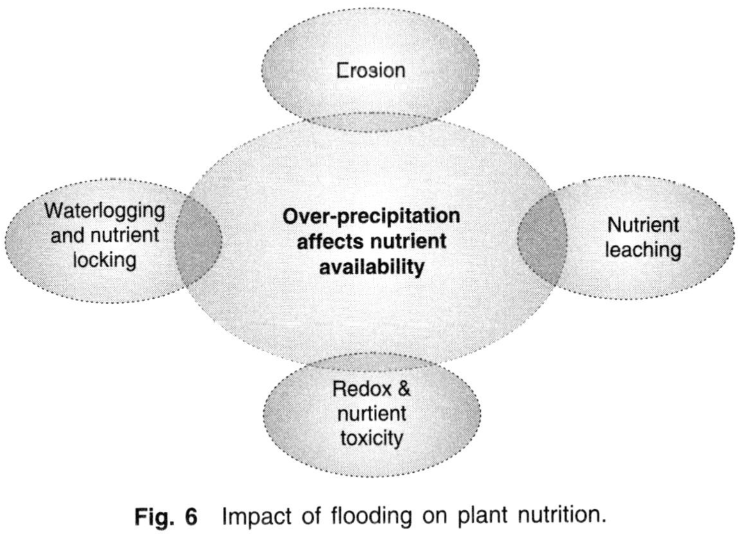

Fig. 6 Impact of flooding on plant nutrition.

- High intensity rainfall events accelerate the erosion leading to loss of nutrient rich top soil and surface applied fertilizers.
- Fastened the nitrate leaching process.
- Waterlogged soils become hypoxic.
- Oxygen dependent active transport of nutrients get slowdown and impaired the plant growth.
- Change in soil redox status can lead to Mn, Fe and Al toxicity.
- Under oxygen limited situation, nitrate is used as alternative electron acceptor, which may lead to N losses through denitrification.

4.1.4 Soil Temperature

Many studies have demonstrated that changes in soil temperature can directly affect plant nutrient acquisition by changing root transport properties for NH_4^+, NO_3^-, PO_4^{3-} and K^+. Soil temperature might alter nutrient uptake capacity and it is largely determined by the degree to which growth and root shoot ratio is affected. If changes in soil temperature lead to an increased root-absorbing surface relative to shoot size, then there will be a down-regulation of ion influx. On the other hand, if changes in soil temperature lead to a relative increase in shoot versus root size, then ion uptake is likely to increase per unit root.

Soil temperature has marked effects on root characteristics such as growth, morphology, longevity, respiration and membrane fluidity, which influence plant nutrient acquisition efficiency (Bassiri Rad, 2000). However, some research evidence showed no interactive effect of elevated CO_2 and temperature on P and N acquisition (Bassiri Rad *et al.*, 1996). Higher temperature does not modulate the nutrient dilution effect often observed in plants exposed to elevated CO_2. Temperature increases lead to higher respiration rates, shorter periods of seed formation, which in turn lowers biomass production. Impacts of temperature on ecosystems cannot be understood without taking nutrient cycling into account. Meta-analysis of soil warming experimental results showed significant positive effects on N mineralization and plant productivity (Rustad *et al.*, 2001). According to Kerkhoff *et al.* (2005) the actual net primary productivity is invariant with temperature and argues that nutrient feedbacks offset temperature effects on kinetics of photosynthesis and respiration. Modelling studies also demonstrated the importance of nutrient cycling in modulating ecosystem response to temperature. Nutrient cycling is strongly affected by precipitation and soil moisture availability (Austin and Vitousek, 1998).

4.1.5 Soil Salinity

Salinity affects the mineral nutrient uptake in plants through its effect on nutrient availability, transport, partitioning in plants and ionic imbalance due to the competition of nutrients such as K^+, Ca^{2+} and NO_3^- with the toxic ions Na^+ and Cl^-. Mineral nutrients play a vital role in determining plant resistance to salinity (Yuncai and Schmidhalter, 2005). Because it causes a similar effect on plant growth and development through a water deficit or soil moisture stress condition, potassium is equally important to maintain the turgor pressure of the plant under salinity stress. Calcium is a key signal messenger for regulating plant resistance to salinity as that of K. High K^+ : Na^+ ratios will also improve the resistance of the plant to salinity. The competition between Cl^- and NO_3^- under saline conditions means that the form of N plays a critical role in determining the growth of salinized plants. Compared to primary and secondary nutrients, micronutrients might be less important with respect to plant resistance to salinity.

4.2 Biotic Factors Affecting Nutrient Use Efficiency

4.2.1 Soil Biology

Soil fauna play a vital role in maintaining nutrient availability to crop plants. Directly below ground herbivory that can reduce fine root absorbing surface area or facilitation of root growth through earthworm channels. Indirectly, it improves the soil physical properties, nutrient transformation, population level interactions with microflora and the comminution and digestion of litter. Experimental evidence indicates that nutrient mineralization increases significantly in soil systems containing soil fauna, compared to those devoid of fauna (Carcamo *et al.*, 2001). In their association with their host plant rizhobacteria may differently colonize their host plant. They may colonize the rhizosphere, the root surface or the intercellular spaces of the host plant. The colonizing ability of rizhobacteria is determined by utilizing organic acids rather than sugars, their chemotaxis response, mobility and production of lypopolysaccharides and proteins (Lugtenberg and Bloemberg, 2004). Microbial respiration and the mineralization of N and sulphur increase over the range of temperatures in different soils. Lipson *et al.* (2000) demonstrated that winter increases microbial biomass in dry alpine meadows were followed by spring declines related to a sustained increase in soil temperature and a corresponding decrease in soluble organic carbon.

4.2.2 Root Morphology

Soil physical and chemical parameters control the soil nutrient concentrations and movement into the root system, besides soil temperature also influences nutrient uptake through direct effects on the growth and physiology of plant root systems. The amount and duration of root growth, root morphology and root spatial distributions can be affected by soil temperature. Crop plants uptake the ions such as iron, NH_4^+, NO_3^- appears to occur at higher rates near the apical portions of roots, while uptake of K^+ occurs at equal rates along the root (Clarkson, 1996). Such variation in ion uptake with root age may have important implications for plant nutrition if the average longevity of populations of fine roots is altered by environmental conditions.

4.2.3 Root Physiology

Changes in soil temperature directly affect root physiology in several ways in combination with soil-mediated controls on nutrient availability (diffusion, mass flow and mineralization) and nutrient uptake.The influx of nutrients into living cells of roots occurs by transport across the plasma lemma through high and low affinity transporter proteins that operate at low and high concentrations, respectively and are thought to be more or less specific to each ionic species (BassiriRad, 2000). In his earlier work he demonstrated that PO_4^{3-} uptake by *Eriophorum vaginatum* increased as soil temperature rose from 5 to 15°C, but decreased with further increases in temperature. In addition, other factors such as plant demand for nutrients, root system, surface area and previous N nutrition are known to modify the response of nutrient uptake to changes in soil temperature. Root respiration is known to increase with rising soil temperature (Atkin *et al.*, 2000) due to higher availability of carbohydrates from enhanced photosynthesis, providing more energy for active transport. Higher rates of root respiration result in higher concentrations of CO_2 in soil solution of which the dissociation products H^+ and HCO_3^-, promote ion exchange reactions at the surface of clay and humus particles, freeing nutrient ions for uptake (Larcher, 1995). Formation of carbonic acid as a result of increased respiration can decrease rhizosphere soil pH, which is widely known to affect the availability and uptake of essential ions, especially micronutrients (Marschner, 1995). The effect of enhanced root and microbial respiration on soil solution chemistry and plant nutrition is determined to a large extent by the buffering capacity of the soil. Properties of cell membranes also change with soil temperature, affecting nutrient uptake. There is a decrease in water uptake at low soil temperatures, caused primarily by increased resistance to water movement within the root due to higher viscosity of water and reduced permeability of cell membranes (Wan *et al.*, 2001). The climatic variables such as excess moisture, moisture stress, and high CO_2 and soil temperature were highly influence the plant metabolitic function, nutrient availability and uptake mechanisms (Clair and Lynch, 2010). Different climate change variables and their interaction on mineral nutrient availability is given in Table 2.

Table 2 Impact of weather variables on plant nutrient availability

Weather variables	*Influence on plant metabolism and nutrient availability*
Excessive moisture (flood, storm, heavy rainfall)	Nutrient leaching, removal of fertile clay and loss of surface nutrients and organic carbon by the erosion process
Deficient moisture (monsoon failure, elevated temperature)	Excessive evapotranspiration altered mass flow of nutrients to roots thereby affect the availability of NO_3, SO_4,Ca, Mg and Si
High soil temperature and CO_2	Water tension changes and influences the root absorption. Increased/ decreased root activities, which affect the availability of all nutrients, especially P and Zn
High CO_2	Root absorption and exudation process of soil microbes especially mycorrhiza, which enhances the availability of P and Zn
Moisture stress and high soil temperature	Alter the microbial diversity, which is directly responsible for biological N fixation and N cycling through symbiotic and non-symbiotic processes
Waterlogging /Flooding	Reduction and oxidation processes affect the nutrient availability mainly Mn, Fe, Al and B
High atmospheric temperature	Alter the plant physiological and metabolism, thereby affect the nutrient uptake from the soil
Excess precipitation and soil temperature	Enhances the solubility and capillary rise of salts from sub surface to surface, which lead to salinity

Plants accumulate nutrients from the soil solution pool and nutrients must be in solution to be mobile in the soil. In the absence of roots, steady-state solution phase concentrations of nutrient ions are controlled by adsorption-precipitation and desorption-dissolution reactions between nutrients and the surface complex of soil, mineralization and immobilization for solutes of organic origin and additions from fertilizer (Barber, 1995). The different processes and parameters controlling nutrient availability is given in Table 3. Biological transformation between organic and inorganic pools is strongly influenced by moisture and temperature and thus global climate change may strongly influence solution concentrations of N as well as S. Some studies proposed that soil C pool size will not change as increased soil respiration and decomposition caused by soil warming will be moderated by the increased C supply below ground (Kirschbaum, 2000). Increasing CO_2 may not exert a significant direct effect on N mineralization but associated warming increased N mineralization and leading to increased solution-phase N. Some studies have examined the impacts of elevated CO_2 on solution-phase K availability is not strongly controlled by biological activity. Rates of adsorption and desorption reactions will accelerate with increased temperature and changes in soil moisture may further modify reactions by altering the ionic strength of the soil solution. However, uncertainties surrounding the magnitude of temperature increases coupled with the spatial and temporal variation in soil moisture make it challenging to predict how climate change will impact plant nutrient availability.

Table 3 Impact of climate change on soil and plant factors controlling nutrient availability

Nutrient acquisition governing factors	*Soil and plant process controlling nutrient availability*	*Impact climate change variables on nutrient use*
Concentration of nutrients in soil solution	Excess moisture, drought, temperature, organic matter and microbial activity; kind, time, rate and placement of fertilizer influence mineralization and immobilization process in the nutrient cycle. Type of clay minerals, buffering capacity, temperature, pH, soil moisture and ionic strength control the adsorption or desorption of applied nutrients in the soil.	Increasing soil temperature has positive and negative impact on soil and plant process controlling nutrient availability; atmospheric rise of CO_2 enhances fine root growth and exudates alter buffering capacity of the soil. Excess or deficit of soil moisture boost or slow down the nutrient availability and uptake processes. Increased temperature and excess soil moisture will fasten the volatilization of surface applied N fertilizers and leaching losses of nutrients.
Nutrient transport /movement	Mass flow or diffusion of nutrients from the solution phase to plants mainly depend on the soil moisture, temperature, bulk density, hydraulic conductivity of soil, soil solution concentration, water movement and plant nutrient uptake potential.	Atmospheric rise of CO_2 reduces transpiration rate, which in turn suppresses nutrient movement to root system but it increases root exudation and fine root growth, thereby enhancing buffering capacity, ionic concentration in solution and nutrient movement through diffusion. Excess moisture or drought may improve or slow down the nutrient movement by mass flow and diffusion process.
Plant nutrient uptake	Root morphological traits such as length, diameter, surface area, branching, spatial distribution and root hairs, ionic concentration and nature of ion present in the solution phase influence the nutrient availability and uptake kinetics.	Atmospheric rise of CO_2 and temperature will increase fine root and root surface area development respectively. Excess or soil moisture stress may increase or depress the nutrient movement by counteracting mass flow or diffusion process.

4.3 Fertilizer Management

Best fertilizer management strategies are based on the '4 R' principle of right source, at the right rate, at the right time and at right placement to improve the nutrient use efficiency of crops, thereby reducing fertilizer requirement and associated GHGs emissions. Managing nitrogen fertilizer levels is challenging because appropriate application rates will differ for each agro ecosystem and growing season. Fertilizer applications that are carefully timed to maximize crop uptake, reduce application rates and N_2O emissions without decreasing crop yield. Fertilizer placement is mainly based on the nature and type of fertilizer and crops grown. By managing the 4 Rs principle properly, we can able to achieve higher nutrient use efficiency with sustainable crop yield and environmental quality.

5. MANAGEMENT STRATEGIES FOR ENHANCING NUTRIENT USE EFFICIENCY

5.1 Adoption of Resource Conserving Technologies

Avoiding excessive fertilizer applications to cropland can reduce farmers' operating costs. A number of resource conserving practices and technologies are available to increase the efficiency of fertilizer use and reduce N_2O emissions.

The ways and means are matching the supply of fertilizer to the demands of specific crops, ensuring the appropriate timing of nutrient applications, using efficient irrigation practices, converting to nitrogen-fixing plants as cover crops in the crop rotation and employing advanced fertilization techniques such as controlled-release fertilizers and nitrification inhibitors, which slow the conversion of ammonium to nitrate nitrogen.

- Conservation agriculture practices involving zero or minimum-tillage with direct seeding, permanent or semi-permanent residue cover and crop rotations have potential to increase soil organic matter, enhance input use efficiency, minimize the GHG emission and improve soil health.
- Minimum tillage and residue mulching maintain the favourable microclimate in the soil thereby enhance the nutrient use efficiency and minimize the greenhouse gas emission.
- Reducing the intensity and frequency of disturbance will help to protect soil carbon sinks. Practising reduced cultivation techniques where appropriate or moving towards plough sole or seeding zone of the soils.
- Site-specific measures of no-till farming, terracing or buffer strips for preventing run-off and erosion losses of P, N and other nutrients.
- The rotation of crops is not only necessary to offer a diverse source to the soil micro-organisms, but their root at different soil depths and are capable of exploring different soil layers for nutrients. Nutrients that have been leached to deeper layers and that are no longer available for the commercial crop can be reutilized by the crops in rotation.
- Practising suitable cover crop concept for protecting the soil from degradation.
- Designing the seeds using suitable bio-inoculants such as azospirillum for cereals, rhizobium for legumes, acetobacter for sugarcane and phosphobacteria for all the crops. It stimulates the germination and growth by production of growth promoting substance like vitamin-B complex, indoleacetic acid and gibberellic acids.

- Band placement of P and K fertilizer in no-till systems to improve nutrient availability during early growth phase.
- Band placement of fluid P fertilizer on calcareous soils overcomes the P fixation capacity and enhances the P use efficiency.
- Need based K application is essential to replenish declining indigenous K supply from minerals and irrigation and maintain the mobilizing non-exchangeable K in the clay minerals.
- Splitting of K applications to minimize leaching, increase stalk strength and resistance to diseases and improve the quality of harvested produce.
- Site-specific management of spatial variability in soil supply and crop removal through variable-rate application of P or K.
- Controlled release of nutrition is recommended on light soils, in rainy areas, where mid-season application is not feasible or where nitrogen application is restricted.
- Foliar feeding can provide the nutrients needed for normal development of crops in cases where absorption of nutrients from the soil is disturbed. As the uptake of nutrients through the foliage is considerably faster than through roots. Foliar sprays are also one of the proven methods of prompt correction of nutrient deficiencies and application of nutrients through foliar nutrition during critical growth stages dramatically increases yields and improve crop quality.
- Fertigation technique enables growers to maximize the use of water resources and increase the efficiency of fertilizer. This technique is particularly appropriate for high value crops under arid and semi-arid conditions. It involves the addition of soluble fertilizers into irrigation systems, preferably using a drip system which allows uniform water distribution and feeding of the crop. The fertilizer can be applied to the crop whenever it is needed at active root site of the plants.

5.2 Climate Smart Soil Management Practices

Good soil management can help to achieve maximum nutrient efficiency and sometimes it regulates emissions of three key greenhouse gases *viz.*, carbon dioxide, methane and nitrous oxide from agriculture. The science behind the role of soil and fertilizer management in reducing greenhouse gas emissions is still relatively young and the relationship between carbon and nitrogen in the soil is complex. More understanding is needed to fine-tune practices further. Following existing features of good soil management practice is generally still advantageous due to the wider production and environmental benefits gained in the climate changing era.

- Reducing compaction will improve nutrient uptake in the soil and help to reduce the release of nitrous oxide.
- The most efficient management practice to reduce nitrous oxide emission is site-specific nutrient management. The emission could also be reduced by usage of nitrification inhibitors such as nitrapyrin and dicyandiamide. There are some plant derived organics such as neem oil, neem cake and karanj seed extract which can also act as nitrification inhibitors, which also reduce CH_4 emissions from fertilized rice fields.

- Efficient nitrogen management can reduce nitrous oxide emissions by meeting crop requirements and minimizing residual nitrogen. Use recognized nutrient management plan and follow soil test based fertilizer recommendations.
- Application of organic enriched P and Zn application especially pulse and oilseed crops improved the crop yield and also enhance the nutrient use efficiency.
- Optimum potassium application is recommended in K deficit soil to maximize other nutrient efficiency, especially nitrogen.
- Soil application of sulphur is advocated especially pulses and oilseeds growing areas and sulphur deficient soil to combat climate change.
- Site specific multi-micronutrient mixture application is advocated to alleviate soil micronutrient deficiency and enhance the crop yield under varying climate.
- Soil application of azosprillum, rhizobium and acetobacter is recommended for cerelas, pulses and sugarcane crop respectively and phosphobacteria application is common for all the crops.
- Biofertilizers will be enriched with 200 kg of compost or organic manure and kept overnight. Enriched biofertilizer is incorporated in the soil at the time of sowing or planting to enhance the nutrient use efficiency.
- Crop specific bio-fertilizers application are known to play a number of vital roles in soil fertility, crop productivity and production in agriculture as they are ecofriendly and cannot substitute for inorganic fertilizers. These are ecofriendly, low cost inputs, which can be used by small and marginal farmers to reduce their chemical fertilizer cost and to sustain the soil health and crop yield.
- Application of biochar with soil test based balanced fertilizer enhanced the fertilizer use efficiency and reduced the prime greenhouse gas emission.
- Application of biochar with biofertilizer enabling the carbon sequestration and reduce greenhouse gases emission with high input use efficiency.
- Practising crop and soil based integrated application organic, inorganic and biofertilizer enhancing the crop yield and sustain the soil health.
- Carbon storage in soils undermost forms of agricultural management is limited, but farmers can potentially improve and maintain organic matter in soils by regular addition of crop residues and manures such as compost and digestate.
- Precise fertilizer application based on the remote sensing and GIS platforms through soil variability map and variable rate applicator techniques. Precision farming systems are already available to ensure farmers can draw up careful plans and the most advanced systems can reduce fertilizer usage by about one-third. For small scale farmers in the developing world, who have no access to modern farming equipment, the best solution for improving fertilizer practices is to increase access to independent advice from local experts such as research institutes.

Nutrient management strategies to combat climate change

- Appraise the nutrient status of the farm using standard soil testing method and prepare the fertility rating chart and categorize the deficient nutrient.

- Assess the different nutrient resources potential available to the farmers.
- An assessment of how best to fill the gap between requirement and availability through decision supporting system.
- Balanced fertilizer recommendation should be made by integrating soil test value, crop requirement and available nutrient resource by using decision supporting systems.

(i) The principles of sound fertilizer management

The basic principle behind the best fertilizer management practices is simple and given in the Fig. 7, that is the 4R using the right source, at the right rate, right time and right place which implies that how fertilizer applications can be managed to achieve economically feasible, socially acceptable and environmentally sustainable crop production.

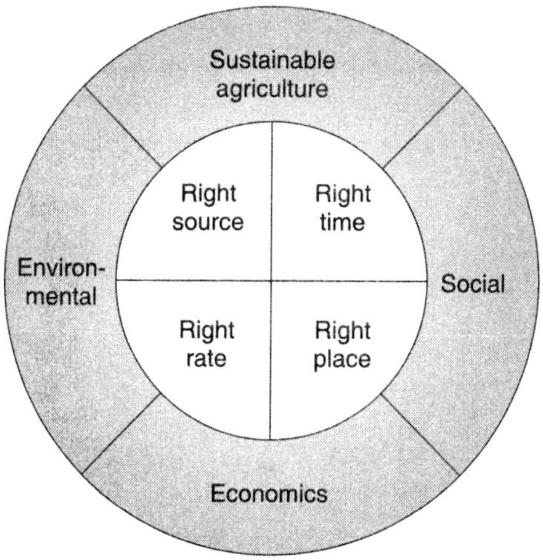

Fig. 7 Principles of fertilizer management and its impact.

According to the Roberts (2006) the following are guiding principles for fertilizer management.

Right source: Match the fertilizer source and product to crop need and soil properties. Be aware of nutrient interactions and balance nitrogen, phosphorus, potassium and other nutrients according to soil analysis and crop needs. Balanced fertilization is one of the keys to increasing nutrient use efficiency.

Right rate: Match the amount of fertilizer applied to the crop needs. Too much fertilizer leads to leaching and other losses to the environment and too little results in lower yields and crop quality and less residue to protect and build the soil. Realistic yield goals, soil testing, omission plots, crop nutrient budgets, tissue testing, plant analysis, applicator calibration, variable rate technology, crop scouting, record keeping and nutrient management planning are the best management practices that will help determine the right rate of fertilizer to apply.

Right time: Make nutrients available when the crop needs them. Nutrients are used most efficiently when their availability is synchronized with crop demand. Application timing (pre-plant or split applications), controlled release technologies, stabilizers, inhibitors and product choice are examples of best management practices that influence the timing of nutrient availability. Advanced tools such as SPAD meter and leaf colour chart to be used to meet out the crop demands at right time.

Right place: Place and keep nutrients where crops can use them. Application method is critical for efficient fertilizer use. Crop, cropping system and soil properties dictate the most appropriate method of application, but incorporation is usually the best option to keep nutrients in place and increase their efficiency. Conservation tillage, buffer strips, cover crops and irrigation management are other best management practices that will help keep fertilizer nutrients where they were placed and accessible to growing crops.

(ii) Ways to foster improved nitrogen use efficiency by manufacturer **(IFA, 2007):**

- Develop and promote fertilizer recommendations that balance soil nutrient supply and mineral fertilizer applications with actual crop needs
- Develop products with improved physical characteristics
- Develop products with chemical compositions that meet the variable nutrient requirements of crops and improve use efficiency
- Make a wide range of smart fertilizer products available to farmers
- Develop and promote nutrient/fertilizer best management practices
- Measure the performance of the recommended products and practices

5.3 Breeding for Nutrient Use Efficiency

5.3.1 Exploiting Genetic Potential to Enhance Nutrient Use Efficiency

Nutrient efficient genotypes are adapted to environments with low nutrient availability. It is useful to screen and exploit the genetic variability present across crop species to enhance nutrient use efficiency. The nutrient use efficiency of the genotypes should be looked into two angles, namely, ability of the plant to extract the nutrient from soil (uptake) and ability of the plant to convert intake nutrients into grain yield (utilization efficiency). These parameters can be measured by nutrient use efficiency and utilization efficiency as given by Ortiz-Monasterio (2001). To satisfy site and management specific needs under changing climatic situation, the genotypes to be evaluated in high and low nutrient status environment and group them for low and high input responder. Simultaneous selection for both nutrient use efficiency and tolerance to nutritional stress is possible if the mechanisms that confer efficiency and tolerance are not competitive (Fritsche-Neto and DoVale, 2012).

Nutrient-efficient genotypes may have an increased capacity (i) to exploit the soil (large root surface area), (ii) to convert non-available nutrient forms into plant available forms and (iii) to take up nutrients across the plasma membrane (Rengel, 2001). Plant species increase their capacity to access nutrients by altering root morphology and by changing the capacity and affinity of plasma membrane-embedded transporters capable of carrying nutrients into the cytosol (Rengel and Marschner, 2005).

The existence of genetic variability had been reported for the nutrient use efficiency (NUE) of cultivars. The efficient use of nutrient is the cumulative effect of following efficiencies that varies with genotypes to genotypes within crop species.

(a) *Uptake efficiency:* It depends on acquiring the nutrients from soil, entry rate into roots and radial transport into roots and all are based on root traits.

(b) *Incorporate efficiency:* Transport of nutrients to shoots and leaves are based on shoot parameters.

(c) *Utilization efficiency:* Based on remobilization, i.e., utilisation absorbed nutrients to produce biomass (Baligar *et al.*, 2001).

The environment factors like solar radiation, temperature and soil moisture also influence the NUE in plant. Hence, in the changing climate situation it is essential to opt balance between the stress adapted genotypes and NUE to sustain the crop yield.

The plant species and cultivars within species differ in uptake and utilization of nutrients. It is mainly attributed to morphological, physiological and biochemical processes in plant and its soil, nutrients, biological and management practices (Baligar *et al.*, 2001). Hence, the NUE improved by the manipulation of plants, soil, fertilizer, biological, environment factors and ideal management practices.

The crop improvement approaches are used to increase nutrient use efficiency (NUE) in crop plants are traditional breeding, marker-assisted selection, targeted breeding through pyramiding and by gene transfer and manipulation (Good *et al.*, 2004; Rengel and Marschner, 2005). The prospects for genetically manipulated plants to enhance nutrient availability under given environmental conditions can be strengthened by (i) understanding the regulation of the exudation process, (ii) elucidating the regulation of root morphology, (iii) characterizing the synthesis and activity of membrane-embedded nutrient transporters, and (iv) understanding the interactions among root exudation, indigenous rhizosphere microorganisms and nutrient availability (Rengel and Marschner, 2005).

5.3.2 Exploiting Variability of Micronutrients

In climate smart management practices, it is important to incorporate crop and edaphic factor specific micronutrient management. The efficient and sustainable strategy is to develop micronutrient efficient plant genotypes that can tolerate low nutrient supply may increase productivity on low fertility soils and reduce the fertilizer requirements (Khoshgoftarmanesh *et al.*, 2010). To increase the micronutrients in edible portion of food crops, it is essential to identify the micronutrient responsive cultivars. The reports prove that there are inherent genetic variability is available in crop species in utilizing micronutrients. Screening for micronutrient responsive genotypes and exploring the available genetic variability in crop improvement programme are vital to the present situation.

6. FUTURE CHALLENGES AND THREATS

- A key priority for future research is to determine whether there will be a sustained increase in nutrient availability due to global warming and how this will affect productivity.
- Till date, experimentation on crop plants has not found conclusive evidence that physiological efficiency is altered in high CO_2 environments.
- Further research could clarify the best fertilizer additions under specific circumstances and develop new nitrogen related products and smart delivery mechanisms that are driven by factors related to temperature, water and host plant.

- Better understanding of the relationship between N_2O emissions and extreme weather events may also become increasingly important to enhance nutrient use efficiency under changing climate.
- Crop specific fertigation through water soluble fertilizer and its use efficiency under changing climate to be studied in different agro-ecological zones.
- The nanotechnology promises slow release fertilizers (nanoporous zeolites) and soil quality and plant health monitoring systems (nanosensors). The new products and tools would go a long way in managing soil fertility and enhancing crop productivity.
- Greater research and development efforts are required to increase shelf-life of biofertilizers by way of isolating location-specific strains, better production technology avoiding contamination and better storage and handling.
- The liquid cultures containing cell protectants maintain high microbial numbers and promote formation of resting cells like cysts and spores having resistance to abiotic stresses.
- Dynamics and response of secondary and micronutrient under elevated CO_2 may be explored in detail under field condition.
- Adaptation to climate change requires multidisciplinary solutions that include the development of appropriate nutrient efficient cultivars and mechanisms to facilitate farmers' access to the cultivars.
- Varieties with increased resilience to abiotic and biotic stresses will play an important role in autonomous adaptation to climate change.
- Identify the candidate genes and nutrient uptake efficient traits valuable for breeding and incorporate these into elite cultivars and to evaluate their performance under real agricultural field conditions under changing climate.

7. CONCLUSION

Climate smart agricultural technologies are available to increase the efficiency of fertilizer use and reduce N_2O emissions. The ways and means are synchronizing the supply of fertilizer to the demands of specific crops, ensuring the appropriate timing of nutrient applications, using efficient irrigation practices, converting to nitrogen-fixing plants as cover crops and crop rotation. Practising advanced fertilization techniques such as controlled release fertilizers, urease inhibitors and nitrification inhibitors, which slow the conversion of ammonium to nitrate nitrogen. Biofertilizer helps in increasing crop productivity by way of increased biological nitrogen fixation, phosphorus solubilization and mobilization, increased availability or uptake of nutrients through solubilization or increased absorption stimulation of plant growth through hormonal action or antibiosis or by decomposition of organic residues. In addition, biofertilizer replaces part of chemical fertilizers, reduces amount and cost of chemical fertilizers, thus prevents the environmental pollution from excess application of chemical fertilizers.

In the context of global environmental changes and other constraints to increase yield further, the efficient use of nutrient, especially N and water have emerged as two key targets. Complementary to bridging the yield gap, tailoring of cultivars with high nutrient use efficiency and yield with abiotic tolerance is required to cope up the climatic variability. Adoption of climate smart soil and crop management technologies and improved cultivars must be explored in future to combat the climate change and sustain the livelihood security of the farming community.

References

Afzal, A., and Bano, A. (2008). Rhizobium and phosphate solubilizing bacteria improve the yield and phosphorus uptake in wheat (*Triticum aestivum* L.). *Int. J. Agri. Biol.* 10: 85-88.

Atkin, O.K., Edwards, E.J., and Loveys, B.R. (2000). Response of root respiration to changes in temperature and its relevance to global warming. *New Phytologist.* 147: 141-154.

Austin, A.T., and Vitousek, P.M. (1998).Nutrient dynamics on a precipitation gradient in Hawaii.*Oecologia* 113: 519-529.

Baligar, V.C., Fageria, N.K., and He, Z.L. (2001) .Nutrient use efficiency in plants, *Commun. Soil Sci. Plant Anal.* 32(7&8): 921-950.

Barber, S.A. (1995). Soil Nutrient Bioavailability: a Mechanistic Approach, second ed. JohnWiley & Sons Inc., New York, USA. pp. 414.

BassiriRad, H., Thomas, R.B., Reynolds, J.F., and Strain, B.R. (1996). Differential responses of root uptake kinetics of NH_4^+ and NO_3^- to enriched atmospheric CO_2 in field grown loblolly pine. *Plant, Cell and Environment.*19: 367-371.

Bassirirad, H. (2000). Kinetics of nutrient uptake by roots: responses to global change. *New Phytol.*147: 155-169.

Biswas, P.P., and Sharma, P.D. (2008). A new Approach for estimating fertilizer response ratio: the Indian scenario, *Indian J. Fert.* 4(7): 59-62.

Brennan, R.F., and Bolland, M.D.A. (2009). Comparing the nitrogen and potassium requirements of canola and wheat for yield and grain quality. *J. Plant Nutr.* 32(12): 2008-2026.

Carcamo, H.A., Prescott, C.E., Chanway, C.P., and Abe, T.A. (2001). Do soil fauna increase rates of litter breakdown and nitrogen release in forests of British Columbia,Canada.*Can J For Res.* 31: 1195-1204.

Clair, St., and Lynch, Jonathan, P. (2010). The opening of Pandora's Box: climate change impacts on soil fertility and crop nutrition in developing countries. *Plant Soil.* 335: 101-115.

Clarkson, D.T. (1996). Root structure and sites of ion uptake. In: Waisel, Y., Eshel, A., Kafkafi, U .(eds) *Plant roots: the hidden half*, 2[nd]edn. Marcel Dekker, New York. pp. 483-510.

Cruz, C., Lips, S.H., and Martins-Loucao, M.A. (1993). The effect of nitrogen source on photosynthesis of carob at high CO_2 concentrations. *Physiologia Plantarum.* 89: 552-556.

Dobermann, A., and Cassman, K.G. (2005). Cereal area and nitrogen use efficiency are drivers of future nitrogen fertilizer consumption. *Science in China Ser. C Life Science*s. 48: 745-758.

Dobermann, A. (2007). Nutrient use efficiency measurement and management.IFA International Workshop on Fertilizer Best Management Practices, Brussels, Belgium, First edition, IFA, Paris, France. pp. 1-28.

DeLuca, T.H., MacKenzie, M.D., Gundale, M.J., and Holben, W.E. (2006). Wildfire-produced charcoal directly influences nitrogen cycling in ponderosa pine forests. *Soil Sci Soc Am J.* 70: 448-453.

Duxbury, J.M., Bouldin, D.R., Terry, R., and Tate, R.L.(1982).Emissions of nitrous oxide from soils. *Nature.* 298: 462-464.

El-Komy, H.M.A. (2005). Co-immobilization of *A. lipoferum* and *B. megaterium* for plant nutrition. *Food Technol Biotech.* 43(1): 19-27.

Fangmeier, A., Gruters, U., Hogy, P., Vermehren B., and Jager, H.J. (1997). Effects of elevated CO_2, nitrogen supply and tropospheric ozone on spring wheat-II.Nutrients (N, P, K, S, Ca, Mg, Fe, Mn, Zn). *Environ. Pollut.* 96(1): 43-59.

Flynn,C. Helen . (2009). The Role of Nutrient Management in Mitigation. Focus 16 brief 7 of May 2009. www.ifpri.org.

Fritsche-Neto, R., and DoVale, J.C. (2012). Breeding for Stress tolerance or resource-use efficiency. In: R. Fritsche-Neto and A. Borém (eds.), Plant Breeding for Abiotic Stress Tolerance, Springer-Verlag Berlin Heidelberg. pp. 13.

Good, A.G., Shrawat, A.K., and Muench, D.G. (2004). Can less yield more? Is reducing nutrient input into the environment compatible with maintaining crop production? *Trends in Plant Science.* 9(12): 597-605.

Guo, J.H., Liu, X.J., Zhang, Y., Shen, J.L., Han, W.X., Zhang, W.F., Christie, P., Goulding, K.W.T., Vitousek, P.M., and Zhang, F.S. (2010). Significant acidification in major Chinese croplands.*Science.* 327: 1008-1010.

Gupta, A.K. (2004). The complete technology book on biofertilizers and organic farming.National Institute of Industrial Research Press. India. pp. 168.

IPCC- Inter-Governmental Panel on Climate Change (2007). Climate change 2007: The physical science Basis, Summary for Policy makers.

IFA- International Fertilizer Association. (2007). Fertilizer Best Management Practices: General Principles, Strategy for their Adoption and Voluntary Initiatives vs. Regulations." Proceedings of the International Workshop on Best Fertilizer Management Practices, Brussels, Belgium, 7-9 March.IFA, Paris, France.

IFA - International Fertilizer Association. (2011). Global fertilizer trade map.www.icis.com/fertilizer.

Johnston, A.E., Poulton, P.R., and Coleman, K. (2009). Soil organic matter: its importance in sustainable agriculture and carbon dioxide fluxes. In: D.L. Sparks (Editor), *Advances in Agronomy, Vol 101.* Elsevier Academic Press Inc, San Diego, pp. 1-57.

Kannan, P., Balasubramaniyan, P., and Prabukumar, G. (2011). Effect of different nutrient management practices on soil health indicators in groundnut system. In Annual report-2011, Dryland Agricultural Research Station, Chettinad, Tamil Nadu Agricultural University, Coimbatore.pp. 20-23.

Kerkhoff, A.J., Enquist, B.J., Elser, J.J., and Fagan, W.F. (2005) Plant allometry, stoichiometry and the temperature-dependence of primary productivity. *Global Ecology and Biogeography.* 14: 585-598.

Khan, K. S., and Joergensen, R. G.(2009). Changes in microbial biomass and P fractions in biogenic household waste compost amended with inorganic P fertilizers. *Bioresour. Technol.*100: 303-309.

Khoshgoftarmanesh, A.H., Schulin, R., Chaney, R.L., Daneshbakhsh, B., and Afyuni, M. (2010). Micronutrient-efficient genotypes for crop yield and nutritional quality in sustainable agriculture: A review. *Agronomy for Sustainable Development.* 30: 83-107.

Kirschbaum, M.U.F. (2000). Will changes in soil organic carbon act as a positive or negative feedback in global warming. *Biogeochemistry.* 48: 21-51.

Larcher, W. (1995). Plant physiological ecology. Springer, Berlin Heidelberg New York.

Lipson, D.A., Schmidt, S.K., and Monson, R.K. (2000). Carbon availability and temperature control the post-snow melt decline in alpine soil microbial biomass. *Soil BiolBiochem.* 32: 441-448.

Loladze, I. (2002). Rising atmospheric CO_2 and human nutrition: toward globally imbalanced plant stoichiometry. *Trends in Ecology & Evolution.* 17(10): 457-461.

Lugtenberg, B.J.J., and Bloemberg, G.V. (2004) .In *Pseudomonas* Vol. 1 ed. Ramos, J.L. Kluwer Academic Plenum Publishers, New York. pp. 403-430.

Manoj-Kumar, Swarup, A., Patra, A.K., Purakayastha, T.J., Manjaiah, K.M., and Rakshit, R. (2011). Elevated CO_2 and temperature effects on phosphorus dynamics in rhizosphere of wheat (*TriticumaestivumL.*) grown in a Typic Haplustept of subtropical India. *Agrochimica. 55*: 14-31.

Marchner, H. (1995). Mineral Nutrition of Higher Plants.Academic, San Diego Samuel B.

Mehrvarz, S., Chaichi, M. R., and Alikhani, H. A. (2008).Effects of phosphate solubilizing microorganisms and phosphorus chemical fertilizer on yield and yield components of Barely (*Hordeum vulgare* L.). *J. Agric. & Environ. Sci.* 3: 822-828.

Mosier, A.R., Bronson, K.F., Freney, J.R., and Keerthisinghe, D.G. (1994).Use of nitrification inhibitors to reduce nitrous oxide emission from urea fertilized soils. *CH_4 and N_2O: Global Emissions and Controls from Rice Fields and other Agricultural and Industrial Sources,* (ed.K Minami, A Mosier and R Sass), Tsukuba: National Institute of Agro-Environmental Sciences.pp. 197-207.

Ortiz-Monasterio, J.I., Manske, G.G.B., and Ginkel, M.V. (2001).Nitrogen and phosphorus use efficiency, In *Application of Physiology in Wheat Breeding,* Reynolds, M.P., Ortiz-Monasterio, J.I. and McNab, A. (eds.), CIMMYT, Mexico.

Pregitzer, K.S., and King, J.S. (2005).Effects of Soil Temperature on Nutrient Uptake. Ecological Studies, Vol.181 Bassiri Rad, H. (Ed.) Nutrient Acquisition by Plants. An Ecological Perspective, Springer-Verlag Berlin Heidelberg.

Rengel, Z. (2001). Genotypic differences in micronutrient use efficiency in crops. *Communications in Soil Science and Plant Analysis.* 32: 1163-1186.

Rengel, Z., and Marschner, P. (2005). Nutrient availability and management in the rhizosphere: exploiting genotypic differences, *New Phytologist.* 168: 305-312.

Roberts, T.L. (2006). Improving nutrient use efficiency. In Proceedings of the IFA Agriculture Conference "Optimizing Resource Use Efficiency for Sustainable Intensification of Agriculture", 27 February - 2 March 2006, Kunming, China. International Fertilizer Industry Association (IFA), Paris, France.

Rustad, L.E., Campbell, J.L., Marion, G.M., Norby, R.J., Mitchell, M.J., Hartley, A.E., Cornelissen J.H.C., and Gurevitch, J. (2001).A meta-analysis of the response of soil respiration, net nitrogen mineralization, and above-ground plant growth to experimental ecosystem warming. *Oecologia.* 126: 543-562.

SamanSeneweera.(2011).Effects of elevated CO_2 on plant growth and nutrient partitioning of rice (*Oryza sativa* L.) at rapid tillering and physiological maturity. *Journal of Plant Interactions.* 6(1): 35-42.

Sardans, J., Penuelas, J., Prieto, P., and Estiarte, M. (2008). Drought and warming induced changes in P and K concentration and accumulation in plant biomass and soil in a Mediterranean shrubland. *Plant Soil.* 306: 261-271.

Seneweera, S. P., and Conroy, J.P. (1997).Growth, grain yield and quality of rice (*Oryza sativa* L.) in response to elevated CO_2 and phosphorus nutrition. *Soil Science and Plant Nutrition.* 43: 1131-1136.

Seufert, V., Ramankutty, N., and Foley, J.A. (2012). Comparing the yields of organic and conventional agriculture. *Nature.* 485(7397): 229-232.

Sharma, J.S., and P.D. Sharma.(2011). Food Security-Indian Scenario. In: Brar, M.S., and Mukhopadhyaya, S.S. (ed.) Potassium role and benefits in improving nutrient management for food production, quality and reduced environmental damages. Volume I IPI, *IPNI.* pp. 15-43.

Sturm, M., Schimel, J., Michaelson, G., and Welker, J.M. (2005). Winter biological processes could help convert Arctic tundra to shrubland. *BioScience.* 55: 17-26.

Tilman David, Kenneth, G., Cassman, Pamela A. Matson, Rosamond Naylor and Stephen Polasky.(2002). Agricultural sustainability and intensive production practices. *Nature.* 418: 671-677.

Van Noordwijk, M., Martikainen, P., Bottner, P., Cuevas, E., Rouland, C., and Dhillion, S.S. (1998).Global change and root function. *Global Change Biology.* 4: 759-772.

Wan, X., Zwiazek, J.J., Liffers, V.J., and Landhausser, S.M. (2001). Hydraulic conductance in aspen (*Populustremuloides*) seedlings exposed to low root temperatures. *Tree Physiol.* 21: 691-696.

Whitelaw, M. A. (2000). Growth promotion of plants inoculated with phosphate solubilizing fungi. *Adv. Agron.* 69: 99-151.

Wilts, A. R., Reicosky, D. C., Allmaras, R. R., and Clapp, C.E. (2004). Long term corn residue effects: Harvest alternatives, soil carbon turnover and root-derived carbon. *Soil Sci. Soc. Am. J.* 68: 1342-1351.

Yuncai, H.U. and Schmidhalter,U.R.S. (2005). Drought and salinity: A comparison of their effects on mineral nutrition of plants, *J. Plant Nutr. Soil Sci.* 168*:* 541-549.

Zaidi, A., and Khan, M. S. (2006). Co-inoculation effects of phosphate solubilizing microorganisms and *Glomus fasciculatum* on green gram – *Bradyrhizobium* symbiosis. *Turk. J. Agric.* 30: 223-230.

7

Crop Diversification and Conservation Agriculture Towards Sustainable Crop Improvement in Global Scenario

Ashim Chakravorty* **and Parthadeb Ghosh**

Cytogenetics and Plant Breeding Section, Biotechnology Research Unit, Department of Botany University of Kalyani, Nadia-741235, West Bengal, India

ABSTRACT

The world's population is anticipated to increase from 5.7 billion in 1995 to 7.7 billion in 2020. Of this, 80% live in developing countries, where the population increases by 1.9% per year. Fast increasing population, changing climate, shrinking land and water resources are the major challenges before us to meet the food and nutritional security of the population at the global level. Improvement in the production and productivity of crops are the immediate need to boost the availability of food and improve the economic conditions of the farmers. Crop diversification is intended to give a wider choice in the production of a variety of crops in a given area so as to expand production related activities on various crops and also to lessen risk. For conservation agriculture to succeed, farmers will need to be in the forefront, helping to identify, develop, and deploy new technologies. Farmers interested in sustainable crop production systems have begun to adopt and adapt improved crop management practices, a step toward conservation agriculture, which may be considered the ultimate solution. If we are to insure that adequate food, fibre, and fuel are produced, farmers must employ appropriate crop management technologies not only to stabilize or increase crop production in a cost-effective manner, but also to conserve the integrity and sustainability of their resource base.

Keywords: Crop improvement, conservation agriculture, diversification, globalization, sustainable crop production.

1. INTRODUCTION

The first food summit of 1974 vowed to 'eradicate world hunger within a decade'. However, after a long gestation period and complacency, a series of World Food Summits starting from 1996, rather in quick succession, took determination to minimise world hunger only to half by 1915. Unfortunately, the number is on the rise and has crossed one billion mark (FAO, 1995).

The global human population scenarios developed by IPCC's SRES depict numbers varying from 8.1 to 14.0 billion in 2080. However, many of us believe that it would cross 9.0 billion in

*Corresponding author's E-mail: ashimbot@gmail.com

Roychowdhury R. (Ed.), Crop Improvement in the Era of Climate Change.

2050 (Fishchetti, 2011). This is so, when we all know that the land which is basic to agriculture is finite and fragile. Water is life but is limited and getting everyday polluted. Also, the climate so crucial for agriculture is changing, rather fast. Thus, our ever-growing demands are on a hot plate of supply-side constraints that are evaporating our popular global slogans like, 'Food for All for Ever' and 'Feed the World Save the Planet' (Fishchetti, 2011).

Now, it has been known that green revolution in agriculture is at the cost of soil fertility degradation, groundwater depletion, biodiversity and soil erosions and water and air pollutions. Consequently, there are declaration in total factor productivity, reduction in the quality of the agricultural produce and increase in the cost of production in association with other growth-retarding and cost-escalating factors (Rai, 2011).

The enhancement of quality agricultural productivity with far greater biotic and a biotic resilience to hostile environments and climate changes would require substantial investments and technological innovations and timely effective interventions. Even on food front alone, 70% increase in grain yield would be required to feed our people by 2050. It should be kept in mind that in our gradual endeavour for progress, agro-ecosystems are to be developed in the natural ecosystem in such a way that finite natural resources and limited land capacities can be uninterruptedly capable of meeting the ever-increasing needs for food, feed, fibre, fuel and timber.

Thus, the course of development necessitates increase in agricultural production with greater environmental protection. Only efficient and advanced technology can help us to reach our goal of materialising conservation agriculture with adequate sustainability. Agriculture must honour healthy ecosystems, by recognising the importance of fertile soil, drawing on natural resources of plant nutrition, and using mineral fertilizer with justified doses.

Improved genetically diverse crop varieties and animal breeds conforming to a range of 'agro-ecosystems' farming conditions, and tolerating climate change can serve our purpose. The pesticides used kill not only pests but their enemies as well. Hence, their overuse can impair all, the producer, consumer and the environment. Thus, agricultural development must anchor itself in fundamental perception, policies and strategies. In our pursuit of sustainable farming system approach, the following developmental efforts may be kept in mind (Rai, 2011):

1. Simultaneous realization of increased agricultural productivity and enhanced or at least maintained ecosystem functions.
2. Input use efficiency enhancement including that of water, nutrients, pesticides, energy, land and labour.
3. Enhanced use of conserved and natural biodiversity to build systems' reliance to biotic, abiotic and economic stresses.
4. Continuous efforts to increase soil organic matter, biological nitrogen fixation and agriculturally important microorganism to ensure soil as a productive living legend would be required.
5. Strengthening linkage between conservation of genetic resources and the use of diversity for plant/breed improvements.
6. Develop commensurate human resource, generate appropriate technology and ensure its extension in continuum.

7. Linking public and private sector support building coordination and reducing transaction costs would be helpful.
8. Incorporating the value of natural and ecosystem services into agricultural input and output price policy.
9. Building regulatory provisions and having research and advisory systems for a wide range of production and marketing conditions.
10. Recognizing and incorporating customary access and management practices into sustainable agricultural production intensification initiatives.
11. Sensor based decision support system development and its wise use.
12. Continuously enhancing capacity, knowledge, understanding and skill of farmers/ entrepreneurs for realizing the desired goal on a changing time scale.
13. Striving for quality input development and supply system for quality outputs, outcomes and impacts.
14. In our endeavours, align subsidy, services and incentives as instrument of positive change for attaining and sustaining continuous gains.

2. DIVERSIFICATION DEFINED IN AGRICULTURE

Diversification of agriculture refers to the shift from the regional dominance of one crop to regional production of a number of crops, to meet ever increasing demand for cereals, pulses, vegetables, fruits, oilseeds, fibres, fodder and grasses, fuel, etc. (Pimentel *et al.*, 2010; Dimmitt, 2009). It aims to improve soil health and a dynamic equilibrium of the agro-ecosystem (Pidiwirny, 2010). Crop diversification takes into account the economic returns from different value-added crops. It is different from the concept of multiple cropping or succession planting in which multiple crops are planted in succession over the course of a growing season. Moreover, it implies the use of environmental and human resources to grow a mix of crops with complementary marketing opportunities, and it implies a shifting of resources from low value crops to high value crops, usually intended for human consumption such as fresh market fruits and vegetables (Harries *et al.*, 2007). With globalization of the market, crop diversification in agriculture means to increase the total crop productivity in terms of quality, quantity and monetary value under specific, diverse agro-climatic situations worldwide.

There are two approaches to crop diversification in agriculture. First is horizontal diversification, which is the primary approach to crop diversification in production agriculture (Singh, 2009). Here, diversification takes place through crop intensification by adding new high-value crops to existing cropping systems as a way to improve the overall productivity of a farm or region's farming economy. The second is the vertical diversification approach in which farmers and others add value to products through processing, regional branding, packaging, merchandising, or other efforts to enhance the product (Tweeton and Flora, 2001). Opportunities for crop diversification vary depending on risks, opportunities and the feasibility of proposed changes within a socio-economic and agro-economic context. Crop diversification may occur as a result of government policies. The adoption of crop technologies is commonly assumed to be influenced primarily by resource-related factors when institutional and infrastructure factors can play as much or more of a role in their adoption.

technology for successful crop diversification. The research system not only needs to address the issues connected with continuance and indulgence and knowledge in the areas of emerging technologies but also create a cadre of scientists through the continuous upgrade of skills and human resource development. The researchers also need to popularize the technologies, impart knowledge and skills to the extension functionaries for the transfer of technologies to the farmers. This knowledge-based farming will call for much more interaction between the researchers, extension workers and farmers. The fruits of the innovative technologies should reach the farmers at the earliest and also spread in the quickest possible time (Olson, 1998).

Production agriculture in the future may also be characterized by increasing diversity. But increased diversity is different from more diversification. "Diversification" refers to expanding the number of activities or enterprises managed and controlled by one firm or business. "Diversity" refers to the differences between the firms and businesses that comprise the industry. In fact, agriculture of the future may exhibit more specialization (i.e., less diversification) within a business, and at the same time more diversity between businesses (Hazra, 2005).

Industrialization of production means the movement to large-scale production units that use standardized technology and management and are linked to the processor by either formal or informal arrangements. Size and standardization are important characteristics in lowering production costs and in producing more uniform crop products and animals that fit processor specifications and meet consumers' needs for specific product attributes, and their food safety concerns (Wolf, 1998, McBride, 1997).

The transformation of crop and livestock production from commodity to differentiated product industries will be driven by consumers' desire for highly differentiated food products; their demands for food safety and trace-back ability; continued advances in technology; and the need to minimize total costs of production, processing, and distribution. Food systems will attempt to differentiate themselves and their products by science and/or through marketing. Ways to differentiate through science include gaining exclusive rights to genetics through patentable biotechnology discoveries; exclusive technology in processing systems; and superior food safety integrity. Marketing may include: branding, advertising, packaging, food safety, product quality, product attributes, bundling with other food products for holistic nutritional packages, and presentation of products in non-traditional formats (McBride, 1997).

In recent decades there has been an increased awareness of the importance of the perspective and practice of ecological agriculture. Proponents of ecological agriculture argue that agriculture cannot function as an isolated system (i.e., as having no exchanges of matter or energy with its environment). They argue that agriculture must consider the limits of the natural resources used to produce agricultural commodities as well as the limits of the sinks needed to dispose of the wastes from agricultural activities (Boehlje, 1992).

An increasing emphasis will be placed on managing and optimizing supply chains from genetics to end-user/consumer. This supply-chain approach will improve efficiency through better flow scheduling and resource utilization, increase the ability to manage and control quality throughout the chain, reduce the risk associated with food safety and contamination, and increase the ability of the crop and livestock industries to quickly respond to changes in consumer demand for food attributes (McBridge, 1997).

Food safety is a major driver in the formation of chains. One way to manage food safety risk is to monitor the production/distribution process all the way from the final product back through the chain to genetics. A trace-back system combined with HACCP (Hazard Analysis Critical Control Points) quality assurance procedure facilitates control of the system to minimize the chances of a food contaminant or to quickly and easily identify the sources of contamination (Rao, 2005; Olson, 1998).

Agricultural production has always been a risky business, but this may become increasingly so in the future. Not only will the traditional variables of price, weather, disease, etc., continue to buffet the industry, new sources of risk may be encountered. Some food distribution channels may require particular quality characteristics that are not available in predictable quantities in open, spot markets. The risk of changing consumer preferences or a food safety scare may be a much more critical and important risk to manage than price or availability of raw materials. One reason for a contractual arrangement to source raw materials from a qualified supplier is to reduce price and availability risk and food safety risks from contamination, and simultaneously to obtain the attributes needed in the final product from the specific-attribute raw material. But this arrangement may reduce flexibility and introduce relationship risk – the risk that the qualified supplier arrangement is terminated (Tyner and Boehlji, 1997).

The transformation of a segment of agriculture from a commodity to a differentiated product industry introduces at least three new risks. First, differentiated products are positioned to respond to unique market segments that value the attribute that is differentiated. Assuming this attribute is measurable (which may be a risk in itself because many food attributes, including quality, are difficult to measure), one risk is that consumers' and end-users' attitudes and willingness to pay for some attributes may change over time. For example, consumer attitudes with respect to food additives, biotechnology, and genetically modified organisms (GMOs) do not appear to be stable or predictable across cultures and across time (Tweeton, 2001).

Second, alternative techniques to accomplish product differentiation may develop over time, and those firms or individuals that can produce the differentiated product could increase. Thus, differentiated products are regularly commoditized over time, and initially higher margins are eroded as new competitors appear. This speed of commoditization is also a source of uncertainty. Finally, differentiated products in the food market, particularly if that product is a branded product, also carry the risk as well as the reward of branding. Brand value can be quickly destroyed by defects or quality lapses, and in the food product markets, food safety is a risk that can quickly destroy brand value (Tweeton, 2001).

The agriculture of the past could reasonably and accurately be characterized by typical or representative farms for various geographic regions and crop or livestock enterprises. Certainly, farms were not identical in terms of the technology used, the size of the business, the financial characteristics, the ownership structure, etc., but in a particular locale and for a particular crop or livestock sector they exhibited many similarities – in fact more similarities than differences. But, increasingly, agriculture is not characterized by similarities among firms, but by differences between firms. Not only do we now have enterprise specialization (e.g., corn/soybean farms or hog farms rather than diversified farms of the past that included corn/soybeans and hogs produced on the same farm), we are further separating production activities in livestock industries such that firms may be involved in only one phase of dairy, pork, or beef production. For example, some firms specialize in breeding, gestation, and farrowing in pork production, and separate firms specialize in finishing or the final feeding phase of pork production (Sharma *et al.*, 1996).

Diversity is also increasingly characterizing the types of products produced even within a segment or sector of agriculture. With increasing diversity in consumer demands and the opportunity for product differentiation at the production level, farmers are no longer just producing commodity crop and livestock products (Perry and Banker, 2000). For example, some farmers are producing high oil corn, while their neighbours are producing white corn or high starch corn. A further form of diversity in the farm sector is in terms of commitment to and dependency on farming as a source of family income. Many farm families combine farm employment with off-farm employment, or home-based businesses, not only as a way to start farming, but also as a permanent and satisfying way of life.

There is more diversity in size of farming operation today than in the past. Even though large-scale businesses are growing rapidly as the dominant size in some parts of the livestock industries, smaller scale production units continue to be a significant part of these industries as well as of other segments of agriculture. This occurs most often in segments focused on niche markets and local customers, such as fresh fruit and vegetable production for local consumption and roadside stands (Perry and Banker, 2000).

Differences in marketing and financial strategies among firms are an important part of the new diversity in agriculture. Farmers are using various methods of raising capital, not just borrowing money and obtaining equity from their own savings. Longer-term lease arrangements for land and machinery and buildings are becoming an important part of the farm capital structure for some businesses and yet are not part of that structure for other businesses. Some farmers are using equity capital from outside investors rather than relying strictly on their own retained earnings or family sources and different farmers use different strategies for marketing their crops. Some use cash markets, whereas others forward contract their production. Some producers sell at the farm gate, while others are in value-added cooperatives that retain ownership further down the distribution channel towards the final food consumer (Dhar, 2004).

An additional dimension of diversity in agriculture is in the production technology used. Some producers are heavily dependent on purchased inputs, whereas others are more focused on holistic sustainable production systems that recycle resources and thus reduce the amount of purchased inputs. Some farmers use highly capital-intensive production systems, whereas others who have more labour than capital to contribute to their business find it much more profitable to use labour-intensive technologies and production systems. Thus, there is increasing diversity in production technology, management practices, ways of doing business, techniques of financing and organizing the business, etc. Even though diversification is declining in production agriculture, diversity is increasing (Wolf 1998; Boehlje 1999).

Thus, to conclude, it can be said that variations have led to the evolution of regional niches for various crops. Historically, regions were often associated with the crops in which they specialize for various agronomic, climatic, hydro-geological, and even, historical reasons. But, in the aftermath of technological changes encompassing biochemical and irrigation technologies, the agronomic niches are undergoing significant changes. With the advent of irrigation and new farm technologies, the yield level of most crops, especially that of cereals, has witnessed an upward shift making it possible to obtain a given level of output with reduced area or more output with a given level of area and creating thereby the condition for inter-crop area shift (diversification) without much disturbance in output level. Besides, as agriculture becomes drought proof and growth becomes more regionally balanced, there has been a reduction in the instability of agricultural output.

In the face of these new changes including the achievement of food self-sufficiency, the area shift that tended towards cereals in the immediate aftermath of the green revolution, has started moving in the opposite direction, i.e., from cereals to non-cereals.

Although these reverse area shifts actually took place in the mid-1970s as a part of the process of commercialization, they became more pronounced since the mid-1980s as a response partly to emerging supply deficit in edible oils and partly to the changing comparative advantage of crops. Since the recent trend in inter-crop area shifts has its origin in the price and trade policy changes of the 1980s, they indicate the increasing market influence on area allocation. The area under commercial crops has almost doubled in the last three decades. Among the foodgrain crops, the area under superior cereals, i.e., rice and wheat, is increasing; while that of coarse cereals (millets) is on decline. The area share of jute and allied fibres has also gone down substantially. Like any other economy, the share of agriculture in the GDP is also declining in India. Increase in income from the agriculture sector, further growth of non-crop sub-sectors within agriculture; faster growth of non-foodgrain crops; and faster growth of superior cereals among the food grains are all happening, but the pace of such change is far too slow. An accelerated pace of diversification to create positive import of higher income, higher employment and conservation and efficient use of natural resources emphasizes the need for efficient policies, especially in technological development, selective economic reforms and institutional change. A strategy of crucial importance is growth enhancing non-farm activities. This calls for investment in rural infrastructure and skill upgradation and it also implies a careful examination and adjustment of macro-policies, which influence the relative profitability of different activities and in turn determine the nature and pace of diversification. In order to ensure social equity, policies on structural adjustment and reforms must pay special attention to the band of marginal and small farmers and agricultural labourers (Boehlje, 1999).

This new agriculture profoundly changes the competitive environment in the industry. In the commodity agriculture of the past, most agribusinesses had to compete only in terms of cost. If you were a low-cost supplier and did not expand beyond the sustainable growth rate of the business, you could expect to be successful – to survive and maybe even thrive in the long run. In the new agriculture that includes differentiated products and more tightly aligned marketing/distribution systems with producers being raw material suppliers for manufacturers and food processors, competition includes quality features and responsiveness or time to market as well as cost. In the agriculture of the future successful companies will need to be better, faster and cheaper to have a sustainable competitive advantage.

4. CONSERVATION AGRICULTURE: TOWARDS SUSTAINABLE, RESOURCE-CONSERVING CROP MANAGEMENT SYSTEMS

In recent years, farmers interested in sustainable crop production systems have begun to adopt and adapt improved crop management practices, a step toward CA, which may be considered the ultimate solution. CA, which focuses on the complete agricultural system, involves major changes in farm cropping operations from the widely used, traditional tillage-based farming practices. Appropriate CA technologies encompass innovative crop production systems that combine the following basic tenets (Sayre 1998; Derpsch 1999):

- **Dramatic reductions in tillage:** Ultimate goal – zero till or controlled till seeding for all crops in a cropping system if feasible.
- **Rational retention of adequate levels of crop residues on the soil surface:** Ultimate goal – surface retention of sufficient crop residues to protect the soil from water run-off and erosion; improve water infiltration and reduce evaporation to improve water productivity; increase soil organic matter and biological activity; and enhance long-term sustainability.
- **Use of sensible crop rotations:** Ultimate goal – employ economically viable, diversified crop rotations to help moderate possible weed, disease, and pest problems; enhance soil biodiversity; take advantage of biological nitrogen fixation and soil enhancing properties of different crops; reduce labour peaks; and provide farmers with new risk management opportunities.
- **Farmer conviction of the potential for near-term improved economic benefits and livelihoods from sustainable CA systems:** Ultimate goal – secure farm level economic viability and stability. To achieve this will involve the development of innovation systems focused on the needs of farmers and will include multiple agents who will use their comparative advantages to adapt the principles of CA to the farmers' various biophysical and socioeconomic conditions.

These basic tenets define an approach to crop and soil management that is not location-specific; i.e., the knowledge, the approach, and the fundamental and strategic principles are applicable to a wide range of crop production systems, from low-yielding, dry rain-fed conditions to high-yielding irrigated conditions. However, the optimum application of these techniques will vary across different agro-climatic situations. Specific and compatible management components (weed control tactics, nutrient management strategies, appropriately-scaled implements, etc.) will need to be developed through adaptive research with active farmer involvement to facilitate farmer adoption of CA under contrasting agro-climatic conditions and production systems, much as specific crop cultivar traits (grain colour, end-use quality characteristics, genetic disease resistance requirements, etc.) vary for specific production situations (Sayre, 1998).

Successful farmer adoption of the first three CA tenets will essentially alter generations of traditional farming practices and implement use (including hand hoes), especially for small- and medium-scale farmers in developing countries who may have had minimal exposure to new farming technologies. In fact, the change in mind-set not only by farmers but also by scientists, extension agents, private sector members, and policy-makers in developing as well as developed countries may be the most difficult aspect associated with the development, transfer, and farmer adoption of appropriate CA-based technologies (Wall, 2007).

In many cases, it may be difficult to explain the importance of CA adoption to farmers beyond its potential to reduce production costs, mainly by tillage reductions (Patino-Zuniga, 2009). It is therefore necessary to educate farmers on the links between excessive tillage and residue removal with soil sustainability problems, and how these problems can be alleviated through CA. Most small holder farmers are risk-averse and cannot afford reductions in farm productivity to try a new system. This makes it very important to show immediate benefits in increased productivity of land or labour from CA, apart from the effects of CA on soil and land regeneration. Increased, immediate profits will be relevant to more farmers than the delayed gratification of improving their land gradually through adoption of CA-based technologies. However, if the development, adaptation, and adoption of CA-based technologies are done properly, they can offer farmers and society both possibilities (Wall, 2007).

5. POTENTIAL CA CONTRIBUTIONS TO ENHANCED RESOURCE PRODUCTIVITY AND EFFICIENCY

For CA to succeed, farmers will need to be in the forefront, helping to identify, develop, and deploy new technologies. If we are to insure that adequate food, fibre, and fuel are produced, farmers must employ appropriate crop management technologies not only to stabilize or increase crop production in a cost-effective manner, but also to conserve the integrity and sustainability of their resource base. To achieve this, farmers, researchers, and policy-makers alike must insure that the following actions are properly implemented (Derpsch 2005; Patino-Zuniga 2009).

5.1 Increase the Efficiency of Water Use for Crop Production

Under rain-fed conditions this may be done by reducing run-off and evaporative losses, as well as reducing other limitations to crop productivity, to generate "more crop per rainfall drop." Figure 1 below illustrates how the use of sound CA-based technologies (zero till seeded maize in rotation with wheat with surface retention of all crops residues) provided consistently superior rain-fed maize yields over 10 years, as compared to the normal, tillage-based, farmer practice for maize production in the central highlands of Mexico. Maize is the predominate crop in these largely rain-fed highlands, which average around 500 mm^{-1} of rainfall annually with a range of 300 – 800 mm^{-1} which is erratically distributed—50 mm in an hour during an afternoon may easily be followed by two weeks without rain.

Improving the productivity of irrigation water is increasingly important: in many developing countries, especially in Asia, irrigated crop production accounts for a major portion of strategic food supplies, insuring food security. The water resources available for irrigation are becoming increasingly scarce and irrigated systems are becoming more fragile, especially with respect to increases in soil salinity from poor irrigation management.

The results contribute to the growing evidence that the use of furrow irrigation with raised bed seeding systems can provide striking increases in irrigation water use efficiency, especially permanent raised bed seeding systems where no tillage is used on top of the beds, but beds are "reshaped" as needed in the furrows. As a consequence, in the southern part of the state of Sonora in northwest Mexico, over the past 25 years nearly all farmers have shifted from flood irrigation to furrow irrigation for all crops, and they indicate that they obtain an average irrigation water savings of 20-25% with equal or higher crop yields (Sayre and Moreno Ramos, 1997).

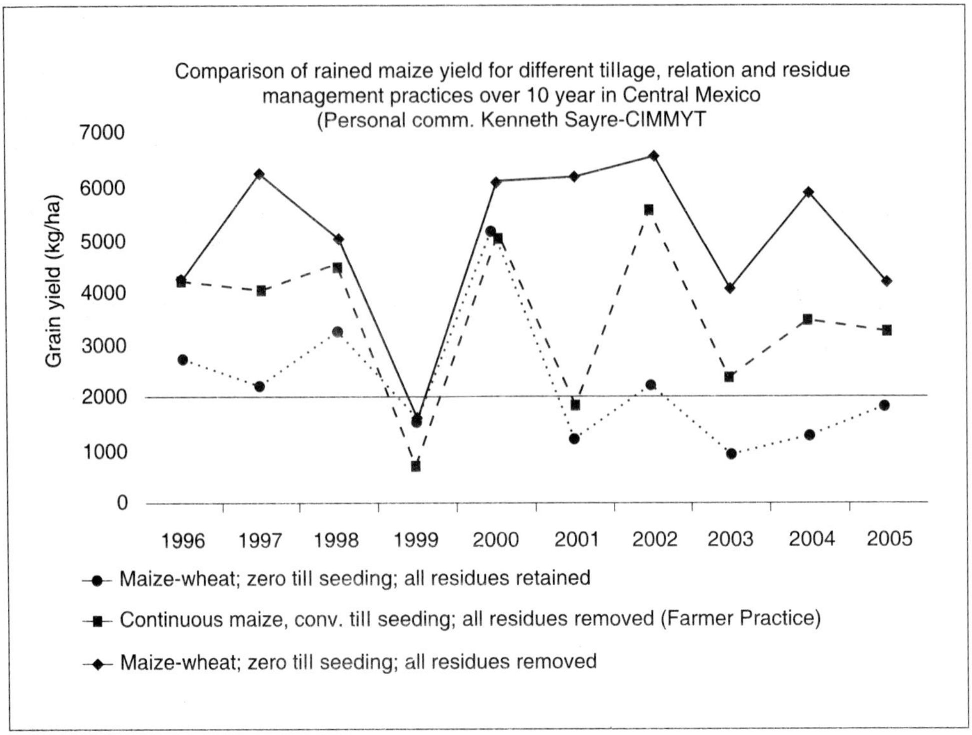

Fig. 1 Use of sound CA-based technologies provided consistently superior rainfed maize yields over 10 years, as compared to the normal, tillage-based, farmer practice for maize production in Mexico.

5.2 Halt and Reverse the Widespread Degradation of the Soil Resource Base

Soil degradation from wind and water erosion, as well as a decline in soil's physical, biological, and chemical properties, can be linked to excessive levels of tillage and extensive removal/burning of crop residues that are associated with many conventional farming systems. Some studies have indicated that conservation agriculture is an effective way of controlling erosion (Packer *et al.*, 1992; Uri *et al.*, 1999). Soil erosion in Brazil fell from 3.4-8.0 t/ha under conventional tillage to 0.4 t/ha under no-till, and water loss fell from approximately 990 to 170 t/ha (Sorrenson*et al.*, 1997, 1998). A watershed study in Brazil showed a 22% reduction in sediment load from 1994 to 1998 because of no-till adoption (Derpsch, 2001b). In Paraguay, 23 t/ha/year of soil are lost on average when using conventional agriculture whereas only 0.53 t/ha/year are lost using the no-till system (Venialgo, 1996).

5.3 Augment Crop and Soil Biodiversity

Augmenting cropping diversity through CA offers farmers' alternative, economically viable crop rotation options that help minimize ever-present economic risks and enables farmers to react to rapid fluctuations in markets. In addition, there is considerable evidence that sound crop rotations can result in positive yield increases for the different crops in a rotation (Sayre and Moreno Ramos, 1997).

5.4 Confront Increasing Input Prices by Boosting Input Use Efficiency to Reduce Production Costs

The increases in the prices of chemical fertilizers, as well as other inputs like herbicides, pesticides, and fuel in the last 18-24 months, have been astounding, especially in many developing countries. In some countries, farmers have access to only restricted amounts of fertilizers through government rationing programmes or have not been able to afford/acquire fertilizer. There is some concern that CA-based technologies using markedly reduced/zero till seeding systems combined with soil surface residue retention may lead to decreased fertilizer N-use efficiency, requiring the use of higher fertilizer rates to obtain similar yields as conventional till systems. When this has occurred, it has likely been due to factors associated with differing production situations. Obviously, suitable fertilizer management practices that are compatible with appropriate CA-based technologies may result in enhanced N-use efficiency (Sayre and Moreno Ramos, 1997). Figure 2 below illustrates how N fertilizer use efficiency for wheat was enhanced in a permanent raised bed furrow irrigated wheat-maize system in northwest Mexico when adequate residues were retained, as compared to the traditional conventional till system. The results show how the removal of residues on the permanent beds dramatically restricted N use efficiency (Sayre, 1998).

5.5 Reduce Agricultural-Related Greenhouse Gas (GHG) Emissions

Soil erosion and leaching of applied agricultural products (fertilizers and pesticides) has long been recognized as detrimental to the environment. However, greenhouse gas (GHG) emissions

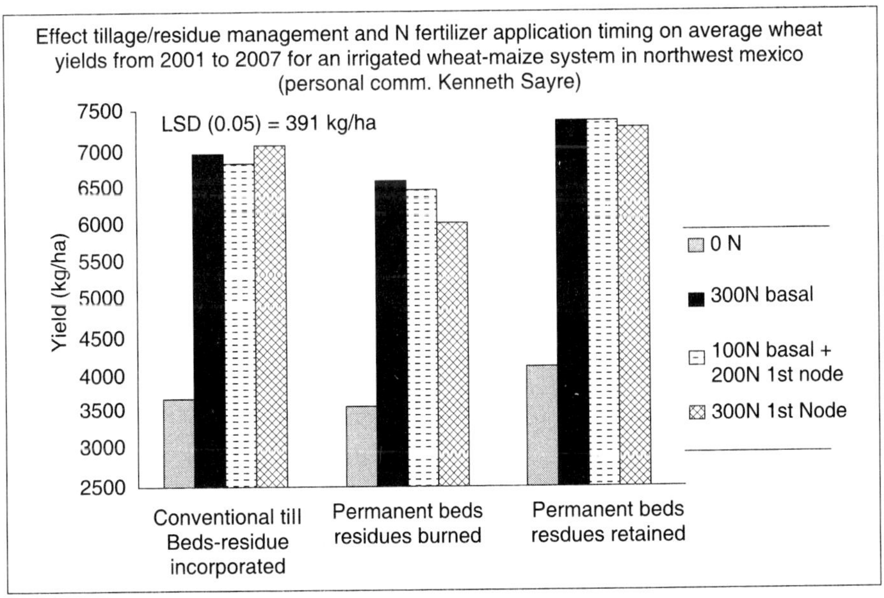

Fig. 2 N fertilizer use efficiency for wheat-maize system in northwest Mexico.

from agriculture, almost certainly linked to climate change, are a more recent concern and certainly more controversial.

Emissions of CO_2 by agriculture can be decreased by reducing tillage and maintaining crop residues on the soil surface to increase carbon sequestration in the soil (Reicosky, 2001), especially when combined with the reduced burning of fossil fuels for field operations associated with reduced/zero till seeding systems. In addition, the field burning of crop residues—yet a widespread practice in developing countries—should be stopped. At the same time NO_X emissions, far more damaging to the environment than CO_2, can be reduced by improving N fertilizer management and cutting methane emissions (also more damaging than CO_2) by limiting the extent of flooded rice cultivation where appropriate (Patino-Zuniga *et al.*, 2009).

If adequate levels of crop residues are retained on the soil surface and combined with reduced or zero till seeding systems, there is good evidence that C sequestration in the soil will occur with corresponding reductions in CO_2 emissions (Reicosky, 2001). Figure 3 illustrates the changes in the soil organic content under rain-fed maize production in the central highlands of Mexico with contrasting management practices. Zero till with full residue retention has resulted in a substantially higher soil organic matter content (SOM) (and therefore C content) as compared to conventional till with full residue removal, in this experiment established in 1991. This figure also illustrates that achieving a new equilibrium in soil properties through the adoption of CA is a long-term process.

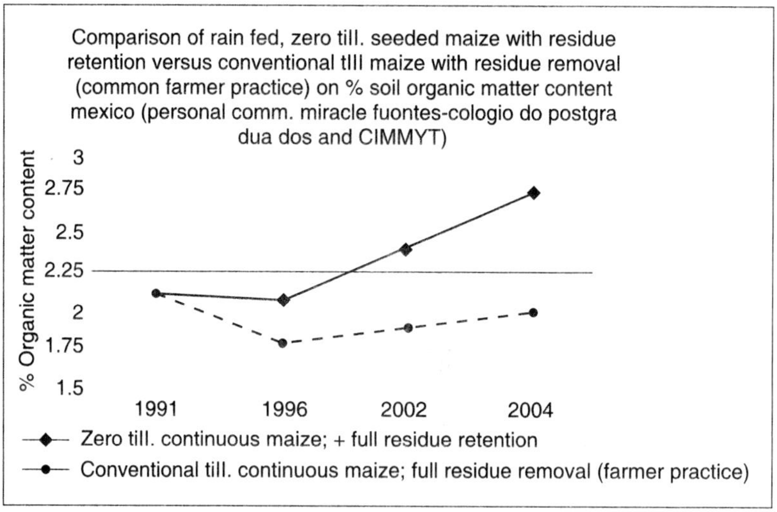

Fig. 3 Changes in the soil organic content under rain-fed maize production in the central highlands of Mexico.

5.6 Confront the Growing Shortages of Agricultural Labour

Agricultural labour shortages are growing even in the two most populous Asian countries—China and India—and this is causing many farmers to consider the adoption of CA-based technologies which, under most situations, can reduce labour requirements. One of the major benefits that small-holder farmers perceive with CA is the labour savings (Wall, 2007). In Asia, for example, hand transplanting of puddled rice after conventionally tilled, irrigated wheat has a high labour requirement that peaks in June and July (especially in northwest India). This creates serious labour shortages during this critical time, and has provoked farmer interest in technologies available for direct seeding rice without puddling. One of these technologies is direct (zero till) seeding of rice into dry soil after zero till wheat. Farmer interest is particularly keen in the lowland rice growing areas, where a major portion of the water used in rice production is provided by irrigation.

Although there are few reliable economic comparisons of dry seeded, zero till rice compared to transplanted, puddle rice, much more is known in relation to wheat. Variable costs are substantially lower for zero till (partly due to reduced labour costs) and gross benefits are higher, generating correspondingly higher net benefits for the zero till wheat. These results are similar to most other examples where farmers have adapted suitable CA-based technologies.

5.7 Reductions in Fuel and Machinery Use

Along with reductions in labour, in mechanized systems CA results in a marked reduction in the use of tractors and equipment, all of which cut fuel use, reducing both farmers' costs and GHG emissions. Generally, CA reduces tractor use by approximately 70%, depending on the intensity of tillage in the conventional system (Wall, 2002). The reduction in tractor use means that a single tractor can provide the required traction for a greater area. In the example of Bolivia (Wall, 2002), this provided for the expansion of the agricultural area using existing tractors, but in the Indo-Gangetic plains (especially in northwest India) it has meant that relatively large-scale farmers could become service providers to the smaller farmers in the community. Anyone who has watched a tractor working dry soil, especially alluvial soil, cannot help but notice the cloud of dust in which the tractor and the operator have to work. In CA systems, where residue covers the soil and is not tilled, dust is almost completely absent, benefitting the operator and the farmer, since tractors and equipment operating in dust-free environments require considerably less maintenance and their useful life is extended, reducing again the costs of production.

5.8 Land Degradation and Its Impact on Agriculture

Land degradation is a serious environmental problem that threatens ecosystem health and food security worldwide. This environmental problem has been a major global issue during the 20th century and will remain high on the international agenda in the 21st century (Eswaran *et al.*, 2001). Though methods to assess land degradation on a large-scale are deficient, there is growing concern that degradation of agricultural soil resources is already seriously limiting production and diminishing ecosystem services (Vlek, 2008). One major hypothesis for the observed improvement in land productivity (vegetation greenness) revealed in recent regional and global studies is the potential impact of atmospheric fertilization (Vlek *et al.*, 2008; Hellden and Tottrup, 2008). In fact, there are many studies that highlight the benefits of CO_2 –or NO_X- fertilization to enhance plant growth and development (Lewis *et al.*, 2004; Ainsworth and Long, 2005; Long *et al.*, 2006; Reay *et al.*, 2008). In an attempt to disentangle the impact of atmospheric fertilization

enable the diffusion of the technology. For its successful implementation in developing regions where it is needed most, the design and dissemination of cost-effective farming tools, access to herbicides and economic incentives will be required in addition to creating awareness.

Acknowledgements

The authors acknowledge Dr. G. C. Mitra, Emeritus Professor, CSIR, who made them aware of the basic aspects of plant biology.

References

Ainsworth, E.A. and Long, S.P. (2005). What have we learned from 15 years of free-air CO_2 enrichment (FACE)? A meta-analytic review of responses to rising CO_2 in photosynthesis, canopy properties and plant production. *New Phytologist.* 165: 351-372.

Boehlje, M. and Schrader, L.F. (1996). Agriculture in the 21st Century. Journal of Production Agriculture. 9(3): 335-340.

Derpsch, R. (1999). Expansiónmundial de la SD y avancestecnológicos. In Proceedings of the 7th National Congress of AAPRESID, 18-20 August 1999. Mar del Plata, Argentina.

Derpsch, R. (2001b). Conservation tillage, no-tillage and related technologies. I World Congress on Conservation Agriculture, Madrid, 1-5 October 2001.

Derpsch, R. (2005). The extent of Conservation Agriculture adoption worldwide: Implications and impact. III World Congress on Conservation Agriculture: Linking Production, Livelihoods and Conservation, 3rd to 7th October 2005, Nairobi, Kenya, p.15.

Devender, T. V., Dimmit, M. and Mcginley, M. (2009). Grasses. In: Encyclopedia of Earth. Eds. Cutler J. Cleveland (Washington, D.C.: Environmental Information Coalition, National Council for Science and the Environment). Accessed December 18, 2009. Available at http://www.eoearth.org/article/Grasses.

Dhar, Biswajit and Murali Kallummal (2004): 'Trade Liberalisation and Agriculture: Challenges Before India' in http://www.macroscan.com government of India (2005): Agricultural statistics at a Glance, Ministry of Agriculture.

Elliott, E.T., (1986). Aggregate structure and carbon, nitrogen and phosphorus in native and cultivated soils. *Soil Science Society ofAmerica Journal.* 50: 627-633.

Eswaran, H., Lal, R. and Reich, P.F. (2001). Land degradation: an overview. In: Bridges, E.M., I.D. Hannam, L.R., Oldeman, F.W.T., Pening de Vries, S.J. Scherr, and S. Sompatpanit (eds.). Responses to Land Degradation. Proc. 2nd.International Conference on Land Degradation and Desertification, KhonKaen, Thailand.Oxford Press, New Delhi, India.

FAO, (2002). The Conservation Agriculture Working Group Activities 2000 – 2001. Food and Agriculture Organization of the United Nations, Rome, pp. 25.

FAO, (2008). Conservation Agriculture Carbon Offset Consultation – West Lafayette, Indiana, USA, 28-30 October 2008.

FAO. (1995). Code of Conduct for Responsible Fisheries. Rome, FAO. (1995). p. 41.

Fishchetti, M. (2011). Observations-opinion, arguments, and analyses from the editors of Scientific American. Accessed from October 12, 2011. Available at www. Scientific American.com.

Glantz, M.H. (1998). Creeping Environmental Problems in the Aral Sea Basin. In Kobori, I., and M.H. Glantz, eds., Central Eurasian Water Crisis: Caspian, Aral and Dead Seas. New York: United Nations University Press.

Gulati, A. and Kelley, T. (1999), "Trade Liberalisation and Indian Agriculture", Delhi: Oxford University Press.

Harris, J. M. and Cleveland, C. (2007). Market. In: Encyclopaedia of Earth. Eds. Cutler J. Cleveland (Washington, D.C.: Environmental Information Coalition, National Council for Science and the Environment). Available at http://www.eoearth.org/article/Market

Hazra, C. R. (2001). Crop diversification in the Asia-Pacific region. FAO Corporate Document Depository. Available at http://www.fao.org/docrep/003/x6906e/x6906e00.htm.

Hazra, C.R. (2005). "Crop Diversification In India" report prepared for ministry of agriculture in India

Hellden, U. and Tottrup, C. (2008). Regional desertification: a global synthesis. Global and Planetary Change. Doi:10.1016/ j.gloplacha.2008.10.006.

Henao, Julio and Carlos Baanante, (2006). Agricultural Production and Soil Nutrient Mining in Africa; International Fertilizer Development Center, Muscle Shoals, Alabama, USA.

Kay, B.D. (1990). Rates of change of soil structure under different cropping systems, *Adv. Soil Sci.* 12, 1-52.

Long, S.P., Ainsworth, E.A., Leakey, A.D.B., Nösberger, J. and Ort, D.R. (2006). Food for Thought: Lower-than-expected crop yield stimulation with rising CO2 concentrations. Science, 312, 1918-1921.

Lumpkin, T. A. and Sayre, K. (2009). Enhancing Resource Productivity and Efficiency through Conservation Agriculture. 4th World Congress on Conservation Agriculture, February 4-7, New Delhi, India. p. 3-9.

Madhavan, R and Renuka, (2003), "Productivity Growth in Indian Agriculture: The Role of Globalisation and Economic Reform, Asia Pacific Development Journal, Vol. 20, No. 2, pp, 57-72.

Martius, C., Tiessen, H. and Vlek, P.L.G. (2001). The management of organic matter in tropical soils: what are the priorities. Nutrient cycling in agroecosystems, 61: 1-6.

Olson, K. (1998). Precision agriculture:current economic and environmental issues. Minneapolis, Minn: Sixth Joint Conference on Food, Agriculture and Environment, August. 31-sept. 2.

Packer, I.J., Hamilton, G.J. and Koen, T.B. (1992). Runoff, soil loss and soil physical properties changes of light textured surface soils from long-term tillage treatments, Aust. J. Soil Res. 30 (1992), pp. 789-806.

Papendick, R.I. and Parr, J.F. (1997). No-till farming: the way of the future for a sustainable dry land agriculture, Ann. Arid Zone 36 (1997), pp. 193-208

Paroda, R.S and Arora, A.K. (1991). Plant Genetic Resources; conservation and Management. IBPGR Publication, Malhotra Publishing House, India, pp. 392.

Paroda, R.S. and Mal, B. (1989). New plant sources for food and industry in India. In:Wickens, G.E., Haq, N. and Dey, P. (eds). New crops for food and industry.Chapman and Hall, London, pp. 135-149.

Paroda, R.S., Arora, A. K. and Chandel, K.P.S. (1987). Plant Genetic Resources-Indian perspective. National Bureau of Plant Genetic Resources, New Delhi, India, pp. 539.

Patino-Zuniga, L., Ceja-Navarro, J.A.B., Govaerts, M., Luna-Guido, Sayre, K. D. and Dendooven, L. (2009). The effect of different tillage and residue management practices on soil characteristics, inorganic N dynamics and emissions of N20, CO2 and CH4 in the central highlands of Mexico: a laboratory study. Plant Soil. 314: 231-241.

Perry, J., and Banker, D. (2000). Contracting changes how farms do business. Rural Conditions and Trends. Washington, D.C.: US. Department of Agriculture Economic Research. 10(2): pp. 50-55.

Pidwirny, M., and Heimsath, A. (2010). Soil. In: Encyclopedia of Earth. Eds. Cutler J. Cleveland (Washington, D.C.: Environmental Information Coalition, National Council for Science and the Environment). Available at http://www.eoearth.org/article/Soi

Pimentel, D., Hogan, C.M., and Bertaglia, M. (2010). Agriculture. In: Encyclopedia of Earth. Eds. Cutler J. Cleveland (Washington, D.C.: Environmental Information Coalition, National Council for Science and the Environment). Accessed May 26, 2010. Available at http://www.eoearth.org/article/Agriculture.

Pimenttel, D., Hepperly, P., Hanson, J., Douds, D. and Seidel, R. (2005). Environmental, Energetic, and Economic Comparisons of Organic and Conventional Farming Systems.BioScience. 55(7): 573-582.

Rai, M. (2011). Sustainable agricultural production intensification-a new paradigm. In: Proceedings of National Symposium on Technological Interventions for Sustainable Agriculture, 3-5 May, Ranichauri, India: p. 1.

Rao, C. H. (2005), "Agriculture, Food Security, Poverty and Environment Essays on Post Reform India, Oxford University Press.

Reay, D.S., Dentener, F., Smith, P., Grace, J. and Feely, R. (2008). Global nitrogen deposition and carbon sinks. Nature Geoscience 1, p. 430-437.

Reichert, J.M. and Norton, L.D. (1994). Aggregate stability and rain-impacted sheet erosion of air-dried and prewetted clayey surface soils under intense rain, Soil Sci. 158 (1994), pp. 159-169.

Reicosky, D.C. (2001). Effects on soil organic carbon dynamics: Field experiments in the U. S. corn belt. In D. E. Scott, R. H. Mohtur and G. C. Steingerdt (eds). Sustaining the Global Farm: Selected papers from the 10th International Soil Conservation Organization Meeting held May 24-29, 1999 at Purdue University and the USDA-ARS National Soil Erosion Laboratory. pp 418-485.

Reicosky, D.C. and Saxton, K.E. (2007). Reduced environmental emissions and carbon sequestration. In: Baker, C.J., Saxton, K.E., editors. No-tillage Seeding in Conservation Agriculture. 2nd edition. Rome, Italy: FAO and CAB International. p. 257-267.

Salonius, P. (2008). Intensive Crop Culture for High Population is Unsustainable. Available at http://www.mindfully.org/Farm/2008/Intensive-Crop-Unsustainable10feb08.htm.

Sayre, K.D. (1998). Ensuring the use of sustainable crop management strategies by small wheat farmers in the 21st century.Wheat Special Report No. 48. Mexico, D.F. CIMMYT.

Sayre, K.D. and Moreno Ramos, O. H. (1997). Applications of raised-bed planting systems to wheat. Wheat Special Report No. 31. Mexico, D. F.: CIMMYT.

Sharma, A.K., Metha, P. and Sharma, S.K. (1996). Evaluation of Risk, in Diversified Crop Faming of Himachal Pradesh-A case study of Vegetable Farms. Indian Journal of Production Agriculture. 9(3):335-340

Singh, A., Boukerrou, L. and Miller, M. (2009). Diversification in agriculture. In: Encyclopedia of Earth. Eds. Cutler J. Cleveland (Washington, D.C.: Environmental Information Coalition, National Council for Science and the Environment). Accessed on Nov. 18, 2009. Available at http://www.eoearth.org/article/Diversification_in_agriculture.

Sorrenson, W.J. (1997). Financial and economic implications of no-tillage and crop rotations compared to conventional cropping systems. TCI Occasional Paper Series No. 9, Rome, FAO.

Sorrenson, W.J., Lopez-Portillo, J. and Nuñez, M., (1997). The economics of no-tillage and crop rotations in Paraguay. Policy and investment implications. MAG/GTZ/FAO, p. 215.

Tyner, W.E. and Boehlje, M. (1997). Food System 21: Gearing up for the New Millenium. EC710, Perdue University Cooperative Extension Service, Dept. of Agricultural Economics and the Faculty of the School of Agriculture. West Lafayette, IN.

Uri, N.D., Atwood, J.D. and Sanabria, J. (1999). The environment benefit and cost of conservation tillage, Environ. Geol. 38 (1999), pp. 111-125.

Venialgo, N. (1996). Efecto de dos sistemas de labranza en el control de la pérdida de sueloporerosiónhídrica. Trabajopresentado en el II EncuentroLatinoamericano de SiembraDirecta en PequeñasPropiedades, Edelira, Paraguay, 11 – 14 de marzo 1966.

Vlek, P.L.G. (2008). The Incipient Threat of Land Degradation. Journal of the Indian Society of Soil Science, Volume: 56, Issue: 1.

Wall P.C. (2007). Tailoring Conservation Agriculture to the needs of small farmers in developing countries: An analysis of issues. Journal of Crop Improvement 19, 137-155.

Wall, P.C. (2002). Extending the Use of Zero Tillage Agriculture: The Case of Bolivia. Paper presented at the International Workshop on Conservation Agriculture for Sustainable Wheat Production in Rotation with Cotton in Limited Water Resource Areas, Tashkent, Uzbekistan, October 13-18, 2002.

8

Contingency Crop Planning for Aberrant Weather Situations

Ayon Alipatra[1], Hirak Banerjee[2*] and Krishnendu Ray[2]

[1]*Mandan Bharti Agriculture College, Agwanpur, Saharsa-852201, Bihar, India*
[2]*Bidhan Chandra Krishi Viswavidyalaya, Mohanpur-741252, West Bengal, India*

ABSTRACT

Second green revolution is urgently needed to feed worlds' growing population. With the rising demand for food in the coming decades, India will have to depend on the rain-fed areas (IFPRI, USA). In India out of 108 mha of total rain-fed area about 47 mha is drylands (nearly 43%) which is more than one-third of the net sown area and their contribution to total foodgrain production is about one-fifth in the country. The main climate-related factors that would affect agricultural productivity in the coming decades are changes in temperature, precipitation and finally short-term weather variability. Indian agriculture is largely rainfall dependent, especially where the quantity and distribution of summer monsoon rain decides the crop production to a great extent. Weather is the most variable factor in rain-fed crop production which can be least controlled. Unfavourable deviated weather from the normal expected one is called aberrant weather. Aberrant weather is a common feature of rain-fed agriculture which includes early and delayed onset of monsoon, normal onset of monsoon followed by immediate prolonged dry spell, early cessation of rains, extended monsoon, variation in quantum of rainfall received, high intensity of rainfall, high atmospheric temperature, low relative humidity, hot dry winds, high atmospheric water demand etc. The adverse effect of these aberrations on crop growth varies with the degree of deviation and the crop growth stage at which such deviations occur. Contingency crop planning has been evolved for efficient weather management which refers to mitigate any unexpected, unfavourable and hence unwanted accidental weather situations occurring at any time without prior knowledge before the crops are sown or even after the crops are sown. Therefore, the contingency crop planning is proposed to mitigate such situation through the choice of appropriate crops and varieties, cropping systems or other necessary relevant farm practices which may have to be developed for different situations like rainfall aberration, drought and flood. Contingency cropping is highly location-specific due to variation in amount and distribution of rainfall. Therefore, proper planning and adoption of location-specific contingency crop planning for varied weather conditions needs to be worked to effectively improve livelihood of vulnerable small holders.

Keywords: Aberrant weather, contingency crop planning, drought, flood, rainfall.

*Corresponding author's E-mail ID.: *hirak.bckv@gmail.com*

Roychowdhury R. (Ed.), Crop Improvement in the Era of Climate Change.

1. INTRODUCTION

Second green revolution is urgently needed to feed worlds' growing population. According to the International Food Policy Research Institute (IFPRI), Washington "with the rising demand for food in coming decades India will have to depend on the rain-fed areas to help increase supply". In India out of 108 mha of total rain-fed area about 47 mha is categorized as to drylands (nearly 43%). These areas cover 84 districts which is more than one-third net sown area and their contribution to total foodgrain production is about one-fifth in the country.

The main climate-related factors that would affect agricultural productivity in the coming decades are changes in temperature, precipitation and finally short-term weather variability. Indian agriculture is largely rainfall dependent, especially where the quantity and distribution of summer monsoon rain decides the crop production to a great extent. Since the food production in the country largely depends on the monsoon behaviour, many efforts have been made to understand and predict the monsoon variability. Yet the variability of summer monsoon is still less predictable (Joshi and Kar, 2007). In recent years some predictions have been made regarding disparity in the changes in distribution of rainfall and temperature (Table 1). It is also predicted that by 2071 the overall summer monsoon in India will increase by 20%, extreme rainfall events would rise sharply especially in parts of Gujarat, Maharashtra, Madhya Pradesh, Karnataka and Andhra Pradesh. The onset of summer monsoon could become more variable. Overall, the changes are likely to increase by 10%. So the predictions clearly indicate that agriculture will be more challenging and thereby threatening the food security and livelihood security of farming communities. Such a scenario warrants careful crop planning and management, as well as development of suitable varieties.

Table 1 Expected changes in temperature and rainfall in India

Year	*Season*	*Change in temperature (°C)*		*Change in rainfall (%)*	
		Min.	*Max.*	*Min.*	*Max.*
2020s	Kharif	0.87	1.17	–1.81	5.10
	Rabi	1.08	1.54	1.95	4.36
2050s	Kharif	1.81	2.37	7.18	10.52
	Rabi	2.54	3.18	–9.22	3.82
2080s	Kharif	2.91	4.62	10.10	15.18
	Rabi	4.14	6.31	–24.83	–4.50

2. WEATHER: MOST VARIABLE FACTOR IN CROP PRODUCTION

Unfavourable deviated weather from the normal expected one is called aberrant weather. Aberrant weather is a common feature of rain-fed agriculture. Monsoon aberration experienced in 2002 has been depicted in Fig. 1.

Generally following aberrations in weather behaviour may occur, *viz.*

2.1 Rainfall Characteristics

Crop production in dry areas is mainly dependent upon rainfall received during the south-west monsoon. The rainfall is seasonal, erratic and highly variable with space and time. Vagaries of

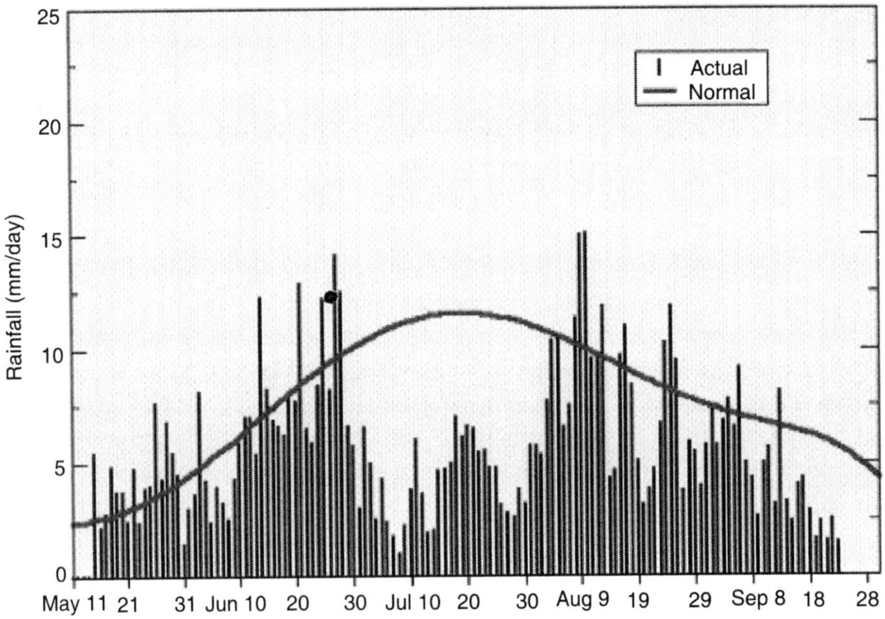

Fig. 1 Monsoon aberration in India during 2002.

monsoon rains and breaks of varying duration are most common in India. Commonly observed important aberrations in the rainfall behaviour are depicted below.

- Early onset of monsoon.
- Normal onset of monsoon followed by immediate prolonged dry spell.
- Delayed onset of monsoon.
- Early cessation of rains.
- Extended monsoon.
- Variation in quantum of rainfall received.
- High intensity of rainfall in association with the movement of cyclones/depressions.

2.2 High Atmospheric Temperature

Because of high temperature the atmospheric demand for moisture increases causing high evapotranspiration (ET) losses resulting in moisture stress.

2.3 Low Relative Humidity

It results in high ET losses causing moisture stress.

2.4 Hot Dry Winds

It causes desiccation of leaves. High turbulent winds especially during summer months cause soil erosion resulting in dust storms and loss of fertile soil.

2.5 High Atmospheric Water Demand

Due to this condition the potential ET exceeds the precipitation during most part of the year.

3. EFFECT OF ABERRANT WEATHER CONDITIONS ON CROPS

In dry farming areas rainfall behaviour is erratic and uncertain. The common deviations in rainfall behaviour include delayed onset, early withdrawal and intermediary dry spells during rainy season. The adverse effects of these rainfall aberrations on crop growth vary with the degree of deviation and the crop growth stage at which such deviations occur (Table 2).

Table 2 Effect of erratic rainfall pattern on crops

Rainfall pattern	*Effect on crops*
Delay in commencement	Reduced length of cropping season or duration; delayed sowing
Early withdrawal or cessation	Moisture stress at maturity leading to more percentage of chaffy grain
Occurrence of dry spells	
a. Immediately after sowing	Germination will be affected thereby reducing plant population
b. At vegetative stage	Stem elongation, leaf area expansion, branching or tillering will be affected
c. At flowering stage	Reduced grain/pod number due to poor anthesis and pollination
d. At ripening stage	Grain filling as well as grain size will be reduced

4. APPROACH TOWARDS EFFICIENT WEATHER MANAGEMENT

Weather is the most variable factor in rain-fed crop production which can be least controlled. According to Clarkson *et al.* (2001), awareness about weather and climate extremes, variability and change will impact on farm operations. Out of various approaches contingent plans have been evolved over the years for efficient weather management. Suitable manipulations in crop management practices are needed to minimize adverse effects of abnormal rainfall behaviour. These management decisions constitute contingency planning. Such management practices done after crop establishment and in the middle of crop growth are called mid-season or mid-term corrections. Some of the management options that can be adopted to reduce the adverse effects of aberrant weather are discussed here.

4.1 Contingency Crop Planning

Contingency planning refers to mitigate any unexpected, unusual, unfavourable and hence unwanted accidental weather situations occurring at any time without prior knowledge before the crops are sown or even after the crops are sown. Therefore, the contingency crop planning is proposed to mitigate such situation through the choice of appropriate crops and varieties varieties, cropping systems or other necessary relevant farm practices (Addy, 2009).

4.1.1 Delayed Onset of Monsoon

In recent years the onset of south-west monsoon gets delayed. So the crops/varieties which are regularly grown in a particular region cannot be sown in time. As a result delayed sowing of the crops can lead to reduced and even uneconomical crop yields. To manage with the possible situation due to late onset of monsoon, farmers of arid, semi-arid and tribal areas of sub-humid regions are being advised to adopt following technologies.

- Watershed approach in conservation of soil and water
- Conservation of run-off water in farm pond
- Preparation of land in advance of monsoon
- Maximum use of farm yard manure
- Application of fertilizer as per recommendations
- Sowing of groundnut, sorghum, cotton in *kharif* season immediately after onset of monsoon.
- Alternate crops/varieties: Certain crops/varieties can perform better even if sown late in the season, depending on the receipt of rains (Table 3).

Table 3 State-specific alternate crops/varieties under late sown condition

State	*Existing crops*	*Alternate crops*
Andhra Pradesh	Pigeonpea	Castor (var. Aruna)
	Groundnut	Sunflower
Rajasthan	Pearl millet	Green gram and cowpea

- Maintenance of optimum plant population per unit area
- Adoption of recommended short duration varieties of suitable crops
- Adoption of intercropping system to reduce the risk of crop failure
- Conservation of moisture by timely inter-culture and mulching
- Complete control of weeds
- Timely adoption of plant protection measures

4.1.2 Prolonged Dry Spell Immediately After Sowing

Breaks of monsoon for 7-10 days may not be a serious concern, but breaks more than of 15 days duration especially at critical stages leads to reduction in yield. Immediately after sowing of crop, dry spell might result in poor germination due to soil crusting, withering of seedlings and poor establishment of crop stand. The adverse effect of moisture stress should be mitigated with top priority to avoid total crop failure. It is necessary to maintain proper plant stand to ensure better yield. Therefore, it is better to resow the crop than to continue with inadequate plant stand.

4.1.3 Mid-Season Correction or Dry Spells During Crop Period

After sowing of crops and during early growth period, any unexpected or unfavourable weather abnormality like long dry spell of 3 to 6 weeks, drought or early cessation of monsoon may occur. Unexpected or accidental occurrence of unfavourable weather after sowing can be overcome by adopting contingency crop plan. To mitigate the damaging effects of droughts after sowing of crop following mid-season corrections are recommended.

(i) **Ratooning:** The rate of soil moisture depletion increases with increasing leaf area. If drought occurs at 40-50 days after sowing (DAS), reduction in leaf area either by ratooning or thinning plant population can mitigate the adverse effects of drought. Sorghum and pearl millet respond well to ratooning. Spraying of 2% urea after drought period is useful for the crops like castor, pigeon pea and groundnut.

(ii) **Mulching:** If the break in monsoon is very brief, soil mulching has been found to be effective measure in extending the period of storage of water in the soil profile due to reduced heat load on the soil and hence the evaporation losses are minimized which in turn leads to extended periods of water availability. Repeated inter-cultivation in black soils results in formation of soil mulch on the surface. It acts as barrier to evaporation loss from soil profile. Deep cracks formed due to drought increases the loss of stored water by evaporation. Soil mulch minimizes deep cracking leading to reduced evaporation losses.

(iii) **Thinning:** If drought occurs during pre-flowering stage thinning is useful. Every third plant or alternate row may be removed to reduce the soil moisture loss and preserve it. Moisture stress of 35-50 days can be lessened by such thinning.

(iv) **Weed control:** Weed control is very essential to save the crop from harmful effect of drought as weeds rob soil moisture and nutrients as well.

(v) **Rainwater harvesting and recycling:** During the periods of moisture stress, *in-situ* water harvesting and/or run-off recycling are the potent measures for crop life saving. There is no control over the receipt of rainwater but it can be managed in a better way after it is being stored. Supplementary irrigation provided to the crop through rainwater harvesting and recycling particularly during the grain formation stage will be very much effective. Run-off recycling holds immense prospects in deep black soil where the seepage losses are very less. High intensity rains are quite common during the rainy season contributing 75% of seasonal rainfall. The inevitable run-off (10-40% of the total rainfall) could be stored in dug-out ponds of appropriate sizes and recycled to the donor area, in the event of severe moisture stress.

(vi) **Stripping of crop leaves:** Generally, photosynthesis occurs in the upper 3-4 leaves. Hence excluding these leaves, lower leaves from the plant may be removed to reduce transpirational losses.

(vii) **Intercropping and risk distribution:** Application of meteorological information in terms of the frequency and probability of breaks in monsoon rains can be made to select a combination of crops of different durations. The choice of intercrop depends upon the pattern of rainfall (Table 4). Three main possibilities are:

- Rainfall more uncertain in the early part of season.
- Rainfall more uncertain in the later part of season.
- Rainfall more or less uniformly distributed

4.1.4 Early Withdrawal of Monsoon

This situation is equally or more dangerous than late onset of monsoon. Rainy season crop will be subjected to terminal stress leading to poor yields. West Bengal, receives rainfall from middle

Table 4 State-specific alternative intercropping system

State	*Alternative intercropping system*
Haryana	Pearl millet + cowpea
Uttar Pradesh	Pearl millet + green gram
Karnataka	Pigeon pea + cowpea
Andhra Pradesh	Sorghum + pigeon pea
Maharashtra	Sorghum + green gram

of June to September, but the distribution of rainfall has been extremely uneven and irregular. The amount of rain and its distribution during the monsoon season also varies to great extent from year to year at all places. Hence, it is difficult to maintain and increase the per hectare production without irrigation especially in case of early withdrawal of monsoon. Under this condition the crop requires one or two protective irrigation to avoid crop failure. As each crop has certain critical stages, it is better to take care of these stages under short supply of water for irrigation to obtain higher water use efficiency. The critical stages of irrigation for *kharif* crop are given in Table 5.

Table 5 Critical stages for irrigation in *kharif* crop

Crops	*Most critical stages*
Paddy	Initiation of ear head and milking stage
Jowar	Boot stage and milking stage
Groundnut	Flowering stage, peg and pod formation
Maize	Silking and milking stage
Cotton	Boll formation
Tobacco	Growth stage
Sugarcane	Tillering and maximum vegetative growth stage
Castor	Flowering, pod formation and seed setting stage
Pulses	Flowering and pod formation
Sesamum	Tillering, flowering and capsule formation

Similarly post-rainy season crops fail due to inadequate available soil moisture (Reddy and Reddy, 2011). Early withdrawal of monsoon before sowing in *rabi* season seeding leads to problem of crop stand establishment. Therefore, sowing should be done in early September to avoid the above two problems. All the suggested measures to mitigate the adverse effect of prolonged dry spells during crop period are necessary to overcome the soil moisture stress during the early withdrawal of monsoon.

4.1.5 Extended Monsoon

Extended monsoon is seldom a problem in present-day agriculture. It benefits long duration *kharif* crops taken as component crops in intercropping system and is also advantageous for *rabi* crops.

5. LOCATION-SPECIFIC CONTINGENCY CROPPING

Contingency cropping is growing of a suitable crop in place of normally sown highly profitable crop of the region due to aberrant weather conditions. Depending upon the date of receipt of rainfall, crops are selected. Contingency cropping is highly location-specific due to variation in amount and distribution of rainfall. For example in arid regions, the spatial distribution of rainfall is highly variable. It is very common that rainfall received varies from field to field in the same location. Temperature gradually falls from August onwards reaching minimum in December. Contingency plan and mid-term corrections vary with the type and time of occurrence of rainfall aberration (Table 6).

Table 6 Contingency crop plan for different rainfall abnormalities

Rainfall abnormality	*Contingency plan and mid-term correction*
1. Delayed onset of rainfall	
a. Delay exceeding 4 weeks	Short duration crops
i. Delay in South-west monsoon (Normal - June)	
Delay upto July 31	
	Finger millet/pearl millet
Delay upto August 31	
	Sama (little millet)/Cowpea
ii. Delay in North-east monsoon (Normal - October)	
Delay upto November 15	
	Sunflower/Pearl millet/Finger millet
Delay upto November 30	
	Coriander/Chick pea
b. Delay of 1 - 2 weeks	Short duration varieties of same crop, e.g., Co 19 (150 days) and Co 25 (110 days) for sorghum; Co 5 (130 days) for redgram.
2. Early withdrawal of rainfall	Anti-transpirant spray, harvesting of millets at physiological maturity for fodder purpose
3. Intermediary dry spell	
a. Immediately after sowing	Gap filling with subsequent rains under reduced crop stand (< 20%). If stand reduction is more than 20% re-sown the crop, mulching especially dust mulch to reduce evaporation
b. At vegetative stage	Mulching, anti-transpirant spray, spraying potassium chloride, thinning of 33-50% population
c. At flowering stage	Anti-transpirant spray, harvesting for fodder and ratooning with subsequent rains in millets (e.g., sorghum)
d. At ripening stage	Anti-transpirant spray, harvesting at physiological maturity for fodder purpose

Crops have to be selected with suitable crop duration to coincide with the length of the growing season. Generally, short duration pulses like green gram, black gram and cowpea may suit the situation. Farmers with economic strength and motivation for high profits with some amount of risk can go for crops of long duration crops with flexibility or elasticity in yield. Pearl millet and sorghum can be ratooned if monsoon extends. Sunflower can be introduced for higher profits with certain amount of risk. Crops like sorghum, pearl millet, etc. can be grown for grain if monsoon extends and if not, fodder can be obtained. Contingency crops for different situations and regions are suggested below (Table 7).

Table 7 Contingent crops for different locations of India

Location	*Onset of monsoon*	*Crops with suitable varieties*
Dantiwada (Gujarat)	July 15	Pearl millet (GHB 32, GHB 39); Sorghum (CSH 5, CSH 6); Castor (GAUCH-1, CAUCH-4)
	July 25	Sorghum (CSH-6, GJ-35); Castor (GAUCH-1, GAUCH-4); Cluster bean (Malosan, HG 75)
	August 15	Castor (GAUCH-1, GAUCH-4); Sorghum (fodder) (Malavan, S-1049); Pearl millet (fodder)
	August 31	Sorghum (fodder); Sesamum (Purva-1); Pearl millet (fodder)
	September 15	Sorghum (fodder); Sesamum; Maize (fodder)
Hisar (Haryana)	Up to July 20	Pearl millet (HHB-50, HHB-45)
	August 15	Cluster bean (FS-227)
	October 15	Bengal gram (G-24, C-214)
	October 30	Mustard (RH-30)
Jodhpur (Rajasthan)	July 20	Pearl millet; Green gram

6. DROUGHT SITUATION

In recent past, agricultural productivity is greatly affected by drought condition in most of the agriculturally sound state of the country. Drought is the single natural calamity which leads to significant loss in foodgrain production, reduction in water resources, ultimately causes threat to livelihood security to rural poor. It also affects the agricultural GDP and the country faces great economic constraint. Impact of drought on agricultural production, water resource and socio-economic status of people in India during 2002 has been summarized in Table 8.

6.1 Contingent Measure Against Drought

The occurrence of drought and flood is common in India. It is an irony that our country receives nearly 1500 mm of rainfall annually of which 80% is received in a span of 4 months (June-September) which is quantitatively enough for most of the crop needs. However, the aberration in temporal and spatial distribution makes the crop vulnerable to drought as well as flood. Such adverse effects on crops can be combated in two ways — preventive measures and ameliorative measures.

Table 8 Impact of drought on agricultural production, water resource and socio-economic status of people in India during 2002

Areas	*Drought impact*
Agricultural production	• 21.53 million hectares remained unsown
	• 47 million hectares sown crops were damaged
	• Foodgrain production was reduced to the tune of 29 million tonnes
Water resources	• 33% less storage in 77% reservoirs (33% less than last 10 years average)
	• In drought affected areas groundwater goes below normal levels table by 2-4 m
	• 1.5 billion litres of water per day transported to drought affected areas
Livestock	• 150 million cattle affected
	• Milk procurement was less than 22% in Rajasthan, 8% in MP and 7% in Tamil Nadu
Livelihood	• 300 million people in 180 million hectare area were affected
	• Loss of job for 1250 person-days
Economic status	• Agricultural GDP was reduced by 3.1%
	• Agricultural income loss was Rs.39,000 crore
	• Rs. 20,000 crore were released by Govt. of India
	• In Punjab, cost of irrigation was increased by Rs. 2000/ha

6.1.1 Drought Preventive Measures

Among the different *kharif* crops the upland rice is most affected by drought. Therefore, diversified land use with low duty non-paddy crops is the best option in these lands. In real sense the technology available to mitigate drought is mostly preventive in nature and requires early planning. Therefore, it is imperative to have a long-term planning at the beginning of the season for judicious use of water, land and crops in a particular locality for obtaining best results. In rain-fed areas the major thrust in drought mitigation should be on rainwater management through *in situ* conservation and water-harvesting through on-farm reservoirs/ capturing run-off from local catchments to recycle at the time of need. Some of the important preventive measures that can be adopted early in the season to mitigate the impact of drought and augment sustainable crop production are elaborated below.

(i) Upland situation

(a) Select efficient crops and cropping systems matching the length of growing season. Some of the promising crops for rain-fed uplands are maize, cowpea, pigeon pea, black gram, rice bean, finger millet, groundnut, sesame, castor, pumpkin and sweet potato.

(b) Choose short duration varieties which possess faster rate of growth, deep and penetrating root system and ability to escape drought.

(c) Store rainwater to use as life-saving irrigation. On-farm water harvesting structures lined with 6:1 soil: cement mortar of 6 cm thickness in 10% land area helps to harvest the rainwater for providing protective irrigation.

(d) Perform off-season ploughing to conserve moisture, reduce pest and weed problem as well as to facilitate early sowing.

(e) Plough and sow the crops across the slope to develop a ridge and furrow type of land configuration for effective soil moisture conservation to overcome drought for longer period.

(f) Adopt intercropping/mixed cropping system in recurrent drought prone areas as mentioned in Table 9.

(g) Adopt integrated farming system: Apart from crop component, inclusion of animal components and some other enterprises like bee keeping, mushroom cultivation, goatery, poultry, planting fruit trees, timber species will diversify production and impart stability to production system.

Table 9 Different intercropping systems in drought-prone areas

Sl. No.	*Inter-cropping*	*Row ratio*	*Set specification*	*Row distance of intercrop (cm)*
Pigeon pea based system				
1.	Pigeon pea + Groundnut	2:6	30-210-30	30
2.	Pigeon pea + Sesame/Niger	2:4	30-150-30	30
3.	Pigeon pea+Green gram/Black gram	2:3	30-120-30	30
4.	Pigeon pea + Finger millet	2:4	30-100-30	20
5.	Pigeon pea + Rice	2:5	30-90-30	15
6.	Pigeon pea + Rice (mixed broadcast)	40:60	Seed rate ratio of individual crop for broadcasting	
7.	Pigeon pea + Radish*	2:2	30-90-30	30
8.	Pigeon pea + Okra	2:2	30-90-30	30
Maize based system				
9.	Maize + Arhar	2:2	30-90-30	30
10.	Maize + Cowpea	2:2	30-90-30	30
11.	Maize + Cowpea (fodder)	2:1	30 cm uniform row	30
12.	Maize + Runner beat*	2:2	30-120-30	40
13.	Maize +Yam*	Two rows of maize grown at 30 cm distance in both sides of Yam planted in mounds at 90 cm x 90 cm to act live staking		
Rice based system				
14.	Groundnut + Rice	1:4	Uniform row	15
15.	Green gram/ Black gram + Rice	1:4	Uniform row	15
16.	Green gram/ Black gram + Rice	1:2	In drought year, if rice fails, pulse crop is maintained and in a normal year pulse is cut for fodder and rice is maintained	
17.	Okra + rice	2:4	30-75-30	15
18.	Radish* + Rice	2:4	30-75-30	15

*Suitable for inland hilly districts

(h) Practice tree-based farming systems (Table 10) and grow different field crops: Growing of annual crops is highly vulnerable to frequent occurrence of drought. Under such situation, agro-forestry system provides multiple benefits in terms of providing food, fodder, fuel, fruits, fiber and industrial products, apart from providing security against risk of drought (Korwar, 1992; Pathak and Solanki, 2002).

Table 10 Various tree-based farming systems

Sl. No.	*System*	*Tree species*	*Companion crops*
1.	Agri-silvi	*Acacia mangium, Dalbergia sissoo, Tectona grandis, Casuarina equisetifolia, Gmelina arborea*	Maize/Cowpea/Green gram/Black gram/ Sesame
2.	Silvi-pastoral	*Leucaena leucocephala, Dalbergia sissoo, Acacia auriculiformis*	Hybrid napier/Guinea/Stylo/Dinanath
3.	Agri-horti	Guava/ Custard apple/ Mango/Pomegranate	Arhar/Cowpea/Niger/Okra

(i) Follow suitable agronomic practices as follows.
- Apply a portion of FYM in the seed furrows at the time of sowing to conserve moisture and thereby prevent seedling mortality from early drought.
- Grow short duration rice varieties such as Annanda, Shankar, Jogesh, Kalinga-III, Heera, Vandana and Anjali.
- Sow crops like ragi, maize, arhar, green gram, black gram, cowpea, sesame, groundnut, castor in place of upland rice.
- Vegetables come up well in drought/low rainfall years. Utilize the ponds, reservoirs and water bodies for growing tomato, cauliflower, radish, brinjal in the inland hilly districts and cowpea, lady's finger and chilli in coastal plains.

(ii) Medium/low land situation

Rainwater management in medium and low lands is crucial for mitigation of drought and improvement in production. A technology for storing excess rainwater in refuges of medium/low land has been standardized by devoting 10% of the cultivable area. The objective is to minimize run-off by encouraging its entry into the soil (*in situ* water conservation) and capturing that which cannot get into the soil. In the first step, the field bund is strengthened by raising the dyke height to 45 cm with provision of weir at 20 cm height for spilling over excess water to run-off collection tank (refuge). The refuge is constructed at the lower reach of the plot with top width 3.0 m, bottom width 2.0 m and depth 1.8 m. The length of the refuge is equal to the width of the plot. This technology is based on the principle that out of total annual rainfall (1500 mm) nearly 50% of the rainfall comes from a few intense showers resulting in higher run-off. On the other hand, in certain years there is a break in rainfall at a stretch for 10-12 days during crop growth period. This long stretch of dry period affects the rice crop adversely. The excess run-off discharged over the weir height during intense showers in the early season if collected in the refuge can provide protective irrigation to mitigate the intermittent drought in rice crop.

6.1.2 Drought Ameliorative Measures

It is difficult to define the exact crop and weather scenario during an anticipated drought or dry spell. Hence, it is really a difficult task to delineate rigid contingent measures well in advance of the cropping season applicable to all situations. However, there are three distinct periods of *kharif* season relating to crop growth stage and associated farm practices. Depending on the rainfall onset and pattern of distribution, seven types of scenarios have been projected and required contingent measures have been suggested below.

(i) Early season drought (June 10 to July 31)

Scenario 1: Early onset and sudden stoppage of monsoon

(a) Upland

- When there is more than 50% mortality, resow the crop up to July after receipt of sufficient rainwater. It is always wise to raise low water requiring crops like finger millet (Suvra, Bhairabi, Dibyasinha, Godavari), green gram (K-851, Sujata, Durga, Kamdev, PDM-54), black gram (T-9, Pant-U-19, Pant-U-30, Ujala and Sarala), cowpea (SEB-2, Pusa Barsati, Utkal Manik), sesame (Uma, Usha, Nirmala, Prachi), rice bean (RBL-6, BRB-1), castor (Jyoti, Kranti, Harita). If mortality is less than 50%, then only gap filling may be done.
- Cultivate vegetables like cowpea, radish, okra, cauliflower, brinjal, tomato wherever possible.
- Niger (Deomali, Alasi-1) and horse gram (Urmi) are to be sown in August.
- In wide as well as close spaced line-sown crops complete hoeing, weeding followed by ridging to the base of the crop rows at 20 days after sowing for *in situ* moisture conservation.

(b) Medium and low land

- In rice cultivation if the population is less than 50%, resow the crop. Select medium duration varieties (125 days) for coastal districts. Sprouted seeds may be direct seeded or fresh seedlings of early varieties may be raised for transplanting. The sprouted seeds can be sown in the lines by seed drill.
- While if the rice population is more than 50%, carry out weeding and adjust the plant population by removing and redistributing the hills.
- Raise community nursery of rice for transplanting at a reliable water source to save time for further delay.
- Sow the seeds at 5-6 cm depth by placing 6-8 seeds at one point with a spacing of 20 cm × 10 cm and cover it with a mixture of FYM:SSP (10:1) to avoid seedling mortality due to moisture stress in lowland. Use a seed rate of 100-120 kg per ha to maintain 400-600 plants/m^2.
- In saline soil use FYM/green leaf manure, sow sprouted seeds and gap fill the crop.

Scenario 2: Late onset, uplands not covered till mid-July

- Sow drought tolerant crops like green gram, black gram, cowpea, sesame, castor in place of upland rice.
- Maize, cowpea may be grown in the first week of August to meet the fodder crisis.

- Niger (Deomali, Alasi-1) and horse gram (Urmi) are to be sown in August.
- Grow sweet potato varieties like Gouri, Shankar, Samrat, Shree Nandini, Shree Bhadra, Shree Ratna in the ridges and allow the furrows to conserve rainwater.
- Grow vegetables like tomato, cauliflower, radish, brinjal in the inland hilly districts and cowpea, lady's finger and chilly in the coastal plains.
- Apply 20% N with full P and K of the recommended dose as basal along with well decomposed organic manure for early seedling vigour.
- Major emphasis should be given on *in situ* rainwater conservation, harvesting excess run-off for its recycling to make provision for life saving irrigation.

(ii) Mid-seasons drought (August 1 to September 15)

Scenario 3: Crops other than paddy in uplands affected

- Complete hoeing and weeding in crop fields to provide dust mulch.
- Spray 2% KCl + 0.1 ppm boron to black gram.
- Foliar application of urea (2%) at pre-flowering and flowering stage of green gram.
- Spray urea (1%) in brinjal. Take up spraying measures against mealy bug and mite which are more prevalent in dry weather.
- Top dress the crops after receipt of rain.
- Remove the borer affected tillers/late formed tillers/dried leaves in sugarcane and follow wrapping and propping in chains. Stripe out the lower 4-5 leaves.
- Spray planofix or celmone 10 ppm (2 ml in 9 litres of water) at 45 days and 20 ppm (4 ml in 9 litres of water) 10 days later to prevent boll shedding in cotton.
- Spray urea (2%) in late planted jute to encourage growth.
- Top dress nitrogen to ginger and turmeric @ 60 and 30 kg/ha, respectively after receipt of rainfall followed by mulching.
- Practice mulching with organics to extend the period of moisture availability.
- Close the drainage holes and check the seepage loss in direct sown medium land rice regularly.

Scenario 4: *Beushaning* of rice delayed

- If the crop is more than 45 days old then do not practise *beushaning* (blind cultivation).
- Weed out the field without waiting for rainfall.
- Go for gap filling using seedling of same age to have a uniform distribution of plant.
- Strengthen the field bunds and close the holes to check seepage loss.
- Withhold N fertilizer application up to receipt of rainfall.

Scenario 5: Transplanting of rice delayed

Generally, in this case rice seedlings are over-aged.

- Seedlings up to 45 and 60-70 days old can be transplanted in case of medium and late duration rice varieties, respectively without much reduction in yield.
- Remove the weeds and follow plant protection measures against blast in the nursery.

- Pulverize the main rice field in dry conditions, if it is not ploughed earlier to save time in final puddling.
- Use tractor/power tiller/tractor mounted rotavator for speedy land preparation/ puddling to cover more area with less time.
- Follow close transplanting using 5-7 seedlings/hill.
- Apply 50% recommended nitrogen at the time of transplanting.
- Apply life-saving irrigation to maintain the nursery seedlings in good health.
- Do not top-dress nitrogen in nursery.

Scenario 6: *Beushaned*/transplanted rice affected at early vegetative stage

- Provide protective irrigation through recycling of harvested rainwater.
- Remove the weeds and follow plant protection measures against blast in the nursery if existing.
- Withhold N fertilizer application up to receipt of rainfall.
- Apply potassic fertilizers wherever soil moisture allows or wait up to receipt of rainfall.
- Strengthen the field bunds and close the holes to check seepage loss.

(iii) Late season drought (September 16 to October 31)

Scenario 7: Medium and low land rice affected at vegetative/reproductive stage

It occurs as a result of early cessation of monsoon rains. The management practices are as follows.

- Provide protective irrigation through recycling of harvested rainwater.
- Provide irrigation at critical stages such as flowering, grain filling, etc. in alternate furrows in wide spaced crops.
- Crops like cowpea, maize, green gram may be harvested for fodder purpose to avoid their failure as grain crops.
- When the soil becomes hard it is difficult to uproot groundnut from the field, sprinkle water from water harvesting structures/nallahs to soften the soil.
- Under situation of complete failure of *kharif* crop dismantle it. In such situation or where land is remaining fallow, sow (dibble) the pre-*rabi* crops. The ideal pre-*rabi* crops with residual moisture condition are horse gram, castor, niger, black gram and sesame in uplands and well drained medium lands.

7. FLOOD SITUATION

Flood is one of the most devastating natural calamities which may cause serious damage to agricultural productivity. Comprehensive planning and adoption of management practices should be done under various flood situations (early and late) for different land situations.

7.1 Early and Late Flood

7.1.1 Early Flood (Up to Mid-August) Under Medium/Low Land Situation

- Select rice varieties like Kanchan, Ramachandi, Durga, Uphar, Sarala, Varshadhan for semi-deep low lands.

- If damage is more than 50% re-transplant rice crop of medium duration group.
- In partially damaged fields, allow the rice plants to stand upright. Do not go for beushaning as it may further reduce the plant population.
- Weed out the rice field; make gap filling and top dress N and K to boost the growth if situation permits.

7.1.2 Late Flood (September)

If flood occurs during mid August to early September, then following measures should be taken:

- Transplant 40-65 days old seedlings after flood water recedes.
- Broadcasting/line sowing of sprouted seeds of relatively short duration rice varieties in soft puddle after flood water recedes.
- Apply moderate dose of fertilizer (40:20:20 N: P_2O_5: K_2O kg/ha).
- Particularly in up-medium land, where there is no scope for revival of rice, go for pulses like black gram, green gram, horse gram etc.

7.2 Plant Population

- Raise nursery by Dapog method to transplant wherever possible.
- Maintain a buffer nursery in the backyard/highland area to ensure adequate plant population in the field after flood damage.

7.3 Fertilizer Application

- Reduce nitrogen application and apply recommended dose of P and K as basal to increase flood resistance.
- Apply moderate dose of NPK at beushaning, if not applied earlier.
- Top dress N and K in flood affected areas if situation permits.

8. SUMMARY AND CONCLUSION

In India, aberrant weather especially vagaries of monsoon rains, rapid fluctuations of temperature, sudden occurrence of drought and/or flood is a great concern in present-day rain-fed agriculture. Around 61% of the cultivated area in India is under rain-fed cultivation and most vulnerable to vagaries of weather. Economic status of farmers of this region is very poor and the ability to overcome the crop failure due to adverse weather is extremely bad (Rao, 2009). As a result of which partial or total crop failure has become very common leading to great economic loss to the farmers in our country. Thus to prevent the farming communities from economic loss due to vagaries of weather, climatic abnormalities and natural calamities such as drought, flood, etc. careful crop planning and management is essential. The age-old adage "prevention is better than cure" thus holds great promise in this regard. Preventive measures such as rainwater harvesting, judicious use of water, land and crops as well as suitable varieties in a particular locality is most important to obtain the best results under any unfavourable or unwanted situations. Therefore, proper planning and adoption of location-specific contingency crop planning for varied weather conditions needs to be worked out to effectively improve livelihood of vulnerable small holders.

References

Addy, K. (2009). Aberrant weather situations and contingency crop planning. http://addykraipur.blogspot.com/2009/04/aberrant -weather-situations-and.html.

Clarkson, N. M., Abawi, G. Y., Graham, L. B., Chiew, F. H. S., James, R. A., Clewett, J. F., George, D. A. and Berry, D. (2001). Seasonal stream flow forecasts to improve management of water resources: Major issues and future directions in Australia. In *Proceedings of the 26th National and third International Hydrology and Water Resource Symposium.* The institution of Engineers. Perth. pp. 653-658.

Joshi, N. L. and Kar, A. (2009). Contingency crop planning for dryland areas in relation to climate change. *Indian Journal of Agronomy*. 54(2): 237-243.

Korwar, G. R. (1992). Alternate land use systems. In *Dryland Agriculture in India – State of Art Research in India*, Somani, L. L. (ed.). Scientific Publishers, Jodhpur. pp. 143-168.

Pathak, P. S. and Solanki, K. R. (2002). *Agroforestry technologies for different agro-climatic regions of India.* ICAR, New Delhi. pp. 42

Rao, V. U. M. (2009). Climate Risk Management for Crop Production. In *Advances in Plant – Atmospheric Interactions*, Rao, V. U. M, Subba Rao, A. V. M, Rao, G. G. S. N., Ramanna Rao, B. V., Vijaya Kumar, P. and Venkateswarlu, B. (eds.). CRIDA, Hyderabad. pp. 156-159.

Reddy, T. Y. and Reddy, G. P. (2011). Problems of crop production in drylands. In *Lecture Notes on Dryland Farming and Watershed management*. pp. 21-23.

9

Effects of Climate Change on Crops: Food Security and Biotechnology

Mohd Mazid[1*], **Taqi Ahmed Khan**[2a,2b] **and Saima Quddusi**[3]

[1]*Advanced Plant Physiology Division, Department of Botany, Faculty of Life Sciences Aligarh Muslim University, Aligarh-202002, India*

[2a]*Department of Biochemistry, Faculty of Life Sciences, Aligarh Muslim University Aligarh-202002, India*

[2b]*Applied Biotechnology Department, College of Applied Sciences, Sur, Oman*

[3]*Amity Institute of Biotechnology, Amity University, Lucknow-227106, India*

ABSTRACT

Climate change is perhaps the biggest challenge facing the world today, and the very existence of man depends on how effectively this challenge is tackled. Agriculture is highly sensitive to climate variability and weather extremes, such as droughts, salinity, floods and severe storms. Agriculture production in many countries, including India will be impacted by climate variability. As these changes take place, agriculture is getting affected adversely, and the threat of decrease in food production is becoming very real. This at a time when our population is still increasing steadily. Developing countries like India would be the worst hit. Already, for several years, the pattern of monsoon in India has become unpredictable, uncertain and erratic. It is estimated that greater loss is expected in Rabi as compared to Kharif crops. By 2020 in some African and Asian countries, yield from rain-fed agriculture could be reduced by up to 50%. The worst affected would be millions of small and marginal farmers and people who are already the most vulnerable. More importantly in developing countries, it will be difficult for formers to carry on farming in the increased temperatures. Since, the environmental and plant's life are interdependent and environmental factors greatly affect the plants so, it is necessary for us to maintain a balanced environment for long a safe and long life. We should not disturb the natural environment because nature is infinitely precious. If the present trend continues, some certain finite threshold limit will be crossed and human civilization's survival will be in jeopardy. All governments have come together to play havoc with the lives of people across the globe. Our health, agriculture, habitation, everything depends on how effectively we are able to tackle this problem. Today more than 21 million Indian farmers tap underground resources of water to their fields, and two-thirds of India's crops are irrigated with groundwater. This water is running out and unlike the rivers, it will not be quickly replaced. All above circumstances create a problem of food security and so therefore biotechnology will appear as a remedial tool to sort out drastic effects of these circumstances in near future.

Keywords: Agriculture, climate change, global warming, productivity and population.

**Corresponding author's email: mazidmohd699@gmail.com*

Roychowdhury R. (Ed.), Crop Improvement in the Era of Climate Change.

1. INTRODUCTION

Studies on inter-annual and long-term variability of monsoon and annual rainfall have found that the variation in rainfall for the Indian subcontinent is within statistical limit (Kumar *et al.*, 1999). However, analysis of past weather data indeed indicates a warming trend at many places in India and changes in rainfall pattern in different parts of the country. Therefore, it is very difficult, at this juncture, to convince the planner and developmental agencies to incorporate the impact of climate change into their projects and agricultural system. However, 60% of the total cropped area is still rain-fed in India and dependent on uncertainties of monsoon. The country's foodgrains production during 2002–03 had slumped to 174.19 Mt, due to widespread salinity, from the record level of 212.02 Mt in 2001-02. Climatic variability and occurrence of extreme events are major concerns for the Indian subcontinents. There is need to quantify the growth and yield responses of important crops and also identify suitable land use options to sustain agricultural productivity under this large range of climatic variations. In India, the analysis of seasonal and annual surface air temperatures has shown a significant warming trend of 0.57°C per hundred years (Arora *et al.*, 2005).

The warming is found to be mainly contributed by the post-monsoon and winter seasons. The monsoon temperatures do not show a significant trend in any major part of the country. Similar warming trends have also been noticed in Pakistan, Nepal, Sri Lanka and Bangladesh. Extreme weather conditions, such as salinity, floods, droughts, heat and cold waves, flash floods, cyclones and hailstorm, are direct hazards to crops. More subtle fluctuation in weather during critical phases of crop development can also have substantial impact on yields. Cultivated areas are subject to a broader range of influences of salinity along with these extreme weather conditions, including changes in commodity prices, costs of inputs and availability of irrigation water. Climate may have indirect and possibly lagged influences on harvested areas.

Moreover, Mall *et al.* (2006) concluded that the negative trends in solar radiation and an increase in minimum temperature, resulting in declining trends of potential yields of rice and wheat in the Indo-Gangetic plains. In Delhi the minimum and maximum temperatures show a rising trend both in summer or 'kharif' growing season (June–October) as well as winter or 'rabi' growing season starts after summer monsoon. There was also a small declining trend in solar radiation during rabi and kharif season after 1980. In addition, climate change is no longer a distant scientific prognosis but is becoming a reality. The anthropogenic increases in emissions of GHGs and aerosols in the atmosphere result in a change in the radiative forcing and a rise in the earth's temperature. The bottom-line conclusion of the Third Assessment Report of the Intergovernmental Panel on Climate Change (IPCC) is that the average global surface temperature will increase by between 1.4°C and 3°C above 1990 levels by 2100 for low emission scenarios and between 2.5 and 5.8°C for higher emission scenarios of GHGs and aerosols in the atmosphere. Since solar radiation is closely related to crop growth, any decrease in this will significantly reduce agricultural productivity.

The accompanied increase in minimum temperatures increases maintenance respiration requirement of the crops and thus further reduces net growth and productivity. Productivity of some of the important cropping systems has either become static or shown a decline in some locations. Recent trends of a decline in the yield of rice-wheat cropping system in

Indo-Gangetic plain and north western India have raised serious concern about the region's food supply (Aggarwal *et al.*, 2004). Moreover, the standard deviation of future projections of area-averaged monsoon rainfall centred around 2050s is not significantly different relative to the present-day atmosphere implying thereby that the year-to-year variability in mean rainfall during the monsoon season may not significantly change in the future. More intense rainfall spells are, however, projected over the land regions of the Indian subcontinent in the future thus increasing the probability of extreme rainfall events in a warmer atmosphere. Providing sufficient food for the world's increasing population is becoming more difficult as land, water, and vegetative resources are progressively degraded through prolonged overuse. In the future, this difficulty will be exacerbated by climate change (Laux *et al.*, 2010).

Moreover, climate change alters the biophysical environment in which crops grow and how crops respond to some factors of climate change, such as CO_2, temperature, precipitation, and evapotranspiration. Atmospheric CO_2 accumulation without changing temperature and precipitation patterns might likely be of benefit for crop production. Plants commonly respond to higher levels of CO_2 with increased rates of photosynthesis, because CO_2 absorption is facilitated by the stronger gradient between the atmosphere and air spaces inside the leaves. C3 plants, such as rice, soybean, and groundnut, exhibit lower rates of net photosynthesis than C4 plants (e.g., maize) at the current CO_2 level ($\approx$ 385 ppm). At elevated levels, C3 plants may become more competitive than C4 plants, due to larger increases in photosynthetic rates. In addition to the enhanced photosynthesis rates, plants respond with a partial closure of their stomata, thus reducing transpiration per unit leaf area and improving their water use efficiency.

However, a significant impact in terms of the direct fertilisation effect of CO_2 cannot be expected before 2050 when CO_2 is likely to reach twice the preindustrial level (Cline, 1991). Initial studies dealing with the climate change impact on crop productivity focussed on the effects of an increased CO_2 level, followed by studies that additionally took the change of average climate conditions into account, such as a rise in the mean global temperature and/or change in rainfall. Additionally, they simulated the economic consequences of the simulated crop yield changes using a world food trade model. They found negative changes to the modelled yield in low latitudes, where many developing countries are located, and is contrary to the increased yield in middle and high latitudes, the predominant location of developed countries.

According to Porter and Semenov (2005), the two largest causes of yield variation are temperature and rainfall. Their independent effects are three to four times larger than those caused by the variation in solar radiation. The increased variation and changes in mean temperature and precipitation are expected to dominate future changes in climate, as they affect crop productivity. Various studies dealing with the effects of climatic variability have pointed to the conclusion that an increased annual variability of weather, as expected due to global warming, causes an increased variation of yields. Genetic engineering provides a means to introduce genes into plants via mechanisms that are different in some respects from classical breeding. A number of commercialized, genetically engineered (GE) varieties, notably canola, cotton, maize and soybean, were created using this technology, and at present the traits introduced are herbicide and/or pest tolerance. In 2007 these GE crops were planted in developed and developing countries on more than 280 million acres worldwide, representing nearly 10% of rain-fed cropland. Although the

United States leads the world in acres planted with GE crops, the majority of this planting is on large acreage farms.

In developing countries, adopters are mostly small and resource-poor farmers. For farmers and many consumers worldwide, planting and eating GE crops and products made from them are acceptable and even welcomed; for others GE crops raise food and environmental safety questions, as well as economic and social issues. When contemplating environmental impacts of GE crops, it is important to consider that the fundamental issues raised are similar in many ways to those encountered with crops created by other genetic modification methods, such as mutation or marker assisted selection (MAS), and cultivated in other ways, such as the use of integrated pest management (IPM) and organic and bio-control methods. Regarding these issues, it was suggested that certain analyses are too crude to allow meaningful assessments of environmental consequences and require geographical data to be collected on a smaller, more defined basis. Although the strategies implemented to delay resistance have helped sustain efficacy of Bt crops longer than many scientists expected, field-evolved resistance to Bt crops was reported (Tabashnik *et al.*, 2003).

2. INNOVATIVE APPROACHES TO STUDY THE RESPONSES OF CROP PLANTS TO THE ELEVATED CO_2

2.1 Open Chamber Carbon Dioxide

Open chamber CO_2 enrichment technology for conducting long term experiments have been developed and designed for south Asian tropical conditions. CO_2 enrichment research network of India, Bangladesh Nepal, Pakistan and Sri Lanka for multicountry multidisciplinary experiments was co-ordinated in the IARI, New Delhi, India.

2.2 Free Air Carbon Dioxide Enrichment

Free Air CO_2 enrichment technology, a PC based system, was established with the help of National Physical Laboratory, New Delhi to generate realistic biological data on the crop responses to the higher CO_2 concentration.

2.3 Global Change and Terrestrial Ecoystems and CO_2 Research Network

This system is active in tackling vulnerable issue and adaptation strategies for meeting the rise in global food demand in the face of global environmental changes. Response of crop plants such as Cicer, Brassica and wheat has been studied in terms of photosynthesis and productivity. These crops plants showed amelioration of the adverse moisture stress effect at high concentration of CO_2 by increasing the water status of plants and greater root growth. Use of biotechnology to understand the role of photosynthesis repression genes and the role of glucose signaling associated with hexokinase located at plasma-membrane or tonoplast to regulate photosynthesis under elevated CO_2 need to be studied. The identification of HIC (high CO_2) gene of Arabidopsis carry a 3-keto-acyl-co-enzyme- A synthase gene encodes a negative regulator of stotmatal development. HIC plants carrying such as gene may be useful for developing plants a possible remedy for every climatic change of future.

3. AGRICULTURAL ECONOMY

In many developing countries, agriculture is the cornerstone of their economy, the basis of economic growth and the main source of livelihood. But agriculture in the developing world is often cited as being one of the sectors most vulnerable to climate change. In Asia, for example, the majority of available fresh water is used for agriculture; farming techniques are relatively simple; and much of the continent is already hot and dry. Any changes in precipitation and temperature patterns will thus have major impacts on the viability and yields in crop production. To exacerbate the situation, several studies warn of an unprecedented confluence of pressures on agriculture with population growth and development driving up global demand for food and competition for land, water and energy intensifying as the impacts of climate change starts to take effect. Any strategy to enhance agricultural productivity in Asia needs to ensure that natural resources are managed sustainably and adapted to climate change.

The constraints on food crop production and distribution differ between regions and, between industrialised and developing countries. Climate change has the potential to exacerbate the stresses on crop plants, potentially leading to catastrophic yield reductions. The elevated CO_2 increased grain yield under potential, irrigated and rain-fed conditions. There was a linear increase in grain yield as the CO_2 concentration increased from 350 to 700 ppm under all the levels. Kumar and Parikh (2009) estimated the functional relationship between farm level net revenue and climate variables, introduced through linear, quadratic, and interaction terms, to understand the climatic sensitivity of Indian agriculture. They found that the overall impacts due to the climate change scenario for a 2°C rise in temperature and a 7% increase in precipitation are negative and about 8.4% of the total farm level net-revenue for India. The temperature increase results in significant negative impacts, while the higher precipitation considered under the scenario increases the net revenue. On the whole the negative impacts due to temperature change more than compensate for the small positive impact due to precipitation change.

Most of the simulation studies have shown a decrease in duration and yield of crops as temperature increased in different parts of India. Such reductions were, however, generally, offset by the increase in CO_2; the magnitude of these crops varied with crop, region and climate change scenario. In north India, irrigated wheat yields decreased as temperature increases, a 2°C increase resulted in 17% decrease in grain yield but beyond that the decrease was very high. These decreases were compensated by increase in CO_2 due to letter's fertilizing effect on crop growth. CO_2 concentration ha to rise up to 450 ppm to nullify the negative effect of 1°C increase in temperature. The effect of climate change scenario of different periods can be positive or negative depending upon the magnitude of change in CO_2 and temperature (King and Neilson, 1992). He has also developed two scenarios based on IPCC as optimistic (low increase in temperature; high increase in CO_2) and pessimistic scenarios (high increase in temperature; low increase in CO_2) for different years. The possible impact of these showed that the irrigated wheat and rice yields in north India will not be significantly affected due to direct effect until 2050. It is only in 2070 when the temperature increases are very large, that the crops show large reduction in yield. Mean decline in yields across different scenarios ranged from 14% in Pune (West India) to 23% in Gwalior.

yields by only 0.5%. This was nullified when the temperature increased by 0.08 °C. Similarly, the small beneficial effect of still higher CO_2 concentrations was nullified by further increase in temperature. The beneficial effect of 700-ppm CO_2 was nullified by an increase of only 0.9°C in temperature. Further increase in temperature always resulted in lower yields than control irrespective of the increase in CO_2.

Pre-anthesis and total crop duration got reduced by 10 and 12 days with 2°C rise. Irrigated yield, which averaged around 58% of the potential yield, in general increased with temperature rise up to 2°C. Total crop duration got reduced only by 4 days with 2°C rise. In most climate change applications, long-term historical weather data are used as input for the crop models. In general, at least 30 years of historical weather data are preferred to represent annual weather variability; different climate change scenarios can then be applied to these data records. The simplest approach is to assume a fixed climate change and to modify the data with a constant number, such as an increase or decrease of 1, 2, 3°C etc. for temperature. Uncertainties involved in such an assessment include: (1) the timing, magnitude and nature of climate change; (2) the ability of ecosystems to adopt either naturally or through managed intervention to the change; (3) future increase in population and economic activities and their impacts on natural resources systems; and (4) how society adapts through the normal responses of individual, businesses and policy changes that after the opportunities and incentives to respond. The uncertainties, the long periods involved and the potential for catastrophic and irreversible impacts on natural resources systems raise questions as to how to evaluate climate impacts, investments, and other policies that would affect or be affected by changes in the climate.

In India, substantial work has been done in the last decade aimed at understanding the nature and magnitude of change in yield of different crops due to possible climate change. The rainfall fluctuations in India have been largely random over a century, with no systematic change detectable in summer monsoon season. However, areas of increasing trend in the seasonal rainfall have been found along the West Coast, North Andhra Pradesh and northwest India and those of decreasing trend over east Madhya Pradesh, Orissa and northeast India during several years. More subtle fluctuation in weather during critical phases of crop development can also have substantial impact on yields. Cultivated areas are subject to a broader range of influences, including changes in commodity prices, costs of inputs and availability of irrigation water.

4. INCLUSIVE AGRICULTURE

Inclusive agriculture means promotion of agriculture in those areas among those farmers who have been grossly excluded from the agriculture; incentives, subsidies and similar other benefits. The focus of inclusive agriculture would largely be on: (a) promotion of agriculture dry land and hilly and sub-mountainous and desert zone; (b) small, marginal and landless agricultural labourers who really take care of agriculture but hardly enjoy the benefit; (c) role of women in agriculture as a cultivar, entrepreneur, extension agent and farm labours; (d) incentives to all kinds of agriculture both food crops and cash crops; livestock sector, fruits and vegetables and horticulture activities and (e) access to credit from the organized institutions and marketing facilities for the farmers in disadvantaged conditions and areas. In 2007, government of India has also emphasized that the growth to be all inclusive, the agricultural strategy must focus on the 85% of farmers who are marginal, increasingly female and who find it too difficult to access to inputs, credits and extension or to market their output. With a strong base for planning, promotion, execution and coordination of agricultural research and education to meet emerging challenge in the country and wide network of institutes spread across the country the ICAR has significantly contributed to overall development of farm and related sectors.

Indian projections under future climate change scenario of increase GHGs concentrations indicate marked increase in both rainfall and temperature into the 21st century, particularly becoming conspicuous after 2040's. Increase in GHG concentration may lead to overall increase in the rainy day intensity by 1-4 mm/day except for small areas in northwest India where the rainfall intensities decrease by 1mm/day. According to India Metrological Department, the monsoon rainfall at all India level does not show any trend but there are some regional patterns. This in turn would impact yield of crops. The processes of climate are slow and its effects on agriculture are projected to be apparent over several decades. The emerging results of analysis of impact of climate change on forest biomass in India seem to be highly vulnerable to the projected change in climate. Majority of the vegetation in India is likely to be less optimally adopted to its existing locations and consequently vulnerable to the adverse climatic changes. Biodiversity is also likely to have adverse impact to this (Gitay *et al*., 2002).

Agriculture plays a key role in overall economic and social well-being of India. Though the share of agriculture in both GDP and employment has declined over time, the pace of decline in its share in employment has been much slower than that of GDP. The share of agriculture in GDP is declined from 39% in 1983 to 24% in 2000–01 compared with much lower rate of decline in its share in total employment from 63% to 57% during the same period. Declines in the share of agriculture in GDP were not commensurate with the fall in dependency in agriculture. Such trends have resulted in fragmentation and decline in the size of land holdings which leads to agronomic inefficiency, a rise in unemployment, a low volume of marketable surplus. These factors could contribute to increase vulnerability to global environmental change (Sharma, 2007).

In India, average food consumption at present is 650 gm per capita per day whereas the corresponding figures in China and USA are 980 gm and 2850 gm respectively. Productivity of some of the important cropping systems has either become static or shown a decline in some locations. Recent trends of a decline or stagnation in the yield of rice-wheat cropping system in Indo-Gangetic plain and north-western India have raised serious concern about the regions food supply (Kumar *et al*., 2004). This trend clearly indicates the reduced factor of productivity in case of the rice-wheat cropping systems. These variations in trends of productivity indicate the effects of other biophysical and socio-economic components, which needs to be eliminated before embarking on assessing the impacts of climate change and its variability on growth and yield of crops.

In few years, with the growing recognition of the possibility of global climate change, an increasing emphasis on world food security in general and its regional impacts in particular have come to forefront of the scientific community. Crop growth, development, water use and yield under normal conditions are largely determined by weather during the growing season. Even with minor deviations from the normal weather, the efficiency of extremely applied inputs and food production is seriously impaired. The CO_2 concentration was in the steady state at 280 ppm till the pre-industrial period (1850). It is rising since then at the rate of 1.5 to 1.8 ppm per year. The concentration of CO_2 is likely to be doubled by the end of 21st century.

Open top chambers and FACE technology are currently being used for the study of the response of crop plants to the elevated CO_2. Results from such studies have shown an increase in plant photosynthetic rate and crop yield (Allen *et al.*, 1992). Global warming may also threaten India food security if there is a negative effect on agriculture. Although, the effect of increasing CO_2 concentrations will increase the net primary productivity of plants, but climate changes, and the changes in disturbance regimes associated with them, may lead either to increased or decreased net ecosystem productivity. In many tropical and subtropical regions, potential yields are projected to decrease for most projected increases in temperature. The impacts of elevated CO_2 should be considered among others, in the reference of: (a) changes in air temperature, particularly nocturnal temperature due to increase in CO_2 and other trace gases and changes in moisture availability and their effect on vegetative versus reproductive growth; (b) need for more farm resources and (c) survival and distribution of pest populations, thus developing a new equilibrium between crops and pests.

5. ENGINEERED CROPS AND PLANT BIODIVERSITY

Molecular and genomic technologies enable identification of genetic variants and development of genetic descriptions of diversity, leading to greater appreciation of these resources. Careful measures are needed when cultivating GE crops (Dahlberg, 1987); however, this situation is not unique to GE plants and can happen with conventionally breed, commercialized crops. Studies of the impact of transgenes moving into wild relatives and the potential to change ecosystem dynamics are currently requested in environmental impact assessments for commercial release of a new GE plant. Another consideration regarding effects of GE crops on diversity of local, adapted crop varieties is that most current, commercial GE varieties were developed by large, mostly international companies.

6. INTERACTION MECHANISMS AND INNOVATION

Several attitudes and practices complicate interactions between the actors in the medicinal plants sector: (a) a long-established mistrust among public research organizations, the private sector, and NGOs; (b) the different philosophical foundations of scientific versus traditional medicine and the resistance to hybridize them; and (c) traditional rivalries between different branches of science concerned with medical plants. The interaction around research and information is also very complex, because the vast knowledge about medicinal plants is held by different groups with dissimilar interests and philosophies. Local health practitioners and rural communities hold the ethanobotanical knowledge and concentrate on traditional healthcare, whereas research organizations in the public and private sector have the scientific knowledge and focus on product development and market share. Ayurveda is a holistic healing system that evolved among the Brahmin sages of ancient India 3000-5000 years ago.

7. CONVENTIONAL MEDICINES

By 1983 the Indian government had formally recognized the tremendous knowledge base and enormous potential of traditional Indian systems of medicine and supported the documentation and registration of this knowledge, especially knowledge of Ayurveda. India has a rich diversity of medically significant plants and very sophisticated indigenous medical traditions such as Ayurveda (Srivastava *et al.*, 1996). Medicinal plants traditionally have assumed great importance in the local healthcare systems of most poor people in India. During the late 1990s, exports of medicinal plants grew in response to increased international demand for a variety of herbal products, including medicines, health foods, and cosmetics. The greatly expanding sector faces many challenges, including: (a) environmental degradation from destructive harvesting of plants; (b) the exploitation of poor plant collectors; (c) the sharing of benefits and profits between small and big companies; (d) regulatory and quality constraints in domestic and export markets; (e) international competition; and (f) intellectual property issues in light of India's World Trade Organization membership.

The scale of the sector is enormous. It has a large number and diversity of sectors with different interests, and policies influencing the sector are dispersed across many sector and ministerial portfolios. Tribal communities in forest areas do most of the plant collection and have few alternative livelihood opportunities. A large number of small, family-owned businesses manufacture and sell traditional herbal medicines locally, but they lack the research, manufacturing, and marketing skills to take advantage of opportunities in the domestic and international markets. In contrast, a small number of large Indian pharmaceutical companies have developed associated research, design, and marketing organizations with strong links to the international market. Their products have some resemblance to traditional herbal healthcare systems but are in fact increasingly developed through a hybrid of local, allopathic, and scientific means (Andrews, 2003).

India's recently approved intellectual property right system recognizes the rural population's entitlement to this traditional knowledge. The government also supported the elaboration of best practices for manufacturing traditional medicines. The challenge for the sector is to create innovation capacity that simultaneously promotes sector growth while integrating economic, poverty reduction, and environmental concerns. Major tasks include: (a) building the network of knowledge by fostering partnerships between different branches of science and between science and traditional healthcare systems; (b) rethinking the approach towards sector and (c) redesigning research and training initiatives based on an exploration of the interfaces between traditional and scientific medicine. The processes that we now think of as "globalization" were central to the environmental cause well before the term "globalization" came into its current usage. Global environmental concerns were born out of the recognition that ecological processes do not always respect national boundaries and that environmental problems often have impacts beyond borders.

Using an evolutionary optimization algorithm, in which partitioning of a fixed quantity of protein-nitrogen among enzymes was allowed to vary, and selecting on photosynthetic rate, resulted after several generations in individuals with a light-saturated photosynthetic rate that was 60% higher. Although the contemporary debate on globalization has been contentious, it has not always been useful. Conditions of insecurity and conflict impose high costs on the pursuit of sustainable development just as they impose hurdles in the way of globalization (Najam *et al.*, 2005). The newly prosperous and the established wealthy will have to come to terms with the limitations of the ecological space in which both must operate, and also with the needs and rights of those who have not been as lucky.

Growth, of course, is a paradox in the context of sustainable development. We need growth in order to meet the needs of people, especially the poorest among us; but permanent global growth is impossible in a finite system. Studies demonstrate that we already exceed the productive capacity of nature by 25 to 30% of the ecosystems are currently overused. Although scares about "limits to growth" have proved less than credible in the past, simple economic logic argues that, as competition for scarce natural resources increases, prices will be driven up and sooner than we might have assumed. In the past, technology has and in the future, it certainly could help to alleviate some of these pressures by developing new solutions and by more widely deploying existing technological solutions. However, the prospects of higher demand, growing prices and dwindling stocks are already propelling new races for control over key resources. The race is now on not just for oil, but for metals, minerals, timber and even for recyclable waste. For many developing countries endowed with critical resources in high demand, this provides an opportunity to harness the power of globalization and pull them out of poverty.

8. HERBICIDE-TOLERANT GENETICALLY ENGINEERED CROPS

The concept of a super weed conjures up the image of a weed taking over entire ecosystems, undeterred by existing herbicides. Although this scenario is not based on fact, problems with herbicide-resistant weeds are real, but not new. These problems have occurred with traditionally bred crops, as well as with (high tolerance) HT GE plants. Although this phenomenon does not lead to so-called "environmental disasters," it reduces the effectiveness of certain weed control strategies and decreases weed management options. Good weed management practices can ameliorate this situation in conventional, organic, and GE cropping systems (Lemaux, 2009). Overuse of single herbicides can lead to this situation and will reduce the effectiveness of the GE crop. Having HT cultivars with resistance genes for herbicides with alternate modes of action that can be used in rotation will slow resistance development in weeds.

Crop tolerance to herbicides is achieved (a) by mutations that render a plant not susceptible to the herbicide or (b) through the introduction of transgenes. An example of the first approach was the identification of varieties that, after treatment with a chemical mutagen, were tolerant to imidazolinone herbicides are tolerant to herbicides. In the second approach to be commercialized, GE crops, tolerant to glyphosate or Roundup, were engineered with a bacterial gene encoding a target enzyme, 5-enol-pyruvylshikimate-3- phosphate synthase (EPSPS), which confers tolerance to the herbicide. When looking at herbicide usage and it is important to note that in addition to use on GE crops, herbicides can be applied directly to conventional crops. Also, depending on weed pressure, multiple applications can be used in the same area during the same season. Taking these facts into consideration, glyphosate use per acre has increased dramatically from 1995 to 2005, coupled with a concomitant dramatic drop in the use of other herbicides (Lieberman and Gray, 2008).

9. ENGINEERED PLANTS AND ENVIRONMENT

Cleaning polluted soil or water using living organisms is called bioremediation or phytoremediation. Phytoremediation when plants are used. Non-engineered bacteria and plants are able to remove heavy metals such as aluminium, selenium, mercury, and organic pollutants from contaminated soil by concentrating them in the cells of their roots, stems, or leaves. One phytoremediation effort has focused on selenium pollution, a worldwide problem arising from refinery effluents, industrial wastewater, and discharges from electric power plants. Removal was achieved in the field using fast-growing Indian mustard (*Brassica juncea* L.), which accumulates selenium to hundreds of parts per million. Engineering Indian mustard with a gene from a hyperaccumulator resulted in plants that produced greater biomass and longer roots, and accumulation and volatilization of selenium was significantly increased. Mercury is one of the most hazardous heavy metals and is particularly problematic in aquatic environments where organic mercury moves from fish to humans. *Arabidopsis thaliana* engineered to express modified bacterial mercuric ion reductase detoxified mercury by converting the more toxic ionic form to a less toxic elemental form (Bizily *et al.*, 1999). Another potential environmental advantage of GE plants is making better use of resources such as land and water. Lowering crop production losses reduces the likelihood that environmentally sensitive areas will be cultivated, potentially allowing maintenance or even expansion of protected forests, lakes, shores, wetlands, and wilderness.

10. REGULATION OF GE CROPS

The United States created a formal regulatory structure for GE organisms establishing the concept that GE foods would be regulated on the basis of product, not process, and would be regulated on a case-by-case basis. GE crops and products made from them are under regulatory control of three federal agencies: the Food and Drug Administration (FDA), the EPA, and the USDA. The EPA's charge also includes GE plants in which a small part of a pest, such as a viral regulatory sequence (e.g., 35S promoter), is used. A division of the USDA oversees environmental safety of planting and field-testing GE plants to ensure GE crop field tests are performed under specified conditions and any unusual occurrences. All these three agencies do not oversee each GE crop; however, all have legal rights to demand immediate market removal of any product if valid scientific data show safety concerns for consumers or the environment. Each event must go through regulatory approval, even if a first event with the same fragment received approval. Regulated articles are evaluated for impact on the environment and on agriculture. A petition for non-regulated status containing these data is formulated by the event's creator and after which an Environmental Assessment can be issued and determination of non-regulated status granted. From June 1992 to January 2009, information on status, requesting institution, genes introduced, phenotype of GE plants, field test data, and environmental assessments of deregulated articles is publicly available (Lieberman and Gray, 2008). This finding raised concerns, particularly about crops that produce pharmaceuticals.

11. USE OF GENETICALLY ENGINEERED CROPS

Sustainability has no single meaning, but one accepted definition is to meet the basic needs of today's inhabitants while preserving resources to enable future generations to flourish. Sustainability has become a goal of the United Nations' Development Group's Millennium project, "to 'Ensure Environmental Sustainability' by integrating principles of sustainable development into a country's policies and programmes to reverse the loss of environmental resources". Although the need for sustainable agricultural systems is now widely accepted, the manner in which to achieve them is not universal and even the precise goals are not well defined. Some estimates are that between 70 and 90% of increases in food production resulted from changes in conventional agricultural practices rather than cultivation increases (Heichel, 1976). This high production does have negative environmental impacts, as well as sizeable consumption of fossil fuels, unsustainable rates of water use and top soil loss, and contributions to environmental degradation, e.g., air pollution, soil erosion, reduced biodiversity, pest resistance, pollution of lakes and streams, and overuse of surface and groundwater.

To achieve agricultural sustainability, causes and cures for these problems must be addressed through all possible means. Numerous agricultural practices or methods, such as integrated pest management (IPM), biological control, organic methods, and use of GE plants, coupled with selected conventional agricultural methods, can play important roles in future sustainable agricultural practices. For example, practitioners of IPM use comprehensive information on the life cycles of pests and their interactions with the environment, in combination with available pest control methods, to manage pest damage with the least possible hazard to people, property, and the environment (Bajwa and Kogan, 1997). Biological control involves the use of a specific living organism to control a particular pest and cause the least harm to beneficial insects.

The use of GE organisms can also contribute to sustainable practices by augmenting and replacing certain conventional practices. For example, plants can be created that increase water use and fertilizer efficiencies, that remediate soil contaminants, increase no-till or low-till practices to help reduce GHGs, and produce higher yields without increasing land usage, particularly in developing countries (Godfray *et al.*, 2010). Although GE plants can contribute to a more sustainable agriculture, their development and availability do not ensure positive contributions. That depends on how they are deployed and whether their use results in changes in farming practices that increase sustainability. To achieve true sustainability agriculture must use the best of all practices.

When considering effects of GE crops and products, just as with those produced by other agricultural methods, it is important to factor in economic and social implications. Perhaps more difficult in these subject areas is the fact that there are not always definitive, factual responses to the issues raised; responses often reflect attitudes rather than information based on peer-reviewed scientific literature. Companies developing the new GE crops invest substantial amounts of time and money in the research, development, and regulatory approvals needed to bring products to market. Patents can also ensure that results and techniques needed for inventions are ultimately made public, although actual use of the IP is restricted to licence holders of the technology. Without this protection, situations could arise where findings are not published and processes are kept secret. To ensure that investments made in creating these crops are recouped, seed producers require purchasers of patented GE seeds to sign agreements stating that they will not reuse or sell the seed, and thus growers must repurchase seed each year.

Although legal agreements were not involved, the hybrid was a type of "biological patent" that prevented replanting because farmers were unable to create hybrid seeds without the inbred parents, which were protected by the companies that produced the hybrid seed. It has often been difficult or impossible to obtain the multiple rights needed to develop and market GE crops. One widely publicized example is Golden Rice, where a large number of IP issues had to be resolved before the engineered genes responsible for provitamin-A production could be introduced into local varieties in developing countries (Herdt *et al.*, 2003). Large agricultural biotechnology companies amassed IP assets through these means or through their own research efforts into the development of new GE crops.

12. EXPORT MARKET AND GE CROPS

Strict rules regarding GE presence in seeds and foods for international markets are a key driver for the need to segregate crops, but lack of standardized, internationally accepted marketing standards, testing methods, and protocols poses significant challenges to the smooth operation of the domestic and international agricultural marketplace. The United States must work toward internationally accepted, science-based standards for trade in GE products that include sampling and testing methods and tolerance levels that result in fair trade practices and that lead to unrestricted shipment of products in international markets. Presently, the European Union has approved only a few GE maize varieties for cultivation in member states. Imported foods, however, can contain GE ingredients that have been approved for food and feed and can be labelled non-GMO in the European Union if they contain less than 0.9% GE content. The WTO ruled that 24 of the 27 approval procedures for GE crops imposed undue delays in the European Union (Lieberman and Gray, 2008). Europe now faces increased pressure to allow planting of GE crops.

The stated industry view was that "adhering to a zero-tolerance rule is not possible in international trade with agricultural commodities" and further that "the German and European food and feed industry will no longer be in the position to obtain input materials on the world market" use of genetic engineering opens the door to improving agricultural crops in ways not possible. But with this capacity comes the responsibility to proceed with caution, investigating possible outcomes carefully. Conversely, there is also a responsibility to utilize the technology where it can provide improvements to human health and the environment and make farmers' efforts more productive. Although no human activity can be guaranteed 100% safe, the commercial GE crops and products available today are at least as safe as those produced by conventional methods. Particularly with regard to environmental safety, we must stay vigilant in our evaluation of GE crops and their impacts to ensure long-term utility, just as with those created using classical methods (Jimenez-Lopez and Hernandez-Soriano, 2012).

The ability of Indian farmers to find pathways out of poverty and to contribute actively to the growth process depends on improving infrastructure and education, distributing key technologies and inputs, and promoting producer and marketing organizations that link small farmers to new market chains. Indian farmers cannot overcome these constraints on their own, and there is a need in the short term for greater public sector involvement in many Asian countries than is currently fashionable. The challenge is therefore to develop new institutional arrangements between the public and private sectors that foster private sector development without leaving small holder farmers isolated during the transition.

Early classical theory viewed economic development as a growth process requiring the systematic reallocation of factors of production from a primary sector characterized by low productivity, traditional technology, and decreasing returns to a modern industrial sector with higher productivity and increasing returns (Kögel and Prskawetz, 2001). However, in reality, the dependence of agriculture on a fixed supply of land meant that its expansion was constrained, implying that agricultural output cannot proportionally increase with increased labour supply under a given technology. On the demand side, the need to satisfy basic needs implied that, at the very least, agricultural growth must match population growth to avoid the "Malthusian trap" and stagnant development. Classical theorists observed that most developing countries comprise "dual' economies. In this view, labour productivity is lower in agriculture than in industry, and hence development requires the movement of agricultural labour into non-agricultural sectors. However, the analysis also suggests that inequality increasingly becomes a constraint to poverty reduction during later stages of development. Most studies show that growth has a significant impact on poverty reduction, but there is substantial variation in the literature about the extent to which poverty declines.

These variations highlight the importance of understanding the structure of growth and its relationship with poverty. A growing econometric literature uses cross-country or time-series data to estimate sectoral and subsectoral growth-poverty elasticity. The distributional effects are, in turn, further broken down into intersectional, direct-distributional, and inter-household transfer linkages. They found that the agricultural sector contributes the most to overall poverty alleviation, followed by the services and informal sectors. The importance of food supply in the growth process, its link to poverty reduction should be understood within the broader context of development.

13. APPROVED PLANTS AND FOODS: TRANSGENIC ANALYSIS

Through the use of the new tools of genetic engineering, genes can be introduced into the same plant or animal species or into plants or animals that are not sexually compatible; the latter is a distinction with classical breeding. This technology has led to the commercial production of GE crops on approximately 250 million acres worldwide. These crops generally are herbicide and pest tolerant, but other GE crops in the pipeline focus on other traits. For some farmers and consumers, planting and eating foods from these crops are acceptable; for others they raise issues related to safety of the foods and the environment (Chassy, 2002). This admonition rings true today when it comes to the impact of information people receive, particularly in the popular press, about GE crops and foods. The intent is to present as accurate a scientific picture as possible, although this does not imply that people possessing the same scientific understanding will necessarily make the same choices about the advisability of GE crops for consumption, because different people have different values.

GE crops, and products made from them, must be evaluated on a case-by-case basis and, although scientific information should be a part of the considerations for using and consuming these crops, issues beyond the technical, science-based facts also need to be considered. Much processed food is produced using enzymes, or whole organisms with enzymes, that are responsible for altering the nature of the food such as bacteria (yogurt), yeast (beer, wine), and multicellular fungi (blue cheese). These enzymes were modified genetically through traditional methods and in some cases recombinant DNA methods as well. In cheese-making, rennet is used to coagulate milk to separate the curds and whey. Rennin or chymosin was the first protein produced through recombinant DNA recombinant DNA means to be used in food (Demain and Vaishnav, 2009). Other products produced by recombinant DNA methods include food supplements, such as vitamin B2, α-amylase (used to produce high-fructose corn syrup and dry beer), and lactase (added to milk to reduce the lactose content for persons with lactose intolerance).

Similarities and differences exist between the modifications made in organisms by classical breeding versus those made via recombinant DNA methods. Both approaches can involve changes in the sequence, order, and regulation of genes in an organism and can utilize many of the same enzymes. Thus, if a change in seed characteristics is desired, a gene can be linked to regulatory signals that result in expression only in the seed or even in a specific compartment of the seed (Debeaujon *et al.*, 2003), an outcome difficult to achieve with classical breeding. But in many crosses, wild relatives and different genera or species are not compatible or crosses can be made but the resulting embryos must be rescued by *in vitro* culture to obtain a plant. The ability to select desirable alleles and eliminate deleterious ones in a fast, reliable manner is critical to the development of improved germplasm through breeding. Genetic markers can speed identification of plants with desired alleles in large populations via a process termed MAS.

MAS is particularly valuable for traits when: a) phenotypic screens in the field are difficult or costly, e.g., drought/frost tolerance or resistance to exotic diseases or pests; b) multiple alleles exist; and c) recessive or low heritability traits exist that require progeny testing. However, MAS can be used to introduce desired traits from wild or related species and these species are acceptable to organic growers. Utilizing MAS to create food crops has led to the term, "Super Organics" crops that are grown under certified organic conditions and were created through classical breeding using genomic information and MAS, not through recombinant DNA methods (Varshney *et al.*, 2006). The first GE plant was tobacco, reported in 1983, but no plants were commercially grown until the FlavrSavrTM tomato was commercialized in 1994. If success is measured by increases in global acreage or farmer acceptance, certainly these GE crops have been successful. Despite sizeable GE crop acreage, the diversity of crop types and traits in commercial production is limited. Few minor acreage GE crops are at present commercially successful, i.e., papaya, certain types of squash and sweet corn.

It is the final product, rather than the means by which it is modified, that is more likely to result in unintended effects. Nonetheless, the nutritional composition of GE foods, including levels of protein, carbohydrate, fat, vitamin, mineral, fibre, moisture, and phytochemicals, is analysed for substantial equivalence, and levels of individual nutrients and anti-nutrients in GE foods are compared with levels in conventional counterparts. Therefore, comparisons of nutritional content of GE foods must be measured against variation in conventional foods grown under comparable conditions. For example, nutrient composition of GE potato tubers was compared with control wild-type and tissue culture–derived non-GE potato tubers of two cultivars, cv. Record and cv. Desiree, grown under the same conditions. One significant difference was detected in defatted, non-toasted soybean meal, the starting material for the production of commercially utilized soybean protein. Peanut allergy was the least common only 9% of the population but it is one of the most severe and durable allergies. The nature of the proteins causing these allergic reactions is well characterized in certain cases, thus making it is possible to engineer the organism to make lower levels of the proteins responsible for the allergies or change their conformation to reduce allergic responses (Mine and Rupa, 2004).

Organic farming is a method of agricultural production that does not allow the use of synthetic pesticides, fertilizers, or growth enhancers. Foods grown under organic certification differ from conventionally produced food by the manner in which they are grown, handled, and processed, but an "organic" label does not guarantee the nature of the product, the food, or ingredient, only its production method. However, organic certification does not imply that foods produced using organic methods are more nutritious or safer than those produced without organic methods. Because of the paucity of scientific data, the UK Food Standards Agency decided in October 2007 to seek a contractor who will evaluate relevant studies and compare the nutrient and non-nutrient content of organic and conventional foods to determine if any compositional differences have nutritional or other health effects in the context of the complete diet (DEFRA, 2006). Differences reported in nutrient composition between organically and conventionally produced foods are interesting but, as seen in the examples given, it is very difficult to control all variables that might affect nutritional quality and ensure that the observed variations are significant and reproducible. For example, in milk no significant differences have been reported in other major nutrients such as calcium, zinc, vitamin B2, or vitamin B12 (Krebs *et al.*, 1995).

Despite these various efforts, ~250,000 to 500,000 children deficient in vitamin-A becomes blind each year; half of them die within twelve months. To develop a bio-fortification strategy to address vitamin A deficiency, researchers developed the first variety of Golden Rice (GR1), a GE variety with increased levels of β- carotene, a precursor to vitamin A, compared with non-GE rice. In 2005 the development of a new Golden Rice variety, GR2, was published; in GR2 a maize gene is substituted for the daffodil genes, boosting β-carotene levels to 37 μg/g estimated to provide 50% of a child's RDA of vitamin A in 72 g of dry GR2 rice. The GR1 and GR2 rice varieties are in use in breeding programmes in the Philippines, India, Bangladesh, China, and Vietnam; the use of Golden Rice is being governed by the Golden Rice Humanitarian Board and is based on full regulatory compliance. Golden Rice might increase vitamin A sufficiency for people in areas difficult to reach with other vitamin A distribution efforts or for people with limited opportunities to grow or purchase sufficient amounts of fresh vegetables or fruits. Golden Rice will not be the single solution to vitamin A deficiency worldwide, but it is another tool that can be used in public health programmes to combat vitamin A deficiency. Researchers using recombinant DNA methods now have the technology to transfer genes, not only within a species, but also from one kingdom to another. This technology opens the door to changing agricultural crops in ways not previously possible. These changes can result in plants that are better able to survive pest attack and abiotic stresses, can be enhanced nutritionally, or can be used to immunize humans and animals. But, with this capacity to change comes the responsibility to proceed with caution, investigating possible outcomes carefully. Conversely, there is also a responsibility to utilize the technology where it can provide improvements to human health and the environment and make farmers' efforts more productive.

14. MITIGATION STRATEGIES

Agriculture is one of the major sources of GHGs emissions. Climate change has been a cause of serious concern if the agricultural sector has to grow in the context of country's overall economic growth, to respond to rural household livelihood, country's food security and poverty alleviation. It may take some time to fully experience the devasting effects of climate change on agriculture but the time is ripe for the governments, private sector and public to have adequate concern, commitment and accountability to mitigate the effects of climate change (Costello *et al.*, 2009).

Significantly, investing expanding, modernizing and equipping agricultural meteorology facilities in all 127 agro-ecological regions to make it world class, thereby continuously improving weather and climate forecasting system. Evolving policy and programmes to manage and mitigate risks due to climate change. Improving early warning systems followed by effective monitoring and evaluating its impact. Developing climate impact modules that give a better understanding of how climate change may affect crop livestock and fish farming and forestry at a local in ordered to be well prepared. Diversifying pattern of livelihoods and adopting agricultural changing and forestry practices to efficient water management and soil conservation practices and growing resilient crops and trees.

Developing a database on climate soil and water use and crop yields to assess, map and monitor land-use performance under given technology conditions. Assessment of how vulnerable our food system is and how we can adopt agriculture, livestock, fisheries and forestry to future climate relates disasters. Increasing coastal inundation, salinization and erosion as a consequence of sea-level rise and human activities may contaminate and reduce the size of productive agricultural lands, thereby threatening household livelihood and country's food security. Steps to mitigate the impact of climate change on agriculture need top priority. Building sufficient resilience of the food systems to avoid enormous future economic losses in agriculture livestock fisheries and forestry. Evolving comprehensive climate resilience strategies comprising risk assessment, development of varieties that can perform well in stressful conditions, better and water and livestock management and bringing about specific changes in agricultural practices that can respond to climate change. While agricultural research institutes and universities have already been engaged in researching drought-resistant and saline-resistant crop varieties for the arid regions and rainfall tolerant and short duration varieties for flood-prone regions, government and private sector will have to invest substantially in agricultural research on one hand and motivate farmers to take better advantage of the dry rabi season in the flood-prone regions and help them supplement their income through non-farm activities on the other (Khan *et al.*, 2012).

15. ADAPTATION

To develop land use plans, food security programmes, fisheries and forestry policies that can help farming community adopt to climate changes, to undertake cost analyses of climate changes risks for irrigation or coastal protection and for investment decision. Promotion of best crop-livestock fish farming practices thoroughfarmers' capacity building and networking. Conceptualization and implementation of national Adaptation Programme of Action on Climate change. Developing contingency plants to cover new and evolving scenarios.

However, the proposed shift in soybean production from the current main season to a second season may necessitate additional planning and change in management practices. Changes in the cropping sequence, irrigation and agriculture land use can be additional alternative options for adaptation in agriculture. Kumar and Parikh (2001) have showed that even with adaptation by farmers of their cropping patterns and inputs, in response to climate change, the losses would remain significant. The loss in farm-level net revenue is estimated to range between 9% and 25% for a temperature rise of 2°C–3.5°C. Agro climatic regions vulnerable to climate change and identify suitable adaptation practices to be followed in order to sustain the productivity of these regions to some extent. These adaptation strategies may include altered crops and cropping systems to maintain soil fertility in sustainable manner and improved management practices. Modern technologies in agriculture could also be beneficial with or without climate change; government should encourage farmers to shift towards newer technologies (Kumar and Parikh, 2001). This change was largely due to very cool weather during January to March 2000, which was favourable to grain formation and filling.

The gluts and shortage of onions and potatoes in recent times besides being caused by policy and management are also manifestation of the effects of climate variability. Such variation in food production would be much larger in rice, pulses and oilseeds where larger portion of the crop area is rain-fed. Any changes in weather during smaller periods would have much larger effect on short-season crops such as vegetables that have less time available to adjust and adopt relative to longer duration crops. Thus, it is very likely that at least in the short run the effects of climatic variability are much larger than the projected impact of global climatic change. Therefore, it appears that if we can evolve strategies for managing climatic variability in agricultural production, adaptation required for climatic change would presumably be automatically taken care of.

Studies have shown that in some crops growing under enhanced CO_2 condition, there is initially a large response, but over time, this response declines and approaches that of crops growing under current CO_2 levels. The impact of the weeds, diseases and insect pests on crop growth, development and final yield formation assume to be controlled. Which need to be incorporated in the future studies for better assessment of final yield. Reducing the wide range of uncertainty inherent in projections of global and regional climate change will require major advances in our scientific understanding on the subject in the years to come. Projections about the probability, frequency, and severity of extreme weather events should be carefully evaluated.

Future long-term projected scenarios of climate change may show the impact after more than 50 years whereas in the mean time several changes in Indian agriculture such as new heat tolerance crop varieties, farm level adaptation, change in demand, market and changes in agricultural technologies is expected to transform agricultural production in India significantly. Whole Indian agricultural system may become changed in coming decades due to the fast technological change, which is expected. These lacking in the studies need to be incorporated in future studies for better-integrated assessment of impact of climate change on Indian agriculture for sustainable development, mitigation and policy planning. Therefore, it is very difficult, at this juncture, to convince the planner and development agencies to incorporate the impact of climate change into their projects and agricultural system. However, 60% of the total cropped area is still rain-fed in India and dependent on uncertainties of monsoon. We must focus on how the possible climate change will affect the intensity, spatial and temporal variability of the rainfall, surface and groundwater availability for irrigation, evaporation rates and temperature in different agro-climatic regions (Medellín-Azuara *et al.*, 2008).

16. CONCLUSION

Climate change is a change in the statistical distribution of weather over periods of time that range from decades to millions of years. It can be a change in the average weather or a climate change in the distribution of weather events around an average. Climate change may be limited to as pacific region, or may occur across the whole earth. The main reasons of climate change including as two factors (a) natural (b) human. Natural factors, viz., as natural fluctuations in the sun intensity, complex motion of the earth around the sun, volcanic eruptions and shorter-term cycles like ElNino; while human factors including increase in GHGs, deforestation and land use changes etc. The expert committee on climate change would be looking into the aspects of measures required for agriculture and to ensure that food security is not threatened due to climate change. Department of Agriculture is in process of preparing a comprehensive action plan on the agriculture agenda for the climate change. The integrated watershed development, a management oriented approach ensures productive and sustainable use of all its natural resources and also to conserve them for the future generations. It takes shortage in supplies especially of water, helps in moving towards higher living standards and provides clan air fresh water leading to a healthy environment. The above analogies, leading to a healthy environment. The above analogies are based on the concept that the watershed framework provides the necessary inter-ecological linkages of the system science and is a logical system for the integration and sustainable management of natural resources.

Acknowledgements

We thank eminent authorities whose works have been consulted and whose ideas and insights have richly contributed to this work, and my research partners who have shared productively my interest in the study. Financial support from the Indian Council of Medical Research, New Delhi to Taqi Ahmed Khan in the form of S.R.F (45/14/2011-BIO/BMS) and Mohd Mazid in the form of J.R.F (0354-40/2008.E.U-39) is gratefully acknowledged.

References

Aggarwal, P. K., and Mall, R. K. (2002). Climate change and rice yields in diverse agro-environments of India. II. Effect of uncertainties in scenarios and crop models on impact assessment. *Climatic Change*, 52(3): 331-343.

Aggarwal, P. K., Joshi, P. K., Ingram, J. S. I., and Gupta, R. K. (2004). Adapting food systems of the Indo Gangetic plains to global environmental change: key information needs to improve policy formulation. *Environmental Science & Policy*, 7(6): 487-498.

Allen Jr, L. H., Drake, B. G., Rogers, H. H., and Shinn, J. H. (1992). Field techniques for exposure of plants and ecosystems to elevated CO2 and other trace gases. *Critical Reviews in Plant Sciences*, 11(2-3): 85-119.

Andrews, T. M. (2003). Choose Ye This Day!!! A study surrounding the dialogue that concerns the use of traditional herbal medicine versus modern scientific medicine in Ghana. *African Diaspora ISPs*, 31: 45-49.

Arora, M., Goel, N. K., and Singh, P. (2005). Evaluation of temperature trends over India/Evaluation de tendances de température en Inde. *Hydrological Sciences Journal*, 50(1): 34-39.

Bajwa, W. I., and Kogan, M. (1997). Compendium of IPM definitions (CID).*Electronic Publication (http:// www. ippc. orst. edu/IPMdefinitions).*

Bizily, S. P., Rugh, C. L., Summers, A. O., and Meagher, R. B. (1999). Phytoremediation of methylmercury pollution: merB expression in Arabidopsis thaliana confers resistance to organomercurials. *Proceedings of the National Academy of Sciences*, 96(12): 6808-6813.

Brown, R. A., and Rosenberg, N. J. (1997). Sensitivity of crop yield and water use to change in a range of climatic factors and CO< sub> 2</sub> concentrations: a simulation study applying EPIC to the central USA. *Agricultural and Forest Meteorology*, 83(3): 171-203.

Chassy, B. M. (2002). Food safety evaluation of crops produced through biotechnology. *Journal of the American College of Nutrition*, 21(suppl 3): 166S-173S.

Cline, W. R. (1991). Scientific basis for the greenhouse effect. *The Economic Journal*, 101(407): 904-919.

Costello, A., Abbas, M., Allen, A., Ball, S., Bell, S., Bellamy, R., and Patterson, C. (2009). Managing the health effects of climate change. *Lancet (London, England)*, 373(9676): 1693-1733.

Dahlberg, K. A. (1987). Redefining development priorities: Genetic diversity and agroecodevelopment. *Bulletin of Science, Technology & Society*, 7(3-4): 367-382.

Debeaujon, I., Nesi, N., Perez, P., Devic, M., Grandjean, O., Caboche, M., and Lepiniec, L. (2003). Proanthocyanidin-accumulating cells in Arabidopsis testa: regulation of differentiation and role in seed development. *The Plant Cell Online*, 15(11): 2514-2531.

Demain, A. L., and Vaishnav, P. (2009). Production of recombinant proteins by microbes and higher organisms. *Biotechnology Advances*, 27(3): 297-306.

Gitay, H., Suárez, A., Watson, R. T., and Dokken, D. J. (2002). *Climate change and biodiversity*. Geneva, CH: IPCC.

Godfray, H. C. J., Beddington, J. R., Crute, I. R., Haddad, L., Lawrence, D., Muir, J. F., and Toulmin, C. (2010). Food security: the challenge of feeding 9 billion people. *Science*, 327(5967): 812-818.

Heichel, G. H. (1976). Agricultural Production and Energy Resources: Current farming practices depend on large expenditures of fossil fuels. How efficiently is this energy used, and will we be able to improve the return on investment in the future?. *American Scientist*, 18:64-72.

Herdt, R., Toenniessen, G., and O'Toole, J. (2007). Plant biotechnology for developing countries. *Handbook of Agricultural Economics*, 3: 2641-2667.

Jimenez-Lopez, J. C., and Hernandez-Soriano, M. C. (2012). Testing Methods for Agriculture and Food Safety. *Biochemical Testing*, 18: 55-82.

Khan, T. A., Mazid, M., and Mohammad, F. (2012). Climate change, sustainable agriculture and future needs: A perspective of parallel re-thinking. *Journal of Industrial Research & Technology*, 2(1): 40-50.

King, G. A., and Neilson, R. P. (1992). The transient response of vegetation to climate change: a potential source of CO 2 to the atmosphere. *Water, Air, & Soil Pollution*, 64(1): 365-383.

Knox, J. W., Hess, T. M., Daccache, A., and Ortola, M. P. (2011). What are the projected impacts of climate change on food crop productivity in Africa and S Asia. *DFID Systematic Review Final Report*.

Kögel, T., and Prskawetz, A. (2001). Agricultural productivity growth and escape from the Malthusian trap. *Journal of Economic Growth*, 6(4): 337-357.

Krebs, N. F., Reidinger, C. J., Hartley, S., Robertson, A. D., and Hambidge, K. M. (1995). Zinc supplementation during lactation: effects on maternal status and milk zinc concentrations. *The American Journal of Clinical Nutrition*, 61(5): 1030-1036.

Kumar, K. K., Rajagopalan, B., and Cane, M. A. (1999). On the weakening relationship between the Indian monsoon and ENSO. *Science*, 284(5423): 2156-2159.

Kumar, K. S., and Parikh, J. (2001). Indian agriculture and climate sensitivity. *Global Environmental Change*, 11(2): 147-154.

Kumar, P., Kumar, A., and Mittal, S. (2004). Total factor productivity of crop sector in the Indo-Gangetic Plain of India: Sustainability issues revisited. *Indian Economic Review*, 9: 169-201.

Laux, P., Jäckel, G., Tingem, R. M., and Kunstmann, H. (2010). Impact of climate change on agricultural productivity under rain-fed conditions in Cameroon—A method to improve attainable crop yields by planting date adaptations. *Agricultural and Forest Meteorology*, 150(9): 1258-1271.

Lemaux, P. G. (2009). Genetically engineered plants and foods: A scientist's analysis of the issues (part II). *Annual Review of Plant Biology*, 60: 511-559.

Lieberman, S., and Gray, T. (2008). The World Trade Organization's Report on the EU's Moratorium on Biotech Products: The Wisdom of the US Challenge to the EU in the WTO. *Global Environmental Politics*, 8(1): 33-52.

Mall, R. K., Lal, M., Bhatia, V. S., Rathore, L. S., and Singh, R. (2004). Mitigating climate change impact on soybean productivity in India: a simulation study. *Agricultural and Forest Meteorology*, 121(1): 113-125.

Mall, R. K., Singh, R., Gupta, A., Srinivasan, G., and Rathore, L. S. (2006). Impact of climate change on Indian agriculture: A review. *Climatic Change*, 78(2): 445-478.

Medellín-Azuara, J., Harou, J. J., Olivares, M. A., Madani, K., Lund, J. R., Howitt, R. E., and Zhu, T. (2008). Adaptability and adaptations of California's water supply system to dry climate warming. *Climatic Change*, 87: 75-90.

Mine, Y., and Rupa, P. (2004). Immunological and biochemical properties of egg allergens. *World's Poultry Science Journal*, 60(3): 321-330.

Najam, A., Runnalls, D., and Halle, M. (2005) Environment and Globalization: The Five Propositions Environment and Globalization: The Five Propositions.

Porter, J. R., and Semenov, M. A. (2005). Crop responses to climatic variation. *Philosophical Transactions of the Royal Society B: Biological Sciences*, 360(1463): 2021-2035.

Rajendran, K., and Kitoh, A. (2008). Indian summer monsoon in future climate projection by a super high-resolution global model. *Current Science*, 95(11): 1560-1569.

Sharma, V. P. (2007). *India's Agrarian Crisis and Smallholder Producers' Participation in New Farm Supply Chain Initiatives: A Case Study of Contract Farming*. Indian institute of management (IIM).

Srivastava, J., Lambert, J., and Vietmeyer, N. (1996). *Medicinal plants: An expanding role in development* (No. 320). World Bank Publications.

Tabashnik, B. E., Carrière, Y., Dennehy, T. J., Morin, S., Sisterson, M. S., Roush, R. T., and Zhao, J. Z. (2003). Insect resistance to transgenic Bt crops: lessons from the laboratory and field. *Journal of Economic Entomology*, 96(4): 1031-1038.

Varshney, R. K., Hoisington, D. A., and Tyagi, A. K. (2006). Advances in cereal genomics and applications in crop breeding. *Trends in Biotechnology*, 24(11): 490-499.

PART C

Biotechnological Approaches to Crop Improvement

10

Crop Hybridization in the Era of Climate Change

Shilpi Paul[1*] and Shrilekha Misra[2]

[1]*G.B. Pant Institute of Himalayan Environment and Development, Kosi-Katarmal Almora-263643, Uttarakhand, India*

[2]*Banasthali Vidyapeeth, Rajasthan, India*

ABSTRACT

The twentieth century is known for new varieties/cultivars and green revolution which was occurred at the time of climate change/stability. Will agricultural production and crop improvement be able to maintain continuous rise under changing climate? A number of agriculture/agri-biotech techniques are introduced to think and cope up with these problems. Hybridization in crop is not new but it can be used as a strategy to increase/develop new varieties in the climate change era. Hybridization is broadly defined as crosses between two genetically diverse populations/species which may further be observed as occurring natural and artificial. Natural hybridization may be affected by numerous factors, both intrinsic and extrinsic factors like inbreeding and local environment compositions which may facilitate hybridization. While in non-natural hybridization, it is easy to discuss the outcome of non-natural hybridization in novel environments. Here intrinsic and extrinsic factors, like inbreeding among parental populations, spatial changes in the local environment like fragmentation, and translocation of elite species, may influence on hybrid fitness under controlled conditions. Such phenomenon can play very important role in crop improvement under climate change. Natural hybridization is tightly controlled by genes and surrounding environment. It can occur in novel environment under relieve selection pressure. In such environment genetic recombination not only destroys beneficial gene complexes but also creates novel complexes with a positive selection pressure. Non-natural hybridization has a serious threat towards the genetic integrity of specific trait under continuous selection pressure. Hybrid varieties of cereals, vegetables, pulse and medicinal plants are good examples and play an important role in production.

Keywords: Agriculture, climate change, environmental effect, gene transfer, hybridization.

1. INTRODUCTION

Today we are facing twin challenges to feed growing population and climate change. The twentieth century has been known for increase in agricultural production through the development of new varieties and fertilizers. Also know as green revolution which helped to keep food prices

*Corresponding author's E-mail: shilpipaul01@gmail.com

Roychowdhury R. (Ed.), Crop Improvement in the Era of Climate Change.

at a low level (Swanson *et al.*, 2009). Based on the changing picture of climatic conditions question arises will agricultural production be able to continue rising in the coming years under a changing climate?

Gene flow and hybridization are one of the major evolutionary processes shaping plant populations according to the environment (Arnold, 1997; Hartl and Clark, 1997; Arnold *et al.*, 1999; Barton, 2001). It may affect by the environmental conditions which results in the introduction of new alleles and changes in the frequencies of resident alleles. This may also promote inter- and intraspecific hybridization to the evolution of invasiveness in plant populations (Ellstrand and Shierenbeck, 2000) with respect to climate change. Simultaneously, with the advent of transgenic crops, gene flow between crop fields and wild populations has emerged as a prime concern (Kareiva *et al.*, 1996; Snow and Mora´n-Palma, 1997; Hails, 2000; Ellstrand, 2003). Gene flow between crop and wild plants can provide a suitable model for hybridization between differentially adapted populations or subspecies. In natural population, crops and their wild relatives are crossed and can contribute to each other's gene pool (Ellstrand *et al.*, 1999). The migration of crop alleles into wild populations depends on their fitness effect based on the environmental conditions (Ellstrand, 2003). If the novel alleles fitness is decreasing in wild conditions that shows, the alleles for such trait values favoured in the domestication process (Stewart *et al.*, 2003). In such situation, decreasing frequency of alleles with time, results the loss of crop alleles in the population.

2. HYBRIDIZATION

The crossing of two plants/lines of dissimilar genotype is known as hybridization. In plants, crossing is done by placing pollen grains from one genotype (male parent), on to the stigma of flowers of the other genotype (female parent). The prevention of self-pollination and chance of cross-pollination in female parent is essential. The pollen from the desired male parent reaches the stigma of female flowers for successful fertilization. The progeny resulting from the hybridization are known as hybrid or F1.

Hybridization is required to create genetic variation to enhance gene pool. When two genotypically different plants are crossed, the genes from both the parents are brought together in progenies. Segregation and recombination produce many new gene combinations in F2 progenies/later generations. The level of variation produced in the segregating generations would depend on the number of heterozygous genes in the F1. If F1 is 100% heterozygous, it means it has only those genes which differ in the two parents. The main aim of hybridization is to transfer one or few qualitative characters, for crop improvement (Sharma, 1996; Singh, 2006). Hybridization can be carried out by following methods:

(a) **Combination Breeding:** The transfer of one or more characters into a single variety from other varieties. These characters may be of oligogenes or polygenes. Using this approach, increased yield of a variety will be obtained by correcting the weaknesses in the yield contributing traits, e.g., tiller number, grains per spike; test weight etc.

(b) **Transgressive Breeding:** Transgressive breeding is the production of plants in F2 progenies which are superior to both the parents for one or more characters. Through this method plants are produced by an accumulation of plus or favourable genes from both the parents.

(c) **Hybrid Varieties:** In most self-pollinated crops, F1 is more vigorous and higher yielding than the parents. It can be used directly as a variety. F1 should be superior to its parents.

2.1 Conditions and Steps Involved in Hybridization

The following steps are involved in hybridization:

1. Choice or selection of parents according to climatic condition
2. Evaluation of parents, i.e., by selfing and studying the progeny
3. Emasculation (for artificial hybridization)
4. Crossing or pollination
5. Bagging and labelling
6. Harvesting of F1 seed
7. Raising F1 generation

2.2 Types of Hybridization

Based on the parental relationship of the two parents, hybridization may be classified into two broad categories:

1. Inter-varietal
2. Distant hybridization

2.2.1 *Inter-varietal Hybridization*

The parents involved in hybridization belongs to the same species, may be two strains, varieties or races of the same species. Inter-varietal hybridization is generally used for crop improvement. The inter-varietal crosses may be simple or complex depending upon the number of parents involve in crossing.

(a) **Simple Cross:** Only two parents are crossed to produce the F1 hybrid. The F1 may selfed to produce F2 or will be used for backcross programme.

(b) **Complex Cross:** More than two parents are involved in hybridization, to produce F2 progenies. This is also known as convergent cross i.e., bringing together of different genotypes/line/accession from several parents into a single hybrid. For example, in the development of variety CIM Arogya of *Artemisia annua;* six high artemisinin yielding genotypes (considered parents) were planted in polycross design for matting themselves to produce high yielding plant/variety (Paul *et al.*, 2010). In view of this, it may be expected that in future complex crosses would become more and more important for climate change. The improved varieties of common crops like wheat and rice are good examples of complex crosses. Such crosses would become routine in near future in the improvement of other self-pollinated crops with the progress in the level of their improvement and environment.

2.2.2 *Distant Hybridization*

Cross between different species of the same genus or of different genera is called distance hybridization. The transfer of one or few simply inherited characters in a crop species is the main objective in such hybridization. Such inter-specific hybridization may be used for developing new varieties, e.g., Clintonoat variety of oat (*Avena sativa* x *A. byzantina*; both hexa-ploid oat species), and CO 31 rice (*Oryza sativa* var. indica x *O. perennis*). Most of the today's sugar cance varieties have been developed from complex crosses between *Saccharum officinarum* (noble canes), *S. barberi* (Indian canes) and other *Saccharum* species, e.g., *S. spontaneum* (Kans. Intergeneric hybridization may also be used to develop a new crop species, e.g., *Triticale* from a cross between *Triticum* sp. and *Secale cereale* (rye). Wild species often provide genes which are not present in the cultivated species. Distant hybridization is likely to become increasingly important in the correction of specific defects of crop species. Some, wild species may contribute valuable 'yield specific genes' as well to the cultivated species (Singh, 2006).

3. GENE TRANSFER

3.1 Transfer of a Dominant Gene

When a high yielding and widely adapted variety A (dominant gene) is crossed with another variety B (recessive gene) then there is a generalized scheme of the backcross programme for transfer of gene from variety B to variety A. The steps involved during dominant gene transfer are: F1 generation (F1 progeny plants are backcrossed to variety A and all the F1 plants will be heterozygous for dominant allele), first backcross generation (half of the plants may be dominant and the remaining half may be recessive and dominant plants will be selected for backcross to BC1 plants), BC2-BC5 generations {segregation may occur for dominant gene in each back cross, selection (BC2 and BC3) for the plant type of variety A}, BC6 generation (approximate 98.4% genes from variety A) and BC6F2 generation (progenies homozygous for dominant allele and similar to the plant type of variety A). Several similar progenies are mixed to constitute the new variety. Yield tests for new variety will be in a replicated yield trial along with the variety A as a check. Evaluation will be based on the plant type, date of flowering, date of maturity, quality etc. (Singh, 2006).

3.2 Transfer of a Recessive Gene

When a desire character is on recessive gene, all the backcrosses cannot be made one after the other. After the first backcross or two backcrosses, F2 must be grown to identify for recessive gene. This process includes hybridization (recurrent parent consider as female plant is crossed with the recessive donor parent), F1 generation (F1 plants are backcrossed to the recurrent parent), BC1 generation (all plants should be self-pollinated), BC1F2 generation (recessive plants are selected and backcrossed with the recurrent parent), BC2 generation (plants are selected for their similarity to the recurrent parent and again backcrossed with the recurrent parent), BC3 generation (plants are self-pollinated to generate F2), BC3F2 generation, BC4 generation (back-crossed to recessive parent), BC5 generation (again self-pollinated to raise F2 generation), BC5F2 generation (plants are subjected to recessive and selection is done for recessive and seeds are collected separately), BC5F3 generation (selection and seeds from several similar recessive homogeneous progenies are mixed to constitute the new variety and yield will be tested; Allard, 1960; Sharma, 1996; Singh, 2006).

4. HETEROSIS

Heterosis (coined by Shull in 1914) may be defined as the superiority of F1 hybrid over both the parents in terms of yield and some other characters. Heterosis (hybrid vigour) is manifested as an increase in vigour, size, growth rate, yield or some other characteristic. But sometimes, it may be inferior to the weaker parent. Heterosis results from the masking of harmful effects of recessive alleles by their dominant alleles. Heterosis may increase yield, reproductive ability, size and general vigour, better quality, early flowering, maturity, greater resistance to diseases, pests, greater adaptability, faster growth rate and number of plant parts. Another form of heterosis is **luxuriance** (increased vigour and size of inter-specific hybrids) and the basic difference between heterosis and luxuriance lies in the reproductive ability of hybrids.

4.1 Physiological Basis of Heterosis

The effect of heterosis on growth regulators and enzymes was reported by Whaley (1952). He suggested that the hybrid embryo has more potential to mobilize stored food materials than their inbreds. Khanna-Chopra *et al.* (1993) also reported that heterotic hybrids showed faster growth rate, higher leaf area index and greater biomass production than do their parents, but their harvest index (HI) is comparable their parental inbreds. Major components of total biomass are net assimilation rate and leaf area index in plants showed effective difference in hybrids. Early flowering mainly due to their faster initial growth rate is also observed in heterotic hybrids. Some physiological parameters which were studied for heterotic plants are: (1) *Net assimilation rate:* Some hybrids showed heterosis for photosynthesis at the seedling stage (Sinha and Khanna, 1975), CO_2 exchange rate (wheat) and for photosynthetic efficiency (rice). (2) *Leaf area index (LAI):* Hybrids in various crops especially during the early growth phases showed distinct advantage over their parents in terms of LAI. (3) *Root growth:* Root growth depends on shoot growth and roots serve as 'sinks' for the photosynthesis. (4) *Hormone balance/imbalance:* Heterosis may be consequence of hormone balance in hybrids as compared to those of their parents. In case of maize, inbreds contain lower levels of GA_3 and respond to exogenous GA_3 application. (5) *Mitochondrial complementation:* The mitochondrial and chloroplast may be the cause of hybrid vigour. It has been observed that mitochondria isolated from seedlings of the hybrids and their parents showed different efficiencies of oxidative phosphorylation and respiration rates.

5. HYBRID ZONES

A hybrid zone is designated as those places where two interbreeding species meet and hybridize. Hybrid zones move, typically at the rate of 0.1–10 metres per year based on the environmental conditions. Such zone is relatively rare and can be of two types. Primary hybrid zones are those primary contact occurs in parapatric speciation. Secondary hybrid zones arise from secondary contact between two populations that were previously allopatric. It is quite difficult to distinguish between primary and secondary contacts.

6. NOVEL ENVIRONMENT FOR HYBRIDIZATION

It is very difficult to define novel environment for hybridization. It may include intrinsic and extrinsic factors, like inbreeding among parental populations, spatial changes in the local environment like fragmentation, and translocation of elite species which may influence on hybrid fitness (Rhode and Cruzan, 2005; Barilani *et al.,* 2007).

A novel environment can be defined as a spatial area, where at least one of the two hybridizing species or populations under normal circumstances are not found; but where admixture is induced by anthropogenic activities. And the fitness expression in hybrids is dependent on the genotypes which are inherited from the parental populations. This may due to recombination between the inherited parental chromosomes. A good example for such condition was recorded after E_1 Nino when smaller seeds were supplied to finches, whereas larger seeds were dominated before the event. Such changes happen due to change in local environment.

7. EFFECT OF CLIMATE CHANGE ON HYBRIDIZATION

7.1 Abiotic and Direct Effects

7.1.1 Water

The major impact in agriculture sector is water supply which becomes scarce in some areas and more subject to extreme variations (Ziska, 2011). According to Doll (2002), two-thirds of current global irrigated areas will need more water by 2070, due to a combination of higher temperatures, lower precipitation and water supply. For example, river run-off is decreasing in some areas (Kundzewicz *et al.*, 2007; 2008). It has also been observed that requirement of water will be increased about 40% due to the impacts of climate change on annual agriculture (Fischer *et al.*, 2007).

7.1.2 Temperature

Crops are very sensitive to temperature. If the temperature rises up to optimal for crop that leading to reduced yields or enforce to develop new hybrids.

7.1.3 Extremes

Under any conditions of climate change (i.e., human or natural), extreme weather events are likely to become more frequent and/or more intense, for temperatures and precipitation (IPCC, 2012). The extreme events to agriculture are droughts, floods and heat waves etc. These extreme events can dramatically reduce or damage crop yields (Rosenzweig *et al.*, 2001; Smith *et al.*, 2007).

The other factors like change in carbon dioxide concentration, precipitation, humidity, light and environmental geography also affect agriculture performance.

7.1.4 Environmental Gradients

Environmental gradients like temperature, humidity and precipitation may vary from one area or region to another. The most suited example of an environmental gradient is the change in temperature from the equator to the poles. The changes in environmental conditions where topography is irregular are troublesome for crop producers especially for hill agriculture. Hence, large-scale agriculture is difficult in mountainous areas.

7.1.5 Tolerance

According to Shelford's concepts:

1. Species with a wide range of tolerance are likely to be widely distributed.

2. Some species may have a wide range of tolerance for one environmental factor and a narrow range for another.
3. When one environmental factor is limiting, the range of tolerance to other factors is likely to be decreased.
4. The range of tolerance is likely to be narrow during the period of reproduction.

Environment can influence the growth and development of organisms only within genetically imposed limits. Species with a wide range of genetic diversity are more likely to respond favourably to new and changed environments than to those with a narrow genetic base. Plants in different stages of growth may have different tolerances.

The developmental phases respond differently to environmental factors. Areas with similar climates are called homoclimes, or with reference to agricultural production, agro-climatic analogues.

7.2 Biotic Factors

The biological community, in their roles as parasite, predator, symbiont, competitor, or vector of pollination and propagation are strongly influenced by distribution of natural plant communities. Such factors may play an important role on the distribution of crops. Many crops depend on biological vectors for pollination, e.g., pollination in Smyrna fig depends on *Blastophaga* wasp. Similarly, many legumes as crops depend on their association with particular nitrogen-fixing bacteria. Sometimes many legumes proved ineffectual in tropical areas until specific forms of bacterial symbionts were introduced.

8. IMPACT OF CLIMATE CHANGE ON CROP YIELD AND BREEDING STRATEGY

Climate change affects agriculture and food production in many complex ways. Food production affected directly through changes in agro-ecological conditions and indirectly by affecting the growth and the distribution of incomes (Schmidhuber and Tubiello, 2007). Changes in temperature and precipitation associated with continued emissions of green house gases (GHGs) are expected to result in long-term changes, including a rise in the global mean surface temperature from 1.8°C to 4.0°C by 2100 and large changes in rainfall, which affect land suitability and crop yields. Current research confirms that while crops would respond positively to elevated CO_2 in the absence of climate change (e.g., Kimball *et al.*, 2002; Jablonski *et al.*, 2002; Ainsworth and Long, 2005), the associated impacts of high temperatures, altered patterns of precipitation and possibly increased frequency of extreme events such as drought and floods, will probably combine to depress yields and increase production risks in many regions, widening the gap between rich and poor countries (e.g., IPCC, 2001).

The Intergovernmental Panel on Climate Change (IPCC, 2007) states that at lower latitudes and in tropical dry areas, crop productivity is expected to decrease "for even small local temperature increases (1-2°C)." At temperate latitudes, higher temperatures are expected to be mostly beneficial to agriculture. The areas potentially suitable for cropping are expected to expand, the length of the growing period will increase, and crop yields may rise. A moderate incremental warming in some humid and temperate grassland may increase pasture productivity and reduce the need for housing and for compound feed (Schmidhuber and Tubiello, 2007). These gains have to be set against an increased frequency of extreme events, for instance, heat waves and droughts in the Mediterranean region, or increased heavy precipitation events and flooding in temperate regions, including the possibility of increased coastal storms (IPCC, 2001; Howden *et al.*, 2007). In drier areas, climate models predict increased evapotranspiration and lower soil moisture levels. As a result, some cultivated areas may become unsuitable for cropping and some tropical grassland may become increasingly arid.

Focusing on the relationship between climate variables and the productivity of food crops with linkage between climate change and crop yields have been largely explored. According to IPCC (2007), in many African countries access to food will be severely affected, "yields from rain-fed agriculture could be reduced by up to 50% by 2020".

8.1 Breeding Strategy During Climate Change

In the case of wheat genetic improvement, the Mediterranean climate is characterized by low and variable annual rainfall which can fluctuate from 200–800 mm and irrigation is not an available option for commercial wheat production (Nachit and Elouafi, 2004). Due to high year-to-year variability, breeding cultivars that combine drought resistance, yield potential, and yield stability are a prerequisite for stable productivity in this region. Some modern techniques like molecular markers have become an essential part for genomic and genetic research which allow discrimination in DNA sequence amongst cultivars and breeding lines and thus offer the scientific community singularly powerful tools to monitor, track, and exploit sequence variation in germplasm. Many types of markers have been developed and are now an essential part of functional, structural genomics and molecular breeding (Varshney *et al.*, 2007). Microsatellite markers or single sequence repeats (SSR) have proved useful in wheat research since they offer reproducibility, multiallelic nature, codominant inheritance, genome specificity, relative abundance, and good genome coverage (Varshney *et al.*, 2005; Ganal and Roder, 2007). The localization of genes to chromosomes (Roder *et al.*, 2004), identification of quantitative trait loci (QTLs) for yield and quality traits (Ganal and Roder, 2007), characterization of wheat varieties and germplasm (Donini *et al.*, 2000; Roder *et al.*, 2004), marker assisted selection (MAS), and backcrossing were carried out through SSRs. Another promising type is single-nucleotide polymorphism (SNP) or biallelic markers because they form the basis for most genetic variation between individuals; they are widely distributed and amenable to high multiplex detection systems (Ganal and Roder, 2007).

9. A CASE STUDY – *CAPSICUM ANNUUM*

Chilli is one of the most versatile vegetable crops in India. Our study on chilli, collected from different parts of India can be a good example of hybridization and effect of different climatic conditions on growth and other parameters. The *Capsicum* is the unique genus of the family Solanaceae finding diverse uses from nutritional and culinary to pharmaceutical uses. In this genus, native to the Americas, more than 30 species have been described, but only five of them, *Capsicum annuum* var. *annuum*, *C. chinense*, *C. frutescens*, *C. baccatum* var. *pendulum*, and *C. pubescens* are considered to be domesticated (Pozzobon *et al.*, 2006; Moscone *et al.*, 2007).

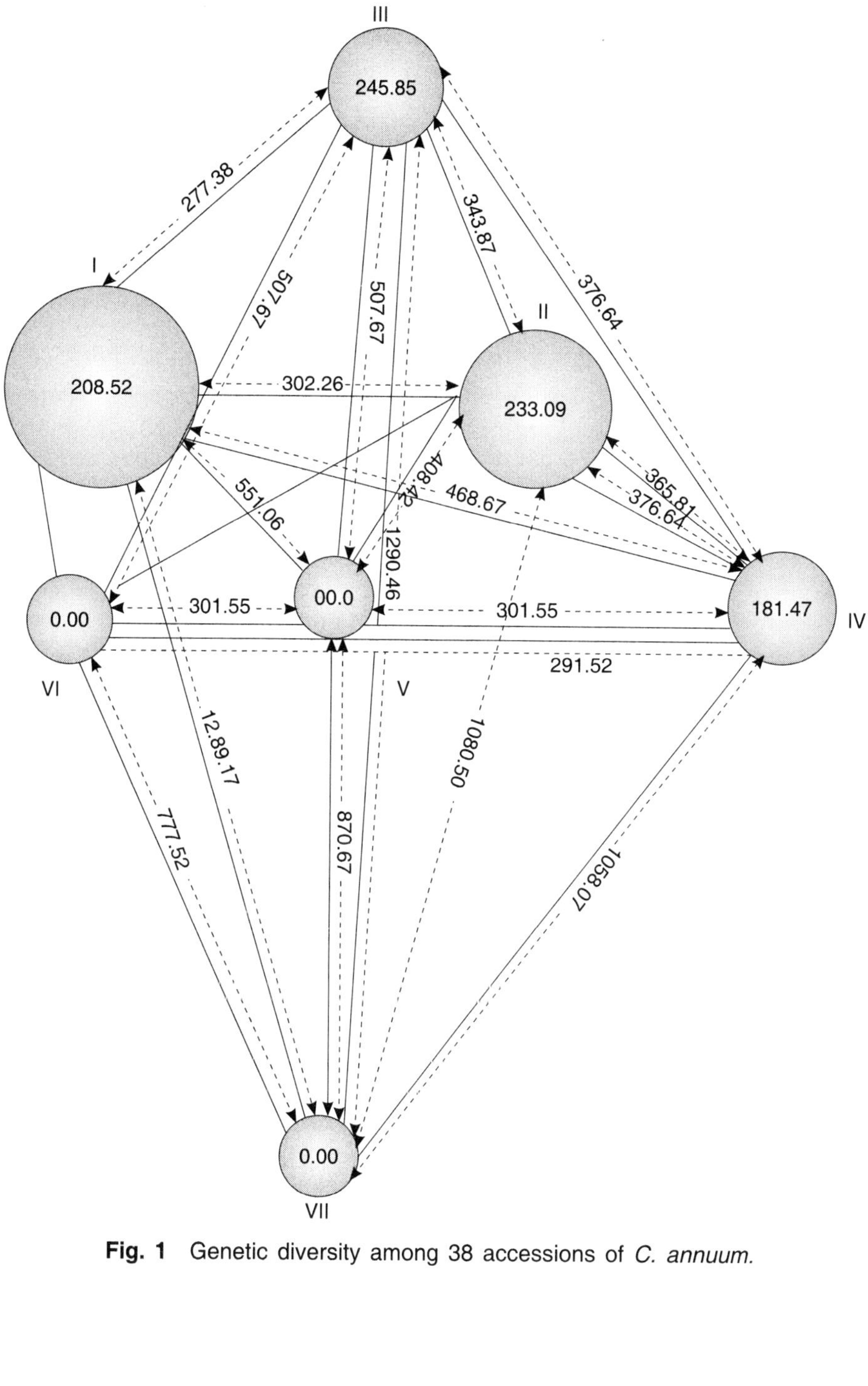

Fig. 1 Genetic diversity among 38 accessions of *C. annuum.*

Pungent peppers commonly known as chilies in India belong to the species *C. annuum* L. which is the most widely cultivated species in the world. Both hot and sweet peppers have been reported to have originated from *C. annuum*. For culinary purposes, its fruits are used for pungency (capsaicin) and also colour (capsanthin). Capsaicin is also used for medicinal purposes particularly in anti-inflammatory formulations. Genetic divergence among 38 accessions of *C. annuum* collected from diverse locations in India (28 from Uttar Pradesh, 5 from Assam—east India, 3 from Maharashtra and 2 from Uttarakhand—high altitude), was estimated from the data pooled over 3 consecutive years for 15 morphological, growth and chemotypic characters that included days to first and second flowering, fruit onset, plant height, primary, secondary and tertiary branches, leaf surface area, fruit length and diameter, fruit surface area, fresh and dry fruit weight, capsaicin and capsanthin content. Based on this characterization the plants could be grouped into 7 clusters wherein substantial diversity among accessions was indicated by the wide range of D^2 values (752.901 - 1918683.00). Accessions with distinct identity were marked, which are likely to be quite suitable for breeding through hybridization combining desirable traits. The accessions which were collected from Assam and Lucknow containing high capsaicin content (%); Maharastra samples contained high capsanthin content (%) and some genotypes of Assam contained both high capsaicin and capsanthin content (Fig. 1). Therefore, the above listed accessions of the *C. annuum* could be utilized, as parents for hybridization in recombination breeding programme followed by repeated selection to obtain maximum gain (Mishra *et al.*, 2011). The inference drawn from such experiments that climatic conditions also plays a very crucial role in secondary metabolite synthesis.

10. CONCLUSION

Major part of the world population depends on agriculture than any other trade in the world and even only 8 to 9% of the total land surface of the earth is under cultivation. Most of this cultivated land is in the middle latitudes, between 30° and 60° north and south of the equator and only four countries like the United States, the Soviet Union, China, and India lie in that area. The distribution of land that is suitable for cultivation is primarily determined by the nature of the environment—climate, soil, and landforms. The process of climate change is constantly going on, although the changes occur gradually. Based on the some reports the climate has since shown a warming trend, and glacial ice in high mountain valleys has been observed to decrease markedly during the past 50 years. Due to gradual climatic change, many crop growers would maintain that it is not realistic to take it into account, but within the large climatic changes and cycles there are smaller changes that can be witnessed within a generation. Advanced agriculture technologies are required to make profitable use of the great areas in tropics and deserts which are now agriculturally unproductive and more hybridization techniques will be required to cope up the problems arising from climate change.

Acknowledgements

Authors are thankful to the director of the institute for library and internet facility and Dr. Shekhar Mallick for their help and support.

References

Ainsworth, E.A., and Long, S.P. (2005). What have we learned from 15 years of free-air CO_2 enrichment (FACE)? A meta-analytic review of the responses of photosynthesis, canopy properties and plant production to rising CO_2. *New Phytologist.* 165: 351-372.

Allard, R.W. (1960). *Principles of Plant Breeding.* John Wiley and Sons, New York.

Arnold, M. L. (1997). *Natural hybridization and evolution.* Oxford Univ. Press, New York.

Arnold, M. L., Bulger, M. R., Burke, J. M., Hempel, A. L., and Williams. J.H. (1999). Natural hybridization: How low can you go and still be important? *Ecology.* 80: 371-381.

Barilania, M., Bernard-Laurentb, A., Muccia, N., Tabarronic, C., Karkd, S., Garridoe, J. A. P., and Randia, E. (2007). Hybridization with introduced chukars (*Alectoris chukar*) threatens the gene pool integrity of native rock. *Bio. conser.* 137: 57-69.

Barton, N. H. (2001). The role of hybridization in evolution. *Mol. Ecol.* 10: 551-568.

Doll, P. (2004). Impact of climate change and variability on irrigation requirements: a global perspective. *Climatic Change.* 54: 269-293.

Donini, P., Law, J. R., Koebner, R.M.D., Reeves, J.C., and Cooke, R.J. (2000). Temporal trends in the diversity of UK wheat. *Theor. App. Gene.* 100: 912-917.

Ellstrand, N. C. (2003). Dangerous liaisons? When cultivated plants mate with their wild relatives. Johns Hopkins Univ. Press, Baltimore, MD.

Ellstrand, N. C., and Schierenbeck, K.A. (2000). Hybridization as a stimulus for the evolution of invasiveness in plants? *Proc. Natl. Acad. Sci. USA* 97: 7043-7050.

Ellstrand, N. C., Prentice, H.C., and. Hancock. J.F. (1999). Gene flow and introgression from domesticated plants into their wild relatives. *Annu. Rev. Ecol. Syst.* 30: 539-563.

Fischer, G., Tubiello, F.N., van Velthuizen, H., and. Wiberg, D.A. (2007). Climate change impacts on irrigation water requirements: Effects of mitigation, 1990-2080. *Technol. Forecasting Soc. Change.* 74: 1083-1107.

Ganal, M.W., and Roder, M.S. (2007). Microsatellite and SNP markers in wheat breeding, Vol. 2. In: Varshney, R.K., Tuberosa, R., (eds). *Genomic assisted crop improvement: genomics applications in crops.* The Netherlands: Springer, 1–24.

Hails, R. S. (2000). Genetically modified plants; the debate continues. *Trends Ecol. Evol.* 15: 14-18

Hartl, D. L., and Clark, A.G. (1997). Principles of population genetics. Sinauer Associates, Sunderland, MA.

Howden, S.M., Soussana, J. F., Tubiello, F.N., Chhetri, N., Dunlop, M and Meinke, H. (2007). Adapting agriculture to climate change. *Proceedings of the National Academy of Sciences*, 104: 19691-19696.

Intergovernmental Panel on Climate Change 2001 http://www.acrim.com/%5C/Reference%20Files/CLIMATECHANGE%202001%20-%20The% 20Scientific%20Basis.pdf IPCC. (2007). Climate change: the physical science basis. In: Solomon S, Qin D, Manning M, Chen Z, Marquis M, Averyt KB, Tignor M, Miller HL, eds. Contribution of Working Group I to the Fourth Assessment Report of the Intergovernmental Panel on Climate Change. Cambridge, UK, New York, USA: Cambridge University Press.

Jablonski, L.M., Wang, X., and Curtis, P.S. (2002). Plant reproduction under elevated CO_2 conditions: a meta-analysis of reports on 79 crop and wild species. *New Phytologist.* 156: 9-26.

Kareiva, P., Parker, I.M., and Pascual, M. (1996). Can we use experiments and models in predicting the invasiveness of genetically engineered organisms? *Ecology.* 77: 1670-1675.

Khanna–Chopra, R., Maheshwari, D., and Gangadhar, R., Sinha, S.K. (1993). Expression of heterosis – Aphysiological analysis. In Verma, M. M., Virk, D.S., Chahal, G.S., and Dhillion, B.S .(eds) *Proc Sym Heterosis breeding in crop Plants- Theory and applications. Crop Imp Sco* India, Ludhiana pp. 20-36

Kimball, B.A., Kobayashi, K. and Bindi, M. (2002). Responses of agricultural crops to free-air CO_2 enrichment. *Adv. Agro.* 77: 293-368.

Kundzewicz, Z. W., Mata, L. J., Arnell, N. W., Döll, P., Jimenez, B., Miller, K., Oki, T., Şen, Z., and Shiklomanov, I. (2008). The implications of projected climate change for freshwater resources and their management. *Hydrol. Sci. J.* 53(1): 3-10.

Kundzewicz, Z. W., Mata, L. J., Arnell, N., Döll, P., Kabat, P., Jiménez, B., Miller, K., Oki, T., Şen, Z. & Shiklomanov, I. (2007). Freshwater resources and their management. *Climate Change: Impacts, Adaptation and Vulnerability.* Contribution of Working Group II to the Fourth Assessment Report of the Intergovernmental Panel on Climate Change (ed. by M. L. Parry, O. F. Canziani, J. P. Palutikof, P. J. van der Linden & C. E. Hanson), 173–210. Cambridge University Press, UK. http://www.ipcc.ch/pdf/assessment-report/ar4/wg2/ar4-wg2-chapter3.pdf.

Kyle, L., Swansona, I., Sugiharab,G., and Anastasios A. (2009). Tsonisa Long-term natural variability and 20th century climate change. *Proc. Natl. Acad. Sci. USA.* 106: 16120-16123.

Misra, S., Lal, R.K., Darokar, M.P., and Khanuja, S.P.S. (2011). Genetic Variability in Germplasm Accessions of *Capsicum annuum* L. *Am. J of Plant Sci.* 2: 629-635.

Moscone, E. A., Scaldaferro, M. A., Grabiele, M., Cecchini, N. M., Sánchez García, Y., Jarret, R., Daviña, J. R., Ducasse, D. A., Barboza G. E., and Ehrendorfer, F. (2007). The evolution of chili peppers (*Capsicum-Solanaceae*): a cytogenetic perspective. *Acta. Hort.* 745: 137-170.

Nachit, M.M., and Elouafi, I. (2004). Durum adaptation in the Mediterranean dryland: breeding, stress physiology, and molecular markers. *In:* Rao, S.C., Ryan, J., *editors. Challenges and strategies for dryland agriculture.* CSSA Special Publication 32. Madison, Wisconsin, USA: Crop Science Society of America Inc., American Society of Agronomy Inc. 203-218.

Paul, S., Khanuja, S.P.S., Shasany, A.K., Gupta, M.M., Darokar, M.P., Saikia, D., Gupta, A.K. (2010). Enhancement of artemisinin content through four cycles of recurrent selection with relation to heritability, correlation and molecular marker in *Artemisia annua* L. *Planta Med.* 76: 1468-1472.

Pozzobon, M. T., Schifino-Wittmann, M. T., and Bianchetti, L. D. B. (2005). Chromosome numbers in wild and semidomesticated Brazilian *Capsicum* L. (*Solanaceae*) species: do x = 12 and x = 13 represent two evolutionary lines? *Botanical J. of the Linnean Soc.* 151(2): 259-269.

Rhode, J.M., and Cruzan, M.B. (2005). Contributions of heterosis and epistasis to hybrid fitness. *Am Nat.* 166(5): 24-39.

Roder, M.S., Huang, X.Q., and Ganal, M.W. (2004). Wheat microsatellites in plant breeding: potential and implications. In: Loerz, H., Wenzel, G., eds. *Molecular markers in plant breeding.* Heidelberg: Springer-Verlag, 255-266.

Rosenzweig, C.E., Iglesias, A., Yang, X.B., Epstein, P.R., and Chivian, E. (2001). Climate change and extreme weather events: Implications for food production, plant diseases, and pests. *Global Change Human Health.* 2: 90-104.

Schmidhuber, J., and Tubiello, F. N. (2007). Global food security under climate change. *Proc. Natl. Acad. Sci. USA.* 104 (50): 19703-19708.

Sharma, J.R. (1994). *Principles and Practice of Plant Breeding.* Tata McGraw Hill, Publishing Company Ltd., New Delhi.

Singh, B.D. 2006. *Plant Breeding: Principles and Methods.* Kalyani Publishers, New Delhi.

Singh, P. (2006). *Essentials of Plant Breeding*. Kalyani Publishers, New Delhi.

Sinha, S. K., and Khanna, R. (1975). Physiological, biochemical and genetic basis of heterosis. *Adv. Agro.* 27: 123-174.

Smith, P., Martino, D., Cai, Z., Gwary, D., Janzen, H.H., Kumar, P., McCarl, B., Ogle, S., O'Mara, F., Rice, C., Scholes, R.J., Sirotenko, O., Howden, M., McAllister, T., Pan, G., Romanenkov, V., Schneider, U., Towprayoon, S., Wattenbach, M., and Smith, J.U. (2007a). Greenhouse gas mitigation in agriculture. *Phil. Transa. of Royal Soc. B.,* 363. doi:10.1098/rstb. 2007-2184.

Snow, A. A., and Mora´n-Palma. P. (1997). Commercialization of transgenic plants: potential ecological risks. *Bioscience.* 47: 86-96.

Stewart, C. N. Jr., Halfhill, M. D., and Warwick. S. I. (2003). Transgene introgression from genetically modified crops to their wild relatives. *Nat. Rev. Gen.* 4: 806-817.

Swansona, K.L., George Sugiharab, I., and Tsonisa, A.A. (2009). Long-term natural variability and 20th century climate change. *PNAS.* 106(38): 16120-16123.

Varshney, R. K., Graner, A., and Sorrells, M. E. (2005). Genic microsatellite markers in plants: features and applications. *Trends Biotechnol.* 23: 48-55.

Varshney, R.K., Beier, U., Khlestkina, E., Kota, R., Korzun, V., Röder, M., Graner, A., and Börner, A. (2007). Single nucleotide polymorphisms in rye: discovery, frequency and applications for genome mapping and diversity studies. *Theo. App. Gene.* 114: 1105-1116.

Vilkoun, J., (2001b). Geographical distribution of wild wheat in their historical setting and current context. In Caligari, P.D.S., and Branfham, P.E. (eds): *Wheat Taxonomy the legacy of John percivial*, The Linnean special issue 3:81-94

Ziska, L.H., (2011). Climate change, carbon dioxide and global crop production: Food security and uncertainty. In: Dinar, A., Mendelsohn, R., editors. *Handbook on Climate Change and Agriculture.* Cheltenham, United Kingdom: Edward Elgar Publishing. p. 9-31.

11

Plant Breeding Tools and Techniques for Crop Improvement Programme: A Conventional and Modern Approach

Viralkumar Bhikhalal Mandaliya and Vrinda Someshwar Thaker*

Centre for Advanced Studies in Plant Biotechnology and Genetic Engineering Department of Biosciences, Saurashtra University, Rajkot-360005 (Gujarat), India

ABSTRACT

Plant breeding encases the existing knowledge of conventional and modern tools and techniques to produce new crop varieties or lines with desirable properties. The current trend in plant breeding is to bridge conventional and modern knowledge. An application of the latest tools and techniques of molecular genetics in crop improvement is a major perspective of plant breeders. The major advanced molecular tools like molecular tagging of genes/QTLs governing complex agronomic traits, identification of candidate genes, and marker assisted breeding for tolerance to biotic and abiotic stresses and quality traits are known to plant breeders. The advances in molecular techniques like transgenic technologies, genome sequencing and functional genomics have offered new dimensions to plant breeding programmes. Mankind, in his quest to improve plant quality, has been looking for variant forms i.e., varieties and hybrids. Hybrid, in general, is the product of a cross between two genetically dissimilar parents. The utilization of hybrids has become one of the approaches for increasing the yield potential. The exploitation and utilization of heterosis is one of the ways for increase in yield potential, enhancement in abiotic and biotic resistances and improvement in quality crop production. Cytoplasmic male sterility (CMS) is known to occur in more than 140 plant species and utilized by plant breeders as a source of production of F1 hybrid seeds. The discovery of usable sources of CMS has made it easier for plant breeders to increase heterosis and incorporate the desired traits into parents. Though conventional breeding techniques has utilized potential of CMS for crop improvements, biotechnological approaches have advantages of molecular genetics for crop improvements in the current era of climate change.

Keywords: Cytoplasmic male sterility, Mitochondrial markers, Molecular markers, Nuclear-cytoplasmic interaction, Restorer-of-fertility.

*Corresponding author's E-mail: casprogramme@gmail.com

Roychowdhury R. (Ed.), Crop Improvement in the Era of Climate Change.

1. INTRODUCTION

Agricultural practices were invented approximate 10,000 years ago by human civilization. It shaped into scientific studies of genetics around 1900 based on the work of Gregor Mendel. An understanding of genetic principles and their application to plant breeding technology has greatly accelerated the rate of improvement of desired crop plants in last century.

The role of breeders is to search for novel genetic combinations to encase plants with superior traits like quality, yield, regional performance, and tolerance to pests and diseases. The conventional breeding methods in plants depend on self-pollinated and cross-pollinated crops. The most often breeding methods for self-pollinated crops are mass selection, pure line selection, hybridization based on pedigree breeding, the bulk population method to develop hybrid varieties, while breeding methods for cross-pollinated species are mass selection, development of hybrid varieties, and development of synthetic varieties. Since cross-pollinated species are naturally hybrid (heterozygous) for many traits and lose vigour as they become pure bred (homozygous), a goal of each of these breeding methods is to preserve or restore heterozygosity. In addition, mutational breeding using various mutagenic agents (like radiation viz X-rays, gamma rays, alpha particles, beta particles, fast and thermal neutrons, and ultraviolet ray) and chemical mutagens (like alkylating agents, base analogues, acridine dyes), and other biotechnologically chosen agents became a source of mutation breeding in crop improvements.

The conventional breeding methods have led to significant increases in yield potential of crops, but they are limited by intra-special mating and the time it takes hybrid varieties to grow and exhibit the desired trait. Pingali and Heisey (1993) has shown the past trends and future aspects of crops with emerging evidence that yield gains of major crops was slowing down, conventional breeding may not be able to deliver the genetic gains required to achieve higher yields and meet rising food demand.

Since conventional approaches have certain limitation an advanced modern approach gained focus in crop improvement programme. Suslow *et al.* (2002) has reviewed biotechnological tools for plant breeding for genetic enhancement and propagation of crop plants. The major tools include: (i) grafting and tissue culture techniques, (ii) marker-assisted breeding and genomics, and (iii) recombinant DNA and transformation techniques.

Jain *et al.* (2002) had brings together information on based molecular marker. The molecular tools allow detection of specific DNA fragments through successive generations, and thus confirm transmission of the selected traits and incorporated genes. A research group established by the European Commission in 2007 has listed out breeding techniques for genetic transformation viz. zinc finger nuclease (ZFN) technology (ZFN-1, ZFN-2 and ZFN-3), oligonucleotide directed mutagenesis, cisgenesis and intragenesis, RNA-dependent DNA methylation, grafting (on GM rootstock), reverse breeding, agro-infiltration (agro-infiltration "sensu stricto", agro-inoculation, floral dip), and synthetic genomics (Lusser *et al.*, 2011).

2. MOLECULAR MARKERS AS TOOLS FOR CONVENTIONAL PLANT BREEDING

The great phenotypic and genomic diversity available in the gene pool can be made use by developing a genetic marker system (Mace *et al.*, 2008) that will assist breeders in efficient crop improvement. Marker in general term is referred as an object used to indicate a position, place, or route, or a thing serving as a standard of comparison or as an indication of what may be expected. A genetic marker is expressed or unexpressed, gene or DNA sequence from a known or unknown location on a chromosome that can be used to identify individuals or species. The genetic marker system can be categorized into three generations and each one of them has advantages and limitations. The first generation markers were based on morphological markers for genetic and breeding research. However, it showed several difficulties such as expression of dominance or epistatic interactions, pleiotropic effects, and incomplete penetrance and expressivity (Sargent *et al.*, 2004). The second generations of genetic markers were based on isozymes, and it became popular during 1970s and early 1980s to analyze the gene product (Royo and Itoiz, 2004). Despite their great advantages, isozyme markers are very limited in number and often are not polymorphic among closely-related genotypes (Foolad *et al.*, 1993). The third generations of genetic markers are the molecular markers, which work based on genotypic traits and not on phenotypic traits (Mehetre *et al.*, 2004).

2.1 Classification of Molecular Markers

Molecular markers are broadly classified into three classes (Mandaliya *et al.*, 2010, 2011): Hybridization-based molecular markers (e.g., RFLP); PCR-based molecular markers (e.g., AP-PCR, CAPS, STS, RAPD, SCAR, AFLP, SSAP, SSR, ISSR, EST); and sequencing-based molecular markers (e.g., SNP).

2.1.1 *Hybridization-based Molecular Markers*

RFLP is hybridization-based molecular marker. This technique is based on restriction enzymes that reveal a pattern difference between DNA fragment sizes in individual organisms. The major strength of RFLP markers are high reproducibility, codominant inheritance, good transferability between laboratories, provide locus specific markers that allow synteny (conserved order of genes between related organisms) studies, no sequence information required, and relatively easy to score due to large size difference between fragments. There are, however, several limitations for RFLP analysis such as (i) it requires the presence of high quantity and quality of DNA (Landau-Ellis *et al.*, 1991); (ii) it depends on the development of specific probe libraries for the species; (iii) the technique is not amenable for automation; (iv) the level of polymorphism is low, and few loci are detected per assay; (v) it is time consuming, laborious, and expensive; and (vi) it usually requires radioactively labelled probes.

2.1.2 *PCR-based Markers*

PCR, polymerase chain reaction, utilizes a polymerase enzyme to amplify the desired piece of DNA. The major advantages of PCR techniques compared to hybridization-based methods include: (i) a small amount of DNA is required; (ii) elimination of radioisotopes in most techniques; (iii) the ability to amplify DNA sequences from preserved tissues; (iv) accessibility of methodology for small labs in terms of equipment, facilities, and cost; (v) no prior sequence knowledge is required for many applications, such as AP-PCR, RAPD, DAF, AFLP and ISSR; (vi) high polymorphism that enables to generate many genetic markers within a short time; and (vii) the ability to screen many genes simultaneously either for direct collection of data or as a feasibility study prior to nucleotide sequencing efforts (Joko *et al.*, 2006).

The various PCR-based techniques are of two types depending on the primers used for amplification: (i) arbitrary or semi-arbitrary primed PCR techniques that developed without prior sequence information (e.g., AP-PCR, DAF, RAPD, AFLP, ISSR), and (ii) site-targeted PCR techniques that developed from known DNA sequences (e.g., EST, CAPS, SSR, SCAR, STS). But, PCR-based markers have three limitations: (i) reproducibility; (ii) dominant inheritance, and (iii) homology. The reason for such limitation could be: (i) the primer may not amplify any PCR product; (ii) the primer may produce very complex, weak or nonspecific amplification patterns, and (iii) the amplification product may not be polymorphic.

2.1.3 *Sequencing-based Molecular Markers*

SNP can be considered as any of the variations in single nucleotides in a DNA sequence. SNP variation occurs when a single nucleotide, such as a C, replaces one of the other three nucleotide letters—A, G, or T. For example a SNP might change the DNA sequence ACGGCTAA to ATGGCTAA.

2.2 DNA and RNA Based Molecular Markers

A molecular marker is typically derived either from genomic DNA or transcript. These makrers are further classified as: (i) DNA based molecular markers, and (ii) RNA based molecular markers. The followings are the list of the markers developed in the last two decades from the various expressed and unexpressed regions of the genome.

2.2.1 *DNA Based Molecular Markers*

- allele specific associated primers (ASAP) (Gu *et al.*, 1995),
- allele specific oligo (ASO) (Beckmann, 1988),
- allele specific polymerase chain reaction (AS-PCR) (Landegren *et al.*, 1988),
- amplified fragment length polymorphism (AFLP) (Vos *et al.*, 1995),
- anchored microsatellite primed PCR (AMP-PCR) (Zietkiewicz *et al.*, 1994),
- anchored simple sequence repeats (ASSR) (Wang *et al.*, 1998),
- arbitrarily primed polymerase chain reaction (AP-PCR) (Welsh and McClelland, 1990),
- cleaved amplified polymorphic sequence (CAPS) (Akopyanz *et al.*, 1992),
- degenerate oligonucleotide primed PCR (DOP-PCR) (Janiak *et al.*, 2008),
- diversity arrays technology (DArT) (Mace *et al.*, 2008),
- DNA amplification fingerprinting (DAF) (Ford-Lloyd *et al.*, 1997),
- insertion deletions (InDel) (Bia *et al.*, 2006),
- inter-simple sequence repeat (ISSR) (Reddy *et al.*, 2002),
- inverse PCR (IPCR) (Digeon *et al.*, 1999),
- microsatellite primed PCR (MP-PCR) (Geistlinger *et al.*, 1997),
- random amplified microsatellites (RAM) (Hantula *et al.*, 2007),
- random amplified microsatellite polymorphisms (RAMP) (Min *et al.*, 2008),
- random amplified polymorphic DNA (RAPD) (Williams *et al.*, 1990),

- restriction enzyme mediated integration (REMI) (Thon *et al.*, 2000),
- restriction fragment length polymorphism (RFLP) (Bligha *et al.*, 1999),
- sequence characterized amplified regions (SCAR) (Adam-Blondo *et al.*, 1994),
- single nucleotide polymorphism (SNP) (Ching *et al.*, 2002),
- single primer amplification reaction (SPAR) (Bhattacharya *et al.*, 2005),
- sequence-related amplified polymorphism (SRAP) (Li and Quiros, 2001),
- single-strand conformation polymorphisms (SSCP) (Martins-Lopes *et al.*, 2001),
- sequence tagged microsatellite site (STMS) (Ford-Lloyd *et al.*, 1997),
- short tandem repeats (STR) (Wang *et al.*, 1994), and
- variable number tandem repeat (VNTR) (Rogstad, 1994).

2.2.2 RNA Based Molecular Markers

(i) cDNA or mRNA as a source of molecular markers:

- cDNA-RFLPs (Kochert, 1994)
- cDNA-AFLPs (Habu *et al.*, 1997)
- RNA fingerprinting differential display (RNAF-DD) (Liang and Pardee, 1992)
- cDNA-SSR (Chin *et al.*, 1996)

(ii) EST databases as a source of molecular markers:

- Virtual AFLP or in silico AFLP (Qin *et al.*, 2001)
- EST-SSRs (La Rota and Sorrells, 2004)
- EST-SNPs (Grivet *et al.*, 2003)
- Insertion and deletions (InDels) (Bhattramakki *et al.*, 2002)
- Expressed sequence tag polymorphisms (ESTPs) (Harry *et al.*, 1998)
- Conserved orthologue set markers (COS) (Fulton *et al.*, 2002)

(iii) Known gene sequences as a source of molecular markers:

- Amplified consensus genetic markers (ACGMs) (Jourdren *et al.*, 1996)
- SSRs, SNPs, SCARs, CAPSs, and RFLPs for known genes (Tartarini, 2003)
- Gene specific tags (GSTs) (Thareau *et al.*, 2001)
- rDNA as markers (Appels and Honeycutt, 1986)
- Resistance gene analogues (RGAs) (Huettel *et al.*, 2002)
- Exon-retrotransposon amplification polymorphism (ERAP) (Pfaff and Kahl, 2003)
- PCR-based markers targeting exons, introns and promoter regions (Holland *et al.*, 2001)

3. DEVELOPING MOLECULAR MARKERS FROM EXPRESSED AND UNEXPRESSED REGION OF GENOME IN CROPS

The molecular markers are used for several purposes in crop improvement programmes. The molecular markers are generally used for germplasm characterization, genetic diversity analysis, molecular tagging and mapping. The following are summarized the utilization of DNA and RNA based molecular markers in various crops plant for a list of applications.

3.1 DNA Based Molecular Markers

The numbers of molecular markers are utilized in various crop improvements programme. This are introduced for several functions: (i) DNA-based markers introduce polymorphisms that are easy to database, and significantly increase the density of the genetic linkage map, e.g., Grapevine (*Vitis vinifera* L.) (Troggio *et al.*, 2007). (ii) A useful tool to quantify linkage disequilibrium (LD), e.g., sunflower elite inbred lines. This includes the estimation of nucleotide diversity, the assessment of linkage disequilibrium structure (LD) and the evaluation of selection processes (Fusari *et al.*, 2008). (iii) Fine-mapping of candidate regions and to determine haplotypes associated with traits of interest, e.g., sugar beet alleles of expressed genes are very frequently organized as robust intragene haplotypes (Schneider *et al.*, 2001). (iv) In order to understand the genetic basis of phenotypic diversity within and between populations. e.g., analysis of phenotypic diversity in tomato (Deynze *et al.*, 2007). (v) Evolutionary studies and genetic diversity analysis, e.g., elite maize inbred lines. The genetic distance between haplotypes is large, indicative of an ancient gene pool and of possible interspecific hybridization events in maize ancestry (Ching *et al.*, 2002).

3.2 RNA Based Molecular Markers

mRNA, cDNA and ESTs are the major sources for further development of RNA based molecular markers. These are introduced for mainly three horizons. (i) *Valuable genetic diversity analysis and genetic mapping:* the efficient development of highly valuable molecular markers as it represents single or low-copy sequences. Often EST-based RFLP markers allow comparative mapping across different species. ESTs also allow a computational approach to the development of SSR (simple sequence repeat) and SNP (single nucleotide polymorphism) markers (Kisiel and Podkowinski, 2005). The available sequence information allows the design of primer pairs, which can be used to screen cultivars of interest for length polymorphisms (Sharma *et al.*, 2009). (ii) *High-throughput transcript profiling:* ESTs also provide the main resource for the construction of cDNA arrays in plants, e.g., Arpat *et al.* (Arpat *et al.*, 2004) reported that genetic characterization of rapid cell elongation in cotton fibres, and approximately 14,000 unique genes were assembled from 46,603 expressed sequence tags (ESTs) from developmentally staged fibre cDNAs of a cultivated diploid species (*Gossypium arboreum* L.). Similarly, Hunt *et al.* (2011) has shown expression difference in wild-type and glabrous soyberans lines. (iii) *Biological interpretation of expression data:* In this, ontogenic changes in gene expression level has given emphasis, e.g., Boyes *et al.* (2001) has given a model on functional genomics in *Arabidopsis*.

4. IMPACTS OF CYTOPLASMIC MALE STERILITY AS BREEDING TOOLS IN CROP IMPROVEMENT PROGRAMMES

Heterosis is an important way to improve yield and quality for many crops. Hybrids often exhibit heterosis, more commonly known as hybrid vigour, whereby hybrid progeny exhibit superior growth characteristics relative to either of the parental lines. Hybrid cultivars have been developed in many crops such as rice, maize, sorghum, sunflower, oil seed rape, and cotton (Eckardt, 2006). The utilization of crop hybrids has become one of the trends for increasing the yield potential (Virmani and Edwards, 1983). Thus, the exploitation and utilization of heterosis

sterility (Hirose and Fujime, 1980) or on the classical method of emasculation and artificial hand pollination. The drawbacks of these methods are that sometimes a high percentage of non-hybrid seeds are produced and a considerable amount of manpower is required (Daskalov and Mihailov, 1988). Daskalov (1976) described a method of hybrid seed production based on utilization of male sterility combined with a lethal gene that can be easily inhibited. This method represented an improvement of the female parent so that 100 % purity of the F1 crop was ensured.

4.1.4 Wheat (*Triticumtimopheevii*)

CMS and fertility restoration in wheat were first discovered by Kihara (1951) and Fukasawa (1955). In 1962, Wilson and Ross (1962) reported complete CMS in common wheat induced by substituting its cytoplasm with *Triticum timopheevii* cytoplasm. Since then CMS has been used widely for the production of hybrid wheat. Seven restorers of fertility (Rf) genes have been reported to restore the fertility against *T. timopheevii* cytoplasm, and their chromosome locations have been determined (Zhou *et al.*, 2005).

4.1.5 Pearl Millet (*Pennisetumglaucum*)

CMS found in wild millet has been used to produce popular high-yielding and disease resistant hybrids. CMS and fertility restorer lines derived from *Pennisetum purpureum* Schum. were used in the first pearl millet grain hybrids and commercial forage hybrids (Hanna, 1989) and the Tifleaf series derived from these are the mostly widely cultivated in North America, popular as far down as Brazil. Then in the Tifleaf series, Tifleaf 3, released in 1997 (Hanna *et al.*, 1997), produces 20% more forage than the disease susceptible hybrids it replaced.

4.1.6 Pigeon Pea (*Cajanuscajan*)

Howard *et al.* (1919) were the first to report natural out-crossing in pigeonpea. In 1980s, various authors recommended various population improvement schemes in pigeonpea using natural out-crossing but up to there no high-yielding population had been developed and it remained as an academic exercise only (Saxena and Kumar, 2010). To then, production of pigeon pea hybrids using CMS found in five different wild relatives was initiated in 2003 (Saxena and Kumar, 2003), but no cultivars were released during that time Hajjar and Hodgkin (2007). In 2008, a group led by Saxena designed to synthesize CMS genotypes by combining the cytoplasmic genome of wild relatives with the nuclear genome of pigeonpea cultivars. This breeding approach has so far yielded the five CMS cytoplasms (A1 through A5) (Saxena, 2008). Primary structure of CMS-based hybrid pigeonpea is developed and is being adapted to suit specific environment as a continuing process. In 2007, the ICRISAT pigeonpea hybridization programme shared parent seeds with 14 private and three public sector seed companies. It is believed that an excellent beginning has been made in hybrid breeding (Saxena, 2008).

4.1.7 Rice (*Oryza sativa*)

With rice, 95% of hybrids grown in China are derived from crosses using CMS from wild *Oryza sativa* f. spontanea L. (Virmani and Shinjyo, 1988). Released in 1976, these hybrids are currently planted on approximately 45% of China's rice-planting area (Hajjar and Hodgkin, 2007). Since then more than 95% of total area planted to commercial hybrids in China and almost cent per cent in India has a single CMS source, the wild abortive (WA) cytoplasm (Brar *et al.*, 1996), cytoplasmic diversification of CMS system in rice is an urgent need to prevent genetic vulnerability of hybrids to diseases and insect pests, which may result from over-dependence on WA cytoplasm (Hariprasanna *et al.*, 2006).

4.1.8 Rapeseed (Brassicanapus)

Polima (pol) CMS line in rapeseed was discovered by Fu in 1972 (Fu *et al.*, 1997). According to the sensitivity of the male sterility to temperature, pol CMS lines can be divided into three categories: high temperature CMS lines, low temperature CMS lines, and stable CMS lines (Fu *et al.*, 1990). The most successfully used in hybrid production is the high temperature lines (Li *et al.*, 2006). Shan2A is another CMS line in *Brassica napus* bred by Li and Xia (1990). Testing results showed that both the pol CMS and Shan2A CMS systems had the same maintainers and restorers (Yang *et al.*, 1998).

4.1.9 Sunflower (Helianthus annuus)

According to Leclercq (1969) it is generally believed that commercial sunflower hybrids are all based on the PET1 cytoplasm, which had been derived from an interspecific cross *Helianthus petiolaris* × *H. annuus*. These hybrids were high yielding since 1972, significantly expanding the sunflower industry (Prescott-Allen and Prescott-Allen, 1986). Total sunflower production in the US, and approximately 60–70% of production worldwide is estimated to be from these hybrids (Hajjar and Hodgkin, 2007).

4.1.10 Soybean (Glycine max)

The first cytoplasmic-nuclear male-sterile soybean line derived from hybridization between two soybean cultivars was reported by Davis (1985) in a US Patent, but no further information is released since then. Palmer *et al.* (2001) reviewed that more than 10 cytoplasmic-nuclear male-sterile soybean lines were obtained in China. Sun *et al.* (1994) and Zhao *et al.* (1998) developed cytoplasmic-nuclear male-sterile lines OA, YAand ZA as well as their maintainers and restorers from a cross between *Glycine soja* and *G. max* at first and later from crosses between two cultivars.

4.1.11 Sorghum (Sorghum bicolor)

Hybrid cultivar development in sorghum has been possible due to the discovery of workable CMS (Stephens and Holland, 1954) designated as A1. Utilization of these non-milo CMS systems at commercial level depends on factors such as stability of male-sterility, restorer gene frequency in the germplasm, effect of male sterility inducing cytoplasm on agronomic traits, and the availability of commercially viable heterosis (Reddy *et al.*, 2005). The A2 system, originally reported in IS 12662C (SC 171) line, from *Caudatumnigricans* group belonging to subseries caffra (Schertz and Ritchey, 1978) is stable and sufficient number of restorers are available on this cytoplasm (Belum *et al.*, 2007).

4.1.12 *Sugar Beet (Beta vulgaris)*

The maintainer of the CMS line (having sterile cytoplasm with recessive restorer genes—xxzz) is called an 'O-type'. Sugar beet is normally allogamous (out-crossing), governed by a complex gametophytic self-incompatibility system, which prevents self-pollination but allows almost any two plants to cross-pollinate (Owen, 1942). Self fertility may be used in combination with genetic male sterility in population improvement programmes based upon selfed-progeny performance and to develop inbred lines for hybrids (Panella and Lewellen, 2007).

5. MOLECULAR MARKERS IN CMS BREEDING: CONVENTIONAL TO MODERN APPROACH

The terms sterility or fertility are used here in the sense of male sterility or fertility, respectively. Cytoplasmic effects on specific combining ability and heterosis were found to be modulated by cytoplasmic-nuclear interactions and influenced by the environmental conditions (Chandra-Shekara *et al.*, 2007). The CMS is induced in most cases by the sterilizing factors encoded by chimeric cytoplasmic genes (Touzet and Budar, 2004). Fertility restoration by dominant nuclear genes is essential for hybrid breeding based on CMS to obtain high yields in seed crops like sunflower, rapeseed, rye and maize (Kusterer *et al.*, 2005). Thus, successful utilization of CMS in hybrid breeding depends on the availability of specific nuclear restorer genes which down-regulate the male sterilizing factors. Information about the potential utility of such lines in breeding programmes, particularly with reference to heterosis and combining ability of cytoplasmic diverse hybrids is critical in improving the precision and efficiency of hybrid breeding (Hallauer and Miranda, 1981).

Knowledge of germplasm diversity and genetic relationships in the breeding material is an invaluable aid in planning appropriate strategies for crop improvement. Classification of germplasm accessions solely based on discrete morphological characters may not provide an accurate indication of their genetic divergence (Menkir *et al.*, 1997), and phenotypic selection based on traits that are conditioned by additive allelic effects can produce dramatic and economically important changes in breeding populations (Dhillon *et al.*, 2006). In a traditional breeding technology, breeders need to test fertility restoration ability of selected lines annually beginning with the F1 in the following generation, the fertility of the testcross hybrids represents the restoration ability of the tested lines, and this should be investigated before selection for other traits (Zhou *et al.*, 1999). The necessity of doing fertility testcross procedures adds more work to R-line selection and impedes the breeding efficiency of R-line development (Zhou *et al.*, 2005).

Use of molecular markers provides a more accurate approach than traditional methods for manipulating genes of agronomic importance such as fertility restoration. Molecular markers have been used for rapid, detailed, and directed genetic manipulation of crop plants, and to identify and characterize quantitative trait loci (QTLs) associated with plant height and days to maturity, plant domestication, resistance to diseases and insects, and tolerance to drought (Dhillon *et al.*, 2006). In general, this molecular marker's studies aimed to (i) evaluate the genetic diversity and relationship among MS systems, (ii) identify and map markers linked to genes responsible for fertility restoration of CMS systems, and (iii) study the effect of genes (alone and in combination) on male sterility in CMS systems.

In this chapter, utilization of molecular markers in CMS is described into three major points: (i) molecular markers for diversity and various traits, (ii) nuclear makers for restorer-of-fertility, and (iii) cytoplasmic markers for male sterility.

5.1 Molecular Markers for Diversity and Various Traits

5.1.1 Phylogenetic Relationship Among CMS Systems

In any hybrid programme, a large number of crosses are made, while only a few good hybrids are obtained. This process is extremely labour intensive, time consuming and tedious, and hence often, alternative methods are explored for analyzing the potential parental material. The level of genetic diversity between the two parents being used for crossing has been proposed as a possible predictor of F1 performance (Zhang *et al.*, 1994). D2 statistic by Mahalanobis (1936) and combining-ability analysis by Griffing (1956) have been extensively used in predicting the hybrid performance on the basis of morphological traits. However, these methods require extensive field tests and crossing and hence, a need is now felt to determine genetic distance at the molecular level and thus to predict hybrid performance (Melchinger *et al.*, 1990). The results obtained using isozyme analysis by Li et al (Li *et al.*, 1982) have revealed that distances computed from isozyme analysis are in some cases significantly correlated to heterosis. However, the correlations are generally too low for the distances to be of practical predictive value (Peng *et al.*, 1991).

The technique has been used to look at the differences between male sterile and their counterpart maintainer lines, and for classification of the different cytoplasmic systems (Pring *et al.*, 1982). The RFLP and RAPD were initial analyses tools used to develop phylogenetic relationship among the cultivated and wild species (Ratnaparkhe *et al.*, 1995). Sivaramakrishnan *et al.* (1997) have carried out a RFLP analysis of different male parents of *Cajanus* species, the wild species *C. sericeus* used as the female parent and the putative CMS progenies.

RAPD have been employed to determine genetic purity (Boury *et al.*, 1992) and to identify female and male parent lines and detect genetic sibs in cabbage and broccoli (Crockett *et al.*, 2000). Ichii *et al.* (2003) reported the screening results of RAPD markers to distinguish the maintainer from CMS plants at seedling stage, and the RAPD markers that can distinguish WA type and DA type rice cytoplasms. Souframanien *et al.* (2003) have used RAPD markers for the identification of two pigeonpea CMS lines derived from crosses between the wild (*Cajanus scarabaeoides* and *C. sericeus*) and the cultivated species of *Cajanus cajan*.

Although polymorphism was observed in mitochondrial DNA between normal and male sterile cytoplasms and among male sterile cytoplasms, the discrimination is too expensive for inspection purpose (Yamaguchi *et al.*, 1983). Nakajima *et al.* (1997) found that RAPD method is most useful for characterization of both mitochondrial and nuclear genomes in carrots. Interestingly, phylogenetic relationships of mitochondrial and nuclear genomes appeared to be independent of each other in eight carrot cultivars.

Dendrogram constructed based on the similarity index provides considerable genetic variation exists between CMS lines, R line and wild species. Genetic variation at the DNA level is of importance in grouping genotypes to different heterotic groups, which can be of great relevance in assessing the combining ability and exploiting maximum heterosis in crop plants (Ratnaparkhe *et al.*, 1995). Thus, the RAPD technique will provide phylogenetic trees for various crops and will be useful of identification of CMS sources.

5.1.2 Seed Purity Identification

The maintenance of CMS lines is tedious because of out-crossing nature of crops. The identification of CMS line will therefore be helpful in assessing the purity and stability of CMS lines. DNA fingerprinting approaches based on polymerase chain reaction have become methods of choice for germplasm characterization and seed purity assays. CMS plants can be selected rapidly and easily by PCR in an early stage of plant regeneration (Akagi *et al.*, 1995). A variety of DNA markers are now available for fingerprinting cultivars and for marker assisted selection. Hybrid plants possessed many of the fragments specific to each of the parents, thus confirming their hybrid nature.

It has been estimated that 1% impurity in the hybrid seed brings down the potential yield of hybrid by about 100 kg per ha (Mao *et al.*, 1996). In a country like India, where contract farming is practised at many places for hybrid seed production, with the active participation of private sector (Mishra *et al.*, 2003), monitoring genetic purity at each stage of seed production is necessary. Assessment of seed purity is one of the most important quality control components in hybrid seed production. Traditionally, it has been the practice to carry out a grow-out test (GOT), based on morphological traits for assessment of purity of seeds. GOT is time consuming (takes one full growing season for completion), space demanding and often does not allow the unequivocal identification of genotypes.

RAPD was used for molecular characterization of male, female and hybrid plants. The sharing of bands in hybrid confirms their hybrid nature (Garg *et al.*, 2007). The use of Simple Sequence Repeat (SSR) and Sequence Tagged Site (STS) markers for rapid assessment of hybrid and parental line seed purity, as an alternative to GOT was started (Yashitola *et al.*, 2002). Nandkumar *et al.* (2004) also showed the utility of SSR markers for genetic purity of rice hybrids based on a single marker and single seed/ seedling based analysis. Dhillon *et al.* (2006) used SSR in genetic purity testing in hybrid populations of crops such as cauliflower for genotype differences for 144 F1 hybrids, and their 36 parents.

Thus, a molecular marker was used for molecular characterization of male, female and hybrid plants. It will be of immense help for hybrid seed industry to select appropriate marker combinations and assess purity at each stage of seed multiplication.

5.1.3 Mapping Genes

With the characteristics of molecular marker technology, DNA markers have potential to be used for the prediction of heterosis prior to actual fieldwork (Joshi *et al.*, 2001). The development of molecular linkage maps has made it possible to detect and analyze the loci underlying heterosis (Xiao *et al.*, 1995). It is envisaged that the use of such markers may have potential to estimate hybrid performance and heterosis. An essential assumption underlying this is that a strong linear correlation exists between heterozygosity and heterosis (Zhang *et al.*, 1996). A number of efforts have been made in various crop plants to investigate the relationship between DNA marker based genotype variation of the parents in a hybrid programme and hybrid vigour, with varying results. For example, in *Zea mays* L., strong correlation between genetic distance and heterosis was observed by Lee *et al.* (1989) and Smith *et al.* (1990), while Godshalk *et al.* (1990) observed a low correlation between marker distance and yield. Studies by Melchinger *et al.* (1990) showed that the correlation depended on the origin of the lines. Smith *et al.* (1990) observed that the relationship between parental genetic distance and F1 performance became significant with an increase in the sample size as well as the number of markers. Stuber *et al.* (1992) found a significant relationship between parental heterozygosity and hybrid yield when the number of parental lines was increased.

Molecular markers like RFLPs have been used to map Rf genes against *T. timopheevii* cytoplasm in wheat (Ma and Sorrells, 1995). SSR markers are more chromosome specific (Bryan *et al.*, 1997) used for identification of markers linked to Rf genes against wheat (*T. timopheevii* cytoplasm) in two elite R-lines, R18 and R9034. Ratnaparkhe *et al.* (1998) identified the ISSR marker UBC-855 500 for *Fusarium* wilt resistance. This will be useful in marker-assisted selection to screen for disease resistance without the need to inoculate the segregating plant material. Astarini *et al.* (2008) identified nine AFLP markers linked to the Bn; Rf gene, which would facilitate and accelerate interim-maintainer breeding in *Brassica napus* and lay a foundation for map-based cloning of the gene. Joshi *et al.* (2001) investigated the genetic diversity between the wild abortive (WA) type cytoplasmic male sterile lines (IR 58025A and IR 62829A) and their compatible 14 restorer lines, selected for a hybrid rice programme intended to produce high yielding varieties. The genetic diversity they determined using both random (RAPD and ISSR) as well as specific (RFLP and STMS) markers linked to 8 different yield-related traits. Thus, the development of PCR based molecular markers would empower breeders across a wide range of crops mapping and breeding populations.

5.2 Nuclear Makers for Restorer-of-Fertility

CMS can be regarded as a consequence of incompatible interaction between nuclear and mitochondrial genomes. Disturbance in mitochondrial gene expression leading to CMS can be overcome by compensating changes in the nuclear genes. Nuclear genes that have the ability to suppress the male-sterile phenotype and restore production of pollen to plants carrying the deleterious mitochondrial cytoplasm are called restorer of fertility (Rf) genes (Bhat *et al.*, 2008; Zhou *et al.*, 2005).

There are several reports on Rf genes worked out for restorer of fertility. For example, Delourme *et al.* (1994) identified four RAPD markers linked to the fertility restorer locus (Rfo) of (ogu) *Brassica napus*. Trendelkamp *et al.* (1999) identified 11 AFLP markers linked to the restorer gene of tour CMS of *Brassica napus* through BSA method. Imai *et al.* (2003) identified four RAPD-STS and six AFLP-STS markers linked to the fertility restorer gene Rfk1 of CMS Kosena radish, and Koizuka *et al.* (2003) cloned the Rfk1 gene by AFLP combined with BSA. Ashutosh *et al.* (2007) identified two AFLP and one SCAR markers linked to the male fertile gene of CMS *Brassica juncea*. Kusterer *et al.* (2005) mapped the restoration locus Rf1 in sunflower cytoplasm. Yasumoto *et al.* (2008) has studied Ogura male-sterile cytoplasm for orf687 homologue Rf gene in Brassicaceae. Zhou *et al.*, (2005) reported seven restorer-of-fertility genes have been reported to restore the fertility against *T. timopheevii* cytoplasm.

In general, each CMS system is unique and restorers of one system do not work in other systems except where they are closely related. Therefore, a common restorer line able to confer male fertility to different CMS sources may point to common underlying molecular mechanisms (Bhat *et al.*, 2008). The specific markers development for Rf gene will be helpful for crop breeding.

5.3 Cytoplasmic Markers for Male Sterility

5.3.1 Mitochondrial Markers

CMS is a maternally inherited trait reported in various crops (Wang *et al.*, 2010). CMS associated with abnormal open reading frames (ORFs) or genes are known to be located in the mitochondrial genome Sattari *et al.* (2008). Mitochondria are the cellular sites for key metabolic pathways such as the tricarboxylic acid cycle, respiratory electron transfer, and ATP synthesis (Chase, 2006). The mitochondrial genome in plants encodes about 50–60 gene products, a small fraction of the gene products needed for mitochondrial functions (Wang *et al.*, 2010). In many cases, male fertility can be restored by fertility restorer (Rf) genes associated with nuclear genes encoding pentatricopeptide repeat (PPR) proteins (Chase and Babay-Laughnan, 2004). It is now well established that mitochondria are dependent on nuclear encoded proteins for their maintenance and function and a compatible nuclear–mitochondrial genome interaction is essential for normal plant development (Bhat *et al.*, 2008).

In recent studies, 14 mitochondrial genomes associated to male sterility have been sequenced in plants including: *Arabidopsis thaliana* L. (Unseld *et al.*, 1997), sugar beet (Kubo *et al.*, 2000), canola (Handa, 2003), *Cycas taitungesis* (Chaw *et al.*, 2008), tobacco (Sugiyama *et al.*, 2005), *Oryza sativa* L. subsp. indica (Tian *et al.*, 2006), *Oryza sativa* L. subsp. japonica (Notsu *et al.*, 2002), sorghum, and *Tripsacum dactyloides* L., (Allen *et al.*, 2007), maize (Allen *et al.*, 2007; Clifton *et al.*, 2004), and wheat (Ogihara *et al.*, 2005). Based on this, the following are the open reading frames (ORFs) or genes found in mitochondrial genomes associated with male sterility:

- atp1 (ATP synthase subunit alpha)
- atp6 (ATP synthase subunit 6)
- apt9 (ATP synthase subunit 9)
- cox1 (Cytochrome c oxidase sub 1)
- cox2 (Cytochrome c oxidase sub 2)
- cox3 (Cytochrome c oxidase sub 3)
- ccb203 (Cytochrome c biogenesis orf203)
- ccb206 (Cytochrome c biogenesis orf206)
- ccb256 (Cytochrome c biogenesis orf256)
- ccb382 (Cytochrome c biogenesis orf382)
- ccb452 (Cytochrome c biogenesis orf452)

- cob (Apocytochrome B)
- nad1 (NADH dehydrogenase sub 1)
- nad2 (NADH dehydrogenase sub 2)
- nad3 (NADH dehydrogenase sub 3)
- nad4 (NADH dehydrogenase sub 4)
- nad4L (NADH dehydrogenase sub 4L)
- nad5 (NADH dehydrogenase sub 5)
- nad6 (NADH dehydrogenase sub 6)
- nad7 (NADH dehydrogenase sub 7)
- nad9 (NADH dehydrogenase sub 9)
- matR (Maturase)
- orfB (ATP synthase subunit 8)
- orfX (Unknown protein)
- orf240 (Unknown protein)
- rpl2 (Ribosomal protein L2)
- rpl16 (Ribosomal protein L16)
- rps3 (Ribosomal protein S3)
- rps4 (Ribosomal protein S4)
- rrn5 (5S ribosomal RNA protein)
- rrn18 (18S ribosomal RNA protein)
- rrn26 (26S ribosomal RNA protein)

5.3.2 Chloroplastic Markers

As both mitochondria and plastids are inherited from the maternal side, plastid gene based markers could also be practically employed for hybrid crop breeding. It is known that chlorophyll deficiency and male sterility were the two most common abnormalities in plant development caused by functional incompatibility between nucleus and cytoplasm (Grun, 1976). Sometimes both of these abnormalities occur simultaneously, and both are maternally inherited. For example, when the Ogura cytoplasm was introduced into *Brassica* species, the male-sterile progenies exhibited severe chlorophyll deficiency (Bannerot *et al.*, 1977). Wu *et al.* (2001) have found some DNA sequence differences in the Cab genes between radish and *Brassica napus*. They found that this putative Cab gene is expressed in young anthers and used putative Cab gene as a landing marker for the positional cloning of the restorer locus.

There were several reports on male sterility mediated by chloroplasts (van der Hulst *et al.*, 2004). For examples, barley-wheat hybrids, both mitochondria and chloroplasts of the paternal origin were associated with fertility restoration of the hybrids (Bogdanova *et al.*, 2007). In *Pisum*, a paternal leakage of chloroplast DNA appears to be the mechanism that compensates for a nuclear-cytoplasmic conflict (Bogdanova and Kosterin, 2006). Nakajima *et al.* (1999) has used RAPD and six sequence-tagged sites (STSs) to classify five carrot CMS lines. They reported that STS4 fragment analysis had complicated gene structure, including retrotransposon-like sequences and small segments of chloroplast genome. Bhat *et al.* (2008) reported CAPS marker for the plastid gene psbB, which could be used for quick identification of Diplotaxis berthautii and D. erucoides based CMS lines. Sang *et al.*, (2006) reported AFLP based cpDNA diversity between rice CMS line Jin2A and its maintainer Jin2B, and sporophyte CMS lines.

Lössl *et al.* (2000) worked out interaction with the novel plastid-mitochondrial factors during accumulation and degradation of polysaccharides constitutes the variable field performance of individual hybrids. They reprted novel mt-cp-configurations may be enrichment for crop breeding in vegetatively and sexually propagated crops. It seems that distinct mt DNA conformations influence mitochondrial performance by their compatibilities with the plastid factors. For example, in maize Newton *et al.* (1990) showed that mitochondrial factors may play a role also for chloroplast function when they found a mutant mt gene which caused an abnormal function of photosystem I. Based on this, the following are listed orf or genes found in chloroplastic genomes associated with male sterility:

- cab (chlorophyll a/b-binding protein)
- psaA/trnS ([PS I (P700 apoproteinA1)] / [tRNA-Ser (GGA)])
- trnC/trnD ([tRNA-Cys (GCA)] / [tRNA-Asp (GUC)])
- trnD/trnT ([tRNA-Asp (GUC)]/ [tRNA-Thr (GGU)])
- tRNAleu-tRNAleu ((tRNAleu intron 1)
- matK (maturase gene)
- trnK gene (tRNAlys (UUU) gene)
- trnL-trnF ([tRNALeu(UAA)/tRNA-Phe(GAA)]
- nadB (intron) (NADH dehydrogenase subunit B)
- atpE (CF1/CF0 epsilon subunit)

6. PLANT BREEDING IN FUTURE: A CONCLUDING REMARK

To cope up with the demands of ever increasing population, an efficient breeding programme is indispensable. With advent of new technologies, the conventional stratagies are changed to more precise excersice. Initial marker based screening leads to identification and isolation of desired traits from the undesired one. Today, a significant pool of markers are available, both at DNA as well as RNA based, have ease selection of higher yielding varieties under biotic and abiotic stress. Identifacation of local varieties for these stresses as CMS lines and addition of desired gene(s) to maintainer and/or restorer lines may lead for efficient crop production in changing environmental conditions.

References

Adam-Blondon, A. F., Sevignac, M., Bannerot, H., and Dron, M. (1994). SCAR, RAPD and RFLP markers linked to a dominant gene (Are) conferring resistance to anthracnose in common bean. *Theor. Appl. Genet.* 88: 865-870.

Akagi, H., Nakamura, A., Sawada, R., and Oka, M. (1995). Genetic diagnosis of cytoplasmic male sterile cybrid plants of rice. *Theor. Appl. Genet.* 90: 948-951.

Akopyanz, N., Bukanov, N. O., Westblom, T. U., and Berg, D. E. (1992). PCR-based RFLP analysis of DNA sequence diversity in the gastric pathogen *Helicobacter pylori*. *Nucleic Acids Res.* 20: 6221-6225.

Allen, O. J., Fauron, C. M., Minx, P., Roark, L., Oddiraju, S., Lin, N. G., Meyer, L., Sun, H., Kim, K., Wang, C., Du, F., Xu, D., Gibson, M., Cifrese, J., Clifon, W. S., and Newton, K. J. (2007). Comparison among two fertile and three male sterile mitochondrial genomes of maize. *Genetics* 177: 1173-1192.

Appels, R., and Honeycutt, R. L. (1986). rDNA: evolution over a billion years. In *DNA systematics*, vol. II, Plants, Dutta, S. K. (ed.). CRC, Boca Raton, Fla., pp. 81-135.

Arpat, A. B., Waugh, M., Sullivan, J.P., Gonzales, M., Frisch, D., Main, D., Wood, T., Leslie, A., Wing, R. A., and Wilkins, T. A. (2004). Functional genomics of cell elongation in developing cotton fibers. *Plant Mol. Biol.* 54: 911-929.

Ashutosh, A., Sharma, P. C., Prakash, S., and Bhat, S. R. (2007). IdentiWcation of AFLP markers linked to the male sterility restorer gene of CMS (*Moricandia arvensis*) *Brassica juncea* and conversion to SCAR markers. *Theor. Appl. Genet.* 114: 385-392.

Astarini, I. A., Plummer, J. A., Lancaster, R. A., and Yan, G. (2008). Identification of 'Sib' plants in hybrid cauliflowers using microsatellite markers. *Euphytica* 164: 309-316.

Bannerot, H., Boulidard, L., and Chupeau, Y. (1977). Unexpected difficulties met with the radish cytoplasm. *Eucarpia Cruciferae Newsl* 2: 16.

Bateson, W., and Gairdner, A. E. (1921). Male sterility in flax subject to two types of segregation. *J. Genet.* 11: 269-275.

Beckmann, J. S. (1988). Oligonucleotide polymorphisms: A new tool for genomic genetics. *Biotechnol.* 6: 161-164.

Belum, V. S., Reddy, S., Ramesh, P., Reddy, S., and Ramaiah, B. (2007). Combining ability and heterosis as influenced by male-sterility inducing cytoplasms in sorghum [*Sorghum bicolor* (L.) Moench]. *Euphytica* 154: 153-164.

Bhat, S. R., Kumar, P., and Prakash, S. (2008). An improved cytoplasmic male sterile (*Diplotaxis berthautii*) *Brassica juncea*: identification of restorer and molecular characterization. *Euphytica* 159: 145-152.

Bhattacharya, E., Dandin, S. B., and Ranade, S. A. (2005). Single primer amplification reaction methods reveal exotic and indigenous mulberry varieties are similarly divers. *J. Biosci.* 30(5): 669-677.

Bhattramakki, D., Ching, A., Morgante, M., Dolan, M., Register, J., Smith, H., Tingey, S., and Rafalski, A. J. (2002). Insertion-deletion polymorphism in 3' regions of maize genes occur frequently and can be used as highly informative genetic markers. *Plant. Mol. Biol.* 48: 539-547.

Bia, I. V, McMullenb, M. D., Sanchez-Villcdac, H., Schroederc, S., Gardinerd, J., Polaccob, M., Soderlunde, C., Wingf, R., Fangc, Z., and Coe, E. H. Jr. (2006). Single Nucleotide Polymorphisms and Insertion–Deletions for Genetic Markers and Anchoring the Maize Fingerprint Contig Physical Map. *Crop Sci.* 46: 12-21.

Bligha, H. F. J., Blackhalla, N. W., Edwardsb, K. J., and McClung, A. M. (1999). Using Amplified Fragment Length Polymorphisms and Simple Sequence Length Polymorphisms to Identify Cultivars of Brown and White Milled Rice. *Crop Sci.* 39: 1715-1721.

Bogdanova, V. S. (2007). Inheritance of organelle DNA markers in a pea cross associated with nuclear-cytoplasmic incompatibility. *Theor. Appl. Genet.* 114: 333-339.

Bogdanova, V. S., and Kosterin, O. E. (2006). A case of anomalous chloroplast inheritance in crosses of garden Pea involving an accession of wild subspecies. *Dokl. Biol. Sci.* 406: 44-46.

Boury, S., Lutz, I., Gavalda, M. C., Guidet, F., and Schlesser, A. (1992). Genetic Wngerprinting in cauliflower by the RAPD method and determination of the level of inbreeding in a set of F1 hybrid seeds. *Agronomie* 12: 669-681.

Boyes, D. C., Zayed, A. M., Ascenzi, R., McCaskill, A. J., Hoffman, N. E., Davis, K. R., and Gorlach, J. (2001). Growth stage-based phenotypic analysis of *Arabidopsis*: a model for high throughput functional genomics in plants. *Plant Cell* 13: 1499-1510.

Brar, D. S., Guo, Z. Y., Ahmed, M. I., Jachuck, P. J., and Virmani, S. S. (1996). Diversification of CMS system to improve sustainability of hybrid rice technology. In *Abstracts, Proc. 3rd Int. Symp. on Hybrid Rice*, Nov. 14–16, 1996. Directorate of Rice Research, Hyderabad, India, pp. 5-6.

Bryan, G. J., Collins, A. J., Stephenson, P., Orry A., and Smith, J. B. (1997). Isolation and characterisation of microsatellites from hexaploid bread wheat. *Theor. Appl. Genet.* 94: 557-563.

Chandra-Shekara, A. C., Prasanna, B. M., Singh, B. B., Unnikrishnan, K. V., and Seetharam, A. (2007). Effect of cytoplasm and cytoplasm-nuclear interaction on combining ability and heterosis for agronomic traits in pearl millet {*Pennisetum glaucum* (L) Br. R}. *Euphytica* 153: 15-26.

Chase, C. D. (2006). Cytoplasmic male sterility: a window to the world of plant mitochondrial-nuclear interactions. *Trends Genet.* 23: 81-90.

Chase, C., and Babay-Laughnan, S. (2004). Cytoplasmic male sterility and fertility restoration by nuclear genes. In *Molecular biology and biotechnology of plant organelles*, Daniell, H., and Chase, C. (eds.). Kluwer, Dordrecht, The Netherlands, pp. 593-622.

Chaw, S., Shih, A. C., Wang, D., Wu, Y., Liu, S., and Chou, T. (2008). The mitochondrial genome of the gymnosperm *Cycas taitungensis* contains a novel family of short interspersed elements, Bpu sequences, and abundant RNA editing sites. *Mol. Biol. Evol.* 25: 603-615.

Chen, Z., Wu, F., Wang, X., and Zhang, G. (2005). Heterosis in CMS hybrids of cotton for photosynthetic and chlorophyll fluorescence parameters. *Euphytica* 144: 353-361.

Chin, E. C., Senior, M. L., Shu, H., and Smith, J. S. C. (1996). Maize simple repetitive DNA sequences: abundance and allele variation. *Genome* 39: 866-873.

Ching, A., Caldwell, K. S., Jung, M., Dolan, M., Smith, O. S., Tingey, S., Morgante, M., and Rafalski, A. J. (2002). SNP frequency, haplotype structure and linkage disequilibrium in elite maize inbred lines. *BMC Genet.* 3: 19.

Ching, A., Caldwell, K. S., Jung, M., Dolan, M., Smith, O. S., Tingey, S., Morgante, M., and Rafalski, A. J. (2002). SNP frequency, haplotype structure and linkage disequilibrium in elite maize inbred lines. *BMC Genet.* 3: 19.

Chittenden, R. J, and Pellow, C. A. (1927). A suggested interpretation of certain cases of anisogeny. *Nature* 119: 10-12.

Clifton, S. W., Minx, P., Fauron, C. M., Gibson, M., Allen, J. O., Sun, H., Thompson, M., Barbazuk, W. B., Kanuganti, S., Tayloe, C., Meyer, L., Wilson, R. K., and Newton, K. J. (2004). Sequence and comparative analysis of the maize NB mitochondrial genome. *Plant Physiol.* 136: 3486-3503.

Correns, C. (1908). Die Roller der mannlichen Keimzellenbei der Geschlechtshe stmmung Der gynodioccisrchen pflanzen (In De.). Bericht der Deutschen Botanischen Gesellschaft 26A: 686-701.

Crockett, P. A., Bhalla, P. L., Lee, C. K., and Singh, M. B. (2000). RAPD analysis of seed purity in a commercial hybrid cabbage (*Brassica oleraceae* var. capitata) cultivar. *Genome* 43: 317-321.

daSilva, F. P., Endrizzi, J. E., and Smith, L. S. (1981). Genetic study of restoration of pollen fertility of cytoplasmic male-sterile cotton. *Brazil. J. Genet.* 3: 411-426.

Daskalov, S. (1976). Seed setting of male sterile mutants in connection with heterosis breeding in pepper *Capsicum annuum* L. *Genet. Agrar.* 30: 407-417.

Daskalov, S., and Mihailov, L. (1988). A new method for hybrid seed production based on cytoplasmic male sterility combined with a lethal gene and a female sterile pollenizer in *Capsicum annuum* L. *Theor. Appl. Genet.* 76: 530-532.

Davis, W.H. (1985). Route to hybrid soybean production. United States Patents No. 4545146.

DeBonte, L. R., Matthews, B. F., and Wilson, K. G. (1984). Variation of plastid and mitochondrial DNAs in the genus *Daucus*. *Amer. J. Bot.* 71: 932-940.

Delourme, R., Bouchereau, A., Hubert, N., Renard, M., and Landry, B. S. (1994). Identification of RAPD markers linked to a fertility restorer gene for the Ogura radish cytoplasmic male sterility of rapeseed (*Brassica napus* L.). *Theor. Appl. Genet.* 88: 741-748.

Deynze, A. V., Stoffel, K., Buell, C. R., Kozik, A., Liu, J., van der Knaap, E., and Francis, D. (2007). Diversity in conserved genes in tomato. *BMC Genomics* 8: 465.

Dhillon, M. K., Sharma, H. C., Folkertsma, R. T., and Chandra, S. (2006). Genetic divergence and molecular characterization of sorghum hybrids and their parents for reaction to *Atherigona soccata* (Rondani). *Euphytica* 149: 199-210.

Dhillon, M. K., Sharma, H. C., Reddy, B. V. S., Singh, R., Naresh, J. S., and Kai, Z. (2005). Relative susceptibility of different male-sterile cytoplasms in sorghum to shoot fly, *Atherigona soccata*. *Euphytica* 144: 275-283.

Digeon, J., Guiderdoni, E., Alary, R., Michaux-Ferri`ere, N., Joudrier, P., and Gautier, M. (1999). Cloning of a wheat puroindoline gene promoter by IPCR and analysis of promoter regions required for tissue-specific expression in transgenic rice seeds. *Plant Mol. Bio.* 39: 1101-1112.

Eckardt N. A. (2006). Cytoplasmic Male Sterility and Fertility Restoration. *Plant Cell* 18: 515-517.

Foolad, M. R., Jones, R. A., and Rodriguez, R. L. (1993). RAPD markers for constructing intraspecific tomato genetic maps. *Plant Cell Reports* 12(5): 293-297.

Ford-Lloyd, B. V., Kaemmer, D. J. W., Kahl, G. J. W., and Lagoda, P. J. L. (1997). In: Molecular marker assisted analysis of the *Musa* genome complex. *INIBAP: Montpellier (FRA)* 24-28.

Ford-Lloyd, B.V., Kaemmer, D. J. W., Kahl, G. J. W., and Lagoda, P. J. L. (1997). In: Molecular marker assisted analysis of the *Musa* genome complex. *INIBAP: Montpellier (FRA)* 24-28.

Fu, T. D., Yang, G. S., and Yang, X. N. (1990). Studies on "three line" polima cytoplasmic male sterility developed in *Brassica napus* L. *Plant Breed.* 104: 115-120.

Fu, T. D., Yang, G. S., Yang, X. N., and Ma, C. Z. (1997). Discovery, study and utilization of polima cytoplasmic male sterility in *Brassica napus* L. *Prog. Nat. Sci.* 5: 169-177.

Fukasawa, H. (1955). Studies on restoration and substitution of nucleus of Aegiolotricum. II. The interrelationships between ovate cytoplasm and fertility restoring factors. *Cytologia* 2: 211-217.

Fulton, T. M., Van der Hoeven, R., Eannetta, N. T., and Tanksley, S. D. (2002). Identification, analysis and utilization of conserved ortholog set markers for comparative genomics in higher plants. *Plant Cell* 14: 1457-1467.

Fusari, C. M., Lia, V. V., Hopp, H. E., Heinz, R. A, and Paniego, N. B. (2008). Identification of Single Nucleotide Polymorphisms and analysis of Linkage Disequilibrium in sunflower elite inbred lines using the candidate gene approach. *BMC Plant Biol.* 8:7.

Garg, H., Banga, S., Bansal, P., Atri, C., and Banga, S. S. (2007). Hybridizing Brassica rapa with wild crucifers *Diplotaxis erucoides* and *Brassica maurorum*. *Euphytica* 156: 417-424.

Geistlinger, J., Weising, K., Kaiser, W. J., and Kahl, G. (1997). Allelic variation at a hypervariable compound microsatellite locus in the ascomycete *Ascochyta rabiei*. *Mol. Gen. Genet.* 256: 298-305.

Godshalk, E. B., Lee, M., and Lankey, K. R. (1990). Relationship of restriction fragment length polymorphism in single- cross hybrid performance of maize. *Theor. Appl. Genet.* 80: 273-289.

Griffing, B. (1956). Concept of general and specific combining ability in relation to diallel crossing system. *Aust. J. Biol. Sci.* 9: 463-493.

Grivet, L., Glaszmann, J. C., Vincentz, M., da Silva, F., and Arruda, P. (2003). ESTs as a source for sequence polymorphism discovery in sugarcane: example of the Adh genes. *Theor. Appl. Genet.* 106: 190-197.

Grun, P. (1976). Cytoplasmic genetics and evolution. Columbia University Press, New York.

Gu, W. K., Weeden, N. F., Yu, J., and Wallace, D. H. (1995). Large-scale, costeffective screening of PCR products in marker-assited selection applications. *Theor. Appl. Genet.* 91: 465-470.

Habu, Y., Fukada-Tanaka, S., Hisatomi, Y., and Iida, S. (1997). Amplified restriction fragment length polymorphism-based mRNA fingerprinting using a single restriction enzyme that recognizes a 4-bp sequence. *Biochem. Biophys. Res. Commun.* 234: 516-521.

Hajjar, R., and Hodgkin, T. (2007). The use of wild relatives in crop improvement: A survey of developments over the last 20 years. *Euphytica* 156: 1-13.

Hallauer, A. R., and Miranda, J. H. (1981). Quantitative Genetics in Maize Breeding. Iowa State University Press, Ames, Iowa, U.S.A.

Handa, H. (2003). The complete nucleotide sequence and RNA editing content of the mitochondrial genome of rape seed (*Brassica napus* L.): comparative analysis of the mitochondrial genomes of rape seed and *Arabidopsis thaliana. Nucleic Acids Res.* 31: 5907-5916.

Hanna, W. W. (1989). Characteristics and stability of a new cytoplasmic-nuclear male sterile source in pearl millet. *Crop Sci.* 29: 1457-1459.

Hanna, W. W., Hill, G. M., Gates, R. N., Wilson, J. P., and Burton, G. W. (1997). Registration of Tifleaf 3 pearl millet. *Crop Sci.* 37: 1388.

Hantula, J., Dusabenyagasani, M., and Hamelin, R. C. (2007). Random amplified microsatellites (RAMS) - a novel method for characterizing genetic variation within fungi. *Europ. J. of Forest Path.* 26(3): 159-166.

Hariprasanna, K., Zaman, F.U., and Singh, A.K. (2006). Influence of male sterile cytoplasms on the physico-chemical grain quality traits in hybrid rice (*Oryza sativa* L.). *Euphytica* 149: 273-280.

Harry, D. E., Temesgen, B., and Neale, D. B. (1998). Codominant PCR-based markers for Pinus taeda developed from mapped cDNA clones. *Theor. Appl. Genet.* 97: 327-336.

Hirose T, and Fujime Y (1980). Studies on the hybrid seed production by using a new male sterile line in pepper *(Capsicum annuum* L.) I. Morphology and mode of inheritance. *J. Jpn. Soc. Hortic. Sci.* 48: 453-458.

Holland, J. B., Helland, S. J., Sharopova, N., and Rhyne, D. C. (2001). Polymorphism of PCR-based markers targeting exons, introns, promoter regions, and SSRs in maize and introns and repeat sequences in oat. *Genome* 44: 1065-1076.

House, L. R. (1985). A Guide to Sorghum Breeding. International Crops Research Institute for the Semi-Arid Tropics, Patancheru, Andhra Pradesh, India, pp. 83-84.

Howard, A., Howard, G. C., and Khan, A. R. (1919). Studies in pollination of Indian crops. I. Mem Dep Indian Crops. *India. Bot. Ser.* 10: 195-200.

Huettel, B., Santra, D., Muehlbauer, F.J., and Kahl, G. (2002). Resistance gene analogues of chickpea (*Cicer arietinum* L.): isolation, genetic mapping and association with a *Fusarium* resistance gene cluster. *Theor. Appl. Genet.* 105: 479-490.

Huntt, M., Kaur, N., Stromvik, M., and Vodkin, L. (2011). Transcript profiling reveals expression differences in wild-type and glabrous soybean lines. *BMC Plant Biol.* 11: 145.

Ichii, M., Hong, D. L., Ohara, Y., Zhao, C. M., and Taketa, S. (2003). Characterization of CMS and maintainer lines in indica rice (*Oryza sativa* L.) based on RAPD marker analysis. *Euphytica* 129: 249-252.

Imai, R., Koizuka, N., Fujimoto, H., Hayakawa, T., Sakai, T., and Imamura, J. (2003). Delimitation of the fertility restorer locus Rfk1 to a 43-Kb contig in Kosena radish (*Raphanus sativus* L.). *Mol. Genet. Genomics* 269: 388-394.

Jain, S. M., Brar, D. S., and Ahloowalia, B. S. (eds.) (2002). In *Molecular Techniques in Crop Improvement.* Kluwer Academic Publishers, The Netherlands.

Janiak, A., Kim, M. Y., Van K., and Lee, S. (2008). Application of degenerate oligonucleotide primed PCR (DOP-PCR) for SNP discovery in soybean. *Euphytica* 162(2): 249-256.

Joko, R. W., Yu, M., and Katsuhiko, K. (2006). Genetic diversity of *Pinanga javana* Blume (Palmae) in six natural populations in Java, Indonesia as revealed by RAPD markers. *Chromosome Botany* 1: 33-39.

Jones, H., and Clarke, A. (1943). Inheritance of male sterility in the onion and the production of hybrid seed. *Proc. Am. Soc. Hortic. Sci.* 43: 189-194.

Joshi, S. P., Lalitha, K., Bhave, S. G., Ranjekar, P. K., Chowdari, K. V., Gupta, V. S., Apte, G. S., and Dhonukshe, B. L. (2001). Use of DNA Markers in Prediction of Hybrid Performance and Heterosis for a Three-Line Hybrid System in Rice. *Bioche. Genet.* 39(5/6): 179-200.

Jourdren, C., Barret, P., Brunel, D., Delourme, R., and Renard, M. (1996). Specific molecular marker of the genes controlling linolenic acid content in rapeseed. *Theor. Appl. Genet.* 93: 512-518.

Justus, N. J., and Leinweber, C.V. (1960). A heritable partial male-sterile character in cotton. *J. Hered.* 51: 191-192.

Kihara, H. (1951). Substitution of nucleus and its effects on genome manifestation. *Cytologia* 16: 177-193.

Kisiel, A., and Podkowinski, J. (2005). Expressed Sequence Tags and their application for plant research. *Acta Physiol. Planta.* 27(2): 157-161.

Kochert, G. (1994). RFLP technology. In *DNA-based markers in plants, advances in cellular and molecular biology of plants*, edn. 1, vol. 1, Phillips, R. L., and Vasil, I. K. (eds.). Kluwer, Dordrecht, pp. 8-38.

Koizuka, N., Imai, R., Fujimoto, H., Hayakawa, T., Kimura, Y., Kohno Murase, J., Sakai, T., Kawasaki, S., and Imamura J. (2003). Genetic characterization of a pentatricopeptide repeat protein gene, orf687 that restores fertility in the cytoplasmic male-sterile Kosena radish. *Plant J.* 34: 407-415.

Kolreuter, D. J. G. (1763). Voolarfige Nachrcht von linigen das geschlet der Pflanzenbetreffen-den versuchen and Beobachtanager. Engelmen, Leipzing.

Kubo, T., Nishizawa, S., Sugawara, A., Itchoda, N., Estiati, A., and Mikami, T. (2000). The complete nucleotide sequence of the mitochondrial genome of sugar beet (*Beta vulgaris* L.) reveals a novel gene for tRNA Cys (GCA). *Nucleic Acids Res.*28: 2571-2576.

Kusterer, B., Horn, R., and Friedt, W. (2005). Molecular mapping of the fertility restoration locus Rf1 in sunflower and development of diagnostic markers for the restorer gene. *Euphytica* 143: 35-42.

La Rota, C. M., and Sorrells, M. E. (2004). Comparative DNA sequence analysis of mapped wheat ESTs reveals complexity of genome relationships between rice and wheat. *Funct. Integr. Genomics* 4: 34-46.

Landau-Ellis, D., Angermüller, S., Shoemaker, R., and Gresshoff, P. M. (1991). The genetic locus controlling supernodulation in soybean (*Glycine max* L.) co-segregates tightly with a cloned molecular marker. *Mol. Gener. Genet.* 228: 221-226.

Landegren, U., Kaiser, R., Sanders, J., and Hood, L. (1988). DNA diagnostics. Molecular techniques and automation. *Science* 241: 1077-1080.

Leclercq, P. (1969). Une sterilite male chez le tournesol. *Ann. Amelior Plantes* 19: 99-106.

Lee, M., Godshalk, E. B., Lamkey, K. R., and Woodman, W. W. (1989). Association of restriction fragment length polymorphism among maize inbreds with agronomic performance of their crosses. *Crop Sci.* 29: 1067-1071.

Li, D. R., and Xia, R. Z. (1990). Genetic researches on the selection of male sterile trilines in *Brassica napus* L. *Crop Res.* 4(3): 20-26.

Li, G., and Quiros, C. F. (2001). Sequence-related amplified polymorphism (SRAP), a new marker system based on a simple PCR reaction: its application to mapping and gene tagging in Brassica. *Theor. Appl. Genet.* 103: 455-546.

Li, Y. C., Hu, Q., Mei, D. S., Li, Y. D., Xu, Y. S., and Zhao, Q. (2006). Breeding of Zhongyouza 8, a canola variety with large seeds and high oil yield. *Agri. Sci. China.* 5: 257-264.

Li, Z. B., Xiao, Y. H., Zhu, Y. H., Li, R. Q., Liu, C. L., and Wang, J. M. (1982). The study and practice of hybrid rice. Shanghai Academic and Technical press, Shanghai, China.

Liang, P., and Pardee, A. B. (1992). Differential display of eukaryotic mRNAs by means of polymerase chain reaction. *Science* 257: 967-971.

Lössl, A., Götz, M., Braun, A., and Wenzel, G. (2000). Molecular markers for cytoplasm in potato: male sterility and contribution of different plastid-mitochondrial configurations to starch production. *Euphytica* 116: 221-230.

Lusser, M., Parisi, C., Plan, D., and Rodríguez-Cerezo, E. (2011). New plant breeding techniques. State-of-the-art and prospects for commercial development. In: *EUR - Scientific and Technical Research series.* European Commission. EUR 24760 EN. Joint Research Centre Institute for Prospective Technological Studies. The European Union, Luxembourg. doi: 10.2791/54761.

Ma, Z. Q., and Sorrells, M. E. (1995). Genetic analysis of fertility restoration in wheat using restriction fragment length ploymorphisms. *Crop Sci.* 35: 1137-1143.

Mace, E. S., Xia, L., Jordan, D. R., Halloran, K., Parh, D. K., Huttner, E., Wenzl, P., and Kilian, A. (2008). DArT markers: diversity analyses and mapping in *Sorghum bicolor*. *BMC Genomics* 9: 26.

Mace, E. S., Xia, L., Jordan, D. R., Halloran, K., Parh, D. K., Huttner, E., Wenzl, P., and Kilian, A. (2008). DArT markers: diversity analyses and mapping in *Sorghum bicolor*. *BMC Genomics* 9: 26.

Mahalanobis, P. C. (1936). On the generalized distance in statistics. *Proc. Nat. Inst. Sci. India* 2(1): 49-55.

Mandaliya, V. B., Pandya, R. V., and Thaker, V.S. (2010). Single Nucleotide Polymorphism (SNP): A trend in genetics and genome analyses of plants. *Gener. Appli. Plant Physiol.* 36(3-4): 159-166.

Mandaliya, V. B., Pandya, R. V., and Thaker, V.S. (2011). CSNP: A tool for harnessing the genetic potential of Cotton. *Cotton Research J.* 2(1): 1-14.

Mao, C. X., Virmani, S. S., and Kumar, I. (1996). Technological innovations to lower the costs of hybrid rice seed production. In *Advances in hybrid rice technology*, Proceedings of Third International Symposium on Hybrid Rice, Virmani, S. S. et al (eds.). Directorate of Rice Research, Hyderabad, India.

Martins-Lopes, P., Zhang, H., and Koebner, R. (2001). Detection of Single Nucleotide Mutations in Wheat Using Single Strand Conformation Polymorphism Gels. *Plant Mol. Bio. Reporter.* 19:159-162.

Mehetre, S. S., Gomes, M., and Eapen, S. (2004). RAPD analysis of hybrid nature of the offspring of *Gossypium hirsutum* × *G. raimondii*. *Curr. Sci.* 87(1): 25-28.

Melchinger, A. E., Lee, M., Lamkey, K. R., and Woodman, W. W. (1990). Genetic diversity for restriction fragment length polymorphisms, relation to estimated genetic effects in maize inbreds. *Crop Sci.* 30: 1033-1040.

Melchinger, A. E., Lee, M., Lamkey, K. R., Hallaue, A. R., and Woodman, W. W. (1990). Genetic diversity for restriction fragment length polymorphisms and heterosis for two diallel sets of maize inbreds. *Theor. Appl. Genet.* 80: 488-496.

Menkir, A., Goldsbrough, P., and Ejeta, G. (1997). RAPD based assessment of genetic diversity in cultivated races of sorghum. *Crop Sci.* 37: 564-569.

Meyer, V. G. (1975). Sterility from *Gossypium harknessii. J. Hered.* 66: 23-27.

Meyer, V. G., and Meyer, J. R. (1955). Cytoplasmically controlled male sterility in cotton. *Crop Sci.* 5: 444-448.

Min, W., Han, J., Kang, W., Lee, H., and Kim, B. (2008). Reverse Random Amplified Microsatellite Polymorphism Reveals Enhanced Polymorphisms in the 3′ End of Simple Sequence Repeats in the Pepper Genome. *Mol. Cells* 26: 250-257.

Mishra, B., Viraktamath, B. C., Ilyas Ahmed, M., Ramesha, M. S., and Vijayakumar, C. H. M. (2003). Hybrid rice research and development in India. In *Hybrid rice for food security, poverty alleviation, and environmental protection*, Proceedings of the 4th International Symposium on Hybrid Rice, 14–17 May 2002, Virmani, S. S., Mao, C. X., Hardy, B. (eds.). Hanoi, Vietnam. Los Baños (Philippines), International Rice Research Institute, pp. 265-283.

Nakajima, Y., Yamamoto, T., and Oeda, K. (1997). Genetic variation of mitochondrial and nuclear genomes in carrots revealed by random amplified polymorphic DNA (RAPD). *Euphytica* 95: 259-267.

Nakajima, Y., Yamamoto, T., Muranaka, T., and Oeda, K. (1999). Genetic variation of petaloid male-sterile cytoplasm of carrots revealed by sequence-tagged sites (STSs). *Theor. Appl. Genet.* 99: 837-843.

Nandakumar, N., Singh, A. K., Sharma, R. K., Mohapatra, T., Prabhu, K. V., and Zaman, F. U. (2004). Molecular Wngerprinting of hybrids and assessment of genetic purity of hybrid seeds in rice using microsatellite markers. *Euphytica* 136: 257-264.

Newton, K. J., Knudsen, C., Gabay-Laughnan, S., and Laughnan, J.R. (1990). An abnormal growth mutant in maize has a defective mitochondrial cytochrome oxidase gene. *Plant Cell* 2: 107-113.

Notsu, Y., Masood, S., Nishikawa, T., Kubo, N., Akiduki, G., Nakazono, M., Hirai, A., and Kadowaki, K. (2002). The complete sequence of the rice (*Oryza sativa* L.) mitochondrial genome: frequent DNA sequence acquisition and loss during the evolution of flowering plants. *Mol. Genet. Genomics* 268: 434-445.

Ogihara, Y., Yamazaki, Y., Murai, K., Kanno, A., Terachi, T., Shiina, T., Miyashita, N., Nasuda, S., Nakamura, C., Mori, N., Takumi, S., Murata, M., Futo, S., and Tsunewaki, K. (2005). Structural dynamics of cereal mitochondrial genomes as revealed by complete nucleotide sequencing of the wheat mitochondrial genome. *Nucleic Acids Res.* 33: 6235-6250.

Owen, F. V. (1942). Inheritance of cross- and self-sterility in *Beta vulgaris* L. *J. Agr. Res.* 64:679-698.

Palmer, R. G., Gai, J. Y., Sun, H., and Burton, J. W. (2001). Production and evaluation of hybrid soybean. *Plant Breeding Rev.* 21: 263-307.

Panella, L., and Lewellen, (R. T. 2007) Broadening the genetic base of sugar beet: introgression from wild relatives. *Euphytica* 154: 383-400.

Peng, J. Y., Virmani, S. S., and Julfiquar, A. W. (1991). Relationship between heterosis and genetic divergence in rice. *Oryza* 28: 129-133.

Peterson, C. E., and Simon, P. W. (1986). Carrot breeding. In *Breeding Vegetable Crops*, Basset, M. J. (ed.). AVI Publishing Co., Westport, Connecticut, pp. 321-356.

Peterson, P. (1958). Cytoplasmically inherited male sterility in *Capsicum. Am. Nat.* 92: 111-119.

Pfaff, T., and Kahl, G. (2003). Mapping of gene-specific markers on the genetic map of chickpea (*Cicer arietinum* L.). *Mol. Genet. Genomics.* 269: 243-251.

Pingali, P. L., and Heisey, P. W. (1993). "Cereal Crop Productivityin Developing Countries: Past Trends and Future Prospects." In *CIMMYT Economics Program Working Paper,* 99(03). Mexico, D.F. CIMMYT.

Prescott-Allen, C., and Prescott-Allen, R. (1986). The First Resource: Wild Species in the North American Economy. Yale University, New Haven.

Pring, D. R., Conde, M. F., and Schertz, K. F. (1982). Organelle genome diversity in sorghum: male sterile cytoplasms. *Crop Sci.* 22: 414-421.

Qin, L., Prins, P., Jones, J. T., Popeijus, H., Smant, G., Bakker, J., and Helder, J. (2001). GenEST, a powerful bidirectional link between cDNA sequence data and gene expression profiles generated by cDNA-AFLP. *Nucleic Acids Res.* 29: 1616-1622.

Ratnaparkhe, M. B., Gupta, V. S., Ven Murthy, M. R., and Ranjekar, P. K. (1995). Genetic fingerprinting of pigeonpea (*Cajanus cajan* (L.) Millsp.) and its wild relatives using RAPD markers. *Theor. Appl. Genet.* 91: 893-898.

Ratnaparkhe, M. B., Santra, D. K., Tullu, A., and Muehlbauer, F. J. (1998). Inheritance of inter-simple-sequence-repeat polymorphisms and linkage with a fusarium wilt resistance gene in chickpea. *Theor. Appl. Genet.* 96: 348-353.

Reddy, B. V. S., Ramesh, S., and Ortiz, R. (2005). Genetic and cytoplasmic-nuclear male sterility in sorghum. *Plant Breed. Rev.* 25: 139-172.

Reddy, M. P., Sarla. N., and Siddiq. E. A. (2002). Inter simple sequence repeat (ISSR) polymorphism and its application in plant breeding. *Euphytica* 128: 9-17.

Rogstad, S. H. (1994). Inheritance in turnip of variable-number tandem-repeat genetic markers revealed with synthetic repetitive DNA probes. *Theor. Appl. Genet.* 89: 824-830.

Royo, J. B., and Itoiz, R. (2004). Evaluation of the discriminance capacity of RAPD, isoenzymes and morphologic markers in apple (*Malus* x *domestica* Borkh.) and the congruence among classifications. *Genet. Reso. Crop Evol.* 51(2): 153-160.

Sang, X., Yang, Z., Zhong, B., Li, Y., Hou, L., Pei, Y., Li, G., and He, G. (2006). Assessment of purity of rice CMS lines using cpDNA marker. *Euphytica* 152: 177-183.

Sargent, D. J., Davis, T. M., Tobutt, K. R., Wilkinson, M. J., Battey, N. H., and Simpson, D. W. (2004). A genetic linkage map of microsatellite, gene - specific and morphological markers in diploid *Fragaria. Theor. Appl. Genet.* 109: 1385-1391.

Sattari, M., Kathiresan, A., Gregorio, G. B., and Virmani, S. S. (2008). Comparative genetic analysis and molecular mapping of fertility restoration genes for WA, Dissi, and Gambiaca cytoplasmic male sterility systems in rice. *Euphytica* 160: 305-315.

Saxena, K. B. (2008). Genetic Improvement of Pigeon Pea - A Review. *Tropical Plant Biol.* 1: 159-178.

Saxena, K. B., and Kumar, R. V. (2003). Development of a cytoplasmic male sterility system in pigeonpea using *C. scarabaeoides* (L.) Thouars. *Indian J. Genet. Pl. Br.* 63: 225-229.

Saxena, K. B., and Kumar, R. V. (2010). Insect-aided natural out-crossing in four wild relatives of pigeonpea. *Euphytica* 173: 329-335.

Saxena, K. B., Kumar, R. V., Srivastava, N., and Shiying, B. (2005). A cytoplasmic-nuclear male-sterility system derived from a cross between *Cajanus cajanifolius* and *Cajanus cajan. Euphytica* 145: 289-294.

Schertz, K. F., and Ritchey, J. M. (1978). Cytoplasmic-genic malesterility systems in sorghum. *Crop Sci.* 18: 890-893.

Schnable, P. S., and Wise, R. P. (1998). The molecular basis of cytoplasmic male sterility and fertility restoration. *Trends Plant Sci.* 3: 175-180.

Schneider, K., Weisshaar, B., Borchardt, D. C., and Salamini, F. (2001). SNP frequency and allelic haplotype structure of *Beta vulgaris* expressed genes. *Mol. Breed.* 8(1): 63-74.

Sharma, H. C., Dhillon, M. K., Naresh, J. S., Singh, R., Pampapathy G., and Reddy, B. V. S. (2004). Influence of cytoplasmic male-sterility on the expression of resistance to insects in sorghum. In *Fourth*

International Crop Science Congress, 25 Sept.–2 Oct. 2004, Fisher, T., Turner, N., Angus, J., McIntyre, L., Robertson, M., Burrell, A., and Llyod D., (eds.). Brisbane, Queensland, Australia, p. 6.

Sharma, V., Bhardwaj, P., Kumar, R., Sharma, R. K., Sood, A., and Ahuja. P. S. (2009). Identification and cross-species amplification of EST derived SSR markers in different bamboo species. Conserv. Genet. 10: 721-724.

Shifriss, C., and Guri, A. (1977). A study of the potential of cytoplasmic-genic male sterility for hybrid seed production in *Capsicum annuum* L. In *Capsicum 77, C.R.3me Congres EUCARPIA*, 5-8 Juillet, pp. 275-284.

Sivaramakrishnan, S., Seetha, K., Nageshwar Rao, A., and Singh, L. (1997). RFLP analysis of cytoplasmic male-sterile lines of pigeonpea [*Cajanus cajan* (L.) Millsp.] developed by interspecific crosses. *Euphytica* 93: 307-312.

Smith, O. S., Smith, J. S. C., Bowen, S. L., Tenborg, R. A., and Wall, S. J. (1990). Similarities among a group of elite maize inbreds as measured by pedigree, F1 grain yield, grain yield heterosis and RFLPs. *Theor. Appl. Genet.* 80: 833-840.

Souframanien, J., Manjaya, J. G., Krishna, T. G., and Pawar, S. E. (2003). Random amplified polymorphic DNA analyses of cytoplasmic male sterile and male fertile Pigeonpea (*Cajanus cajan* (L.) Millsp.). *Euphytica* 129: 293-299.

Stephens, J. C., and Holland, P. F. (1954). Cytoplasmic male sterility for hybrid sorghum seed production. *Agron. J.* 46: 20-23.

Stephens, J. C., and Holland, R. F. (1954). Cytoplasmic male sterility for sorghum seed production. *Agron. J.* 46: 20-23.

Stuber, C. W., Lincoln, S. E., Wolff, D. W., Helentjaris, T., and Lander, E. S. (1992). Identification of genetic factors contributing to heterosis in a hybrid from two elite maize inbred lines using molecular markers. *Genetics* 132: 823-839.

Sugiyama, Y., Watase, Y., Nagase, M., Makita, N., Yagura, S., Hirai, A., and Sugiura, M. (2005). The complete nucleotide sequence and multipartite organization of the tobacco mitochondrial genome: comparative analysis of mitochondrial genomes in higher plants. *Mol. Genet. Genomics* 272: 603-615.

Sun, H., Zhao, L., and Huang, M. (1994). A study on the cytoplasmnuclear male sterile soybean line. *Chin. Sci. Bull.* 38(16): 1535-1536.

Suslow, T.V., Bruce R. T., and Bradford, K. J. (2002). Biotechnology Provides New Tools for Plant Breeding. In *Agricultural Biotechnology in California Series*, Publication 8043 at the University of California, Division of Agriculture and Natural Resources. Seed Biotechnology Center, UC Davis.

Tartarini, S. (2003). Marker-assisted selection in pome fruit breeding. Marker assisted selection: a fast track to increase genetic gain in plant and animal breeding? Session I: MAS in plants. In *(Conference 10 of the FAO Biotechnology Forum) Molecular marker assisted selection as a potential tool for genetic improvement of crops, forest trees, livestock and fish in developing countries*, 17 November to 12 December 2003, Italy.

Thareau, V., Dehais, P., Rouze, P., and Aubourg, S. (2001). Automatic design of specific gene tags for transcriptome studies. In *JOBIM 2001*, Duret, L., Gaspin, C., and Schiex, T. (eds.). Toulouse, pp. 195-196.

Thompson, D. J. (1961). Studies on the inheritance of male-sterility in the carrot, *Daucus carota* L. var. sativa. *Proc. Am. Soc. Hortic. Sci.* 78: 332-338.

Thon, M. R., Nuckles, E. M., and Vaillancourt, L. J. (2000). Restriction enzyme-mediated integration used to produce pathogenicity mutants of *Colletotrichum graminicola. Mol. Plant Microbe Interact.* 13(12): 1356-65.

Tian, X., Zhang, J., Hu, S., and Yu, J. (2006). The rice mitochondrial genomes and their variations. *Plant Physiol.* 140: 401-410.

Touzet, P., and Budar, F. (2004). Unveiling the molecular arms race between two conflicting genomes in cytoplasmic male sterility? *Trends Plant Sci.* 9: 568-570.

Trendelkamp, H., Uzunova, M. I., and Ecke, W. (1999). Mapping a restorer gene for CMS tour 25-143 cytoplasm in rapeseed (*Brassica napus* L.). In *Proceedings of the 10th international rapeseed congress*, Canberra, Australia.

Troggio, M., Malacarne, G., Coppola, G., Segala, C., Cartwright, D. A., Pindo, M., Stefanini, M., Mank, R., Moroldo, M., Morgante, M., Grando, M. S., and Velasco, R. (2007). A Dense Single-Nucleotide Polymorphism-Based Genetic Linkage Map of Grapevine (*Vitis vinifera* L.) Anchoring Pinot Noir Bacterial Artificial Chromosome Contigs. *Genetics* 176: 2637-2650.

Unseld, M., Marrienfeld, J. R., Brandt, P., and Brennicke, A. (1997). The mitochondrial genome of *Arabidopsis thaliana* contains 57 genes 366, 924 nucleotides. *Nature Genet.* 15: 57-61.

van der Hulst, R. G., Meirmans, P., van Tienderen, P. H., and van Damme, J. M. (2004). Nuclear-cytoplasmic male-sterility in diploid dandelions. *Heredity* 93: 43-50.

Virmani, S. S., and Edwards, I. B. (1983). Current status and future prospects for breeding hybrid rice and wheat. *Adv. Agron.* 36: 145-214.

Virmani, S. S., Sun, Z. X., Mou, T. M., Jauhar Ali, A., and Mao, C. X. (2003). Two-line Hybrid Rice Breeding Manual. International Rice Research Institute, Los Banos, Philippines.

Virmani, S., and Shinjyo, C. (1988). Current status of analysis and symbols for male sterile cytoplasms and fertility restoring genes. *Rice Genet. Newsl.* 5: 9-15.

Vos, P., Hogers, R., Bleeker, M., Reijans, M., Van De Lee, T., Hornes, M., Frijters, A., Pot, J., Peleman, J., Kuiper, M., and Zabeau, M. (1995). AFLP: a new technique for DNA fingerprinting. *Nucleic Acids Res.* 23: 4407-4414.

Wang, F., Feng, C., Stewart, J. McD., O'Connell, M. A., and Zhang, J. (2010). RFLP analysis of mitochondrial DNA in two cytoplasmic male sterility systems (CMS-D2 and CMS-D8) of cotton. *Euphytica* 172: 93-99.

Wang, G., Mahalingan, R., and Knap, H. T. (1998). (C-A) and (GA) anchored simple sequence repeats (ASSRs) generated polymorphism in soybean, *Glycine max* (L.). *Theor. Appl. Genet.* 96: 1086-1096.

Wang, X. D., and Li, Y. Y. (2002). Development of transgenic restorer of cytoplasmic male sterility in upland cotton. *Sci. Agric. Sinica.* 35: 137-141.

Wang, Z., Weber, J. L., Zhong, G., and Tanksley, S. D. (1994). Survey of plant short tandem DNA repeats. *Theor. Appl. Genet.* 88(1): 1-6.

Weaver, J. B. Jr. (1968). Analysis of a double recessive completely male-sterile cotton. *Crop Sci.* 8: 597-600.

Weaver, J.B. Jr., (1986). Performance of open-pollinated cultivars, F2's and CMS Upland × Upland restorer strains. In *Proceedings beltwide cotton production research conference*. pp. 98-100.

Welsh, J., and McClelland, M. (1990). Fingerprinting genomes using PCR with arbitrary primers. *Nucleic Acids Res.* 18: 7213-7218.

Williams, J. G. K., Kubelik, A. R., Livak, K. J., Rafalski, J. A., and Tingey, S. V. (1990). DNA polymorphisms amplified by arbitrary primers are useful as genetic markers. *Nucleic Acids Res.* 18(22): 6531-6535.

Wilson, J. A., and Ross, W. M. (1962). Male sterility interaction of the *Triticum aestivum* nucleus and *Triticum timopheevi* cytoplasm. *Wheat Inform. Serv.* 14: 14-29.

Wu, Y., Tulsieram, L., and Rothstein, S. J. (2001). Identification and characterization of a putative light-harvesting chlorophyll a/b-binding protein gene encoded at a fertility restorer locus for the Ogura CMS in *Brassica napus* L. *Theor. Appl. Genet.* 102: 759-766.

Xiao, J., Li, J., Yuan, L., and Tanksley, S. D. (1995). Dominance is the major genetic basis of heterosis in rice as revealed by QTL analysis using molecular markers. *Genetics* 140: 745-754.

Yamaguchi, H., and Kakiuchi, H. (1983). Electrophoretic analysis of mitochondrial DNA from normal and male sterile cytoplasms in rice. *Jpn. J. Genet.* 58: 607-611.

Yang, G. S., Fu, T. D., and Brown, G. G. (1998). The genetic classification of cytoplasmic male sterility systems in *Brassica napus* L. *Sci. Agri. Sin.* 31: 27-31.

Yashitola, J., Thirumurugan, T., Sundaram, R. M., Naseerullah, M. K., Ramesha, M. S., Sarma, N. P., and Sonti, R. V. (2002). Assessment of purity of rice hybrids using microsatellite and STS markers. *Crop Sci.* 42: 1369-1373.

Yasumoto, K., Nagashima, T., Umeda, T., Yoshimi, M., Yamagishi, H., and Terachi, T. (2008). Genetic and molecular analysis of the restoration of fertility (Rf) genes for Ogura male-sterility from a Japanese wild radish (*Raphanus sativus* var. hortensis f. raphanistroides Makino). *Euphytica* 164: 395-404.

Zhang, Q. F., Gao, Y. J., Yang, S. H., Ragab, R. A., Saghai Maroof, M. A., and Li, Z. B. (1994). A diallele analysis of heterosis in elite hybrid rice based on RFLPs and microsatellites. *Theor. Appl. Genet.* 89: 185-192.

Zhang, Q. F., Zhou, Z. Q., Yang, G. P., Xu, C. G., Liu, K. D., and Saghai Maroof, M. A. (1996). Molecular marker heterozygosity and hybrid performance in Indica and Japonica rice. *Theor. Appl. Genet.* 93: 1218-1224.

Zhao, L., Sun, H., and Huang, M. (1998). The development and preliminary studies on cytoplasmic male sterile soybean line ZA. *Soybean Sci.* 17(3): 268-270.

Zhou, W., Kolb, F. L., Domier, L. L., and Wang, S. (2005). SSR markers associated with fertility restoration genes against *Triticum timopheevii* cytoplasm in *Triticum aestivum*. *Euphytica* 141: 33-40.

Zhou, W.C., Y.H. Zhao, M.L. Zou and S.W. Wang, (1999). Selection and genetic improvement of pollen fertility restorer lines with *Triticum timopheevii* cytoplasm in common wheat. *Plant Breed.* 118: 221-224.

Zietkiewicz, E., Rafalski, A., and Labuda, D. (1994). Genome fingerprinting by simple sequence repeats (SSR)-anchored PCR amplification. *Genomics* 20: 176-183.

12

Molecular Marker-Assisted Technologies for Crop Improvement

Rajib Roychowdhury[1], Abdelmoumen Taoutaou[2], Khalid Rehman Hakeem[3], Mohamed Ragab Abdel Gawwad[4] and Jagatpati Tah[5*]

[1]Department of Biotechnology, Visva-Bharati, Santiniketan-731235, West Bengal, India

[2]Department of Molecular Biology and Biotechnology, University Babes-Bolyai, Clinicilor 5-7, Cluj-Napoca, Romania

[3]Faculty of Forestry, Universiti Putra Malaysia, Serdang-43400, Selangor, Malaysia

[4]Natural Sciences Dept., Faculty of Engineering and Natural Sciences, International University of Sarajevo, Bosnia and Herzegovina

[5]Department of Botany (UGC-CAS), The University of Burdwan, Burdwan-713104 West Bengal, India

ABSTRACT

In recent years, the development and use of molecular markers for the detection and exploitation of DNA polymorphism has become one of the most significant developments in the field of molecular genetics and biotechnology for a variety of applications including examination of genetic relationships between individuals, mapping of useful genes, construction of linkage maps, marker assisted selections, backcrosses, population genetics and phylogenetic studies. The presence of various types of molecular markers, differences in their principles, methodologies, and applications require careful consideration in choosing one or more of such methods. No molecular markers are available yet that fulfill all requirements needed by researchers. According to the kind of study to be undertaken, one can choose among the variety of molecular techniques, each of which combines at least some desirable properties. Thus, recognizing the enormous potential of DNA markers, it has been adopted the capacity for marker development and marker-assisted selection (MAS) mainly for crop breeding and improvement. However, due to rapid developments in marker technology, statistical methodology for identifying quantitative trait loci (QTLs) has been utilized. This chapter provides an introduction to molecular (DNA) markers and the concept of polymorphism, their technical advantages, major types, the principles of QTL analysis and how such markers may be applied in breeding programmes using MAS for crop improvement.

Keywords: Crop improvement, DNA, molecular marker, marker-assisted selection, QTL.

**Corresponding author's E-mail: jt_botbu2012@yahoo.in*

Roychowdhury R. (Ed.), Crop Improvement in the Era of Climate Change.

1. MARKER AND MOLECULAR MARKER: AN INTRODUCTION

Plant breeding is one of the most important pillars of modern agriculture. It allows resolving major problems in both modern and conventional agriculture (Roychowdhury *et al.*, 2013), for instance breeding for biotic and abiotic stress resistance, and important agronomical traits. Breeders hybridize different genotypes in the scope to introduce the desired character into the genotype of interest. As a result, we have actually a multitude of cultivars of potato, wheat, corn, etc. This enormous created biodiversity, with the natural one (wild relative species) makes selection for the desired traits (such as disease and insect resistance) a hard and time consuming labour beside the high costs. Before the molecular technique apparition, plant breeders had no choice without the phenotypic morphological characters for selection. The development of molecular markers has increased the possibilities of an efficient application of marker based selection procedures in plant breeding (Friedt and Ordon, 2007).

A genetic marker (molecular marker) can be defined in one of the following ways: (a) a chromosomal landmark or allele that allows for the tracing of a specific region of DNA; (b) a specific DNA portion with a known position on the genome, or (c) a gene whose phenotypic expression is usually easily discerned, used to identify an individual or a cell that carries it, or as a probe to mark a nucleus, chromosomes, or locus (King and Stansfield, 1997). Since the markers and the genes are close together on the same chromosome, they tend to stay together as each generation of plants is produced. As researchers learn where markers occur on a chromosome, and how close they are to specific genes, they can create a genetic linkage map that can serves for several purposes, including detailed analysis of associations between economically important traits and genes or quantitative trait loci (QTLs) and facilitate the introgression of desirable genes or QTLs through marker-assisted selection (MAS). Genetic markers fall into three major classes: those based on visually assessable, morphological and agronomic traits (morphological/classical/visible marker), those based on protein products from genes or biochemical components (biochemical markers), and DNA (or molecular) markers which reveal sites of variation in DNA and relying on a DNA assay. The idea of using genetic markers appeared in very early literature (Wexelsen, 1933), but the development of electrophoretic assays of isozymes (Markert and Moller, 1959) and molecular markers (Botstein *et al.*, 1980; Williams *et al.*, 1990; Vos *et al.*, 1995; Jaccoud *et al.*, 2001) have greatly improved our understanding in biological sciences to find out the hidden mystery of genome as well as life. Morphological markers are usually visually characterized phenotypic characters such as flower colour, seed shape, growth habits or pigmentation. Biochemical markers are differences in gene products/ proteins/enzymes (isozymes) that are detected by electrophoresis and specific staining. The major disadvantages of morphological and biochemical markers are that they may be limited in number and are influenced by environmental factors or the developmental stage of the plant (Winter and Kahl, 1995). However, despite these limitations, morphological and biochemical markers have been extremely useful to plant breeders (Eagles *et al.*, 2001). Molecular markers should not be regarded as normal genes, as they usually do not have any biological effect, and instead can be thought of as constant landmarks in the genome. They are identifiable DNA sequences, found at specific locations of the genome, and transmitted by the standard laws of inheritance from one generation to the next. Essentially, DNA markers may reveal genetic differences that can be visualized by using a technique called gel electrophoresis and staining with chemicals (ethidium bromide or silver nitrate) or detection with radioactive or colorimetric probes. DNA markers are particularly useful if they reveal differences between individuals of the same or different species. These markers are called polymorphic markers, whereas markers that do not discriminate between genotypes are called monomorphic markers (Fig. 1A and 1B). Polymorphic markers may also be described as codominant or dominant. This description is based on whether markers can discriminate between homozygotes and heterozygotes (Fig. 2). Codominant markers indicate differences in size whereas dominant markers are either present or absent. Strictly speaking, the different forms of a DNA marker (e.g., different sized bands on gels) are called marker 'alleles'. Codominant markers may have many different alleles whereas a dominant marker only has two alleles.

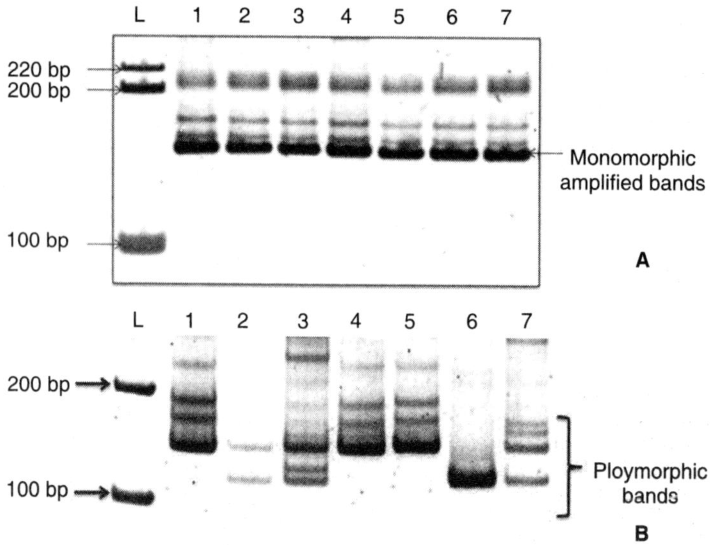

Fig. 1 Diagram representing DNA markers between genotypes 1, 2, 3, 4, 5, 6 and 7. Polymorphic markers are indicated by arrows. Monomorphic marker does not discriminate between genotypes (A) and the polymorphic marker (B) reveals size differences for the marker alleles. Often, the sizes of these markers in nucleotide base pairs (bp) are also provided; these sizes are estimated from a known molecular weight DNA ladder (L).

Molecular markers have demonstrated a potential role to detect tolerance or susceptibility for abiotic stresses in crop plants, genetic diversity and to aid the management of plant genetic resources (Song *et al.*, 2003), resistance to pathogens, and agronomical characters. The existence of various molecular techniques and differences in their principles and methodologies requires careful consideration in choosing one or more of such marker types. This scientific chapter deals on the basic principles, requirements, and advantages and disadvantages of the most widely used molecular markers for genetic diversity studies, genetic mapping, marker-trait association studies and marker assisted selection programmes.

2. MAIN ADVANTAGES OF MOLECULAR MARKERS

Molecular markers for trait selection have numerous advantages over morphological and biochemical markers used in conventional plant breeding. The application of molecular markers in crop improvement has been reviewed by scientists addressing the application of DNA markers

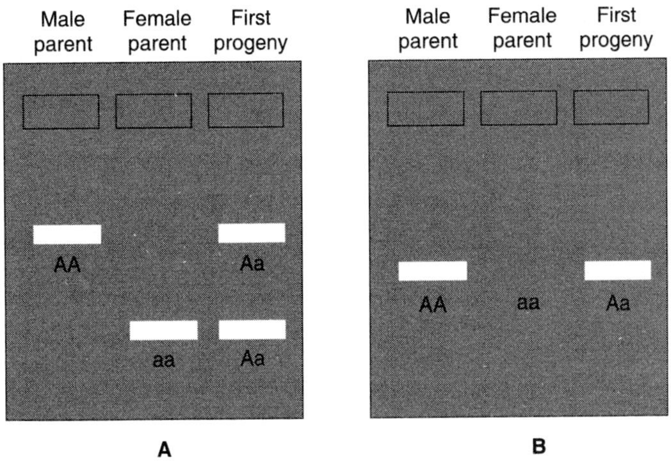

Fig. 2 Hypothetical gel diagram showing comparison between (A) codominant and (B) dominant markers. Codominant markers can clearly discriminate between homozygotes and heterozygotes whereas dominant markers do not. Genotypes at two marker loci (A and B) are indicated below the amplified bands.

in improvement of crop plants in general and emphasized the marker application in heterosis breeding (Collard and Mackill, 2007; Mackill, 2007). They also described the limitations of marker used in crop improvement. The main advantages of molecular markers include the followings (Jena and Mackill, 2008):

Time Saving

Genomic DNA can be isolated from any part of the plant tissue at every stage of its development, and target trait information can be obtained with linked DNA markers before pollination, thus allowing breeders to carry out more informed genetic crosses. Also, it is possible for breeder to perform many rounds of selection in a year (Mohan *et al.*, 1997).

Consistency

Phenotypic evaluation of genetic traits is often complicated by environmental factors. However, DNA markers are mostly neutral to environmental variation. The breeder can evaluate their material independently of the environmental conditions (environmental conditions can be favourable or unfavourable for morphologic and/or biochemical marker expression).

Biosafety

Diagnostic tests for the presence or absence of traits for disease resistance can be conducted by DNA markers tightly linked to the target gene without resorting to pathogen inoculation in the field or greenhouse. Additionally, molecular markers facilitate introgression of genes into elite cultivars in advance of the occurrence of certain races of diseases or biotypes of insects.

Efficiency

Evaluation of breeding lines in early generations of the breeding process with DNA markers can allow breeders to reject progenies from the programme and improve the genetic quality of breeding materials.

More accurate selection of complex traits

Polygenic traits are often difficult to select for using conventional breeding approaches. DNA markers linked to QTL allow them to be treated as single Mendelian factors. Beside analyzing and selecting the interesting characters, molecular markers allow the researchers also to analyze the wild species with potential interest for the breeding programme.

3. MAJOR TYPES

DNA markers may be broadly divided into three classes based on the method of their detection: (1) hybridization-based; (2) polymerase chain reaction (PCR)-based and (3) DNA sequence-based (Jones *et al.*, 1997; Joshi *et al.*, 1999; Gupta *et al.*, 1999). Till date, numerous types of molecular markers have been reported (Semagn *et al.*, 2006) that are as follows: allele specific associated primers (ASAP), allele specific oligo (ASO), allele specific polymerase chain reaction (AS-PCR), amplified fragment length polymorphism, (AFLP), anchored microsatellite primed PCR (AMP-PCR), anchored simple sequence repeats (ASSR), arbitrarily primed polymerase chain reaction (AP-PCR), cleaved amplified polymorphic sequence (CAPS), degenerate oligonucleotide primed PCR (DOP-PCR), diversity arrays technology (DArT), DNA amplification fingerprinting (DAF), expressed sequence tags (EST), inter-simple sequence repeat (ISSR), inverse PCR (IPCR), inverse sequence-tagged repeats (ISTR), microsatellite primed PCR (MP-PCR), multiplexed allele-specific diagnostic assay (MASDA), random amplified microsatellite polymorphisms (RAMP), random amplified microsatellites (RAM), random amplified polymorphic DNA (RAPD), restriction fragment length polymorphism (RFLP), selective amplification of microsatellite polymorphic loci (SAMPL), sequence characterized amplified regions (SCAR), sequence specific amplification polymorphisms (S-SAP), sequence tagged microsatellite site (STMS), sequence tagged site (STS), short tandem repeats (STR), simple sequence length polymorphism (SSLP), simple sequence repeats (SSR), single nucleotide polymorphism (SNP), single primer amplification reactions (SPAR), single stranded conformational polymorphism (SSCP), site-selected insertion PCR (SSI), strand displacement amplification (SDA), and variable number tandem repeat (VNTR). Although some of these markers are very similar (e.g., ASAP, ASO and AS-PCR), some synonymous (e.g., ISSR, RAMP, RAM, SPAR, AMP-PCR, MP-PCR, and ASSR) and some are identical (e.g., SSLP, STMS, STR and SSR), there are still a wide range of techniques for researchers to choose from as per their experimental design. Out of the vast lists of molecular markers, microsatellite or SSR (simple-sequence repeat) markers (Roychowdhury *et al.*, 2012), RAPD (random amplified polymorphic DNA) (Williams *et al.,* 1990), AFLP (amplified fragment length polymorphism) (Vos *et al.,* 1995) RFLP (restriction fragment length polymorphism) (Botstein *et al.,* 1980) are efficiently used in today's research field related to stress tolerance in cereals, floricultural crops, legumes, biodiesel plants, horticultural crops (fruits and vegetables), etc. (Table 1). Amongst these, SSRs and RFLPs are codominant in nature and their genetic map location on crop genome is publicly disclosed; in case of AFLP and RAPD, they produce random amplification and are largely dominant markers in nature (Ram *et al.*, 2007).

PCR based SSR markers are cost-effective, high diversity in crop plants due to their frequently available nature, codominant and greater efficient to others (Temnykh *et al.,* 2000). Compared to RFLPs, microsatellite markers detect a significantly higher degree of polymorphism in rice and are especially suitable for evaluating genetic diversity among closely related cultivars (Akagi

Table 1 A comparison of the properties of six most widely used molecular markers in crop plants

Properties	*RFLP*	*SSR*	*ISSR*	*RAPD*	*AFLP*	*SNP*
Principle	Restriction digestion	DNA amplification	DNA amplification	DNA amplification	DNA amplification	DNA amplification
PCR needed	No	Yes	Yes	Yes	Yes	Yes
Genomic abundance	High	Medium	Medium	Very high	Very high	Very high
Part of genome surveyed	Low copy coding regions	Whole genome	Whole genome	Whole genome	Whole genome	Whole genome
Amount of invested DNA	High	Low	Low	Low	Medium	Low
Quality of invested DNA	Relatively pure	Crude	Crude	Crude	Relatively pure	Crude
Type of polymorphism	Single base changes, insertion, deletion	Changes in length of repeats	Single base changes, insertion, deletion	Single base changes, insertion, deletion	Single base changes, insertion, deletion	Single base changes, insertion, deletion
Level of polymorphism	Medium	High	High	High	Very high	Super high
Primer requirement	None	Yes (selective primer)	Yes (selective primer)	Yes (random primer)	Yes (selective primer)	Yes
Probe requirement	Yes (specific probe)	None	None	None	None	None
Use of radio-isotopes	Yes	No	No	No	Yes	No

Contd...

the AFLP primer combination, the selective nucleotide motif, GC-content, and physical genome size and complexity. The AFLP technique generates fingerprints of DNA from any source, and without any prior knowledge of DNA sequence. Most AFLP fragments correspond to unique positions on the genome and hence can be exploited greatly in genetic and physical mapping (Agarwal *et al.*, 2008).

Single nucleotide polymorphism (SNP) constitutes the most abundant and widely distributed molecular markers in the genome, which varies greatly among species. For example, maize has 1 SNP per 60–120 bp. The SNPs are usually more prevalent in the non-coding regions of the genome. Within the coding regions, an SNP is either non-synonymous resulting in an amino acid sequence change, or it is synonymous not being able to alter the amino acid sequence. The latter can result in phenotypic differences due to mRNA splicing. Direct analysis of genetic variation at the DNA sequence have been made possible by improvements in sequencing technology and availability of an increasing number of EST sequences level. Majority of SNP genotyping assays are based on one or two of the following molecular mechanisms: allele specific hybridization, primer extension, oligonucleotide ligation and invasive cleavage (Sobrino *et al.*, 2005). High throughput genotyping methods, including DNA chips, allele-specific PCR and primer extension approaches make SNPs attractive genetic markers. They are suitable for automation, rapid identification of crop cultivars and construction of ultra-high-density genetic maps. Sophisticated strategies have been developed and applied in various crop species for unlocking the genetic potential (allelic selection) from its non-domesticated genetic stock. One such very recent promising strategy is 'allele mining' which dissects naturally occurring allelic variation at abiotic stress tolerance governed gene loci controlling key agronomic traits and exploits the DNA sequence of one genotype to isolate useful alleles from related genotypes (Latha *et al.*, 2004). As the map-based complete crop genome sequence is now freely available to researcher worldwide, the identification and isolation of novel and superior allele for agronomically important genes became a popular research area in modern crop improvement programme (Ram Kumar *et al.*, 2010) for their stress tolerance property, particularly in drought, salinity and oxidative stress conditions.

4. MARKER-ASSISTED SELECTION (MAS)

As principle, marker-assisted selection (MAS) is based on the possibility to infer the presence of a gene from the presence of a tightly linked marker (to the gene of interest) (Kumar, 1999). The development of genomics using the modern biotechnological tools and techniques provide ample opportunities for scientists to improve trait(s) of interest in a wide range of crop cultivars. Molecular markers for trait selection have numerous advantages over morphological markers that are used in conventional breeding. The use of molecular marker in such efficient trait-specific crop improvemental aspect would help to accelerate the application of marker-assisted selection or MAS (Holland, 2004; Salvi and Tuberosa, 2005). Selecting plants in a segregating progeny that contains appropriate combinations of genes is a critical component. Moreover, researchers typically work with hundreds or even thousands of populations, which often contain large numbers of individuals (Witcombe and Virk, 2001). Marker-assisted selection (also named as 'marker-assisted breeding' or 'marker-aided selection') may greatly increases the efficiency and effectiveness in molecular breeding compared to conventional plant breeding methods.

Once markers that are tightly linked to genes or quantitative trait loci (QTLs) of interest have been identified, prior to field evaluation of large numbers of plants, breeders may use specific DNA marker alleles as a diagnostic tool to identify plants carrying the genes or QTLs (Ribaut *et al.*, 1997). Although, the advantages of marker-assisted selection (MAS) include:

(i) Indirect selection of desirable plant free of environmental, pleiotropic or epistasis effects,

(ii) Selection of genotypes at seedling stage,

(iii) Time saving from the substitution of complex field trials (that need to be conducted at particular times of year or at specific locations, or are technically complicated) with molecular tests,

(iv) Elimination of unreliable phenotypic evaluation associated with field trials due to environmental effects,

(v) Selecting for traits with low heritability (Roychowdhury and Tah, 2013),

(vi) Avoid the transfer of undesirable or deleterious genes ('linkage drag'; this is of particular relevance when the introgression of genes from wild species is involved),

(vii) The ability to discriminate between homo- and heterozygote,

(viii) Pyramiding of genes or combining multiple genes simultaneously,

(ix) Monitoring the introgression,

(x) Identification of recombinants possessing least amount of linkage drag and donor DNA flanking the gene of interest and,

(xi) Testing for specific traits where phenotypic evaluation is not feasible (e.g., quarantine restrictions may prevent exotic pathogens to be used for screening).

Collard and Mackill (2008) established a list of five criteria for a marker to be used efficiently in MAS:

1. Adaptability to flexible and high-throughput detection methods,
2. Efficiency with low quantity and quality DNA,
3. Low cost per assay,
4. Linked tightly to target loci, and
5. Display a high level of polymorphism in breeding material.

However, the development of a second generation of molecular markers based on the sequence information is more amenable to application in plant breeding programmes (Paux *et al.*, 2012). On the other hand, breeding by MAS was carried out with success in the case of several legume crops (Torres *et al.*, 2010), transgenic *Bt* plants (Ribaut and Hoisngton, 1998), etc.

4.1 QTLs Selection for MAS

Although, MAS has proven success in the case of simple trait controlled by major genes, the availability of molecular markers and genetic maps make it feasible also for QTL (Torres *et al.*, 2010). Theoretically, all markers that are tightly linked to QTLs could be used for MAS. However, due to the cost of utilizing several QTLs, only markers that are tightly linked to no more than three QTLs are typically used (Ribaut and Betran, 1999), although there have been reports of up to 5 QTLs being introgressed into tomato via MAS (Lecomte *et al.*, 2004). Even selecting for a single QTL via MAS can be beneficial for crop improvement strategies; such a QTL should account for the largest proportion of phenotypic variance for the specific trait. Furthermore, all QTLs selected for MAS should be stable across the environments (Hittalmani *et al.*, 2002).

4.2 Effective Marker–Trait Association (MTA) and Marker Validation

The identification of the new QTL has been increasing tremendously and this was very much clear from the past two decades' publication scenario in PubMed. Now this involves almost all crop plants and all types of agronomic traits (Dwivedi *et al.*, 2007). However, reports of QTL mapping to date have tended to be based on individual small to moderately sized mapping populations screened with a relatively small number of markers, providing relatively low resolution of marker–trait association or MTA (Salvi and Tuberosa, 2005). Very few of the QTLs reported have been used for MAS. Most MTA reports to date have been based on segregating populations generated, in most cases, from two inbred lines. Genetic variation detected in the mapping population (particularly recombination patterns in the region of the target gene) may not be shared by other genetic and breeding populations because of allelic diversity. Thus, QTL markers identified using a single mapping population may not be automatically used directly in unrelated populations without marker validation and/or fine mapping (Nicholas, 2006). The MTA must be validated in representative parental lines, breeding populations, and phenotypic extremes before it can be used for routine MAS, although this process may be incorporated into genetic mapping programmes. In a portion of cases, markers will lose their selective power during this validation step. In these cases, the most suitable approach is to identify new markers (through fine mapping or candidate gene analysis) in the genomic region around the target locus to find MTAs that are shared across different breeding populations. By developing several markers within or around a single gene, it is much more likely that the parents of any breeding population will be polymorphic for at least one of them. This will then allow breeders to track the alleles donated from each parent throughout the breeding process, speeding up MAS and marker-assisted breeding (MAB) in any cross.

4.3 Multilocational High-throughput Phenotyping

The quantity and quality of phenotyping is becoming the most significant factor affecting the accuracy of genetic mapping and thus the power of the resultant MAS, particularly for complex traits. However, precision and global phenotyping of a large number of plant samples are very expensive and time-consuming. The level of heritability of measured traits depends in part on whether the phenotyping can be repeated across different seasons, locations, and environments. Clustering target locations into mega-environments and comparing these with the success of selection at different locations has been used to understand how breeding programmes can optimize their selection processes to generate germplasm with the best yield and other agronomic characters for specific target environments (Lillemo *et al.*, 2004). Cross-population and environment comparison of phenotyping will determine how the MTAs identified under one environment can be used for selection under another. In this case, well-characterized environments and well-established selection criteria are essential prerequisites for the development of a reliable precision phenotyping system. Precision and high-throughput multilocational phenotyping, together with effective sampling and data acquisition systems being developed for many traits, provide the potential to develop a phenomics-based protocol for trait-specific improvemental programmes. This will not only help our understanding of the phenotypic profile a plant possesses but also improves the precision of genetic mapping and thus MAS for the target phenotype.

4.4 Marker-Assisted Backcrossing (MAB)

Using conventional methods, it typically takes 6–8 backcrosses to fully recover the recurrent parent genome. The theoretical proportion of the recurrent parent genome after n generations of backcrossing is given by: $(2n+1-1)/2n+1$ (where n = number of backcrosses; assuming an infinite population size). The percentages are usually lower in smaller population sizes that are typically used in actual plant improvement programmes. Although the average percentage of the recurrent parent genome is 75% (for the entire BC1 population), some individuals possess more of the recurrent parent genome than others. Therefore, if tightly-linked markers flanking QTLs and evenly spaced markers from other chromosomes (i.e., unlinked to QTLs) of the recurrent parent are used for selection, the introgression of QTLs and recovery of the recurrent parent may be accelerated. This process is called marker-assisted backcrossing. The use of additional markers to accelerate cultivar development is sometimes referred to as 'full MAS' or 'complete line conversion' (Morris *et al.*, 2003). Simulation studies using PLABSIM–a computer programme that simulates recombination during meiosis–indicate that efficiency of recurrent parent recovery using markers is far greater compared to conventional backcrossing. This is an important consideration for researchers because the accelerated release of an improved variety may translate into more rapid profits by the release of new cultivars in the medium to long-term.

4.5 Selective Genotyping and Pooled DNA Analysis

There have been two broad types of approaches of identifying MTA. The first is based on genotyping an entire segregating population with markers densely covering the entire genome, and then testing for associations between phenotypic differences and marker genotypes. The genotyping for this approach is extensive, time-consuming, and expensive, while generating precision phenotype data at this scale may be logistically difficult or even impossible. The second approach is based on genotyping only that part of the population exhibiting extreme phenotypes for the target trait (selective genotyping), association is then inferred by finding allelic frequency differences between the groups of plants with contrasting phenotypes. Combining DNA pool analysis with selective genotyping (so-called bulked segregant analysis) is then the simplest approach to identifying markers for major genes, because it requires analysis of only two DNA pools representing the two phenotypic extremes. Pooled DNA analysis has been very successful in genetic mapping in plants using RFLP and SSR markers with numerous reports for single major genes (van Treuren, 2001; Zhang *et al.*, 2002) and even two to three major QTLs. Developments in single nucleotide polymorphism (SNP) genotyping technologies and methodologies offer a vision of future possibilities for molecular plant breeding. It is possible to carry out genome-wide association mapping by using an integrated technology package including selective genotyping, pooled DNA analysis, and microarray-based SNP genotyping with numerous markers. This system has the power to estimate allele frequencies and identify unique alleles from a pooled DNA sample cf several individuals. When this approach can be successfully implied, it will resolve many of the constraints of bulked segregant analysis described above (Xu and Crouch, 2007). Allele frequencies can be estimated either by collating individual genotypic scores when genotyping is based on individuals or by signal strength comparison when genotyping is based on pooled DNA samples. As high density genome-wide SNP markers will soon be available in many crop species, it is now possible to narrow down the target locus to less than 1 cM in one step using selective genotyping alone. However, optimizing SNP genotyping systems for pooled DNA analysis is considerably more complicated than for SSR markers and suffers a much higher level of redundancy. The high density of SNP markers will soon be available in rice and maize (amongst cereals). Pooled DNA analysis will also be very useful for large-scale analysis of landraces that are highly heterogeneous and thus till date not well characterized.

With recent advances in genomics, the MTA analysis is increasing the phenotyping and not the genotyping. Thus, it may be more efficient to genotype the whole population first to identify the most informative subset of individuals in terms of minimum level of relatedness between individuals plus optimum subpopulation structure and allele representativeness; then, carry out precision phenotyping of this subset, particularly for the traits that are difficult or expensive to evaluate. A final refinement to achieve maximum efficiency would be to combine selective phenotyping with selective genotyping thereby focusing on a very small proportion of the total number of genotypes for the full analysis but ensuring that these are the most informative plants or families. In this approach, the total number of individuals to be phenotyped and genotyped may not change, but the power of the analysis will be dramatically increased. This would be most effective for traits where phenotypic extremes can be easily identified using a simple screening method. For example of abiotic stress tolerance, where a large number of plants or families can be eliminated easily under stress conditions through visual scoring. Although the level of stress and the selection threshold need to be carefully optimized to maximize the probability of finding genes that confer economic tolerance (with little or no yield penalty) as opposed to survival traits (with a substantial negative effect on yield). Following selective genotyping of these individuals with extreme phenotypes, precision phenotyping of the resultant subset of individuals can be performed using physiological component and surrogate traits. High-density planting and selection at early stages of plant development, combined with selective phenotyping and genotyping should also be investigated as a potential option for some traits to allow one to work with more plants or families. Where the target trait is influenced by planting density or strong selection pressure this will clearly confound the ability to make genetic gain. However, many major gene-controlled traits can be investigated in this way without much disturbance. It can be inferred that phenotypic extremes or extremely stress-tolerant plants are those with an accumulation of favourable alleles from multiple loci, each with small to large effects, so that genetic mapping, particularly with relatively small populations, will identify the genetic regions with relatively large accumulative effect on the target trait. This is supported to some extent by results on drought tolerance in rice (Li *et al.*, 2005). When allele frequencies can be estimated from the selected individuals or the DNA pools on which genotyping is based, the putative gene locations can be identified based on the comparison of allele frequencies, even if markers have distorted segregation ratios. However, when only one extreme (only survivors or most resistant individuals) is available, which might be true in the case that no susceptible individuals survive, or reliable estimation of allele frequencies is not possible, a control population is needed to distinguish selection effects from segregation distortion. Genome-wide linkage disequilibrium (LD)-based association mapping may provide a shortcut to discovering functional alleles and allelic variations that contribute to agronomic traits of interest. The Generation Maize is being screened via resequencing and precision phenotyping to test the feasibility of the whole "population" based approach for simply inherited traits. Thus, the selective genotyping and pooled DNA analysis discussed here can be extended for using extremes of inbreeds selected. For association mapping of quantitative traits governed by a large number of minor genes which interact with each other and the environment, through either whole population analysis or selective genotyping, we will face the same challenges as experienced with linkage-based QTL mapping. Trait-specific genetic and breeding materials, with novel properties including phenotype extremes, eternal or fixed segregating populations (e.g., recombinant inbred lines, doubled haploids, near isogenic lines (NILs), introgression lines), genetic stocks (e.g., single segment substitution lines), and mutant libraries, have been developed and maintained by many groups across the world spanning most crops. These are valuable directly for the purpose they were developed but also offer a powerful opportunity when used collectively. In many cases, they have been phenotyped in multiple environments by taking advantage of their permanent property and fixed segregation pattern. By collecting phenotypic extremes from currently available genetic and breeding materials, and using selective genotyping and pooled DNA analysis, one 384 plate could be used for genetic mapping of almost all important major gene- or QTL-controlled traits. This assumes that each trait needs only two wells and that in one plate it is possible to carry out 192 pairwise DNA pool-based comparisons. When such existing materials are used collectively and combined with LD mapping (Mackay and Powell, 2007), they will also provide a shortcut toward one-step genome-wide association mapping for all target traits.

4.6 Cost-benefit in MAS

The cost of using MAS compared to conventional approaches varies considerably between studies. Dreher *et al.* (2003) indicated that the cost-effectiveness needs to be considered on a case-by-case basis. Factors that influence the cost of utilizing markers include: inheritance of the trait, method of phenotypic evaluation, field/glasshouse and labour costs, and the cost of resources. In some cases, phenotypic screening is cheaper compared to marker-assisted selection (Dreher *et al.*, 2003). However, in other cases, phenotypic screening may be time-consuming and expensive assays, and the use of markers will then be preferable. Some studies involving markers for disease resistance have shown that once markers have been developed for MAS, it is cheaper than conventional methods. In other situations, phenotypic evaluation may be time-consuming and/or difficult and therefore using markers may be cheaper and preferable. An important consideration for MAS is that while markers may be cheaper to use, there is a large initial cost in their development.

5. CONCLUSION AND FUTURE LOOK

Various difficulties that a user can encounter during the ontogeny of marker's application have also been discussed. Marker mediated varietal fingerprinting and germplasm characterization appeared the most common and most pervasive application with AFLP and SSR markers. Being cost effective, easy to handle and devoid of any radioisotope requirement, SSR and SNP markers are considered the most suitable and reliable system for DNA fingerprinting. Marker-assisted selection has been successful for introgressing and pyramiding major-effect genes, however many challenges remain to be resolved before MAS can routinely provide added value for breeding very complex traits. Marker-assisted selections for qualitative traits appeared most successful after DNA fingerprinting while for quantitative characters, major disease resistance genes and genes controlling QTL for abiotic stress tolerance, the success is limited. It is anticipated that application of markers will remain restricted in these areas till the allele-specific markers are available and the cost of marker analysis is reduced significantly. Although there have been numerous QTL mapping studies for a wide range of traits in diverse crop species, relatively few markers have actually been implemented in breeding programmes for crop improvement. This lack of adoption is due to the non-reliable nature of used markers to predicting the desired phenotype. In many cases, this would be attributable to a low accuracy of QTL mapping studies or inadequate validation. The rate, scale, and scope of uptake of MAS in public crop breeding programme has continually lagged behind expectations. There are many technical and logistical factors that have hindered the speed and scope of MAS uptake. Steady progress and advancement in DNA markers will make it more attractive for molecular breeding and plant genetics and ultimately help in major crop improvement.

References

Adam-Blondon, A. F., Sevignac, M., Bannerot, H., and Dron, M. (1994). SCAR, RAPD and RFLP markers linked to a dominant gene (are) conferring resistance to anthracnose in common bean. *Theor. Appl. Genet.,* 88: 865-870.

Agarwal, M., Shrivastava, N., and Padh, H. (2008). Advances in molecular marker techniques and their applications in plant sciences. *Plant Cell Rep.,* 27: 617-631.

Akagi, H., Inagaki, Y. A., Fujimura, T. (1997). Highly polymorphic microsatellites of rice consist of AT and a classification of closely related cultivars with these microsatellite loci. *Theor. Appl. Genet.,* 94: 61-67.

Botstein, D., White, R. L., Skolnick, M., and Davis, R. W. (1980). Construction of a genetic linkage map in man using restriction fragment length polymorphisms. *American J. Hum. Genet.,* 32: 314-331.

Brondani, C., Borba, T. C. O., Rangel, P. H. N., and Brondani, R. P. V. (2006). Determination of genetic variability of traditional varieties of Brazilian rice using microsatellite markers. *Genet. Mol. Biol.,* 29(4): 676-684.

Collard, B. C. Y., and Mackill, D. J. (2007). Marker-assisted selection: An approach for precision plant breeding in the twenty-first century. *Philos. Trans. Royal Soc. Lond. B Biol. Sci.,* 17: 1-16.

Collard, B. C., and Mackill, D. J. (2008). Marker-assisted selection: an approach for precision plant breeding in the twenty-first century. *Philos. Trans. Royal Soc. Lond. B. Biol. Sci.,* 363: 557-572.

Dwivedi, S. L., Crouch, J. H., Mackill, D. J., Xu, Y., Blair, M. W., Ragot, M., Upadhyaya, H.D., and Ortiz, R. (2007). The molecularization of public sector crop breeding: Progress, problems, and prospects. *Adv. Agron.,* 95: 163-318.

Dreher, K., Khairallah, M., Ribaut, J., and Morris, M. (2003). Money matters (I): Costs of field and laboratory procedures associated with conventional and marker-assisted maize breeding at CIMMYT. *Mol. Breed.,* 11: 221-234.

Eagles, H., Bariana, H., Ogbonnaya, F., Rebetzke, G., Hollamby, G., Henry, R., Henschke, P., and Carter, M. (2001). Implementation of markers in Australian wheat breeding. *Aust. J. Agric. Res.*, 52: 1349-1356.

Friedt, W., and Ordon, F. (2007). Molecular markers for gene pyramiding and disease resistance breeding in barley. In *Genomics assisted crop improvement*, Vol. 2: Genomics applications in crops, Varshney, R. K., and Tuberosa, R. (eds.). Springer, Dordrecht, The Netherlands, pp. 81-101.

Garris, A., Tai, T., Coburn, J., Kresovich, S., and McCouch, S. R. (2005). Genetic structure and diversity in *Oryza sativa* L. *Genetics*, 169: 1631-1638.

Gupta, P., Varshney, R., Sharma, P., and Ramesh, B. (1999). Molecular markers and their applications in wheat breeding. *Plant Breed.*, 118: 369-390.

Hittalmani, S., Shashidhar, H. E., Bagali, P. G., Huang, N., Sidhu, J. S., Singh, V. P., and Khush, G. S. (2002). Molecular mapping of quantitative trait loci for plant growth, yield and yield related traits across three diverse locations in a doubled haploid rice population. *Euphytica*, 125: 207-214.

Holland, J. B. (2004). Implementation of molecular markers for quantitative traits in breeding programmes—Challenges and opportunities. In *New directions for a diverse planet*, Fischer, T. (ed.). Proceeding of the 4th International Crop Science Congress, Brisbane, QLD, Australia, 26 Sept. 1 Oct. 2004, The Regional Institute Ltd., Gosford, NSW, Australia.

Jaccoud, D., Peng, K., Feinstein, D., and Kilian, A. (2001). Diversity arrays: a solid state technology for sequence information independent genotyping. *Nucleic Acids Res.*, 29: e25.

Jena, K. K., and Mackill, D. J. (2008). Molecular markers and their use in marker-assisted selection in Rice. *Crop Sci.*, 48: 1266-1276

Jones, N., Ougham, H., and Thomas, H. (1997). Markers and mapping: We are all geneticists now. *New Phytol.*, 137: 165-177.

Joshi, S., Ranjekar, P., and Gupta, V. (1999). Molecular markers in plant genome analysis. *Curr. Sci.*, 77: 230-240.

King, R. C., and Stansfield, W. D. (1990). A dictionary of genetics. 4th Ed., Oxford University Press, New York-Oxford, pp. 188.

King, R. C., and Stansfield, W. D. (1997). A dictionary of genetics. 5th Ed., Oxford University Press, New York. pp. 439.

Kumar, L. S. (1999). DNA markers in plant improvement: an overview. *Biotechnology Advances*, 17: 143-182.

Latha, R., Rubia, L., Bennett, J., and Swaminathan, M. S. (2004). Allele mining for stress tolerance genes in *Oryza* species and related germplasm. *Mol. Biotechnol.*, 27: 101 108.

Lecomte, L., Duffe, P., Buret, M., Servin, B., Hospital, F., and Causse, M. (2004). Marker-assisted introgression of five QTLs controlling fruit quality traits into three tomato lines revealed interactions between QTLs and genetic backgrounds. *Theor. Appl. Genet.*, 109: 658-668.

Li, Z. K., Fu, B.Y., Gao, Y. M., Xu, J. L., Ali, J., Lafitte, H. R., Jiang, Y. Z., Rey, J. D., Vijayakumar, C. H. M., Maghirang, R., Zheng, T. Q., and Zhu, L. H. (2005). Genome-wide introgression lines and their use in genetic and molecular dissection of complex phenotypes in rice (*Oryza sativa* L.). *Plant Mol. Biol.*, 59: 33-52.

Lillemo, M., van Ginkel, M., Trethowan, R. M., Hernandez, E., and Rajaram, S. (2004). Associations among international CIMMYT bread wheat yield testing locations in high rainfall areas and their implications for wheat breeding. *Crop Sci.*, 44: 1163-1169.

Mackay, I., and Powell, W. (2007). Methods for linkage disequilibrium mapping in crops. *Trends Plant Sci.*, 12: 57-63.

Mackill, D. J. (2007). Molecular markers and marker-assisted selection in rice. In *Genomics assisted crop improvement*, Vol. 2: Genomics applications in crops, Varshney, R. K., and Tuberosa, R. (eds.). Springer, New York, pp. 147-168.

Markert, C. L., and Moller, F. (1959). Multiple forms of enzymes: tissue, ontogenetic, and species specific patterns. *Proc. Natl. Acad. Sci.*, 45: 753-763

Martin, G. B., Williams, J. G. K., and Tanksley, S. D. (1991). Rapid identification of markers linked to a Pseudomonas resistance gene in tomato by using random primers and near-isogenic lines. *Proc. Natl. Acad. Sci.*, 88: 2336-2340

Mohan, M., Nair, S., Bhagwat, A., Krishna, T. G., Yano, M., Bhatia, C. R., and Sasaki, T. (1997). Genome mapping, molecular markers and marker-assisted selection in crop plants. *Molecular Breed.*, 3: 87-103.

Morris, M., Dreher, K., Ribaut, J. M., and Khairallah, M. (2003). Money matters (II): Costs of maize inbred line conversion schemes at CIMMYT using conventional and marker-assisted selection. *Mol. Breed.*, 11: 235-247.

Nicholas, F. W. (2006). Discovery, validation, and delivery of DNA markers. *Aust. J. Exp. Agric.*, 46: 155-158.

Paran, I., Kesseli, R., and Michelmore, R. (1991). Identification of restriction fragment-length-polymorphism and random amplified polymorphic DNA markers linked to downy mildew resistance genes in lettuce, using near isogenic lines. *Genome*, 34: 1021-1027

Paux, E., Sourdille, P., Mackay, I., and Feuillet, C. (2012). Sequence-based marker development in wheat: advances and applications to breeding. *Biotechnol. Advances*, 30: 1071-1088.

Rahman, M. S., Sohag, M. K. H., and Rahman, L. (2010). Microsatellite based DNA fingerprinting of 28 local rice (*Oryza sativa* L.) varieties of Bangladesh. *J. Bangladesh Agric. Univ.*, 8 (1): 7-17.

Ram Kumar, G., Sakthivel, K., Sundaram, R. M., Neeraja, C. N., Balachandran, S. M., Shobha Rani, N., Viraktamath, B. C., and Madhav, M. S. (2010). Allele mining in crops: Prospects and potentials. *Biotechnol. Advances*, 28: 451-461.

Ram, S. G., Thiruvengadam, V., and Vinod, K. K. (2007). Genetic diversity among cultivars, landraces and wild-relatives of rice as revealed by microsatellite markers. *J. Appl. Genet.*, 48(4): 337-345.

Ravi, M., Geethanjali, S., Sameeyafarheen, F., and Maheswaran, M. (2003). Molecular marker based genetic diversity analysis in rice (*Oryza sativa* L.) using RAPD and SSR markers. *Euphytica*, 133: 243-252.

Ribaut, J.-M., Hu, X., Hoisington, D., and Gonzalez-De-Leon, D. (1997). Use of STSs and SSRs as rapid and reliable preselection tools in marker-assisted selection backcross scheme. *Plant Mol. Biol. Reporter*, 15: 156-164.

Ribaut, J. M., and Betran, J. (1999). Single large-scale marker-assisted selection (SLS-MAS). *Mol. Breed.*, 5: 531-541.

Ribaut, J. M., and Hoisington, D. (1998). Marker-assisted selection: new tools and strategies. *Trends in Plant Sci.*, 3: 236-239.

Roychowdhury, R., Karmakar, J., and Dey, N. (2012). PCR-compatible genomic DNA isolation from different tissues of rice (*Oryza sativa*) for SSR fingerprinting. *Eurasian J. Bioscience*, 6: 85-90.

Roychowdhury, R., Banerjee, U., Sofkova, S., and Tah, J. (2013). Organic farming for crop improvement and sustainable agriculture in the era of climate change. *OnLine Journal of Biological Sciences*, 13(2): 55-70.

Roychowdhury, R., and Tah, J. (2013). Mutagenesis - A potential approach for crop improvement. In *Crop improvement: New approaches and modern techniques*, Hakeem, K.R., Ahmad, P., and Ozturk, M. (eds.). Springer-Verlag, New York, USA, pp. 149-187.

Salvi, S., and Tuberosa, R. (2005). To clone or not to clone plant QTLs: Present and future challenges. *Trends in Plant Sci.*, 10: 297-304.

Semagn, K., Bjornstad, A., and Ndjiondjop, M. N. (2006). An overview of molecular marker methods for plants. *African J. Biotechnol.*, 5(25): 2540-2568.

Sobrino, B., Briona, M., and Carracedoa, A. (2005). SNPs in forensic genetics: a review on SNP typing methodologies. *Forensic Sci. Int.*, 154: 181-194.

Song, Z. P., Xu, X., Wang, B., Chen, J. K., and Lu, B. R. (2003). Genetic diversity in the northernmost *Oryza rufipogon* populations estimated by SSR markers. *Theor. Appl. Genet.*, 107: 1492-1499.

Temnykh, S., Park, W. D., Ayres, N., Cartinhour, S., Hauck, N., Lipovich, L., Cho, Y. G., Ishii, T., and McCouch, S. R. (2000). Mapping and genome organization of microsatellite sequences in rice (*Oryza sativa* L.). *Theor. Appl. Genet.*, 100: 697-712.

Torres, A. M., Avila, C. M., Gutierrez, N., Palomino, C., Moreno, M. T., and Cubero, J. I. (2010). Marker-assisted selection in fababean (*Vicia faba* L.). *Field Crop Res.*, 115: 243-252.

van de Wiel, C., Arens, P., and Vosman, B. (1999). Microsatellite retrieval in lettuce (*Lactuca sativa* L.). *Genome,* 42: 139-149.

van Treuren, R. (2001). Efficiency of reduced primer selectivity and bulked DNA analysis for the rapid detection of AFLP polymorphisms in a range of crop species. *Euphytica,* 117: 27-37.

Vos, P., Hogers, R., Bleeker, M., Reijans, M., Van De, T. L., Hornes, M., Frijters, A., Pot, J., Peleman, J., Kuiper, M., and Zabeau, M. (1995). AFLP: a new technique for DNA fingerprinting. *Nucleic Acids Res.*, 23: 4407-4414.

Wexelsen, H. (1933). Linkage between quantitative and qualitative characters in barley. *Hereditas,* 17: 323-341.

Williams, J. G. K., Kubelik, A. R., Livak, K. J., Rafalski, J. A., and Tingey, S. V. (1990). DNA polymorphism amplified by arbitrary primers are useful as genetic markers. *Nucleic Acids Res.,* 18: 6531-653.

Winter, P., and Kahl, G. (1995). Molecular marker technologies for plant improvement. *World J. Microbiol. Biotechnol.,* 11: 438-448.

Witcombe, J. R., and Virk, D. S. (2001). Number of crosses and population size for participatory and classical plant breeding. *Euphytica,* 122: 451-462.

Xu, Y., and Crouch, J. H. (2007). Marker-assisted selection in plant breeding: From publications to practice. In *From genomics to plant improvement,* Li, Z. K., and Zhang, Q. F. (eds.). Proceedings of the 2nd International Conference on Plant Molecular Breeding, Sanya, Hainan, China, 23–27 March 2007, Beijing Kexin Printing House, Beijing, pp. 46.

Zhang, J., Xu, Y., Wu, X., and Zhu, L. (2002). A bentazon and sulfonylurea sensitive mutant: Breeding, genetics, and potential application in seed production of hybrid rice. *Theor. Appl. Genet.,* 105: 16-22.

13

'Omics': Technique in Modern Agriculture for Crop Improvement

Rajni Shukla[1], Arvind K. Shukla[2] and Yogesh K. Sharma[1*]

[1]*Botany Department, University of Lucknow, Lucknow-226007, India*

[2]*Indian Institute of Soil Science, Bhopal, M.P., India*

ABSTRACT

The complete sequencing of an organism even the human genome has ushered in a new era of system biology is referred to as omics. Fundamental biological processes can now be studied by applying the full range of OMICS technologies (genomics, transcriptomics, proteomics, metabolomics) and beyond to the same biological sample. A variety of omics' sub-disciplines have begun to emerge, each with their own set of instruments, techniques, reagents and software. The omics technology that has driven these new areas of research consists of DNA and protein microarrays, mass spectrometry and a number of other instruments that enable high-throughput analyses. However, new technology is developing constantly and quickly, so it is important that researchers keep up to date with the latest protocols, commercial products and other sources of information. Omics world was developed as a portal to link investigators to the wide variety of resources that are currently available in specific omics fields. In this review, we have outlined some important technologies such as separation method like High Performance Liquid Chromatography (HPLC), Capillary Electrophoresis (CE), Detection Method, Substractive Hybridization, Expressed Sequence Tag Sequencing, Microarray Hybridization, Differencial Display, c-DNA-AFLP, Microarray, Serial Analysis of Gene Expression (SAGE), Multiple Reaction Monitoring for which these technologies are expected to advance crop genetics and breeding, leading to crop improvement. Developments in Omics' technologies have posed new challenges in Omics data visualization and interpretation. For example, proteomics research has matured in plant research, allowing the quantification of thousands of proteins. Emerging metabolomics technologies such as Fourier Transform Mass Spectrometry (FT-MS) allow the identification and quantification of many more metabolites than that which was previously possible. The modularity and flexibility offered by MapMan (a software tool) will be useful in keeping up with these changes allowing it to continue to provide novel ways to organize, evaluate and display data. It will also be adopted to more

*Corresponding author's E-mail: yogesh_s26@yahoo.com

Roychowdhury R. (Ed.), Crop Improvement in the Era of Climate Change.

plants relying on increasingly sophisticated methods for knowledge transfer. Thus, MapMan will aid in data visualization, analysis and interpretation in the next 5 years of omics research.

Keywords: Genomics, metabolomics, omics techniques, proteomics, transcriptomics.

1. INTRODUCTION

With a rising population, growing more food at affordable prices becomes even more important. This challenge requires integrated approaches to increase agricultural production and cope with environmental challenges (Fig. 1). It is the general tendency of human beings to know something about the past and something about the future. The means or the techniques required for the purpose have evolved a systematic system which comprises the science. Various descriptions and narrations are the output of these developed techniques. Trial and error other than need of time have been greatly helpful in evolving different techniques on the date. Further short-comings and demerits in the existing techniques; open the door for developing more accurate and highly sensitive techniques. In view of ever growing population and decreasing natural resources there is need to enhance food production that can possibly be achieved by improving the qualitative and quantitative traits of crop plants by adopting new analytical tools and technologies. *Arabidopsis thaliana*, rice etc. have greatly influenced the discipline of plant and crop science to open the door in modern science of genomics and proteomics (Alavadi *et al.*, 2001). Gene discovery and gene expression profiling technologies are creating an unprecedented opportunity for plant breeder who can now apply molecular markers to assess and enhance the diversity in their germplasm collection to introduce valuable traits from new sources and identity genes that control trait (Setia and Setia, 2008).

Fig. 1 Interaction of external and internal factors in human.

Advances in molecular biology are now providing the tools to study the genetical make-up of plants which allow us to unravel the inheritance of all traits whether they are controlled by single (major) gene or many genes (of smaller effect) acting together, known as the quantitative trait loci (QTI). The emergence of the novel 'OMICS' technologies, such as genomics, proteomics and metabolomics is now permitting researchers to identify the genetic underpinning of crop improvement namely the genes, that contribute to the improved productivity and quality of modern crop variety.

These technologies also help in thorough investigation of the biology behind agronomic trait at the physiological, biochemical and molecular level, and permit the elucidation of molecular circuitry of the crop plants, ultimately, paying way for improved crop production. Indeed the various "OMICS" have become a staple of plant physiology (Raikhel, 2005). Risk assessment frameworks, such as those used for GM crops, have detailed comparative analysis with appropriate non-GM counterparts as their cornerstone. Opinions have been voiced that current analytical approaches are too specific, and need to be complemented by more unbiased, larger scale analysis of gene and protein expression using transcriptomics and proteomics, respectively. In parallel, the use of metabolomics has been advocated as an approach to expand significantly the range of metabolites that can be measured to assess more stringently the potential for any unintended effects. Transcriptomics, proteomics and metabolomics have been termed collectively "omics" technology (Fig. 2 and Fig. 3).

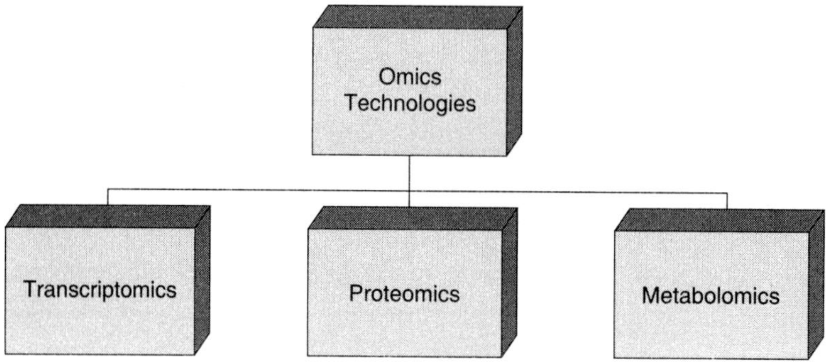

Fig. 2 Components of OMICS technologies.

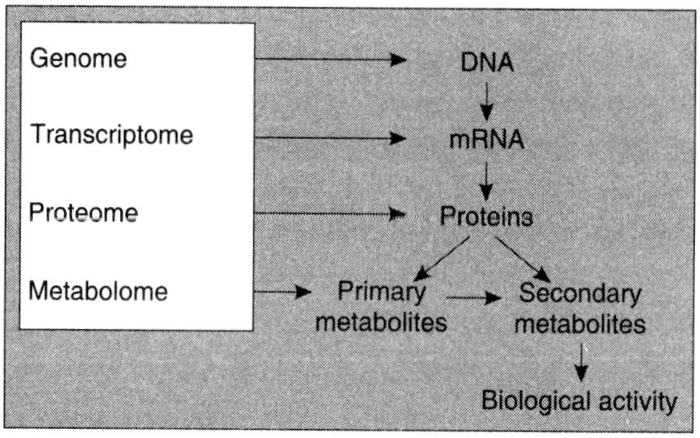

Fig. 3 Control level for cellular process.

1.1 Earliest Research in Plants

Agriculture probably began with the spontaneous growth of plants in the kitchen middens from seeds that had been thrown away. Gardens were made for their beauty by the Assyrians as early as 2000 B.C. The Hanging Gardens of Babylon, built about 612 B.C., were famous for their aesthetic character. Aristotle in the 4th century B.C., who was a teacher of Alexander the Great and later, through him, had access to much of the plant life of the known world. One of these students, Theophrastus of Eresus, is generally regarded as the father of botany. He classified plants by their habit of growth, dividing them into trees, shrubs, undershrubs, and herbs. This paragraph presents a brief historical review of progress in the field of plants because the authors feel that it is impossible to fully understand the present without some knowledge of the past. As the Danish philosopher Kierkegaard wrote, *"Life can only be understood backward, but it can only be lived forward,"* and this also is true for science. According to Aristotle, plants absorb their food ready for use from the soil, and plant nutrition was controlled by a soul or vital principle that allowed plants to absorb only those substances useful in growth. An understanding of plant mineral nutrition did not develop until the 19th and early 20th century. Probably the most important contributions of the 17th century were the anatomical observations of Nehemiah Grew as described in the book "The Anatomy of Plants" published in 1682 and of Marcello Malpighi in "Anatome Plantarum" 1675. Both were interested in physiological processes, while Grew in particular speculated about water absorption and the ascent of sap. Both seemed to think that the sap underwent some change in the leaves that increased its nutritive value, after which it moved to fruits or down to roots. In 1843 the farmers Lawes and Gilbert at Rothamstead turned their farm into a laboratory for long-term studies in soils and pasture management which is still a world leader. And so by 1850 agricultural productivity in Britain was increased by 0.5% per year, unprecedented in history and the period 1850 to 1875 was called the Golden Age of British Agriculture.

Agricultural practices have changed dramatically in the last fifty years as global population has increased. Over the next fifty years, global demand for food is expected to double (Tilman, 2002). Traditions of subsistence farming and organic methods have been increasingly replaced with large scale methods that have coincided with an increased use of nitrogen, phosphorus and pesticides. While new practices have allowed a dramatic increase in food production. They have also reduced the ability of ecosystems to provide vital services. The average land use of a single farm has been more than doubled over the last century (Arth *et al.*, 2007). Mono-crop systems have become more economically rewarding, partly because government agricultural subsidies tend to favour mono-cropping (Lean, 2006). Additionally, mono-cropping increases the susceptibility of plants to disease as a single pathogen has the potential to destroy an entire crop. Rice farms in China have successfully combated diseases that coincide with mono-cropping by planting genetically dissimilar strains of rice in a field (Tilman, 2002).

Changing priorities are putting stress on the world agricultural systems. Enough food is produced globally to adequately feed eight to ten billion people, yet many still suffer in hunger (Tilman, 2007). The problem lies with distribution; not only are agricultural products not reaching starving populations, but they are simultaneously being pulled in other directions. Agriculture is closely linked with other environmental challenges, especially the chemical industry, climate change, water quality and deforestation. Worldwide, a growing human population increasingly relies on unsustainable agricultural practices. As world population approaches ten billion, global food demand is expected to at least double in the next fifty years (Lean, 2007).

2. SOME TERMS IN OMICS

Word	*Definition*
Genomics	The analysis of the genes contained in an organism.
Proteomics	The analysis of protein profiles.
Proteome	The complement of proteins present in a cell.
Transcriptomics	The analysis of transcript profiles.
Transcriptome	The complement of transcripts (mRNA) in a cell.

Metabolomics: The net outcome of metabolism (the metabolites)

3. METABOLOMICS

Metabolomics is a term sometimes used to describe the emerging science of net outcome of the metabolism, the metabolites. The field of metabolomics overlaps with biology, chemistry, physics etc. Metabolomics is the scientific study of chemical processes involving metabolites. Metabolomics is the global analysis of all or a large number of cellular metabolites. Like other functional genomics research, metabolomics generates large amount of data. Handling, processing and analysis of these data are a clear challenge and require specialized mathematical, statistical and bioinformatics tools. Metabolomics needs for bioinformatics span through data and information management, raw analytical data processing, metabolomics standards and ontology, statistical analysis and data mining, data integration and mathematical modelling of metabolic network within a framework of systems biology. Metabolomics is the study of cells by measuring profiles of all, or a large number, of their metabolites. Metabolomics was originally proposed as a method of functional genomics. Metabolomics is the "systematic study of the unique chemical fingerprints that specific cellular processes leave behind the study of their small-molecule metabolite profiles".

Metabolomics is the scientific study of chemical processes involving metabolites. The metabolome represents the collection of all metabolites in a biological cell, tissue, organ or organism, which are the end products of cellular processes (Jordan *et al.*, 2009). Thus, while mRNA gene expression data and proteomic analyses do not tell the whole story of what might be happening in a cell, metabolic profiling can give an instantaneous snapshot of the physiology of that cell.

There is little doubt that improving the nutritional properties of food will be of considerable value to both the public and the agricultural industry, however, there is no consensus on how to achieve this goal, particularly on how to identify which specific target molecules to change. Research into secondary plant metabolism primarily took off in the latter half of the 19th century, however, there was still much confusion over what the exact function and usefulness of these compounds were. All that was known, that secondary plant metabolites were "by-products" of primary metabolism and were not crucial to the plant's survival. Early research only succeeded as far as categorizing the secondary plant metabolites but did not give real insight into the actual function of the secondary plant metabolites. The study of plant metabolites is thought to have started in the early 1800s when Friedrich Wilhelm Serturner isolated morphine from opium poppy, and after that new discoveries were made rapidly. In the early half of the 1900s, the main research around secondary plant metabolism was dedicated to the formation of secondary metabolites in plants, and this research was compounded by the use of tracer techniques which made deducing metabolic pathways much easier.

However, secondary plant metabolites are useful in the long term, often for defence purposes, and give plants characteristics such as colour, fragrance and even obnoxious tastes. Secondary plant metabolites are also used in signalling and regulation of primary metabolic pathways. Plant hormones, which are secondary metabolites, are often used to regulate the metabolic activity within cells and oversee the overall development of the plant (Fig. 4).

There has been some disagreement over the exact differences between 'metabolomics' and 'metabonomics'. The difference between the two terms is not related to choice of analytical platform: although metabonomics is more associated with NMR spectroscopy, and metabolomics

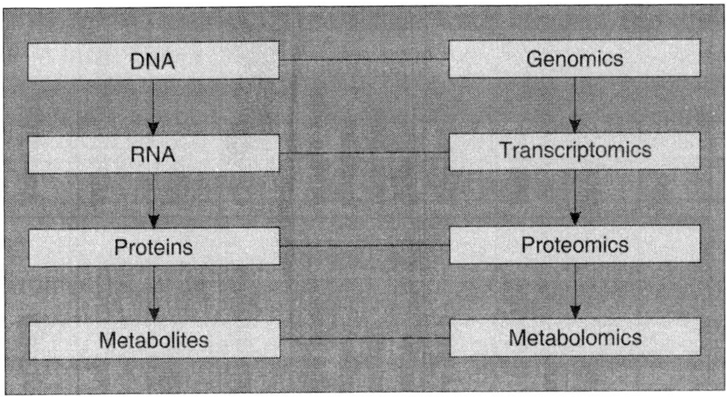

Fig. 4 Metabolic network model (*E. coli*).

with mass spectrometry-based techniques, this is simply because of usages amongst different groups that have popularized the different terms. While there is still no absolute agreement, there is a growing consensus that 'metabolomics' places a greater emphasis on metabolic profiling at a cellular or organ level and is primarily concerned with normal endogenous metabolism.

3.1 Technologies for Analyses

3.2 Separation Methods

3.2.1 Gas Chromatography

Gas chromatography, especially when interfaced with mass spectrometry (GC-MS), is one of the most widely used and powerful method for separation. It offers very high chromatographic resolution, but requires chemical derivatization for many biomolecules. Only volatile chemicals can be analysed without derivatization. (Some modern instruments allow '2D' chromatography, using a short polar column after the main analytical column, which increases the resolution still further.) Some large and polar metabolites cannot be analysed by GC.

3.2.2 High Performance Liquid Chromatography (HPLC)

High performance liquid chromatography (HPLC), compared to GC, HPLC has lower chromatographic resolution, but it does have the advantage that a much wider range of analytes can potentially be measured (Gika *et al.*, 2007).

3.2.3 Capillary Electrophoresis (CE)

Capillary electrophoresis (CE), has higher theoretical separation efficiency than HPLC, and is suitable for use with a wider range of metabolite classes than is GC. As for all electrophoretic techniques, it is most appropriate for charged analytes (Soga *et al.*, 2003).

3.3 Detection Methods

3.3.1 Mass Spectrometry (MS)

Mass spectrometry (MS) is used to identify and to quantify metabolites after separation by GC, HPLC (LC-MS), or CE. GC-MS is the most 'natural' combination of the three, and was the first to be developed. In addition, mass spectral fingerprint libraries exist or can be developed that allow identification of a metabolite according to its fragmentation pattern. MS is both sensitive (although, particularly for HPLC-MS, sensitivity is more of an issue as it is affected by the charge on the metabolite, and can be subjected to ion suppression artifacts) and can be very specific. There are also a number of studies which use MS as a stand-alone technology: the sample is infused directly into the mass spectrometer with no prior separation, and the MS serves to both separate and to detect metabolites.

3.3.2 Surface-based Mass Analysis

Surface-based mass analysis has seen resurgence in the past decade, with new MS technologies focused on increasing sensitivity, minimizing background, and reducing sample preparation. The ability to analyze metabolites directly from biofluids and tissues continues to challenge current MS technology, largely because of the limits imposed by the complexity of these samples, which contain thousands to tens of thousands of metabolites. Among the technologies being developed to address this challenge is Nanostructure-Initiator MS (NIMS), a desorption/ ionization approach that does not require the application of matrix and thereby facilitates small-molecule (i.e., metabolite) identification. MALDI is also used, however, the application of a MALDI matrix can add significant background at <1000 Da that complicates analysis of the low-mass range (i.e., metabolites). In addition, the size of the resulting matrix crystals limits the spatial resolution that can be achieved in tissue imaging. Because of these limitations, several other matrix-free desorption/ionization approaches have been applied to the analysis of biofluids and tissues. Secondary ion mass spectrometry (SIMS) was one of the first matrix-free desorption/ ionization approaches used to analyze metabolites from biological samples. SIMS uses a high-energy primary ion beam to desorb and generate secondary ions from a surface. The primary advantage of SIMS is its high spatial resolution (as small as 50 nm), a powerful characteristic for tissue imaging with MS. However, SIMS has yet to be readily applied to the analysis of biofluids and tissues because of its limited sensitivity at >500 Da and analyte fragmentation generated by the high-energy primary ion beam. Desorption electro-spray ionization (DESI) is a matrix-free technique for analyzing biological samples that uses a charged solvent spray to desorb ions from a surface. Advantages of DESI are that no special surface is required and the analysis is performed at ambient pressure with full access to the sample during acquisition. A limitation of DESI is spatial resolution because "focusing" the charged solvent spray is difficult. However, a recent development termed laser ablation ESI (LAESI) is a promising approach to circumvent this limitation.

3.3.3 Nuclear Magnetic Resonance (NMR) Spectroscopy

Nuclear magnetic resonance (NMR) spectroscopy is the only detection technique which does not rely on separation of the analytes, and the sample can thus be recovered for further analyses. All kinds of small molecule metabolites can be measured simultaneously — in this sense, NMR is close to being a universal detector. The main advantages of NMR are high analytical reproducibility and simplicity of sample preparation. Practically, however, it is relatively insensitive compared to mass spectrometry-based techniques. Although NMR and MS are the most widely used techniques, other methods of detection that have been used include ion-mobility spectrometry, electrochemical detection (coupled to HPLC) and radiolabelling (when combined with thin-layer chromatography).

3.3.4 Metabolic Engineering

Metabolic engineering is the practice of optimizing genetic and regulatory processes within to increase the cells' production of a certain substance. These processes are chemical networks that use a series of biochemical reactions and enzymes that allow cells to convert raw materials into molecules necessary for the cell's survival. Metabolic engineering specifically seeks to mathematically model these networks, calculate a yield of useful products, and pin point parts of the network that constrain the production of these products. Genetic engineering techniques can then be used to modify the network in order to relieve these constraints. Once again this modified network can be modeled to calculate the new product yield. Genetic engineering techniques can then be used to modify the network in order to relieve these constraints. Once again this modified network can be modeled to calculate the new product yield. The ultimate goal of metabolic engineering is to be able to use these organisms to produce valuable substances on an industrial scale in a cost effective manner. Current examples include producing beer, wine, cheese, pharmaceuticals, and other biotechnology products. Since cells use these metabolic networks for their survival, changes can have drastic effects on the cells' ability to survive. Therefore, trade-offs in metabolic engineering arise between the cells ability to produce the desired substance and its natural survival needs. Therefore, instead of directly deleting and/or over-expressing the genes that encode for metabolic enzymes, the current focus is to target the regulatory networks in a cell to efficiently engineer the metabolism (Vemuri and Aristidue, 2005).

At the industrial scale, metabolic engineering is becoming more convenient and cost effective. According to the Biotechnology Industry Organization, "more than 50 biorefinery facilities are being built across North America to apply metabolic engineering to produce biofuels and chemicals from renewable biomass which can help reduce greenhouse gas emissions". Potential biofuels include short-chain alcohols and alkanes (to replace gasoline), fatty acid methyl esters and fatty alcohols (to replace diesel), and fatty acid-and isoprenoid-based biofuels (to replace diesel) (Keasling, 2010) (Fig. 5).

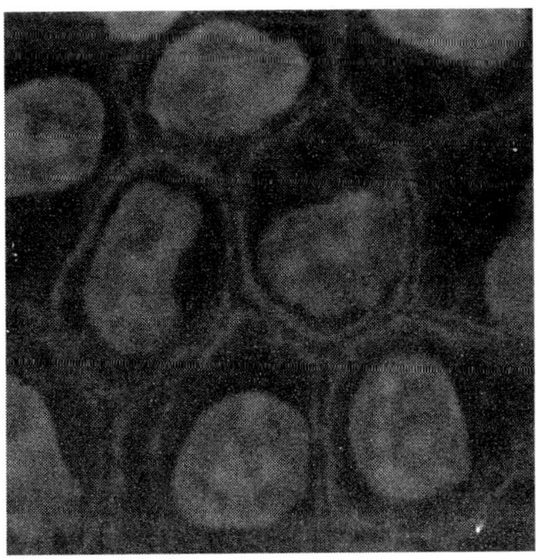

Fig. 5 Optimization of cellular metabolism for industrial use: a necessity.

3.4 Applications of Metabolomics

There are three major approaches used in as applications of metabolomics studies (Table 1): (i) targeted analysis, (ii) metabolite profiling, and (iii) metabolic fingerprinting (Table 1). Metabolomics is used for comparing mutants, assessing responses to environmental stress, studying global effects of genetic manipulation, comparing different growth stages, toxicology, drug discovery, nutrition, cancer, diabetes and natural product discovery. Metabolite profiling, whether targeting specific metabolite classes or untargeted, can also be applied as a tool in systems biology, where metabolite snapshots are used to study cellular dynamics through mathematical models.

Table 1 Major approaches of metabolomics

Approach	*Advantage*	*Disadvantage*
Targeted analysis	Quantitative	Limited number of compound can be targeted
	Low limit of detection	Does not detect compounds that were not targeted
	High throughput	Targetted compounds must be available purified for calibration
Metabolic profilling	Global (not targeted)	Semi-quantitative
	Directly applicable to pattern recognition	Difficult informatics
		Majority of peak not identifiable
		Medium throughput
Metabolic fingerprinting	Global (not targeted)	No compound identification
	Directly applicable to pattern recognition	
	Highest throughput	

4. GENOMICS

Genomics is the study of all of the nucleotide sequences, including structural genes, regulatory sequences, and noncoding DNA segments, in the chromosomes of an organism. The ability to sequence entire genomes has stimulated research directed not only at producing DNA sequence, but also at defining the function of genes on a genome-wide level. Given that genes with related functions are likely to be regulated together, techniques that evaluate global gene expression provide a mechanism for the initial identification and clustering of novel gene sequences with related functions. In 1995, when the entire DNA genome sequence of the first self-replicating organism, *Haemophilus influenzae*, was described, the age of genomics was initiated. In fact, within the last years the complete sequence of *Arabidopsis thaliana* was published (Arabidopsis Genome Initiative, 2000), and that of rice was reported by the commercial sector (Butler and Pockley, 2000; Davenport, 2001; Dickson and Cyranoski, 2001; Dunwell, 2000a; Eckardt, 2000). In Latin America the first successfully sequenced organism has been the prokaryote *Xylella fastidiosa* (Simpson *et al.*, 2000). In the coming months and years the DNA sequences for other different plant species will also be delivered.

The organization of sequences into functional groups based on expression provides a basic framework to direct additional experiments aimed at defining the precise function of the final gene product. Subsequently, when the complete genome sequences were considered a basis for systematic functional analysis, the genomics was divided into two disciplines:

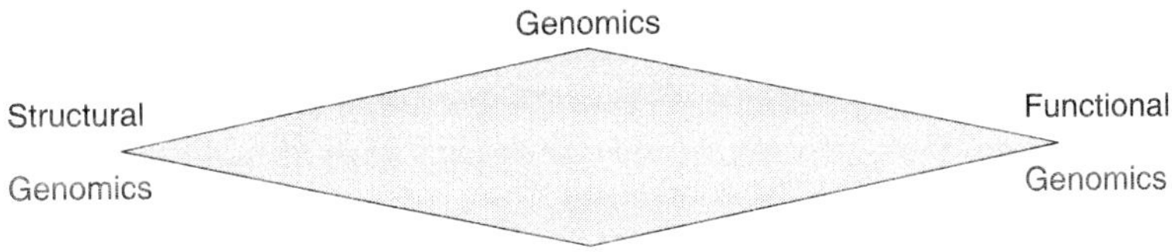

Structural genomics correspond to the initial phase of genome analysis resulting ultimately in the revolution of the complete DNA sequence of an organism. Functional Genomics makes the use of genome sequence to assess, on large scale the functions of genes (Leister, 2005).

4.1 Technologies for Analysis

4.1.1 Subtractive Hybridization

Subtractive hybridization is the first technique to be widely used for the purpose of identifying differentially expressed genes on a global scale. Advantages of the technique include the ability to isolate genes with no prior knowledge of their sequence or identity and the use of common molecular biology techniques that do not require specialized equipment or analyses. Subtractive hybridization methods were first described in the early 1980s for the purposes of creating cDNA libraries (Sargent and Dawid, 1983) and generating probes (Davis *et al*., 1984) of differentially expressed genes. Differentially expressed genes were isolated by hybridizing cDNA representing one sample (tester) to an excess of mRNA representing a second sample (driver). Transcripts expressed in both the tester and driver would form a mRNA/cDNA hybrid molecule, whereas a cDNA sequence unique to the tester would remain single-stranded. Single- and double-stranded molecules were separated using hydroxylapatite chromatography, and the eluted single-stranded cDNA represented genes expressed only in the tester. Differentially expressed cDNA could then be cloned or used directly as probes for screening libraries. Two significant limitations of the original protocol were the requirement of large quantities of mRNA (e.g., 270 μg were used by Sargent and Dawid, 1983) and a bias against the identification of rare transcripts.

Modifications to the original protocol include the production of cDNA with biotin tags (Welcher *et al*., 1986) or oligo(dT)30-latex (Hara *et al*., 1991) to improve the separation of single- and double-stranded molecules. Amplification of selected cDNA by PCR has also been incorporated to decrease the initial amount of mRNA required and to improve the cloning efficiency of selected transcripts (Hara *et al*., 1991). Additional modifications have been developed to alleviate bias against rare transcripts (Gurskaya *et al*., 1996), and Diatchenko *et al*., 1996 described similar protocols, termed *suppression subtractive hybridization* (**SSH**). The SSH method is designed to selectively amplify differentially expressed transcripts while suppressing the amplification of abundant transcripts, thus eliminating the need to separate single- and double-stranded molecules. In addition, SSH normalizes target transcripts to approximately equal abundance. The SSH method was originally described by Diatchenko *et al*. 1996.

Several limitations of the original protocols, such as requirements of large quantities of RNA and bias toward abundant genes, have been overcome by incorporation of PCR into the SSH technique. However, SSH remains applicable only to pair-wise treatment comparisons and must be replicated with the tester and driver reversed to identify gene expression changes in both directions. Additionally, subtractive hybridization does not provide a quantitative measure of expression differences and is most efficient at identifying genes that are completely absent, rather than expressed less abundantly, in the driver sample. Differential Display Techniques known in general as RNA fingerprinting include differential display (Liang and Pardee, 1992) and RNA fingerprinting by arbitrary primed PCR (**RAP-PCR**) (Welsh *et al.*, 1992). Both methods were introduced in 1992 and are based on PCR amplification of random subsets of genes from two or more RNA samples.

4.1.2 Expressed Sequence Tag Sequencing

Expressed sequence tag sequencing is the technological advance facilitating high-throughput sequencing led to the concept of expressed sequence tag (EST) libraries in the early 1990s (Adams *et al.*, 1991; Okubo *et al.*, 1992). The EST sequences are generated by randomly picking clones from a cDNA library and performing a single sequencing reaction to produce 300 to 500 bp of sequence per clone. Differences in gene expression may be identified by counting the number of times a particular sequence appears in EST libraries of genes from different sources. However, EST sequences are often generated from cDNA libraries that have been normalized to equalize the abundance of clones representing different transcripts (Soares *et al.*, 1994). The EST sequenced from normalized cDNA libraries may be compared to identify transcripts that are expressed in one library and absent in another, but accurate quantitative data describing the relative abundance of genes within a library can only be obtained from EST generated from non-normalized cDNA libraries. Despite this restriction, the power of discovering differences in gene expression by comparing the abundance of sequences in EST databases increases with growth of the databases.

4.1.3 Analysis of Gene Expression

Analysis of gene expression is the serial analysis of gene expression (SAGE), a technique for the analysis of gene expression that is essentially an accelerated version of EST sequencing (Velculescu *et al.*, 1995). The SAGE technique is based on the concept that short sequence tags are sufficient to identify a gene transcript, provided the tags represent a known location within the gene (Velculescu *et al.*, 1995). A SAGE tag is generated from an mRNA transcript by extracting a short sequence from a defined location. Typically, a SAGE tag includes nine bases of sequence downstream from the last endonuclease recognition site of a transcript. Multiple SAGE tags are ligated together in a cloning vector so that a typical sequencing reaction of 300 to 500 bp generates the sequences of 20 to 30 SAGE tags. Because each unique SAGE tag theoretically represents a unique mRNA transcript, an overview of all genes expressed in the original RNA sample is generated by SAGE tag sequences. Differences in gene expression among experimental samples can then be identified by comparing the relative abundance of specific SAGE tags in different libraries. Genes or EST sequences represented by the SAGE tags are identified by searching sequence databases to identify gene or EST sequences that contain the SAGE tag in the appropriate location. A thorough description of the bioinformatics involved in the analysis of SAGE data is provided by Lash *et al.* (2000). Since the original description of SAGE, modified protocols have been proposed to reduce the amount of mRNA required (Datson *et al.*, 1999; Peters *et al.*, 1999) and improve the efficiency of cloning tags (Powell, 1998; Kenzelmann and Muhlemann, 1999).

4.1.4 Microarray Hybridization

Microarray hybridization is the evaluation of gene expression using microarray technology that was originally described by Schena *et al.,* (1995) is proving to be an effective approach for a variety of applications. Although this technology has been largely inaccessible to animal scientists in the past, the resources needed to create livestock microarrays are being generated, and specialized equipment for making and reading arrays is becoming more accessible. Method to evaluate gene expression. Three general types of microarrays include oligonucleotide chips made by the synthesis of short oligos directly on a glass wafer oligonucleotide arrays made by spotting presynthesized oligos onto glass slides or nylon membranes, and cDNA arrays, made by spotting PCR amplified inserts of cDNA library clones onto glass slides or nylon membranes. The continued refinement of genomics techniques and their related bioinformatics will provide researchers with efficient tools to investigate gene expression and gain a better understanding of the genetic regulation of physiological processes important to the improvement of agriculture.

4.2 Applications of Genomics

The evaluation of gene expression is an effective way to identify genes important in the regulation of traits of economic importance in livestock production. Several techniques have been developed for this purpose, and important resources are currently in development that will facilitate continued research and discovery in this area. The application of these research methods in animal sciences will contribute to a greater understanding of the genetic regulation of traits important to livestock production. Application of nutritional genomics for biofortification of food staples has great potential to reduce hunger and malnutrition on the continent. To attain these benefits, increased educational and research efforts are needed by government agencies, charities and foundations for development, enhancement and/or adoption of genomics, computational biology and bioinformatics tools in agriculture, food and health-care delivery systems.

The molecular maps for legume crops of the semi-arid tropics (*i.e.,* chickpea, cowpea, groundnut and pigeonpea) are not very dcnsc compared to the other related legume crops (*i.e.,* soybean) but data from related species could be mined to develop maps and markers. For example, gene synteny research in *Vigna* species provides a means to detect orthologous genes by using common sets of chromosomes to probe their genomes (Fatokun *et al.,* 1992). Cowpea (*V. unguiculata*) and mung bean (*V. radiata*) have QTL for seed weight in the same conserved region, which are arranged in identical order (Menancio-Hautea *et al.,* 1992). This result was not surprising because it reflects the selection of chromosome regions controlling the most important character selected by farmers in these crop species: seed size. In cereal crops (*e.i.* maize and sorghum) similar findings for plant height were reported (Lin *et al.*, 1995; Pereira and Lee, 1995), which reveals same domestication evolution pattern among crop species (Paterson *et al.,* 1995). Hence, we envisage the successful integration of data mining, gene discovery, marker development and tactful gene deployment to become routine practice in crop improvement programmes in the next decades.

To fortify food staples to enhance levels of essential and non-essential micronutrients and macronutrients, such as vitamins (e.g., A, C, E, folate), minerals (e.g., iron and zinc) and proteins. Examples include cereals and sweet potato with enhanced levels of vitamins and/or proteins in their seeds and tubers, respectively (Beyer and Potrykus, 2001; Beyer *et al.*, 2002; Potrykus, 2001; Wu *et al.*, 2003; Ye *et al.*, 2000). This field of nutritional genomics takes advantage of the many genes that have been cloned for vitamin pathways and for the synthesis of many other "nonessential" compounds and macronutrients (Dellapena, 1999; 2001).

5. TRANSCRIPTOMICS

Transcriptomics deals with the analysis of gene expression pattern across a wide array of cellular responses, phenotype and conditions. The identification and characterization of differential gene expression from tissues subjected to stress has gained much attention in plant research. The recognition of elements involved in the response to a particular stress enhances the possibility of promoting crop improvement through direct genetic modification. However, the performance of some of the 'first generation' of transgenic plants with the incorporation of a single gene has not always been as expected. Transcriptomics studies often rely on partial reference transcriptomes that fail to capture the full catalogue of transcripts and their variations. Recent advances in sequencing technologies and assembly algorithms have facilitated the reconstruction of the entire transcriptome by deep RNA sequencing (RNA-seq), even without a reference genome. However, transcriptome assembly from billions of RNA-seq reads, which are often very short, poses a significant informatics challenge. The identification of candidate genes influencing any important trait can be approached through an analysis of complementary DNA (cDNA), copies of messenger RNA (mRNA). Devoid of intronic and intergenic sequences, whose biological significance is still obscure, these mRNAs represent only a small percentage of the total genome (about 1-3% in eukaryotes).

5.1 Technologies for Analysis

The analysis of transcripts expressed in different cells (transcriptomics) and tissues has been enormously facilitated by the development of Polymerase Chain Reaction (PCR). Several different strategies have been established to reveal expressed sequence tags (ESTs); their principles and applications are discussed below. In general, the identification of a DNA encoded product is based on the synthesis of a copy DNA followed by PCR amplification.

5.1.1 Differential Display

Differential display is a technique which is based on the synthesis of cDNA, using an oligo dT primer (3′) and an arbitrary oligonucleotide primer for the 5′-end (Liang and Pardee, 1992). This specific condition makes it difficult to conduct analysis using the available databases (Bachem *et al.*, 1996). In an alternative protocol, arbitrary primers are used both for cDNA synthesis and PCR amplification (Welsh *et al.*, 1992). Both variants of RNA fingerprinting require a low annealing temperature during PCR amplification in order to achieve visualizable products. Consequently, the quantity of individual amplification products is not only a function of the initial concentration of that cDNA, but also is dependent upon the quality of a particular match between primer and template (McClelland *et al.*, 1995). As a result, abundant cDNAs with poor matches to the primers used are likely to outperform rare species with perfect matches during the course of PCR amplification. Despite this disadvantage, there are several reports of the description of genes expressed differentially in plants. These include the cloning of: the last enzyme (1 aminocyclopropane-1-carboxylate oxidase) involved in the synthesis of ethylene, a hormone that plays a role in fruit ripening and senescence in tomato (*Barry et al.*, 1996); specific genes involved in the synthesis of flavonoids (Saito *et al.*, 1999; Yamazaki *et al.*, 1999); genes differentially expressed in carpel development (Yung *et al.*, 1999) or in floral transition (Yu and Goh, 2000); modulators such as nitrate transporters concerned with plant/environment interaction (Filleur and Daniel-Vedele, 1999); genes regulated by the light photoreceptor, phytochrome (Kuno *et al.*, 2000); and proteins involved in the response to abiotic stress (Brosche and Strid, 1999; Baldi *et al.*, 1999; Kim *et al.,* 2000). The technique has also been used to discriminate between individuals within a species (Lapopin *et al.*, 1999; Ni *et al.*, 2000).

5.1.2 cDNA-AFLP

cDNA-AFLP, this fingerprinting method relies on the selective amplification of a subset of DNA molecules from a more complex pool. The development of an amplified fragment length polymorphism (AFLP) provides a reliable tool to reveal the few differences between two close individuals (Vos *et al.*, 1995). The system is based on the use of high stringency conditions, facilitated by adding double-stranded adaptors on the ends of restriction fragments, which serve as primer sites during amplification. Selective fragment amplification is achieved by adding one or more bases to the PCR primers which will only then be successfully extended if the complementary sequence is present in the fragment flanking the restriction site, thereby reducing the number of visualized bands. The advantage of this strategy to reveal DNA fingerprinting has been presented in numerous reports more than 260 in the last five years (PubMed database). Because of the advantage of this technique in generating a good individual polymorphism, cDNA has been used as template to amplify transcripts in plants. The use of any pair, of four and six base restriction enzymes, has been proposed as a method to generate differentially expressed products. In fact, a combination of Pst I/Mse I or Ase I/Taq I has been successfully applied (Bachem *et al.*, 1996; Money *et al.,* 1996, Habu *et al.,* 1997). This method seems to be more efficient as a way to discriminate and identify a particular RNA fingerprint; the cloning and subsequent characterization of a specific band can easily be done, particularly if silver stain is used to develop the fingerprinting pattern (Dubos unpublished results). The description of several ESTs have been reported (Suarez *et al.*, 2000; Durrant *et al.*, 2000; Qin *et al.*, 2000; Bachem *et al.*, 2000). All these reports describe the modulation of biochemical pathways responding to different stimuli.

5.1.3 Microarray

Microarray, recently this DNA chip technology has been developed to study gene expression profiles (Schena *et al.,* 1995). Such techniques are based on the capacity to bind either DNA fragments or previously characterized oligonucleotides on a microscope slide. The plates are built by depositing specific DNA fragments at indexed positions using a computer controlled high speed robot with the specific DNA fragment sequences determined from any database originated including: cDNA clones, EST clones, anonymous genomic clones or DNA amplified from open reading frames (ORFs). Today's technology can display 409,000 spots in an area of 1.28 cm^2 (Fodor, 1997). In theory, all 20,000-25,000 *Arabidopsis* genes could be displayed on a single slide (Kehoe *et al.,* 1999). In fact, a slide with the *Arabidopsis* cDNA has been prepared for a large transcript analysis (Deprez *et al.,* 1998). The disposition of genes from the conifer *Pinus taeda* has also been prepared for use in physiological and molecular studies for wood formation. The technique is very sensitive, detecting mRNAs at level of 1/100,000 or 1/500,000 (Gerhold *et al.*, 1999). Despite the advantage of this precise technique, several problems have arisen and are expected to be resolved in the near future, including the high cost, identifying appropriate software for analysis of results, and standardizing methods to allow comparison between results from different labs in light of the fact that in some systems the slide can be used only once (for review, see Richmond and Somerville, 2000; Van Hal *et al.*, 2000). Studies of genetic expression patterns have been done principally in humans and yeast. The determination of differentially expressed genes in cancer cells has been the primary goal of array efforts ("Transcriptome 2000: From functional genomics to systems biology," a conference held in the Pasteur Institute, Paris). Results from *Saccharomyces cerevisiae* include a large study on the modulation of differentially expressed genes involved in carbohydrate metabolism; 6400 different genes were analyzed simultaneously on a slide with the data providing valuable conclusions about the method of regulation of anaerobic respiration (De Risi *et al.*, 1997). Recent examples of such technology applied to plants include those on *Arabidopsis* (Ruan *et al.*, 1998; Maleck *et al.,* 2000; Petersen *et al.,* 2000; Reymond *et al.*, 2000; Seki *et al.*, 2001), soybean and maize (McGonigle *et al.*, 2000), lima bean (Arimura *et al.*, 2000), and strawberry (Aharoni *et al.*, 2000).

5.1.4 Serial Analysis of Gene Expression (SAGE)

Serial analysis of gene expression (SAGE), among the various techniques used to assess transcript abundance, the most powerful is probably SAGE. This technique (Powell, 2000; Velculescu *et al.*, 2000), invented first to quantify gene expression in yeast (Velculescu *et al.*, 1995; 1997), comprises the production of a short 10-14 nucleotide tag, with each tag representing a unique transcript present in a cell (all possible combinations of 4 bases in a 10 bp sequence, 4^{10}, gives more than 1 million different sequences). Determination of the sequence of a tag allows identification of the corresponding gene (Chen *et al.*, 2000; Van den Berg *et al.*, 1999), and the frequency of a tag represents the steady state level of the mRNA from which it was derived. The unique advantages of SAGE over alternative techniques for transcript analysis are: high sensitivity, threshold detection is one transcript in three cells (Ishii *et al.*, 2000); scalability, the technique can be used on any size of sample, from a few cells upwards; detection of all genes, including those of unknown function; avoidance of amplification bias; the data are digital and not derived (e.g., from an analogue fluorescent signal); the data are immortal and can be used at any time in a comparative study; data sets generated by one lab can be related directly to those produced by another.

5.2 Applications of Transcriptomics

The plants genomics provided a platform for analyzing and understanding the genetic and molecular basis of all biological process in plants that are relevant to the species. The knowledge of genomics is valid for development of new plant diagnostic tools. The next challenge for the application of these technologies to mycobacterial research is the investigation of gene and protein expression in vivo.

6. PROTEOMICS

In the past decade, a technological revolution has been witnessed that has culminated in the sequencing of the human genome and the dawning of the post-genomic era. The terms proteomics and proteome were coined by Marc Williams and colleagues in the early 1990s and have since been adopted by the research community at large. The proteome refers to full complement of protein encoded by the genome. Biochemical studies of protein activity have historically focused on the analysis of single molecular species. The rapid pace of discovery of new gene products via large scale genomic and proteomic initiative has necessitated the development of alternative strategies to evaluate protein function. The challenge in recent years has been to develop high throughput approaches to facilitate systematic protein analysis in biological samples map functional interaction between proteins on a global scale and place them in a biological context. Every organism possesses one genome and multiple proteomes cells express genes which encodes protein with encompassing a wide variety of reversible and irreversible chemical reactions, which influence the diversity affinity, function, cellular abundance and transport of proteins. The proteome comprises all of this form which influences the diversity affinity, function, cellular abundance and transport of proteins. The proteome comprises all of these forms, which influence the diversity, affinity function, cellular abundance and transport of proteins.

Proteomics could lead to the discovery of novel markers for diseases and novel targets for drug development. However, there are many more potential applications in the different disciplines of life sciences, and this is reflected by the number of different organisms whose proteomes are currently being investigated. The most extensive database available is the yeast proteome database (YPD by Costanzo *et al.*, 2000) which currently contains more than 6100 proteins encoded by the 6800 genes of the yeast genome. The function of 56% of these proteins is known, and for 13% the function is predicted based on homology analysis. *A priori*, studying plant proteomes is not an easy task: for example, the genome of *Arabidopsis thaliana* (now being sequenced and expected to be completed this year) is 120 Mb and is estimated to contain 20,000 genes (Meinke *et al.*, 1998), which is about five times more than the genome of *E. coli*, but relatively small compared to the estimated number of 140,000 genes of the human genome. However, it is also estimated that the number of human proteins will approach 1,000,000, i.e., every gene encodes for seven proteins, such extended diversity being made possible by specific splicing and, especially, by a wide range of post-translational modifications (PTMs) of the encoded polypeptide chains. Extrapolating these data, the *Arabidopsis* proteome may contain 140,000 proteins, and hence the mapping of the proteins encoded by even a relatively small plant genome requires a technique with very high resolving power. However, the power of proteomics is not necessarily in the mapping of complete proteomes, but rather in the comparison of proteomes, aiming at the identification of differentially expressed proteins. For example, the proteomes of healthy and diseased cells can be compared in order to identify disease-related proteins. Similarly, the proteomes of resistant and susceptible plant tissues can be compared to identify resistance-related proteins, or the proteomes of red and white flowers can be compared to identify proteins involved in flower pigmentation. For separation of complex mixtures of proteins the most powerful technique available is two-dimensional polyacrylamide gel electrophoresis (2D-PAGE), developed in 1975 (O'Farrell, 1975). The proteins are separated according to their isoelectric point (pI; pH at which the net charge of a protein is zero) and then their molecular weight (Mw) by isoelectric focusing (IEF) and by sodium dodecylsulphate (SDS)-PAGE, respectively. This technique allows the separation of thousands of proteins in one experiment. The obtained peptide masses and sequences are used for database searches, allowing identification of known proteins (Wilkins *et al.*, 1998; Hoogland *et al.*, 1999). The amino acid sequence information, obtained by either MS or by the classical Edman degradation, is also of use for unknown proteins (Fig. 6).

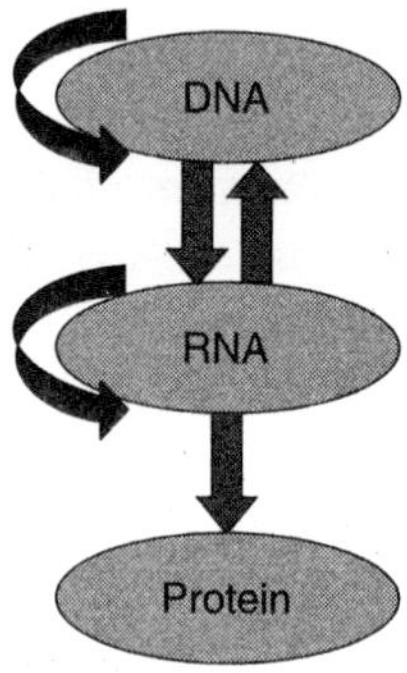

Fig. 6 Central dogma.

Proteomics, the study of the proteins encoded within the genome, is complementary to studies at the transcript level. In addition to describing the protein complement of the genome, the proteome can provide information on proteins expressed at defined points in time or under defined conditions, their location within the cell, and the existence of post-translational modifications. Expression proteomics is conventionally based on a combination of two-dimensional (2D) gel electrophoresis, image analysis, mass spectrometry (MS), and database searching. Protein identification by MS is commonly approached in a hierarchical manner, first utilizing matrix assisted laser desorption-ionization (MALDI) MS to provide a peptide mass fingerprint. This approach is often sufficient to identify proteins where the complete genome sequence is available. However, if identification is not achieved, techniques such as electrospray tandem mass spectrometry (ES-MS/MS) are used to generate peptide sequence data.

6.1 Technologies for Analysis

6.1.1 Electrospray Ionization

Electrospray ionization is a technique used in mass spectrometry to produce ions. It is especially useful in producing ions from macromolecules because it overcomes the propensity of these molecules to fragment when ionized.

6.1.2 Matrix-Assisted Laser Desorption/Ionization (MALDI)

Matrix-assisted laser desorption/ionization (MALDI), a soft ionization technique used in mass spectrometry, allowing the analysis of biomolecules (biopolymers such as proteins, peptides and sugars) and large organic molecules (such as polymers, dendrimers and other macromolecules), which tend to be fragile and fragment when ionized by more conventional ionization methods. A multi-organism, publicly accessible compendium of peptides identified in a large set of tandem mass spectrometry proteomics experiments (http://www.peptideatlas.org/).

6.1.3 Multiple Reaction Monitoring (MRM)

Multiple reaction monitoring (MRM) experiments, using a triple quadrupole instrument, are designed for obtaining the maximum sensitivity for detection of target compounds. This type of mass spectrometric experiment is widely used in detecting and quantifying drug and drug metabolites in the pharmaceutical industry.

6.1.4 FT-ICR Mass Spectrometry

FT-ICR mass spectrometry, Fourier transform ion cyclotron resonance mass spectrometry, also known as Fourier transform mass spectrometry, is a type of mass analyzer (or mass spectrometer) for determining the mass-to-charge ratio (m/z) of ions based on the cyclotron frequency of the ions in a fixed magnetic field. This source is about non-targeted mass spectrometry and targeted approaches which are important methods in the identification of proteins (an important step in proteomics).

6.2 Applications of Proteomics

Proteomics may be used for the elucidation of biosynthetic pathways leading to secondary metabolites. Many proteins contribute in cascades of reactions leading to these metabolites, in addition to the enzymes, transport and regulatory proteins are also involved. In the plant sciences 2D-PAGE has been used for many various applications, e.g., differential gene expression, hormone effects, mutant analysis, plant development, variety identification, identification of organ specific proteins, subcellular localization, plant-microbe interactions, plant defence studies, allergen identification, and identification of proteins in biosynthetic pathways. Proteomics provides a promising approach for studying secondary metabolism in plants and plant cells. Many of these metabolites are used as fine chemicals such as drugs, antioxidants, flavours, fragrances, dyes and insecticides, but unfortunately the yields of such commercially important compounds are generally low and purification is costly.

7. PRESENT VIEW

Development of new analytical techniques is largely focused on increasing resolution and comprehensiveness, increasing speed and throughput of analytical assays and equipment miniaturization. Examples of new promising technologies for metabolomics include micro-fluidic devices (Wang and Chen, 2003; Bilitewski *et al.*, 2003; Figeys *et al.*, 1998; Garcia *et al.*, 2003) UPLC (Bendahl *et al*, 2006; Castro *et al.*, 2005; Church *et al.*, 2005; Johnson and Plumb, 2005; Leandro *et al.*, 2006; O'Conner *et al.*, 2006; Plumb *et al.*, 2004; Sherma, 2005; Wilson *et al.*, 2005)], multidimensional chromatography with various detectors (i.e., GC_GC_TOF–MS(Dalluge *et al.*, 2003; Adabchour, 2005; Adahchour *et al.*, 2003) GC_GC_Quadropole MS (Adahchouretal, 2003)) or LC–LC–MS or LC–CE–MS (Evansetal,1952; Kammerer *et al.*, 2005; Mondello *et al.*,

2005; Sheldon, 2003; Straink *et al.*, 2005), LC–NMR–MS (Bobzin *et al.*, 2000; Burton *et al.*, 1997; Corcoranand Sprael, 2003; Keifer,2003; Fritsche *et al.*, 2002; Holt *et al.*, 1997; Spraul *et al.*, 2003; Yang, 2006) and metabolite arrays. To meet growing challenges in metabolomics, a concerted community-wide effort is required. Several recent community papers (Jnkins *et al.*, 2004; Bino *et al.*, 2004) as well as activities of the recently created Metabolomics Society are major steps in this direction and pave the way for further steady progress of the discipline.

Along the present review the gradient of knowledge existing about filamentous fungi Proteomics is clearly presented, which is perfectly correlated with the protein isolation difficulty. Thus, several studies have been carried out over the intracellular proteome that helped to the development of a huge number of procedures for protein extraction and analysis. Anyway, the Proteomics approach applied to the filamentous fungi is opening a new world of strain improvement, drug resistance analysis, antigen detection, biomarker discovery and environment applications, which was difficult to predict only ten years ago. Proteomics research has matured in plant research, allowing the quantification of thousands of proteins Baerenfaller *et al.*, (2008). Emerging metabolomics technologies such as Fourier transform mass spectrometry (FT-MS) allow the identification and quantification of many more metabolites than was previously possible (Giavalisco *et al.*, 2008).

The Principle of Substantial Equivalence has received comments from all types of stakeholders (producers, regulators, consumers, evaluators, etc.) (Millstone *et al.*, 1999), (Levidow and Murphy 2002), the basic idea behind the principle remains untouched.

8. CONCLUSION

Main focus of the article summarizes recent improvements as well as some principal limitations of shotgun tandem mass spectrometry based proteomics. Furthermore, it also briefly introduces steps of targeted driven quantitative proteomics. Metabolomics is a relatively young discipline, but it has already developed as an important integrative part of genomics and systems biology research. Further developments in this area require improvements in both analytical science and bioinformatics.

Exciting advances have been posted in proteomics. These advances have mainly been observed in the technical field with the development and improvement of technologies such as mass spectrometry, SELDI, and protein microarrays. However, these technologies are still limited in many areas including specificity and sensitivity of detection. In addition, data mining can also see significant attention for better accuracy of protein analysis. It is the integration of these technologies as well as the development of bioinformatics tools that can speed up the discovery of protein targets for therapeutics and lead to more accurate and safer drugs.

The concept of "proteome" more than 12 years ago, large-scale studies of protein expression, localization, activities and interactions have gained an exponential increase of interest, leading to extensive research and technology development. Proteomics is expansively applied in many areas, ranging from basic research, various disease and malignant tumors diagnostic and biomarker discovery to therapeutic applications. Several proteomics approaches have been developed for protein separation and identification, and for the characterization of protein function and structure. Tremendous progress has been made in the past few years in generating large-scale data sets for protein. Although DNA microarrays and proteomics alone will not bring new drugs and vaccines to fruition, they have become powerful tools in the discovery process. With continued technology development, their impact on driving research towards new therapies and vaccines for tuberculosis is likely to increase interactions, organelle composition, protein activity patterns and protein profiles in cancer patients. But further technological improvements, organization of international proteomics projects and open access to results are needed for proteomics to fulfill its potential.

Proteomics is a powerful tool, which has emerged as a result of summarized methodological and technical findings in mass spectrometry, protein visible and fluorescent staining, bidimensional electrophoresis, isobaric labelling, etc. The genome sequence facilities generated in the last years have supported this current trending topic that proteomes and subproteomes analyses represent. These analyses collect from technical updates to multiple strain comparison. As far as the protein isolation became more and more complex the numbers of published analyses diminished due to the complexity of the organelle isolation as well as hydrophobic membrane protein analysis.

When evaluating a new or GM crop variety, comparison with available data on the nearest comparator, as well as with similar varieties on the market, should form the initial part of the assessment procedure. The term 'substantial' has provoked interesting discussions, but has also led to misinterpretations. Therefore, the principle should be rephrased as the 'Comparative Safety Assessment (CSA)' approach. This phrase better outlines the comparative nature of the assessment, while avoiding the idea that it is a safety assessment in itself. Nutritional and toxicological assessment should be performed on the basis of the CSA, and might require additional safety tests.

Developments in Omics technologies will pose new challenges for Omics data visualization and interpretation. The modularity and flexibility offered by MapMan will be useful in keeping up with these changes allowing it to continue to provide novel ways to organize, evaluate and display data. It will also be adapted to more plants relying on increasingly sophisticated methods for knowledge transfer. Thus, MapMan will aid in data visualization, analysis and interpretation in the next 5 years of omics research.

References

Abbas, C. A., and Cheryan, M. (2002). Biorefinery opportunities. *Appl.Biochem. Biotechnol.* 98: 1147.

Achnine, L., Huhman, D. V., Sumner., L. W., Blount., J. W., and Dixon., R. A. (2005) Genomics-based selection and functional characterization of triterpene glycosyltransferases from the model legume *Medicago truncatula. Plant J.*, 41: 875-887.

Adams, M. D., Kelley, J. M., Gocayne, M., Dubnick, M H., Polymeropoulos, H., Xiao, C. R., Merril, A., Wu, B. O., and Moreno, R. F. (1991) *Complementary DNA sequencing: expressed sequence tags and human genome project.* Science (Wash DC).

Aharoni, A., Keizer, L.C., Bouwmeester, H.J., Sun, Z., Alvarez-Huerta, M., Verhoeven, H.A., Blaas, J., Van Houwelingen, A.M., Devos, R.C., Van Der Voet, H., Jansen, R.C., Guis, M., Mol, J., Davis, R.W., Schena, M., Van Tunen, A.J., and O'Connell, A.P.(2000) Identification of the SAAT gene involved in strawberry flavour biogenesis by use of DNA microarrays. *Plant Cell* 12: 647-662.

Arimura, G.I., Tashiro, K., Kuhara, S., Nishioka, T., Ozawa, R., and Takabayashi, J. (2000). Gene responses in bean leaves induced by herbivory and by herbivore-induced volatiles. *Biochem. Biophys. Res. Commun* 277: 305-310.

Aziz, N., Paiva, N. L., May, G. D. and Dixon,R.A. (2005). Profiling the transcriptome of alfalfa glandular trichomes. *Planta,* 221: 28-38.

Bachem, C.W, Oomen, R.J.F., Kuyt, S., Horvath, B.M., Claassens, M.M., Vreugdenhil, D. and Visser, R.G. (2000). Antisense suppression of a potato alpha-SNAP homologue leads to alterations in cellular development and assimilate distribution. *Plant Mol Biol.* 43: 473-82.

Bachem, C.W., Van Der Hoeven, R.S.,De Bruijn, S. M., Vreugdenhil, D., Zabeau, M., and Visser, R.G. (1996).Visualization of differential gene expression using a novel method of RNA fingerprinting based on AFLP: Analysis of gene expression during potato tuber development. *Plant J* 9: 745-753.

Baerenfaller, K., Grossmann, J., Grobei, M.A., Hull, R., Hirsch-Hoffmann, M., Yalovsky, S., Zimmermann, P., Grossniklaus, U.,Gruissem W., and Baginsky S.(2008).Genome-scale proteomics reveals *Arabidopsis thaliana* gene models and proteome dynamics. *Science* 320: 38-41.

Baldi, P., Grossi, M., Pecchioni, N., Vale, G.,and Cattivelli, L., (1999). High expression level of a gene coding for a chloroplastic amino acid selective channel protein is correlated to cold acclimation in cereals. *Plant Mol Biol* 41: 233-243.

Barry, C.S., Blume, B., Bouzayen, M., Cooper, W., Hamilton, A. J., and Grierson, D. (1996). Differential expression of the 1-aminocyclopropane-1-carboxylate oxidase gene family of tomato. *Plant J.* 9: 525-5.

Bobzin, S. C., Yang, S., and Kasten, T.P. (2000**).** LC-NMR: a new tool to expedite the dereplication and identification of natural products. *J Ind.Microbiol Biotechnol.* 25: 342-5.

Broeckling, C. D., Huhman, D. V., Farag, M., Smith, J. T., May, G. D., Mendes, P., Dixon, R. A., Sumner, L.W.(2005**).** Metabolic profiling of *Medicago truncatula* cell cultures reveals effects of biotic and abiotic elicitors on metabolism. *J. Exp. Bot. 56:* 323-336.

Brosche, M., and Strid, A. (1999).Cloning, expression, and molecular characterization of a small pea gene family regulated by low levels of ultraviolet B radiation and other stresses. *Plant Physiol* 121: 479-487.

Bungay, H. (1992). Product opportunities for biomass refining. *Enzyme Microb. Technol., 14:* 501-507.

Butler, D., and pockley, P., (2000). Monsanto makes rice genome public. *Nature* 404: 534.

Chen, F., Duran, A. L., Blount, J. W., Sumner, L. W., Dixon, R. A. (2003) Profiling phenolic metabolites in transgenic alfalfa modified in lignin biosynthesis. *Phytochemistry, 64*: 1013-1021.

Corcoran, O., and Spraul, M. (2003). LC-NMR-MS in drug discovery. *Drug Discov Today*; 8: 624-631.

Costanzo, M.C., Hogan J.D., Cusick, M.E., Davis B.P., Fancher, A.M.,Hodges, P.E., Kondu, P., Lengieza, C., Lew-Smith JE, Lingner,C., Roberg-Perez K.J., Tillberg, M., Brooks, J. E., Garrels, J.I., (2000). The yeast proteome database (YPD) and Caenorhabditis elegans proteome database (WormPD): comprehensive resources for the organisation and comparison of model organism protein information. *Nucleic Acids.Res* 28: 73-76.

Dalluge, J., Beens, J., and Brinkman, U.A. (2003). Comprehensive two dimensional gas chromatography: a powerful and versatile analytical tool. *J Chromatogra.* 1000: 69-108.

Datson,N.A., Jong, J. De, Vanden Berg, M. P., de Kloet, E. R., and Vreugdenhil, E. (1999). MicroSAGE: a modified procedure for serial analysis of gene expression in limited amounts of tissue. *Nucl.Acids Res.* 27: 1300-1307.

Davenport, R.J.(2001).Rice genome. Syngenta finishes, consortium goes on. *Science* 291: 807.

Davis, M. M., Cohen, D. I.,Nielsen, E. A., Steinmetz, M., Paul, W. E., and Hood. L. (1984). Cell-type-specific cDNA probes and the murine 1 region: the localization and orientation of Ad. *Proc. Natl. Acad. Sci.* USA 81: 2194-2198.

Deavours, B. E.,and Dixon, R. A. (2005). Metabolic engineering of isoflavonoid biosynthesis in alfalfa (*Medicago sati*V*a* L.). *Plant Physiol. 138*: 2245-2259.

Deprez, T., Amselem, J., Caboche, M., and Hofte, H. (1998).Differential gene expression in *Arabidopsis* monitored using cDNA arrays. *Plant J.* 14: 643-652.

Derisi, J., Iyer, V.R., and Brown P.O. (1997). Exploring the metabolic and genetic control of gene expression on a genomic scale. *Science.* 278: 680-686.

Dickson, D., and Cyranoski, D. (2001). Commercial sector scores success with whole rice genome. *Nature* 409: 551

Dixon, R. A. (2005). Engineering plant natural product pathways. *Curr.Opin. Plant Biol.*, *8*: 329-336.

Dixon, R. A., Achnine, L.,Deavours,B. E.,and Naoumkina,M. (2005). Metabolomics and gene identification in plant natural product pathways. In *Biotechnology in Agriculture and Forestry, "Plant Metabolomics"*; Saito, K., Dixon, R. A., Willmitzer, L., Eds.;Springer: Berlin, Germany, pp. 243-259.

Dixon, R. A., and Sumner, L. W. (2003) Legume natural products. Understanding and manipulating complex pathways for human and animal health. *Plant Physiol.* 131: 878-885.

Durrant, W.E., Rowland, O., Piedras, P., Hammond-Kosack, K.E., and Jones, J.D. (2000). cDNA-AFLP reveals a striking overlap in race specific resistance and wound response gene expression profiles. *Plant Cell* 12: 963-977.

Eckardt, N.A. (2000). Sequencing the rice genome. *Plant Cell* 12: 2011-2017.

Evans, C.R., and Jorgenson, J.,W. (2004). Multidimensional LC-LC and LC-CE for high-resolution separations of biological molecules. *Anal Bioanal Chem* 378: 1952-61.

Fatokun, C.A., Menancio-Hautea, D.I., Danesh, D., and Young, N.D. (1992). Evidence for orthologous seed weight genes in cowpea and mung bean based on RFLP mapping. *Genetics* 132: 841-846.

Fdoro, S. (1997). Massively parallel genomics. *Science.* 277: 393.

Fiehn, O. (2002). Metabolomics the link between genotypes and phenotypes. *Plant Mol. Biol. 48*, 155-171.

Filleur, S., and Daniel-Vedele F. (1999). Expression analysis of a high affinity nitrate transporter isolated from*Arabidopsis thaliana* by differential display. *Planta.* 207: 461-469.

Garcia, C.D., and Henry, C.S. (2003). Direct determination of carbohydrates,amino acids, and antibiotics by microchip electrophoresis with pulsed amperometric detection.*Anal Chem.* 75: 4778-83.

Giavalisco, P., Hummel, J., Lisec, J., Inostroza, A.C., Catchpole, G., and Willmitzer, L. (2008). High-resolution direct infusion-based mass spectrometry in combination with whole (13)c metabolome isotope labeling allows unambiguous assignment of chemical sum formulas. *Analytical Chemistry* 80: 9417-9425.

Gika, H.G., Theodoridis, G.A, Wingate, J.E., and Wilson, I.D., (2007)."Within-day reproducibility of an LC-MS-based method for metabonomic analysis: application to human urine". *J. Proteome Res.* **6** (8): 3291-303.

Gika, H.G., Theodoridis G.A., Wingate J.E., and Wilson I.D., (2007)."Within-day reproducibility of an LC-MS-based method for metabonomic analysis: application to human urine". *J. Proteome Res.* **6** (8): 3291-303.

Griffin, J.L.(2003). Metabonomics: NMR spectroscopy and pattern recognition analysis of body fluids and tissues for characterisation of xenobiotic toxicity and disease diagnosis.*Curr Opin Chem Biol* **7** (5): 648-654.

Gurskaya, N. G., Diatchenko, L. Chenchik, A., Siebert, P. D., Khaspekov, G. L, Lukyanov, K. A., Vagner, L. L., Ermolaeva, O. D., Lukyanov, S. A., and Sverdlov. E. D. (1996) Equalizing cDNA subtraction based on selective suppression of polymerase chain reaction: cloning of Jurkat cell transcripts induced by phytohemaglutinin and phorbol 12-myristate13-acetate..*Anal. Biochem.* 240: 90-97.

Habu, Y., Fukada-Tanaka, S., Hisatomi, Y., and Iida, S. (1997). Amplified restriction fragment length polymorphism based mRNA finger restriction enzyme that recognizes a 4-bp sequence. *Biochem Biophys Res Commun* 234: 516-521.

Hara, E., Kato, T., Nakada, S., Sekiya, S., and K. Oda. (1991). Subtractive cDNA cloning using oligo (dT)30-latex and PCR: isolation of cDNAclones specific to undifferentiated human embryonal carcinoma cells. *Nucl. Acids Res.* 19: 7097-7104.

Ishii, M., Hashimoto, S., Tsutsumi, S., Wada, Y., Matsushim,a K., Kodama, T., and Aburatani, H. (2000). Direct comparison of GeneChip and SAGE on the quantitative accuracy in transcript profiling analysis. *Genomics* 68: 136-143.

Jordan, K.W., Nordenstam, J., Lauwers, G.Y., Rothenberger, D.A, Alavi, K., Garwood, M., Cheng, L.L. (2009). "Metabolomic characterization of human rectal adenocarcinoma with intact tissue magnetic resonance spectroscopy". *Diseases of the Colon & Rectum* 52 (3): 520-525.

Keasling, D. J. (2010). Advanced Biofuel production in microbes. *Biotechnol. J.* 5: 147-162.

Kehoe, D., Villard, P., and Sommerville, S. (1999) DNA microarrays for studies of higher plants and other photosynthetic organisms. *Trends Plant Sci* 4:38-41.

Keifer, P.A., (2003) Flow NMR applications in combinatorial chemistry. *Curr Opin Chem Biol.* 7: 388-394.

Kenzelmann, M. and Muhlemann, K. (1999) Substantially enhanced cloning efficiency of SAGE (serial analysis of gene expression) by adding a heating step to the original protocol. *Nucl. Acids Res.* 27: 917-918.

Kim, C., Lee, S., Park, H., Bae,, C. Cheong., Y. Choi, Y., Han, C., Lee, S, Lim, C., and Cho, M. (2000) Identification of rice blast fungal elicitor responsive genes by differential display analysis. *Mol Plant Microbe Interact* 13: 470-474.

Koutinas, A. A., Wang, R., and Webb, C. (2004) Restructing upstream bioprocessing: technological and economic aspects for production of a generic microbial feedstock from wheat. *Biotechnol. Bioeng. 85: 524-538.*

Kuno, N., Muramatsu, T., Hamazato, F., and Furuya M (2000). Identification by large scale screening of phytochrome regulated genes in etiolated seedlings of *Arabidopsis* using a fluorescent differential display technique. *Plant Physiol* 122: 15-24.

Kutchan, T., and Dixon, R. A.(2005) Secondary metabolism: nature's chemical reservoir under deconvolution. *Curr. Opin. Plant Biol. 8*: 227-229.

Lapopin, L., Gianinazzi-Pearson, V., and Franken, P. (1999) Comparative differential RNA display analysis of arbuscular mycorrhiza in *Pisum sativum* wild type and a mutant defective in late stage development. *Plant Mol Biol* 41: 669-677.

Levidow, L. and Murphy, J. (2002) The decline of substantial equivalence: how civil society demoted a riskyconcept Proceedings of the Science and Citizenship in a Global Context: Challenges from New Technologies',Institute of Development Studies, University of Sussex, Brighton, (http://technology.open.ac.uk/cts/bpg.htm).

Liang, P., and A. B. Pardee.(1992). Differential display of eukaryotic messenger RNA by means of the polymerase chain reaction. *Science* (Wash DC) 257: 967-971.

Lin, Y.R., Schertz, K.F., and Paterson, A.H. (1995).Comparative analysis of QTLs afftecting plant height and maturity across the poaceae, in referecence to aninterspecific sorghum population. *Genetics* 141: 391-411.

Maleck, K., Levine, A., Eulgem, T., Morgan, A., Schmid, J., Lawton, K.A., Dang,l J.L., and Dietrich, R.A., (2000) The transcriptome of *Arabidopsis thaliana* during systemic acquired resistance. *Nat Genet* 26: 403-10.

Mcclelland, M., Mathieu-daude, F., and Welsh, J. (1995) RNA fingerprinting and differential display using arbitrary primed PCR. Trends Genet 11: 242-246.

Mcgonigle, B., Keeler S.J., Lau, S.M., Koeppe, M.K., and O'keefe, D.P., (2000) A genomics approach to the comprehensive analysis of the glutathione S-transferase gene family in soybean and maize. *Plant Physiol* 124: 1105-20

Mehrotra, B., and Mendes, P. (2006) Bioinformatics approaches to integrate metabolomics and other systems biology data. In *Biotechnology in Agriculture and Forestry, Vol. 57, Plant Metabolomics*; Saito, K., Dixon, R. A., Willmitzer, L., Eds.; Springer-Verlag: Berlin,Germany,; pp. 105-115.

Meinke, D.W., Cherry, J.M., Dean, C., Rounsley, S.D., and Koornneef, M.(1998). Arabidopsis thaliana: a model plant for genome analysis. *Science* 282: 662-682.

Menancio-Hautea,, D.I., Fatokun, C.A., Kumar, L., Danesh, D., and Young, N.D., (1992). Comparative genome analysi of mung bean (*Vigna radiata*) and cowpea (*V. unguic ulata*)using RFLP mapping data. *Theoretical and Applied Genetics* 86: 797-810.

Ni Z, Sun Q, Liu Z, Wu L and Wang, X. (2000) Identification of a hybrid specific expressed gene encoding novel RNA binding protein in wheat seedling leaves using differential display of mRNA. *Mol Gen Genet* 263: 934-938.

Northen, T.R, Yanes, O, Northen, M, Marrinucci, D, Uritboonthai, W, Apon, J., Golledge, S, Nordstrom, A., and Siuzdak, G., (2007). "Clathrate nanostructures for mass spectrometry". *Nature* 449 (7165): 1033-1036.

O'Farrell PH. (1975). High resolution two-dimensional electrophoresis of proteins. J Biol Chem 250: 4007-4021.

Okubo, K., N. Hori, R. Matoba, T. Niyama, A. Fukushima, Y. Kojima, and K.Matsubara. 1992. Large scale cDNA sequencing for analysis of quantitative and qualitative aspects of gene expression.Nat. Genet. 2: 173-179.

Paterson, A.H., Lin, Y.R., Li, Z., Schertz, K.F.,Doebley, J.F., Pinson, S.R.M.,Liu, S.C., Stansel, J.W., and Irvine, J.E. (1995). Convergent domestication of cereal crops by independent mutations at corresponding loci. *Science* 269: 1714-1718

Pereira, M.G., and Lee, M., (1995). Identification of genomic regions affecting plant height in sorghum and maize.Theoretical and Applied Genetics 90: 380-388.

Petersen, M., Brodersen, P., Naested, H., Andreasson, E., Lindhart, U., Johansen, B., Nielsen, H.B, Lacy, M., Austin, M.J, Parker, J. E., Sharma S.B., Klessig D.F, Martienssen, R., Mattsson, O, Jensen, A.B., Mundy, J. (2000)*Arabidopsis* map kinase 4 negatively regulates systemic acquired resistance. Cell 103: 1111-1120.

Powell, J., (2000) SAGE. The serial analysis of gene expression. Methods *Mol Biol* 99: 297-319.

Powell, J. (1998). Enhanced concatemer cloning-a modification to the SAGE (serial analysis of gene expression) technique. *Nucl. Acids Res.* 26:3445–3446.

Qin, L., Overmars, H., Helder, J., Popeijus, H., Van Der Voort, J.R., Groenink, W., Van Koert, P., Schots, A., Bakker, J., and Smant, G. (2000) An efficient cDNA-AFLP-based strategy for the identification

14

Transgenesis via Genetic Engineering: A Crop Improvemental Approach

Ashok Kumar*

Manager-EIA, En-Vision Enviro Engineers Pvt. Ltd. 208, G-Tower Shankeshwar Complex, Above Girish Group of Hospital, Sagrampura Surat-395002, Gujarat (India)

ABSTRACT

The genetic engineering offers many opportunities for agriculture and provides the means to address many of the limitations in crop improvement. The development of genetically engineered plants by the introduction and over-expression of selected genes seem to be a viable option to speed up the breeding of improved plants. Spontaneously, the genetic engineering is a faster way to insert beneficial genes than through conventional or molecular breeding. It is the only option when genes of interest originate from cross-barrier species, distant relatives or from non-plant sources. There are several traits whose correlative association with resistance has been tested in transgenic plants. Genetic engineering allows controlling the timing, tissue specificity and expression level of the introduced genes for their optimal function. The rapidly expanding information from genomics and genetics combined with improved genetic engineering technologies offer a wide range of possibilities for enhanced crop yield. The concept of utilizing a transgenic approach to host plant resistance was realized in the mid-1990 with the commercial introduction of transgenic maize, potato and cotton plants expressing genes encoding the insecticidal δ-endotoxin from *Bacillus thuringiensis.* Transgenic crops expressing foreign gene, which confer, added protection against insect pests and viruses could make a significant contribution to provide sustainable agricultural in the future. It is clear that biotechnology is paving the way to develop transgenic varieties with enhanced productivity as well as improved nutritional quality. The modern approaches therefore will hasten the process of agricultural development in our country. The predominant transgenic crops are maize (corn), soybean, cotton and potatoes. There are many field trials of transgenics for all major field crops to improve the productivity including the stress tolerance. This chapter summarizes recent data concerning to application of genetic engineering for transgenesis, transgenic approaches including stress tolerance and its significance for crop improvement concluding to biosafety of GM application.

Keywords: Biosafety, gene cloning, genetic engineering, transformation, transgenics.

**Corresponding author's E-mail: asokumr@gmail.com*

Roychowdhury R. (Ed.), Crop Improvement in the Era of Climate Change.

1. INTRODUCTION

Plants are the main source of micro- and macronutrients for most of the living systems. Their capacity to photosynthesize allows them to transform solar energy into chemical energy stored in organic and inorganic molecules and use it for their own necessities. For millions of years, humans have selected outstanding characteristics of crop plants like vigour, yield, flavour, resistance to pests and higher nutritional value. The development of genetic engineering offers the possibility of the improving the desired characteristics of crop plants through the direct introduction of specific gene with advantage of being faster and more selective. Genetic engineering is simply an additional tool for improving crop plants in combination with traditional breeding (Chavez-Barcenas *et al.*, 2000).The genetic engineering was first developed in 1973, when restriction enzyme was commercially produced. One of the most prominent developments has been the development of novel transgenic crop plant varieties. Many millions of hectares of commercially produced transgenic crops such as soybean, cotton, tobacco, potato and maize have been grown annually in a number of countries including USA (28.7 million hectares in 1999), Canada (4 million), China (0.3 million) and Argentina (6.7 million) (James, 1999). India made record increase in its adoption with almost tripling of area under Bt cotton from 1,300,000 ha in 2005. Genetic engineering offers exceptional opportunities to meet the growing needs of food and food security by enhancing productivity, profitability and environmental sustainability of farming systems. The rapidly increasing area being covered by genetically modified (GM) crops in India, China, Iran and Philippines signifies that such opportunities are being realized in the developing countries of Asia-Pacific. Recent advances in genetic engineering allowed greater molecular level understanding of many biochemical pathways, particularly for several model organisms and agricultural crops including rice and maize. Modification of pathways and products holds great potential for enhancing agriculture (Davis Lawrence, 2006).

The role of herbicides in agriculture entered a new era with the introduction of glyphosate resistant soybeans in 1995. Currently, the commercial area planted to transgenic crops is in excess of 90 million hectares with approximately 77% expressing herbicide tolerance, 15% expressing insect resistance genes and approximately 8% expressing both traits. Technologies of plant cell biology and molecular biology have attracted much attention because they provide a powerful and novel means to supplement as well as complement the traditional methods of crop improvement. The transgenic approach of plant genetic engineering provides access to an unlimited gene pool for the transfer of desirable genes between any two species of interest, irrespective of their evolutionary or taxonomic relation. The successful generation of transgenic insect resistant crop plants by transferring the insecticidal crystal protein (ICP) gene of *Bacillus thuringiensis* (Perlak *et al.*, 1991; Sutton *et al.*, 1992; Koziel *et al.*, 1993; Adang *et al.*, 1993; Wunn *et al.*, 1996) a Gram-positive, spore forming bacterium, is one of the major breakthroughs in crop science that the new genetics has provided. There are half a dozen different species of lepidopteran insects whose larvae turn off into the tissue of rice plants and resultant damage cause massive yield losses (Herdt *et al.*, 1991).

2. GENETIC ENGINEERING

The genetic engineering is the addition, deletion or manipulation of a single trait in an organism or a molecule to create a desired change. Genetic engineering is the creation of new DNA, usually by linking DNA from different organisms together by artificial means using enzymes known as restriction enzymes. Cloning is the production of many copies of the newly engineered DNA. The amplification of a specific cloned gene or genes, coupled with a marked increase in production of their protein products, makes it relatively easy to extract and purify these proteins in the laboratory. A suitable plasmid (vector) is selected in which to insert a desired gene (donor DNA). Both donor DNA and vector are digested with the same restriction enzyme and then incubated together with ligase to join the donor DNA fragments with the plasmid. The result is a recombinant plasmid that contains the desired DNA fragment. The recombinant plasmid is used to transform a host bacterial cell, creating a new genetic strain of bacteria that stably maintains the recombinant plasmid. The goal of cloning is to isolate a desired gene or segment of DNA from an organism and introduce it into a suitable host cell to obtain large quantities of the DNA. Frequently, this donor DNA is used for the large scale production of important proteins, but the DNA may also be used in the detection of infectious agents or abnormal cells. Normally, the donor DNA is a small portion of the genome of a cell and it is present as one or two copies in each cell. Therefore, before the donor DNA can be extracted, a sufficient number of cells containing the desired DNA must be obtained, either from a small segment of tissue or by culturing the cells. The cells must be disrupted and the genetic material either in chromosomes or in plasmids extracted (Kumar and Srivastava, 2012). Over the last 30 years, the field of agricultural biotechnology has developed rapidly due to the greater understanding of DNA as the chemical double helix code from which genes are made. Genetic engineering is one of the modern agricultural biotechnology tools that are based on recombinant DNA technology. The term genetic engineering, interchanged with terms such as gene technology, genetic modification or gene manipulation, is used to describe the process by which the genetic makeup of an organism can be altered using recombinant DNA technology. This involves using laboratory tools and specific enzymes to cut out, insert and alter pieces of DNA that contain one or more genes of interest. The ability to manipulate individual genes and to transfer genes between species that would not readily interbreed is what distinguishes genetic engineering from traditional plant breeding. With conventional plant breeding, there is little or no guarantee of obtaining any particular gene combination from the millions of crosses generated. Undesirable genes can be transferred along with desirable genes or while one desirable gene is gained, another is lost because the genes of both parents are mixed together and re-assorted more or less randomly in the offspring. These problems limit the improvements that plant breeders can achieve, eating time and funds along the way. Karenlampi *et al.*, (2000) reported the genetic engineering as a tool for improvement of plants for phytoremediation of metal polluted soils. The genetic engineering has tools to cut the DNA into pieces known as restriction enzymes.

(i) **Restriction Enzymes:** Restriction enzymes are enzymes that have the ability to cleave double-stranded DNA. The term restriction comes from the capacity of prokaryotes to defend against or restrict the possibility of takeover by foreign DNA that might gain entry into their cells. Prokaryotes degrade foreign DNA by using their unique restriction enzymes to chop it into relatively large but non-infective fragments. Restriction enzymes are classified into two: exonuclease and endonuclease. The exonucleases are not so important for use in genetic engineering because they are non-specific to restrict the dsDNA. The endonucleases are specific to restrict the dsDNA, they are classified further into three types, I, II or III. Types I and III require ATP to hydrolyze DNA and can also catalyze chemical modification of DNA through addition of methyl groups to specific bases. Type I restriction endonucleases cleave DNA randomly, while type II recognizes the specific nucleotide sequences within dsDNA and cut the DNA within the recognition sites (Kumar and Srivastava, 2012). The type III cut the DNA away from the recognition site, thus they are not much useful in genetic engineering.

(ii) Restriction Endonucleases: Restriction endonucleases are bacterial enzymes that recognize specific nucleotide sequences within a double-stranded DNA molecule and cleave at those locations. These enzymes cut DNA into fragments of various lengths, depending on the number of times the enzyme's recognition site is repeated within the molecule. Most restriction endonucleases used in cloning recognize base pair sequences four to eight nucleotides long and cut within these sequences. Many restriction enzymes have recognition sequences known as palindromes. Palindromes are sequences that are identical when read in the 5→3direction on both strands of the DNA molecule. Restriction endonucleases may cut the DNA to produce fragments with cohesive (sticky) ends or blunt ends (Kumar and Srivastava, 2012). Restriction endonuclease type II is widely used in genetic engineering to modify the genetic material.

3. GENETIC ENGINEERING FOR CROP PRODUCTION

Genetic engineering is used for crop production when the trait to be introduced is not present in the germplasm of the crop of crop or the trait is very difficult to improve by conventional breeding methods and also when it will take a very long time to introduce and improve such trait in the crop by conventional breeding methods. Modern plant breeding is a multidisciplinary and coordinated process where a large number of tools and elements of conventional breeding techniques, bioinformatics, biochemistry, molecular genetics, molecular biology and genetic engineering are utilized and integrated.

3.1 Trasgenic Crops

A transgenic crop is a genetically modified organism (GMO). Transgenic indicates that a transfer of genes has occurred using recombinant DNA technology (Chawla, 2000). Generally, a transgenic crop contains one or more genes that have been inserted artificially either from an unrelated plant or from different species altogether. Even before the creation of transgenics, the alteration of crops to improve their production was performed through selection. In fact, this selection has been going on for thousands of years and only in the past few centuries has it become a dedicated science onto itself. Simply put, to manipulate plants through selection takes many generations', i.e., large investments of time and does not always work. Through the use of transgenics, one can produce plants with desired traits and even increase yields to allow for more crops that last longer and withstand pests and disease (James, 2008). The idea being that we could produce the traits we want without throwing the dice as with selection techniques, allowing us to better feed a growing population and produce more desirable products. A transgenic crop plant contains a gene or genes which have been artificially inserted instead of the plant acquiring them through pollination. The inserted gene sequence known as transgene may come from another unrelated plant or from a completely different species, e.g., transgenic Bt corn which produces its own insecticide, contains a gene from a bacterium. Plants containing transgenes are called genetically modified or GM crops, although in reality all crops have been genetically modified from their original wild state by domestication, selection and controlled breeding over long periods of time (USDA, 2008).

3.2 Need of Transgenic Crop Plants

A plant breeder tries to assemble a combination of genes in a crop plant which will make it as useful and productive as possible. Depending on where and for what purpose the plant is grown, desirable genes may provide features such as higher yield or improved quality, pest or disease resistance or tolerance to heat, cold and drought. Combining the best genes in one plant is a long and difficult process, especially as traditional plant breeding has been limited to artificially crossing plants within the same species or with closely related species to bring different genes together, e.g., a gene for protein in soybean could not be transferred to a completely different crop such as corn using traditional techniques. Transgenic technology enables plant breeders to bring together in one plant useful genes from a wide range of living sources, not from within the crop species or from closely related plants. This technology provides the means for identifying and isolating genes controlling specific characteristics in one kind of organism and for moving copies of those genes into another quite different organism, which will then also have those characteristics. This powerful tool enables plant breeders to do what they have always done, generate more useful and productive crop varieties containing new combinations of genes, but it expands the possibilities beyond the limitations imposed by traditional cross pollination and selection techniques (Anonymous, 2004; Gupta, 2004).

Plants have been modified for many reasons over the years but the most common purpose is to produce the best products possible. Before recombinant DNA technology, it was possible to produce some improved products, although it was very difficult to do and many things could not be improved or changed. An improved product may involve changing the colour or the size of a plant, making it more appealing to buyer's eye. It may be more practical, allowing increased tolerance to cold, frost or drought; all making a crop easier to grow in a constantly changing environment (Gasser and Fraley, 1989). Goals like these are unattainable with traditional selection programmes and thus transgenics is now preferred when trying to come up with new products to improve sales. One of the hottest areas right now is the modification of crops to increase the resistance of plants to insects and diseases they may carry (Chawla, 2000). Improving resistance to these diseases and insects also reduces the need for herbicides and pesticides. This makes the plants safer for the consumer and allows the farmer to save money on chemicals. In conclusion, transgenic crops are an economically safer method of producing crop products, which makes them appealing and potentially profitable. For many years plant breeding entailed the selection of the finest plants to get the best crops. In those days, variation occurred through induced mutation or hybridization where two or more plants were crossed. Selection occurred through nature, using a selection of the fittest concept, where only the seeds best adapted to that environment succeeded, e.g., farmers selected only the biggest seeds with non-shattering seed heads, assuming these to be the best (Pighin Jamie, 2003; Dahiya, 2009).

3.3 Transgenesis

Transgenesis is the process of inserting a gene from one source into a living organism that would not normally contain it. The gene may be transferred from within the same species sometimes known as cisgenesis or come from another species entirely. Many GM organisms are transgenic; transgenesis involves a gene being transferred (transgene). Other ways to create GM organisms include knocking out genes and altering existing genes. The gene must be isolated and amplified, using gene cloning. The gene is then added into a vector designed to best introduce it into the target. To create a transgenic microorganism, the vector can be easily introduced using transformation (Ferber, 1999).

4. PRODUCTION OF TRANSGENIC CROP

The plant breeding involved the selection of the supreme plants to get the best crops. The variation occurred through induced mutation or hybridization where two or more plants were crossed. Selection occurred through nature, using a selection of the fittest concept, where only the seeds best adapted to that environment succeeded, e.g., farmers selected only the biggest seeds with non-shattering seed heads, assuming these to be the best. Now, we can not only select, but also create crops by inserting genes to make a seeds bare any trait desired. In order to make a transgenic crop, there are six main steps, i.e., isolation of DNA, cloning a gene of interest, designing the gene for plant infiltration, transformation, detection of inserted gene and finally plant breeding. For understanding the transgenesis, one must first know about DNA (deoxyribonucleic acids). DNA is the universal programming language of all cells and stores their genetic information. It contains thousands of genes, which are discrete segments of DNA that encode the information necessary to produce and assemble specific proteins. The main role of DNA molecules is the long term storage of information and DNA is compared to a set of blueprints, since it contains the instructions needed to construct other components of cells, such as proteins and RNA molecules. The DNA segments that carry this genetic information are called genes, but other DNA sequences have structural purposes or are involved in regulating the use of this (Kumar and Srivastava, 2012). All genes require specific regions in order to be utilized by a cell. These regions include:

(a) A promoter region, which signals where a gene begins and it used to express the gene;
(b) A termination sequence, which signals the end of a gene;
(c) The coding region, which contains the actual gene to be expressed.

All these regions together allow a gene to create a protein. Once a gene is transcribed into a protein, it can function as an enzyme to catalyze biochemical reactions or as a structural unit of a cell, both of which will contribute to the appearance of a particular trait in that organism. All species are capable of turning DNA into protein through a process known as translation. This capability makes it possible to artificially put genes from one organism into another. This process is known as transgenics. Genes that are determined to contribute to certain traits need to be obtained in a significant amount before they can be inserted into another organism. In order to obtain the DNA comprising a gene, DNA is first extracted from cells and put into a bacterial plasmid. A plasmid is a molecular biological tool that allows any segment of DNA put into a carrier cell and replicated to produce more of it. A bacterial cell, e.g., *E. coli* that contains a plasmid can put aside and used over and over again to produce copies of the gene by cloning the gene. The word cloning referred to production of identical copies of the original gene. Plasmids containing the gene can be used to modify the gene in any way the researcher sees fit, allowing novel effects on the gene trait to be produced. The nucleus of the plant cell is the target for the new transgenic DNA. There are many methods of doing this but the two most common methods include the Gene Gun and Agrobacterium method. The Gene Gun method, also known as the micro-projectile bombardment method, is most commonly used in species such as corn and rice. This procedure involves high velocity micro-projectiles to deliver DNA into living cells using a gun. It involves sticking DNA to small micro-projectiles and firing these into a cell. It enables scientists to transform organized tissue of plant species and has a universal delivery system common to many tissue types from many different species. It can give rise to undesired side effects, such as the gene of interest being rearranged upon entry or the target cell sustaining damage upon bombardment (Chawla, 2000).

The Agrobacterium method involves the use of soil dwelling bacteria known as *Agrobacterium tumefaciens*, which has the ability to infect plant cells with a piece of its DNA. The piece of DNA that infects a plant is integrated into a plants chromosome through a tumour inducing plasmid (Ti plasmid), which can take control of the plant's cellular machinery and use it to make many copies of its own bacterial DNA. The Ti plasmid is a large circular DNA particle that replicates independently of the bacterial chromosome (Chawla, 2000). The importance of this plasmid is that it contains regions of transfer DNA (tDNA), which can be transferred to a plant cell through a process known as a floral dip. A floral dip involves dipping flowering plants into a solution of Agrobacterium carrying the gene of interest, followed by the transgenic seeds being collected directly from the plant. Agrobacterium is capable of transferring large fragments of DNA very efficiently without substantial rearrangements, followed by maintaining high stability of the gene that was transferred (Chawla, 2000). The major steps in production of transgenic crop plants through the genetic engineering are following in detail:

(i) Isolation of Nucleic acid: The isolation of nucleic acid, either DNA or ribonucleic acid (RNA) is the first step in the genetic engineering process. It is a reliable method for isolating nuclear components from the cell. In any isolation procedure, the initial step is the disruption of the cell of the desired organism, which may be viral, bacterial or plant cells, in order to extract the nucleic acid. After a series of chemical and biochemical steps, the extracted nucleic acid can be precipitated to form thread-like pellets of DNA/RNA (Freifelder, 1986).

(ii) Gene Cloning: After genomic DNA isolation, the DNA is treated with specific enzymes called restriction enzymes cutting it into smaller fragments with defined ends to allow it to be cloned into bacterial vectors. Copies of the vector will anchorage many different inserts of the genome. These vectors are transformed into bacterial cells and thousands of copies are produced desired phenotype, the vector harboring the desired sequence is detected, selected, isolated and clones are produced. Restriction enzymes are utilized to determine if the desired gene insert is cloned completely and correctly (Ye *et al*., 2000).

Basic Steps of Gene cloning (Kumar and Srivastava, 2012)

(a) A suitable cloning vector is liberalized by cleavage with a suitable restriction enzyme.

(b) The cellular DNA containing the gene of interest is also cleaved into fragments with this enzyme.

(c) For various reasons the cellular DNA is cleaved in a way yielding N overlapping fragments with a uniform length of 20 kb.

(d) Linearized vector DNA and genomic DNA fragments are then ligated *in vitro*, yielding a population of N re-circularized cloning vector molecules, each of which contains an insert of cellular genomic DNA.

(e) These molecules are introduced into suitable host cells for multiplication.

(iii) Designing and Packaging of Gene: When the gene of interest has been cloned, it has to be linked to pieces of DNA that will control its expression inside the plant cell. These pieces of DNA will switch on (promoter) and off (terminator) the expression of the gene inserted. Gene designing/packaging can be done by replacing an existing promoter with a new one, incorporating a selectable marker gene and reporter gene, adding gene enhancer fragments, introns and organelle localizing sequences, among others. Promoters allow differential expression of genes, e.g., some promoters cause the inserted genes to be expressed all the time, in all parts of the plant (constitutive) whereas others allow expression only at certain stages of plant growth, in certain plant tissues or in response to external environmental signals. The amount of the gene product to be expressed is also controlled by the promoter. Some promoters are weak, whereas others are strong. Controlling the gene expression is an advantage in developing GM plants (Robinson, 2001).

Selectable marker genes are usually linked to the gene of interest to facilitate its detection once inside the plant tissues. This enables the selection of cells that have been successfully incorporated with the gene of interest, thus saving considerable expense and effort. Genetic engineers used antibiotic resistance and herbicide resistance marker genes to detect cells that contain the inserted gene. Cells that survive the addition of marker agents to the growth medium indicate the presence of the inserted gene. Although increase in antibiotic resistance in humans and animals is unlikely to occur using antibiotic resistance marker, genes coding for resistance to non-medically important antibiotics are preferred. Furthermore, alternative types of marker genes have been developed which are related to plant metabolism such as phosphor-mannose isomerase, xylose isomerase and others. Reporter genes are cloned into the vector in close proximity to the gene of interest, to facilitate the identification of transformed cells as well as to determine the correct expression of the inserted gene. Reporter genes that have been used include: the beta glucuronidase gene (gusA gene) which acts on a particular substrate producing a blue product, hence making the transformed cells blue; the green fluorescent protein (gfp) which allows transformed cells to glow under a green light and luciferase gene that allows cells to glow in the dark among others. Several genetic sequences can also be cloned in front of the promoter sequences (enhancers) or within the genetic sequence itself (introns) to promote gene expression. An example is the cloning of the cauliflower mosaic virus promoter enhancers in front of the plant promoter. Once the gene of interest is packaged together with the promoter, reporter and the marker gene, then it is introduced into a bacterium to allow for the creation of many copies of the gene package. The DNA isolated from the bacterial clones can be used for plant cell transformation using particle bombardment. If the use of bacteria *Agrobacterium tumefaciens* is preferred in the plant transformation, the whole gene package should be cloned in between two border sequences, i.e., left and right border of a binary vector. This will allow processing of the Agrobacterium so that only transfer DNA (T-DNA) will be incorporated into the plant genome (Lopez-Bucio *et al*., 2000).

(iv) **Transformation:** The introduction of gene into the plant cells is called transformation or gene insertion. The most common methods include biolistic transformation using the gene gun and Agrobacterium mediated transformation (Goto *et al*., 1999; Lemaux and Peggy, 2008).

(a) **Particle Bombardment (Gene Gun Method):** Particle bombardment is a mechanical method of introducing the desired gene. The gene gun was created as a new way of proceeding with gene transformation of grains in response to the difficulty of getting bacteria across grain cell walls. It was designed by John Sanford at Cornell University in 1987 and made introducing new genetic material into plant cells much easier than previous methods, such as the use of viruses or agrobacterium. The desired genetic sequence is cloned into a plant DNA vector and introduced into the plant using the gene gun or particle gun (Christou, and McCabe, 1992). As in the common gun, the gene gun uses minute particles of tungsten or gold as the bullet. These particles are coated with the DNA solution and fired to the plant cells through the force of the Helium gas inside a vacuum filled chamber. The DNA and the tungsten/gold particles get inside the cell and within 12 hours, the inserted DNA gets inside the nucleus and integrated with the plant DNA (Helenius *et al*., 2000). The tungsten/gold particles are sequestered to the vacuole and eliminated later. Transformed cells are cultured *in vitro* and induced to form small plants (regeneration) that express the inserted gene (de la Riva *et al*., 1998).

(b) ***Agrobacterium tumefaciens* Mediated Transformation:** The genes have moved from one organism to another, e.g., soil bacterium *Agrobacterium tumefaciens* has the natural ability to transfer the genes. It causes crown gall disease in a wide range of broad leaved plants such as apple, pear, peach, cherry, almond, raspberry and roses (Gelvin, 2003). The disease gains its name from the large tumour-like swellings (galls) that typically occur at the crown of the plant above soil level. The bacterium transfers part of its DNA to the plant and this DNA integrates into plants genome causing tumours and associated changes in plant metabolism. The genes that cause the galls are removed and replaced with genes coding for desirable traits. Plant cells infected with the bacterium will not form galls but produce cells containing the desired gene. When these cells cultured in a special medium will regenerate into plants and manifest the desired trait. If introduced gene is functional and gene product is synthesized, then the plant is said to be transformed. Once the inserted gene is stable, inherited and expressed in subsequent generations, the plant is considered a transgenic (Lemaux, 2009).

(v) **Detection of Inserted Genes:** Polymerase chain reaction (PCR) is a quick test to determine if the regenerated transgenic cells or plants contain the gene. It uses primers i.e., forward and backward primers, whose nucleotide sequences are based on the sequence of the inserted gene. The primers and single nucleotides are incubated with the single stranded genomic DNA and several cycles of DNA amplification is conducted in a PCR machine (Supli *et al.*, 2001). Analysis of the PCR products in agarose gel shows that the plants are really transformed, when DNA fragments equivalent in size with the inserted gene is present and amplified. Southern blot analysis determines the integrity of the inserted gene whether the gene is complete and not fragmented at the correct orientation and with one copy number (Anklam *et al.*, 2002). The DNA coding sequence is the probe binding to the single stranded genomic DNA of the transgenic plant which is implanted on a nitrocellulose paper (Ahmed, 2002). Autoradiography can reveal the transgenic status of the plant. Northern blot analysis determines whether the transcript or mRNA of the introduced DNA is present and correctly transcribed in the transgenic plant. The mRNA of transgenic plants are isolated and processed to bind to the nitrocellulose membrane. Labelled DNA is used to bind to the mRNA and can be visualized through autoradiography. Western blot analysis or protein immunoblotting is an analytical technique used to detect whether the transgenic plants produce the specific protein product of the introduced gene. Protein samples are extracted from the transgenic plants, processed into denatured proteins and transferred to a nitrocellulose membrane. The protein is detected using specific antibodies to the target protein (Trapmann, 2002).

(vi) **Backcross Breeding:** Genetic transformation is usually conducted in selected or commercial varieties which already possess the desired agronomic traits but lacks the important trait of the transgene. Thus, once successfully conducted, the genetically modified plant will be easily recommended for commercialization if it shows stability in several generations and upon successfully passing and fulfilling varietal registration requirements. In some cases, such as country restriction to grow for biosafety and sterility problem of transgenic plants, the conventional plant breeding is performed where the transgenic plant becomes the pollen source in the breeding programme and commercial varieties as the recurrent parent. Backcross breeding enables the combination of the desired traits of the recurrent parent and the transgenic line in the offspring. The length of time in developing transgenic plant depends upon the gene, crop species, available resources and regulatory approval. It varies from 6 to 15 years before a new transgenic plant or hybrid is ready for commercial release (Agbios, 2001; Acquaah, 2007; Vogel, 2009).

5. COMMERCIAL BT TRANSGENIC CROPS

Transgenic crops planted commercially are herbicide tolerant soybean, maize, canola, cotton; insect resistant maize and cotton and virus resistant squash and papaya. With the help of genetic engineering, more than one trait can be incorporated into the plant and are called stacked/combined traits, e.g., corn and cotton crops with both herbicide tolerant and insect resistant traits (Metahelix, 2007; USDA, 2008).

Agricultural crops suffer losses due to diseases and pests. Chemical and biological insecticides have been in use for a long time to minimize yield losses due to insect pests. The extensive use of DDT (dichloro-diphenyl-trichloroethane) in 1940 onwards as the first modern insecticide played a key role in protection of crops, forests and controlling insect vectors of human diseases. As a contact poison against several arthropods, it was effectively used to combat mosquitoes spreading malaria, typhus and other insect-borne diseases. The following move towards indiscriminate use of synthetic chemical insecticides led to the contamination of water and food sources, poisoning of non-target beneficial insects and development of insect populations resistant to the chemical insecticides. Increased public concerns about the potential adverse environmental effects associated with the heavy use of chemical insecticides prompted search of alternative methods for insect-pest control. One of the promising alternatives is the use of biological pesticide. The most common biological pesticides used include Cry proteins from *Bacillus thuringiensis* (Bt), which is a soil bacterium. Bt has a vast collection of cry genes. The encoded proteins vary widely with respect to their insect control efficacy and target insect specificity. It produces crystalline inclusions containing these proteins during sporulation. When ingested by insect larvae, these crystalline inclusions dissociate into monomers in the alkaline insect gut. The monomers further undergo proteolytic cleavage which results in an activated insecticidal protein. This protein binds to the larval gut lining and damages it inducing an anti-feeding behaviour in the larvae and eventually death. The cry protein binds to the larval gut lining at specific host-encoded receptors. One of the first genes to be commercially exploited for trait development in crops is the Bt cry. This was the advent of plant produced pesticides. A few genes like cry1Ac, cry1F, cry2Ab, have been effectively incorporated into plants for control of insect pests (Metahelix, 2007).

Farmers have adapted to green revolution technology characterized by the use of high yielding varieties (HYV), chemical fertilizers and pesticides. This technology has considerably increased the productivity, but has created problems of land degradation, pesticide residues in farm produces, gene erosion, environmental pollution, etc. Continuous use of HYVs without proper crop rotation has resulted in enhanced pest incidences. Insect-pest management in HYVs by extensive use of chemical pesticides has certainly brought considerable protection to crops over the past decades. The need of today is to produce more food and livelihood opportunities from less per capita arable land and irrigation water to feed the burgeoning population. Providing ample foodgrain at affordable price for the ever growing global population is only the first part of the challenge. The second and more important part is to produce this in a way that does not destroy the natural resources. Intensive agriculture, though provides sufficient foodgrains, yet treads heavily in the environment. An integrated crop management technology needs to be deployed to counteract the degradation of the agro-ecosystem due to the ongoing intensive agriculture, e.g., use of biological fertilizers, soil and water conservation practices, biodiversity conservation, integrated pest management, etc. If transgenic technology is integrated into the traditional system of crop husbandry, it holds great promise in augmenting agricultural production and productivity, while conserving biodiversity, natural resources and the environment for future generations (Kumar, 2001). The tools and techniques of molecular biology and biotechnology can facilitate harnessing this in crop plants in a safe and sustainable manner. There are an estimated 6,7000 pest species that damage agricultural crops, of which approximately 9000 species are insects and mites (Ross and Lembi, 1985). Insect-pests are the major cause of these losses. Pest associated losses in major crops varies from 52% in wheat, 58% in soybean, 59% in maize, 74% in potato, 83% in rice to 84% in cotton (Oerke, 1994). Insects not only cause direct losses to the agricultural production but also cause losses indirectly due to impaired quality of the produce and their role as vectors of various plant pathogens (Kumar, 2006). In addition to these losses, they increase cost of production by adding costs of pesticides and their application. Not only this, an extensive use of pesticides results in adverse effects on the beneficial organisms, pesticide residues in food and fodder and environmental pollution. People are becoming more health conscious and aware of the pesticide residues in food, fodder and water. There is urgent need of replacing agrochemicals having harmful effects on human health and environment with safer alternatives and adopting integrated pest management (IPM) practices to provide adequate crop protection for sustainable agriculture (Hazra *et al.*, 1998). It is in this context that the use of insect resistant Bt transgenic crops will increase productivity as well as provide benefits to farmers, consumers and environment. *B. thuringiensis* as biopesticide, commonly known as Bt, is a naturally occurring, Gram-positive, spore-forming soil bacterium. Bt has been known to be reservoir of several insecticidal proteins, such as δ-endotoxins, cytolytic proteins, vegetative insecticidal proteins, etc. Among these, δ-endotoxins have been more efficiently utilized for protection of a variety of crops from various insect-pests. A number of commercial Bt formulations have been used for a long time as effective biopesticides to food crops ornamentals, forest trees and stored grains (Meadows, 1993). Compared to chemical pesticides, Bt has certain advantages: it is highly toxic to insects and yet highly specific. Bt toxins are harmless to human beings, animals and a wide array of non-target pests. Therefore, it is ideally suited for incorporation into IPM programmes (Nester *et al.*, 2002). The biopesticide needs to be applied repeatedly and it is effective only against immature stages of target insects feeding on exposed plant surfaces. This limits its usefulness against insect-pests that feed inside the plants. These limitations can be removed by expression of Bt genes in transgenic plants (Krattiger, 1997).

Bt δ-endotoxins constitute a family of related proteins for which more than 150 genes have been identified (Crickmore *et al.*, 1998). The crystalline (Cry) proteins are inactive until they get solubilized by proteases (Tojo and Aizawa, 1983; Milne and Kaplan, 1993) in the insects midgut at high pH (>9.5), releasing proteins called δ-endotoxins. The toxicity is derived from the N-terminal half of the protein, which is composed of seven antiparallel alpha-helices. When insects ingest the crystalline inclusions, they are solubilized under alkaline conditions of insect midgut (Hofmann *et al.*, 1988a), releasing the δ-endotoxins which upon proteolytic activation exhibit a highly specific insecticidal activity. The activated toxin binds readily to specific receptors on the apical brush border of the midgut microvillae of susceptible insects (Hofmann *et al.*, 1988b). Then, α-helices penetrate the membrane to form an ion channel in the apical brush border membrane (Knowles and Dow, 1993). The formation of toxin-induced pores in the membrane allows rapid flux of ions (Sacchi *et al.*, 1986; Wolfersberger, 1989). This disruption of gut integrity leads to death of the insect due to starvation and/or septicaemia. The δ-endotoxins are highly insoluble in normal digestive conditions, therefore not harmful to mammals, birds and fish or to most of the beneficial insects. Moreover, mammals, including human, do not have δ-endotoxin receptors in their guts and all the δ-endotoxins tested. Therefore they are unstable when heated and degrade within 20 seconds by the mammalian digestive enzymes (EPA, 1998). Although, the Cry9C protein is stable under stomach conditions, it is not glycosylated and does not cause any adverse effects characteristic of immune response in mammalian system. Safety of Bt toxin in terms of toxicity and allergenicity towards mammals and other non-target organisms is well documented (Glare and O'Callaghan, 2000). Lack of receptors that bind to Bt toxins and instant degradation of Bt toxin in human digestive system makes them innocuous to human beings. Community exposure to Bt toxins/spray formulations over a period of six decades has not resulted in any adverse effects. Human volunteers consumed Bt toxin at very high concentration without any undesirable effects. Lack of homology to any allergenic protein/epitope sequence makes Bt toxin non-allergenic (Kumar, 2006).

Bt crystals referred as insecticidal crystal proteins (ICP), are protein crystals formed during sporulation in some Bt strains. Bt produces proteins that aggregate to form a crystal. These crystal proteins are toxic to very specific species of insects yet harmless to humans and the natural enemies of many crop pests (benenificial insects). There are more than 150 insects that are known to be susceptible in some way to *Bt*. The crystal proteins bind specifically to certain receptors in the insect's intestine. Not all insects carry the same receptors allowing for high species specificity. Humans and other vertebrates do not have these receptors in their bodies, so the toxin is unable to affect us. Expression of Bt gene in tobacco and tomato provided the first example of genetically engineered plants for insect resistance (Barton *et al.*, 1987; Vaeck *et al.*, 1987). Subsequently, several Bt genes have been expressed in transgenic plants, including tobacco, potato, tomato, cotton, brinjal, rice, etc. Delannay *et al.* (1980) reported for the first time field performance of transgenic tomato plants expressing δ-endotoxin gene. Though Cry1Ab protein was effective against tobacco hornworm, higher level of gene expression was needed for the control of tomato fruit worm (Helicoverpasp). The transgenic cotton plants were reported to provide excellent protection against lepidopteran insects throughout the season and resulted in significantly higher yields (Artim, 2003). Some of the potential benefits of transgenic insect resistant crops include reduced applications of broad spectrum insecticides, increased or protected yields due to season-long control of the target insect pest, protection of stored grain from insect pests and lower mycotoxin levels due to reduction in fungal pathogens associated with *Ostrinianubilalis* (Hubner) feeding on corn (Rice and Pilcher, 1998; Munkvold *et al.*, 1999; Giles *et al.*, 2000; Carpenter *et al.*, 2002).

The disadvantages of Bt transgenic crops may be:

(a) Potential impact on nontarget species;
(b) Increase in toxin levels in the soil may affect soil microflora;
(c) Exchange of genetic material between the transgenic crop and related plant species leading to the development of so called Super weed;
(d) Evolution of new and more virulent biotypes of the pests (Kumar, 2002).

5.1 Bt Cotton

Cotton and other monocultured crops require an intensive use of pesticides as various types of pests attack these crops causing extensive damage. Over the past 40 years, many pests have developed resistance to pesticides. So far, the only successful approach to engineering crops for insect tolerance has been the addition of Bt toxin. On March 26th 2002, Bt cotton was officially approved in India for commercial cultivation, in the six states belonging to southern and central cotton cultivation zones of the country (Barwale, 2002). Bt cotton, with the trade name Bollgard-I, developed by Mahyco-Monsanto Biotech Ltd is genetically engineered with *Bacillus thuringiensis* (Bt) gene with a protein that is toxic to the bollworm pest. The Bt toxin contained by the Bt crops is no different from other chemical pesticides, but causes much less damage to the environment. These toxins are effective against a variety of economically important crop pests but pose no hazard to non-target organisms like mammals and fish. Three major Bt crops are commercially now available corn, cotton and potato. The cotton is the most popular of the Bt crops, it was planted on about 1.8 million acres (7,28,437 ha) in 1996 and 1997 (Barwale *et al.*, 2004). The Bt gene was isolated and transferred from a bacterium *Bacillus thurigiensis* to American cotton. The American cotton was subsequently crossed with Indian cotton to introduce the gene into native varieties. The Bt cotton variety contains a foreign gene obtained from *Bacillus thuringiensis*. This bacterial gene protects the plants from bollworm (*A. lepidoptora*), a major pest of cotton. The worm feeding on the leaves of Bt cotton plant becomes lethargic and sleepy, thereby causing less damage to the plant. Field trials have shown that farmers who grew Bt variety obtained 25%–75% more cotton than those who grew the normal variety. Also, Bt cotton requires only two sprays of chemical pesticide against eight sprays for normal variety. According to the director general of the Indian Council of Agricultural Research, India uses about half of its pesticides on cotton to fight the bollworm menace (Bennett *et al.*, 2006). The number of Bt hybrids released for commercial cultivation till date has crossed 600 with more than 35 seed companies and public sector institutions currently engaged in their development. In addition, the first true breeding variety has also been released by Indian Council of Agricultural Research (ICAR). This provides an opportunity to the farmers to save their own seed without losing the efficacy of Bt gene. The area under Bt cotton reached 7.6 million hectares in 2008-09 constituting nearly 81% of the total cotton area in India. As a result, the production also reached 4.9 million tonnes. All these are indicators of the extraordinary impact and acceptance of Bt technology in cotton by the Indian farmers. The Ministry of Environment and Department of Biotechnology simplified the regulatory procedures leading to expeditious commercial release, especially of events with well established biosafety record (Karihaloo and Kumar, 2009).

5.2 Bt Brinjal

The cultivation of Bt brinjal is proposed in some countries across Asia, including India, where there is currently a moratorium on commercialization and the Philippines, where field trials are going on. The brinjal (*Solanum melongena*) is said to have originated in India and is known to have been cultivated for over 4000 years. Second only to the potato in terms of the total quantity produced, the brinjal continues to be an important domestic crop cultivated across the country accounting for 9% of total vegetable production and covering 8.14% of the land under vegetable cultivation. There are approximately 2500 varieties of brinjal in India of various shapes extending from oval or egg shaped to long or club shaped and colours ranging from white, yellow, green and purple to nearly black. Many popular commercial varieties of brinjal available today are derivatives of older varieties from India and China. The brinjal is low in calories and fats and contains mostly water, some protein, fibre and carbohydrates. It is also an excellent source of minerals and vitamins and is rich in water soluble sugars and amide proteins among other nutrients. The brinjal is a popular component of the Indian diet across the country. The brinjal is known to be consumed both cooked and raw. Some of the most well known brinjal dishes in India include the *begun bhaja* of Eastern India, the *guttivankayakura* of Andhra Pradesh, *katharikaikozhambu* of Tamil Nadu, *upperi* in Kerala, *vangi bath* in Karnataka, *wangyachabharit* in Maharashtra, *bharatu* in Gujarat and *bainganjhonga* in Bihar. There are many concerns with Bt brinjal, which has been engineered to be resistant to certain insect pests using *Bt* genes from the soil bacterium *Bacillus thuringiensis*. These concerns include food safety and possible effects on organisms other than the pest insect (non-target organisms), such as beneficial insects and butterflies. The Bt brinjal is a suite of transgenic brinjals created by inserting a crystal protein gene (*Cry1Ac*) from the soil bacterium *Bacillus thuringiensis* into the genome of various brinjal cultivars. The insertion of the gene, along with other genetic elements like promoters, terminators and an antibiotic resistance marker gene into the brinjal plant is accomplished using Agrobacterium mediated genetic transformation. The Bt brinjal has been developed to give resistance against lepidopteron insects, in particular the Brinjal Fruit and Shoot Borer (*Leucinodesorbonalis*) (FSB). Mahyco, an Indian seed company based in Jalna, Maharashtra has developed the Bt brinjal. The genetically modified brinjal event is termed Event EE1 and Mahyco have also applied for approval of two brinjal hybrids. The Event EE1 was introgressed by plant breeding into various local varieties by University of Agricultural Sciences, Dharwad and Tamil Nadu Agricultural University, Coimbatore. Some of the cultivars of brinjal include: Malpur local, Manjarigota, Kudachi local, Udupi local and Pabkavi local (Choudhary and Gaur, 2008). It was approved for commercialization in India in 2009, but after a public disagreement, Indian Government applied a suspension on its release. The first agreement to develop Bt Brinjal was signed in 2005 between India's leading seed company-Maharashtra Hybrid Seeds Company, better known as Mahyco and two agricultural universities—Universities of Agricultural Sciences (UAS) in Dharwad and Tamil Nadu Agricultural University (TNAU) in Coimbatore. An expert committee (EC-I) was set up in 2006 to examine the biosafety data presented by Mahyco. They concluded that while the current data demonstrated that Bt brinjal was safe and equivalent to its non-Bt counterpart, more studies were required to re-affirm the findings and further trials were needed to ascertain the benefits from Bt brinjal with respect to existing methods for pest management and pesticide reduction. They recommended that large scale trials be allowed to go ahead. In 2009, a second expert committee (EC-II) examined the data from these trials. They concluded that adequate safety tests had been performed, stating that the benefits of Bt brinjal event EE-I developed by M/s Mahyco far outweigh the perceived and projected risks and advised the Genetic Engineering Appraisal Committee (GEAC) to recommend commercialization of the Bt brinjal (Mahyco, 2009).

The GEAC cleared Bt brinjal for commercialization on 14 October 2009. Following concerns raised by some scientists, farmers and anti-GM activists the government of India officially announced on 9 February 2010 that it needed more time before releasing Bt brinjal, with Indian

Environment Minister Jairam Ramesh saying that there is no overriding urgency to introduce Bt brinjal in India (Ghosh *et al.*, 2003). On 17 February 2010, Jairam Ramesh reiterated that the Centre had only imposed a moratorium on the release of transgenic brinjal hybrid and not a permanent ban, saying that until we arrive at a political, scientific and societal consensus, this moratorium will remain. Companies with any seeds of Bt brinjal will have to register the details with the government and the National Bureau of Plant Genetic Resources (NBPGR) was made responsible for storage of all the Bt brinjal seeds in India. Independent testing labs are currently being set up (Jain, 2010). An irregularity was also brought to the notice of the Karnataka Biodiversity Board (KBB) by Environment Support Group (ESG), a charitable trust in Bengaluru, in February 2010. It found that agencies accessed at least 10 brinjal varieties from Karnataka and Tamil Nadu without seeking prior consent of the National Biodiversity Authority and state biodiversity boards. Mahyco became India's first commercial entity to be accused of bio-piracy, or misappropriation, of local germplasm (Sood, 2012).

5.3 Bt Corn

Bt-corn is a type of genetically modified organism, termed GMO. A GMO is a plant or animal that has been genetically modified through the addition of a small amount of genetic material from other organisms through molecular techniques. Currently, the GMOs on the market today have been given genetic traits to provide protection from pests, tolerance to pesticides, or improve its quality. Examples of GMO field crops include Bt-potatoes, Bt-corn, Bt-sweet corn, Roundup Ready soybeans, Roundup Ready Corn and Liberty Link corn. Genetically modified foods are foods derived from GMO crops, e.g., corn produced through biotechnology is being used in many familiar foods, including corn meal and tortilla chips. In addition, corn is used to make high fructose corn syrup, which is used as a sweetener in many foods such as soft drinks and baked goods. While the FDA regulates genetically modified foods, it considers Bt-corn to be nutritionally equivalent to traditional corn. To transform a plant into a GMO plant, the gene that produces a genetic trait of interest is identified and separated from the rest of the genetic material from a donor organism. Most organisms have thousands of genes; a single gene represents only a tiny fraction of the total genetic makeup of an organism. A donor organism may be a bacterium, fungus or even another plant. In the case of Bt corn, the donor organism is a naturally occurring soil bacterium, *Bacillus thuringiensis* and the gene of interest produces a protein that kills Lepidoptera larvae, in particular, European corn borer. This protein is called the Bt delta endotoxin. Growers use Bt corn as an alternative to spraying insecticides for control of European and southwestern corn borer (Witkowski *et al.*, 2002). The Bt δ-endotoxin is highly effective at controlling Lepidoptera larvae, caterpillars. It is during the larval stage when most of the damage by European corn borer occurs. The protein is very selective, generally not harming insects in other orders such as beetles, flies, bees and wasps. For this reason, GMOs that have Bt gene are compatible with biological control programmes because they harm insect predators and parasitoids much less than broad-spectrum insecticides. The Bt endotoxin is considered safe for humans, other mammals, fish, birds and the environment because of its selectivity. Bt has been available as a commercial microbial insecticide since 1960 and is sold under many trade names. These products have an excellent safety record and can be used on many crops until the day of harvest. To kill a susceptible insect, a part of the plant that contains Bt protein must be ingested. Within minutes, the protein binds to the gut wall and the insect stops feeding. Within hours, the gut wall breaks down and normal gut bacteria invade the body cavity. The insect dies of septicaemia as bacteria multiply in the blood. Even among Lepidoptera larvae, species differ in sensitivity to the Bt protein. To add a trait to a crop plant, the gene must be inserted along with some additional genetic material. This additional genetic material includes a promoter sequence that determines how the new trait is expressed in the plant, e.g., the promoter may cause the protein to be expressed in certain parts of the plants or only during a particular period of time. There is a marker gene that allows plant breeders to easily determine which plants have been transformed. Herbicide and antibiotic tolerance promoters are commonly used to identify transformed plants. There may also be a plasmid or vector sequence that allows for rapid multiplication of the gene of interest in a bacterial host prior to insertion in the crop plant (Griffiths *et al.*, 2002).

5.4 Bt Rice

Rice is the major staple crop for India and many other Asian countries. Yellow stem borer (YSB - *Scirpophagaincertulas*) is the major lepidopteran pest causing significant yield losses in rice. It is very difficult to control this pest using contact insecticides since it bores into the stem and feeds on tissue affecting the growth and productivity of the tiller. Early infestation results in dead hearts while later stage infestation leads to generation of white heads or chaffy panicles. Both the symptoms cause yield losses eventually. It is a vital requirement to develop transgenic rice varieties expressing appropriate proteins which could control the pest and prevent yield losses. The added benefit will be the saving in terms of cost and environmental protection due to reduced chemical insecticide sprays. Metahelix has developed rice plants expressing its proprietary, truncated cry1Ac based gene (Metahelix, 2007). These plants have been tested under containment conditions for their resistance to YSB. The unique proprietary promoter that has been used to drive the expression of the insecticidal gene facilitates high levels of expression in the culm region, through which the pest larvae bore into the stem. They are due to enter regulatory trials. Rice plant is a proprietary functional genomics platform using largescale Agrobacterium mediated transformation of rice, *Oryza sative* L. with the growing body of sequence information, there is also a parallel and compelling need to assign functions to predicted gene sequences through experimental validation (Anonymous, 1997).

5.5 Bt Soybeans

The herbicide resistant soybeans containing the active agent glyphosate was developed by US Company's Roundup Ready. Additionally, the new soybean variety, which was approved by CTNBio on August 19, is resistant to insect pests. The soy plants produce a substance called Bt protein that specifically targets a certain group of insects. The gene encoding for the Bt protein is also used in GM cotton plants. Bt cotton is widely grown in the US, India and China. According to Monsanto, the Bt protein is also effective against insects causing massive damage to soy plantations in Brazil. The cultivation of Bt Roundup Ready 2 Yield soybeans will presumably begin in two to five years from now. First of all, soy producers in Brazil are probably waiting for the approval of the soybeans as food and feed on the most important export markets. The respective application for approval in the EU was submitted in 2009. In the insect gut, the protein breaks down to release a toxin known as δ-endotoxin. This toxin binds to and creates pores in the intestinal lining, resulting in ion imbalance, paralysis of the digestive system and after a few days, insect death. Different versions of the Cry genes, also known as Bt genes, have been identified. They are effective against different orders of insects, or affect the insect gut in slightly different ways. Bt-based insecticides are considered safe for mammals and birds and safer for non-target insects than conventional products. What is new in Bt crops is that a modified version of the bacterial Cry gene has been incorporated into the plant's own DNA, so that the plant's cellular machinery produces the toxin. When the insect chomps on a leaf or bores into a stem of a Bt-containing plant, it ingests the toxin and will die within a few days (Lal and Lal, 1993; Nordlee *et al.*, 1996).

Table 1 Cry gene from *Bacillus* sp. and toxicity

S. No.	*Cry gene*	*Toxic to insect orders*
1.	CryIA(a), CryIA(b), CryIA(c)	Lepidoptera
2.	Cry1B, Cry1C, Cry1D	Lepidoptera
3.	CryII	Lepidoptera, Diptera
4.	CryIII	Coleoptera
5.	CryIV	Diptera
6.	CryV	Lepidoptera, Coleoptera

6. TRANSGENESIS FOR PLANT'S STRESS TOLERANCE

Initial attempts to develop transgenics for stress tolerance involved single action genes, i.e., genes responsible for modification of a single metabolite that would confer increased tolerance to salt or drought stress. Stress induced proteins with known functions such as water channel proteins, key enzymes for osmolyte (proline, betaine, sugars such as trehalose and polyamines) biosynthesis, detoxification enzymes and transport proteins were the initial targets of plant transformation. Indeed metabolic traits, especially pathways with relatively few enzymes, have been characterized genetically and appear more amenable to manipulations than structural and developmental traits. Many genes involved in stress response can be simultaneously regulated by a single gene encoding stress inducible transcription factor (Kasuga *et al.*, 1999), thus offering possibility of enhancing tolerance towards multiple stresses including drought, salinity and freezing. The genetic engineering allows controlling the timing, tissue specificity and expression level of the introduced genes for their optimal function. This is an important consideration if the action of a given gene or transcription factor is desired only at a specific time, in a specific organ or under specific conditions of stress. The basic findings on stress promoters have led to a major shift in the paradigm for genetically engineering stress tolerant crops in recent years (Katiyar-Agarwal *et al.*, 1999). The most widely used promoters in generating transgenic plants are constitutively expressed, i.e., they are turned on all the time and throughout the plant life cycle. In some cases, where the gene expression needs to be tailored to a specific organ or a specific time, such constitutive promoters may not be a suitable choice, especially for the stress induced genes because the constitutive expression of some stress induced genes may have serious deleterious effects on the plant. With an increasing number of stress genes becoming available and genetic transformation becoming more or less a routine procedure, characterization of stress induced promoters has taken a firm footing (Katiyar-Agarwal *et al.*, 1999).

In stress tolerant transgenic plants, many genes involved in the synthesis of osmoprotectants organic compounds such as amino acids, e.g., proline, quaternary and other amines, e.g., glycinebetaine and polyamines and a variety of sugars and sugar alcohols, e.g., mannitol, trehalose and galactinol that accumulate during osmoticadment have been used to date (Vinocur and Altman, 2005). Many crops lack the ability to synthesize the special osmoprotectants that are naturally accumulated by stress tolerant organisms. The osmoregulation is the best strategy for stress tolerance, especially if osmoregulatory genes could be triggered in response to drought, salinity and high temperature. Therefore, a widely adopted strategy has been to engineer certain osmolytes or by over-expressing such osmolytes in plants, as a potential route to breed stress tolerant crops. Various strategies are being pursued to genetically engineer osmoprotection in plants. The first step involved in obtaining stress tolerant transgenic plants has been to engineer genes that encode enzymes for the synthesis of selected osmolytes (Bray, 1993). In most of the aerobic organisms, there is a need to effectively eliminate reactive oxygen species (ROS) generated as a result of environmental stresses. Depending on the nature of the ROS, some are highly toxic and need to be rapidly detoxified. In order to control the level of ROS and protect the cells from oxidative injury, plants have developed a complex antioxidant defence system to scavenge the ROS. These antioxidant systems include various enzymes and non-enzymatic metabolites that may also play a significant role in ROS signaling in plants (Vranova *et al.*, 2002).

Many genes that respond to multiple stresses like dehydration and low temperature at the transcriptional level are also induced by abscisic acid (ABA), which protects the cell from dehydration (Mundy and Chua, 1988; Dure *et al.*, 1989; Skriver andMundy, 1990). In order to restore the cellular function and make plants more tolerant to stress, transferring a single gene encoding a single specific stress protein may not be sufficient to reach the required tolerance levels (Yamaguchi-Shinozaki *et al.*, 1994). Genes involved in stress signal sensing and a cascade of stress signalling in *A. thaliana* has been of recent research interest (Winicov and Bastola, 1997; Shinozaki andYamaguchi-Shinozaki, 1999). Components of signal transduction pathway may be shared by various stress factors such as drought, salt and cold (Shinozaki and Yamaguchi-Shinozaki, 1999). A number of signaling components are associated with the plant response to high temperature, freezing, drought and anaerobic stresses (Grover, 2001). The development of tolerant crops by genetic engineering requires the identification of key genetic determinants underlying stress tolerance in plants and introducing these genes into crops. Drought triggers a wide array of physiological responses in plants and affects the activity of a large number of genes. Gene expression experiments have identified several hundred genes which are either induced or repressed during drought (Sahi *et al.*, 2006). The conventional breeding for drought tolerance is a slow process that is limited by the availability of suitable genes for breeding. Some examples of conventional breeding programmes for drought tolerance are the development of rice, wheat and Indian mustard varieties tolerant to salt and to alkali soils by Central Soil

Salinity Research Institute, Karnal, India (CSSRI, 2001); the development of maize hybrids with increased drought tolerance (Bruce *et al.*, 2002), efforts to incorporate salt tolerance to wheat from wild related species (Colmer *et al.*, 2006) and the incorporation of drought tolerance as a selection trait in the generation of new maize and wheat germplasm by the International Maize and Wheat Improvement Center (Ribaut and Poland, 2004).

7. MAJOR CONCERNS ABOUT TRANSGENIC CROPS

There is public interest accompanying the technology of genetic manipulations. Different groups of the society are active in the field, spending an enormous amount of energy in different activities according to their roles, aims and perceptions (Domingo, 2000). An important consideration is ethical do we have the right or we are doing the proper thing when engaged in genetic manipulations (Carr and Levidow, 1997). There are concerns relating to the safety of transgenic food for human consumption (Ewen and Pusztai, 1999). Gene spread and resistance management is also major concerns (Gould, 1998). The first transgenic crops were commercially planted in 1995. By 2000, a total of 44.2 million ha of transgenic plants were grown in 13 countries (James, 2001). While this is only an 11% increase over the area in 1999, the spread of this technology has so far been very rapid. Most transgenic crops are grown in developed countries (76% of total area) and 24% in developing countries. Four countries, USA, Argentina, Canada and China grew 99% of the global total of transgenic crops in 2000. Most of this area is divided among four crops, soybean (58%), maize (23%), cotton (12%) and oilseed rape (7%). The area devoted to transgenic soybean and cotton increased since 1999, but that of maize and oilseed rape decreased (James, 2001). While a wide array of traits has been inserted into many species of plants, there are only little in commercial cultivation, mostly herbicide resistance and insect resistance (Davis Lawrence, 2006). The 74% of all transgenic crops in 2000 were herbicide resistant, 19% insect tolerant and a further 7% contained both these traits. Herbicide resistant soybean was the most widely planted transgenic crop (59%), with insect resistant maize (15%) at second (Shrivastava, 2011). Since 1999, the area devoted to herbicide tolerant plants has increased, while that of insect resistant (Bt-) crops has decreased. The global share of transgenic crops is already considerable; 36% of all soybean, 16% of cotton, 11% of oilseed rape and 7% of maize was transgenic in 2000 (James, 2001). The regulations required for commercialization and field growing of transgenic plants are diverse. The latest directive of the European Union requires an environmental impact assessment and post release monitoring, but specifies no methods. Generally, there is a lack of precise, justified legal framework that hinders assessment, potential commercial application and evaluation of this technology. When the industry developing transgenic plants has been required to assess the potential environmental consequences of their products, the results have not been impressive (Purrington and Bergelson, 1995).

The use of transgenic crops has been an issue for many years. A concern about what effect genetically modified material could have on human health, e.g., transgenic crops have been suggested to cause allergies in some people, although it is uncertain whether transgenic crops are the source of this reaction. Furthermore, the antibiotic resistance genes placed in these crops has been suggested to cause resistance to antibiotics leading to super bugs that cannot be killed with antibiotic treatments (Ferber, 1999). The idea of a population being uncomfortable with ingesting

DNA that originated from another source, such as a virus or bacteria, must also be considered when thinking about producing transgenic crops. However, there is no evidence of the DNA from transgenic crops being any different from the DNA ingested from conventional crops to date. A concern about whether transgenic crops cause damage to the natural environment. One example includes pollen from transgenic corn, which has been suggested to kill the Monarch butterfly larvae. It has been shown that hybrid corn expresses a bacterial toxin in its pollen, which is dispersed over 60 metres by wind. In this range, the corn pollen is deposited on other plants near cornfields where it can be ingested by non-target organisms including the monarch butterfly. These butterflies have been found to eat less, have a slower growth rate and higher death rate (Losey *et al.*, 1999). A second example is the hybridization of crops with nearby weeds. This could cause these weeds to attain resistance to herbicides or other things that we have been trying to avoid for many years. Genes that provide resistance to viral disease or other traits allowing them to survive in their environment could end up benefiting weed populations around a crop field. This trait could make that population more difficult to control. There has been little evidence to support this theory (Crawley *et al.*, 2001). On other side of the coin are the concepts that support the use of transgenic crops. The potential benefits of which are quite obvious, including such things as increased yields, decreasing the use of pesticides, i.e., to save the environment and the cost of pesticides and the production of novel crops such as providing crops with increased nutritional value (Ferber, 1999). It is very important to know the biochemical and physiological mechanisms of action, regulation of gene expression and safety of gene and gene product to be utilized (Tabien, 2000). Some other potential risk of using Bt crops are:

- **Invasiveness:** Genetic modifications, through traditional breeding or by genetic engineering can potentially change the organism to become invasive. Few introduced organisms become invasive, yet it is a concern for the users.
- **Resistance to Bt:** The biggest potential risk to using Bt-crops is resistance. Farmers have taken many steps to help prevent resistance.
- **Cross-contamination of genes:** Although unproven, genes from GM crops can potentially introduce the new genes to native species.

8. SIGNIFICANCE OF TRANSGENIC APPROACHES

Genetic engineering employs a very different method to produce improved crops. Instead of relying on sexual recombination to thoroughly stir the parental genes, genetic engineering preserves the integrity of the parental genotype, inserting only a small additional piece of information that controls a specific trait. This is done by splicing a well characterized chunk of foreign DNA containing a known gene into a chromosome of the host species using restriction enzymes. Restriction enzymes cut the long DNA strand that makes up a chromosome at very specific places and in a very repeatable way, so that foreign DNA fragments can be inserted and integrated into the host chromosome at the restriction site. There are many different restriction enzymes in use today, each recognizing specific, but different, sites in DNA molecules, providing great versatility in snipping out and inserting specific genes. Restriction enzymes are also employed in the sophisticated biochemical procedures that engineer the foreign gene, enabling the host organism to recognize the new information and use it at the proper time, in the proper cellular location and to the proper extent. The advantages of genetic engineering result mainly from the ability to circumvent the shortcomings of sexual reproduction. Hence, the gene pool is unbounded. Improvement affects only the targeted trait (no segregation), so there is less need for large populations and multiple generations of selection. The sterile and vegetatively propagated crops are as readily treatable by this approach as fertile crops. Likewise, the limitations of genetic engineering are complemented by the strengths of conventional methods, in that the new technology can usually target only simple, single gene traits; it is expensive and technically demanding; and it is regulated by government agencies (Manshardt, 2004; Tadele, 2009; Anonymous, 2012).

There are two common ways to transfer an engineered gene into a plant chromosome. *Agrobacterium tumefaciens*is a plant-pathogenic bacterium that has the ability to transfer a portion of its own genetic information into many plant species through a process called transformation, thereby causing the crown gall disease. This natural plant transformation agent has been modified by molecular geneticists in ways that enable it to move any engineered gene into host plants, without the associated disease symptoms. This method has the advantage of simplicity, but it is not well suited to transformation of the economically and nutritionally important cereal crops. Largely because of this limitation, a second method of plant transformation was invented that literally shoots the engineered genes into plant cells using tiny DNA coated tungsten or gold particles as fine as dust. Although somewhat more expensive in terms of equipment requirements, the gene gun approach has the advantage of unlimited range of applicability (Manshardt, 2004).

Both procedures are typically applied to minimally differentiated cells cultured in test tubes, rather than to organized tissues, because plants that regenerate from individual transformation events will consist entirely of genetically engineered cells. Neither transformation method is very efficient in terms of the percentage of cells that initially incorporate the engineered gene into a chromosome, so most of the plantlets regenerated via tissue culture lack the target gene entirely. In order to identify the rare successes, a selection system has been devised to eliminate all but the transformed plants. This is accomplished by including in the tissue culture medium an antibiotic that inhibits growth of typical plant tissues. Another genetically engineered gene linked to the inserted target gene detoxifies the antibiotic and allows transformed tissues to grow normally on the selective medium. The power to transfer traits across sexual barriers between species increases the potential for introduction of compounds that may have unsuspected secondary allergenic, toxic, or anti-nutritional properties. The potential for unintended side effects must be carefully evaluated in every new case. The Food and Drug Administration (FDA) has responsibility to evaluate the human health and safety risks of genetically engineered foods before and after commercialization, although the process is technically voluntary on the part of developers at the present time (Baker and Burnham, 2002; Manshardt, 2004; Pelletier David, 2006; Tadele 2009; Anonymous, 2012; American Radio Works, 2012).

9. BENEFITS OF TRANSGENIC CROP PLANTS

The genetic engineering has been used to produce a variety of crop plants to date, primarily with market led traits, some of which have become commercially successful. Developments resulting in commercially produced varieties in countries such as USA and Canada have centred on increasing shelf life of fruits and vegetables, conferring resistance to insect pests or viruses and producing tolerance to specific herbicides. While these traits have benefits for farmers, it has been difficult for the consumers to see any benefit other than, a decreased price owing to reduced cost and increased ease of production (Tribe, 2012; The California Right to Know Genetically Engineered Food Act. (2012). A possible exception is the development of transgenesis that delays ripening of fruit and vegetables, thus allowing an increased length of storage. Farmers would benefit from this development by increased flexibility in production and harvest. Consumers would benefit by the availability of fruits and vegetables such as transgenic tomatoes modified to soften much more slowly than traditional varieties, resulting in improved shelf life and decreased cost of production, higher quality and lower cost. It is possible that farmers in developing countries could benefit considerably from crops with delayed ripening or softening as this may allow them much greater flexibility in distribution than they have at present. In many cases small-scale farmers suffer heavy losses due to excessive or uncontrolled ripening or softening of fruit or vegetables. The real potential of genetic engineering to help address some of the most serious concerns of world agriculture has only recently begun to be explored (The California Right to Know Genetically Engineered Food Act. (2012).

9.1 Pest Resistance

There is clearly a benefit to farmers if transgenic plants are developed that are resistant to a specific pest, e.g., papaya ring spot virus resistant Papaya has been commercialized and grown in Hawaii since 1996 (Gonsalves, 1998). There may also be a benefit to the environment if the use of pesticides is reduced. Transgenic crops containing insect resistance genes from *Bacillus thuringiensis*have made it possible to reduce significantly the amount of insecticide applied on cotton in thc USA. The pest biotypes are different in various regions, e.g., insect resistant crops developed for use in the USA and Canada may be resistant to pests that are of no concern in developing countries and this is true both for transgenic plants and those developed by conventional breeding techniques. Even where the same genes for insect or herbicide resistance are useful in different regions, typically these genes will need to be introduced into locally adapted cultivars. There is need for more research on transgenic plants that have been made resistant to local pests to assess their sustainability in the face of increased selection pressures for ever more virulent pests (Daniell, 1999).

9.2 Tolerance to Biotic and Abiotic Stresses

The development of crops that have an inbuilt resistance to biotic and abiotic stress would help to stabilize annual production, e.g., Rice Yellow Mottle Virus (RYMV) devastates rice in Africa by destroying the majority of the crop directly, with a secondary effect on any surviving plants that makes them more susceptible to fungal infections. As a result this virus has seriously threatened rice production in Africa. Conventional approaches to the control of RYMV using traditional breeding methods have failed to introduce resistance from wild species to cultivated rice. Researchers have used a novel technique that mimics genetic immunization by creating transgenic rice plants that are resistant to RYMV (Pinto *et al.*, 1999). Resistant transgenic varieties are currently entering field trials to test the effectiveness of their resistance to RYMV. This could provide a solution to the threat of total crop failure in the sub-Saharan African rice growing regions. Numerous other examples could be given to illustrate the range of current scientific research including transgenic plants modified to combat papaya ring spot virus (Souza, 1999), blight resistant potatoes (Torres *et al.*, 1999) and rice bacterial leaf blight (Zhai *et al.*, 2000), or as an example of an abiotic stress, plants modified to overproduce citric acid in roots and provide better tolerance to aluminum in acid soil (de la Fuente *et al.*, 1997).

9.3 Improved Yield

The development of high yielding semi-dwarf wheat varieties is a major concern of green revolution. The genes responsible for height reduction were Japanese NORIN 10 genes introduced into western wheat in 1950 (Gibberellin insensitive dwarfing genes). These genes had two benefits; they produced a shorter, stronger plant that could respond to more fertilizer without collapsing and they increased yield directly by reducing cell elongation in the vegetative plant parts, thereby allowing the plant to invest more in the reproductive plant parts that are eaten. These genes have recently been isolated and demonstrated to act in exactly the same way when used to transform other crop plant species (Peng *et al.*, 1999). This dwarfing technique can potentially be used to increase productivity in any crop plant where the economic yield is in the reproductive rather than the vegetative parts.

9.4 Nutritional Benefits

Vitamin A deficiency causes half a million children to become partially or totally blind each year (Conway and Toenniessen, 1999). Traditional breeding methods have been unsuccessful in producing crops containing a high vitamin A concentration and most national authorities rely on expensive and complicated supplementation programmes to address the problem. Researchers have introduced three new genes into rice; two from daffodils and one from a microorganism. The transgenic rice exhibits an increased production of β-carotene as a precursor to vitamin A and the seed is yellow in colour (Ye *et al.*, 2000). Such yellow, or golden, rice may be a useful tool to help treat the problem of vitamin A deficiency in young children living in the tropics. Iron fortification is required because cereal grains are deficient in essential micro-nutrients such as iron. Iron deficiency causes anaemia in pregnant women and young children. About 400 million women of child bearing age suffer as a result and they are more prone to still born or underweight children and to mortality at childbirth. Anaemia has been identified as a contributing factor in over 20% of maternal deaths (after giving birth) in Asia and Africa (Conway, 1999). Transgenic rice with elevated levels of iron has been produced using genes involved in the production of an iron binding protein and in the production of an enzyme that facilitates iron availability in the human diet (Lucca, 1999; Goto *et al.*, 1999).

9.5 Reduced Environmental Impact

Water availability and efficient usage have become global issues. Soils subjected to extensive ploughing for controlling weeds and preparing seed beds are prone to erosion and there is a serious loss of water content. Low ploughing systems have been used for many years in traditional communities. There is a need to develop crops that thrive under such conditions, including the introduction of resistance to root diseases currently controlled by ploughing and to herbicides that can be used as a substitute for tillage (Cook, 2000).

9.6 Use of Marginalized Land

A vast landmass across the globe, both coastal as well as terrestrial has been marginalized because of excessive salinity and alkalinity. A salt tolerance gene from mangroves (*Avicennia marina*) has been identified, cloned and transferred to other plants. The transgenic plants were found to be tolerant to higher concentrations of salt. The gutD gene from *E. coli* has also been used to generate salt tolerant transgenic maize plants (Liu *et al.*, 1999). Such genes are a potential source for developing cropping systems for marginalized lands.

10. TRANSGENIC RESEARCH STATUS IN INDIA

Present R&D in crop biotechnology in India is focused on the biotechnology crops that can contribute to higher and more stable yields and enhanced nutrition. The current research and development activities in transgenic research in India have led to the generation of a number of transgenic plants harbouring genes of agronomic as well as quality traits. The transgenics in cotton (Bt-cotton hybrids), engineered for tolerance to insect damage have been already released for commercial cultivation in 2002. In the year 2002, first transgenic crop planting of Bt cotton was cultivated in about 50,000 ha in India. Currently, about 6.2 million hectare (2008) area is estimated to be under approved Bt cotton hybrids. The institutions under the ICAR, DST, DBT and CSIR have established a large number of facilities where most advanced transgenic research are being done (Table 2). In the identification of new genes relating to biotic and abiotic stress, quality trait, development of new drug delivery systems, diagnostics, recombinant vaccines, computational biology, and many other related areas, considerable success has been achieved.

Table 2 Major transgenic crops development activities

Crops/plants	*Transgene inserted*	*Source of transgene*	*Targeted traits*
Rice	1. *Cry1A (b)* and *Cry1A (c)* 2. *Xa-21,cry1A(b)* 3. Pyruvate decarboxylase and alcohol dehydrogenase 4. *Cod A* 5. Chitinase, glucanase and osmotin gene 6. *HVA1*	Bacteria Bacteria Plant *A.globiformis* Plant/virus Barley Plant	Generate plants resistant to yellow stem borer. Generate plants resistant to lepidopteron pests. Generate plants tolerant to flooding Tolerance against salt stress. Resistance against sheath blight and fungal infection
Mustard/ Rapeseed	1. Bar, Barnase and Barstar 2. *Osmotin* 3. *CodA* 4. Annexin gene 5. Chitinase, glucanase and RIP gene 6. Chickpea lectin gene	Plant Tobacco *A. globiformis* Arabidopsis Fungi/bacteria/ plant Plant	Generate herbicide tolerant plants, male sterile and restorer lines for hybrid seed production. Enhanced drought and salt tolerance. Tolerance against salt and water stress. Generate stress tolerant plants Generate plants resistant to Alternaria blight disease. Resistance against aphids

Contd...

Contd...

Tomato	1. *Cry1Ab, Cry1Ac* 2. ACC synthase gene 3. OXDC 4. Chitinaseandglucanase 5. Rep antisense	Bacteria Tomato Virus/plants Fungi/plant Plant	Generate plants resistant to tomato fruit borer. Delayed ripening. Generate plants resistant to wilt disease. Generate plants resistant to fungal diseases. Develop resistance to leaf curl virus
Cotton	1. Cry1Aa genes 2. Cry1E & Cry1C	Bacteria Bacteria	Generate plants resistant to cotton boll worm. Generate resistance to *Spodopteralitura* and *Heliothisarmigera.*
Chickpea	1. Bean α amylase inhibitor 2. Lectin	Plant Plant	Generate plants resistant to bruchids. Aphid resistance
Black gram	Coat protein and replicase genes	Virus	Develop virus resistant plants
Pigeonpea	1. Protease inhibitor genes 2. *Cry1 Aa3*	Cowpea Bacteria	Generate plants resistant to pod borer. Generate plants resistant to pod borer.
Wheat	*HVA1* *PIN2, Bar* *tAPx*and*TaCBF2* gene DREB 1a	Barley Plant Wheat Plant	Improved biomass under water deficit Resistance against biotic and abiotic stress Drought tolerance Drought tolerance
Tobbacco	1. *Cry1Ab and Cry1c* gene 2. Chitinase, glucanase and RIP genes	Bacteria Fungi/ plant	Generate plants resistant to insects-pests Generate resistance against brown eye leaf spot and Black shank disease
Potato	1. Osmotin gene 2. *Cry1Ab* 3. *Ama-1*	Tobacco Bacteria Virus/plant	Enhanced tolerance to drought & salt stress Resistance against potato tuber moth Develop nutritionally enriched plants
Brinjal	1. *Cry1 Ab* 2. Chitinase, glucanase and thaumatin encoding genes	Bacteria Fungi/bacteria/ Plant	Resistance against fruit and shoot borer Generate plants resistant to diseases
Cabbage and Cauliflower	*Cry1Ab*	Bacteria	Generate plants resistant to *Plutellasylestella*
Castor	1. *Cry1Ac* 2. Antisense ricin gene	Bacteria Plant	Develop resistance against semilooper Reduction of ricin and RCA endosperm in castor.

11. BIOSAFETY

The boisafety is necessary to protect human health and environment from the possible adverse effects of the products of genetic engineering. Biosafety is one of the issues addressed by the convention on biological diversity (CBD), which was finalized in Nairobi in May 1992 and come into force on 29 December 1993 (Anonymous, 2004). Now, the convention is the main international instrument for addressing biodiversity issues. It provides a comprehensive and holistic approach to the conservation of biological diversity, sustainable use of natural resources, fair and equitable sharing of benefits deriving from the use of genetic resources. In the second meeting of the convention, held in November 1995, an open ended ad hoc working group on biosafety was established to develop a draft protocol on biosafety, focusing specifically on transboundary movement of any genetically modified organism (GMO) resulting from modern biotechnology that may have adverse effect on the conservation and sustainable use of biological diversity. After several years of negotiations, the protocol, known as the Cartagena Protocol on biosafety, was finalized and adopted in Montreal on 29thJanuary 2000 in a meeting of the conference of the parties (Anonymous, 2000). The protocol created an enabling environment for the environmentally sound application of biotechnology, making it possible to derive maximum benefit from modern biotechnology, while minimizing the possible risks to the environment and to human health. The protocol promotes biosafety by establishing rules and procedures for the safe transfer, handling and use of GMO, with specific focus on transboundary movements of GMO. It features a set of procedures including one for GMO that are to be intentionally introduced into the environment and one for GMO that are intended to be used directly as food, fodder or processing. Parties to the protocol must ensure that GMOs are handled, packaged and transported under conditions of safety (Anonymous, 2000). The protocol has been signed by 103 countries. The Biosafety protocol was signed by India on 23rd January 2001. Subsequent to the cabinet's approval on 5th September, 2002, India has acceded the Biosafety protocol on 17th January 2003 (Kumar 2002).

12. FURTHER RESEARCH

The commercial transgenic crops have delivered benefits in crop production, but there are also a number of products in the pipeline which will make more direct contributions to food quality, clean environment, pharmaceutical production and livestock feeds, e.g., rice with higher levels of iron and beta carotene; long life banana that ripens faster on the tree and can be harvested earlier; maize with improved feed value; delayed ripening papaya; papaya ringspot virus resistant papaya; tomatoes with high levels of flavonols, which are powerful antioxidants; drought tolerant maize and wheat; maize with improved phosphorus availability; arsenic tolerant plants; insect resistant eggplant and rice; edible vaccines from fruit and vegetables; low lignin trees for paper making among others.

The transgenic crop research and development should focus on plants that will

(i) Improve production stability.

(ii) Give nutritional benefits to the consumer.

(iii) Reduce the environmental impacts of intensive and extensive agriculture.

(iv) Increase the availability of pharmaceuticals and vaccines.

(v) Developing protocols and regulations that ensure that transgenic crops designed for purposes other than food, such as pharmaceuticals, industrial chemicals, etc., do not spread or mix with either transgenic or non-transgenic food crops.

13. CONCLUSION

Agricultural crops suffer losses due to diseases and pests. Chemical and biological insecticides have been in use for a long time to minimize yield losses due to insect pests. The development of genetic engineering offers the possibility of improving the desired characteristics of crop plants through the direct introduction of specific gene with advantage of being faster and more selective. Genetic engineering employs a very different method to produce improved crops. It is an additional tool for improving crop plants in combination with traditional breeding. The rapidly increasing area being covered by genetically modified crops in India, China, Iran and Philippines signifies that such opportunities are being realized in the developing countries of Asia-Pacific. Recent advances in genetic engineering allowed greater molecular level understanding of many biochemical pathways, particularly for several model organisms and agricultural crops including rice and maize. Genetic engineering is the creation of new DNA, usually by linking DNA from different organisms together by artificial means using enzymes known as restriction enzymes. Cloning is the production of many copies of the newly engineered DNA. The amplification of a specific cloned gene or genes, coupled with a marked increase in production of their protein products, makes it relatively easy to extract and purify these proteins in the laboratory. Transgenic indicates that a transfer of genes has occurred using recombinant DNA technology. Plants have been modified for many reasons over the years but the most common purpose is to produce the best products possible. In order to make a transgenic crop, there are six main steps, i.e., isolation of DNA, cloning a gene of interest, designing the gene for plant infiltration, transformation, detection of inserted gene and finally plant breeding. Transgenic crops planted commercially are herbicide tolerant soybean, maize, canola, cotton; insect resistant maize and cotton and virus resistant squash and papaya. With the help of genetic engineering, more than one trait can be incorporated into the plant, e.g., corn and cotton crops with both herbicide tolerant and insect resistant traits. The genetic engineering preserves the integrity of the parental genotype by inserting only a small additional piece of information that controls a specific trait. This is done by splicing a well characterized chunk of foreign DNA containing a known gene into the chromosome of host species using restriction enzymes. Developments resulting in commercially produced varieties have centred on increasing shelf life of fruits and vegetables, conferring resistance to insect pests or viruses and producing tolerance to specific herbicides. Transgenic crops that need commercialization have to undergo and pass extensive safety trials with regard to potential for food toxicity, food allergenicity, cross pollination and effect on non-target beneficial organisms including biological control agents. Transgenic crops are being adopted rapidly at the global level, but only a few developing countries are growing them in significant quantities. The transgenesis is the most excellent approach to improve the crop plants and boost food security.

Acknowledgements

Author is grateful to K.P. Rathoure (Mrs.) for her technical support and valuable comments.

References

Acquaah George, (2007). Principles of Plant Genetics and Breeding. 1st Edi. Blackwell Publishing USA pp. 163

Adang, M. J., Brody, M. S., Cardineau, G., Eagan, N., Roush, R. T., Shewmaker, C. K., Jones, A., Oakes, J. V. and McBride, K. E., (1993) Plant Mol. Biol. 21: 1131-1145.

Agbios (2001). Principles and Practice of Environmental Safety Assessment of Transgenic Plants. Part No. AGBESAM-01-099B Agriculture and Biotechnology Strategies (Canada) Inc. Available online at http://www.lacbiosafety.org/wp-content/uploads/2011/09/principles-and-practice-of-environmental-safety-assessment-of-transgenic-plants-agbios1.pdf

Ahmed, F.E., (2002). Detection of genetically modified organisms in foods. Trends in Biotechnology, 20(5): 215-223.

American Radio Works, (2012). History of GMOs. Retrieved June 2012, from American Public Media: http://americanradioworks.publicradio.org/features/gmos_india/history.html

Anklam, E., Gadani, F., Heinze, P., Pijnenburg, H., EedeGVD., (2002). Analytical methods for detection and determination of genetically modified organisms in agricultural crops and plant-derived food products. European Food Research and Technology, 214(1): 3-26.

Anonymous, (1997). Rice environments In: Rice Almanac. 2nd ed. IRRI, Philippines. pp. 17

Anonymous, (2000). Cartagena Protocol on Biosafety to the Convention on Biological Diversity. Montreal, Canada.

Anonymous, (2004). Agricultural biotechnology: Safe and responsible use. Curr. Sci., 87: 425-426.

Anonymous, (2004). What Are Transgenic Plants? Available online at http://cls.casa.colostate.edu/transgeniccrops/what.html

Anonymous, (2012). Agricultural Biotechnology in India. Available online at http://loveable-eccentric.blogspot.in/2012/04/agricultural-biotechnology-in-india.html

Artim, L., (2003). Application for determination of non-regulated status for lepidopteran insect protected VIP3A cotton transformation event COT102. Submitted by Syngenta Seeds, Inc., Research Triangle Park, NC 27709-2257 to the Biotechnology Regulatory Services, Riverdale, MD

Baker, G. A. and Burnham, T. A., (2002). The market for genetically modified foods: consumer characteristics and policy implications. Inte F AgribusManag. Rev., 351-360.

Barton, K., H. Whitely and N.S. Yang, (1987). *Bacillus thuringiensis* δ-endotoxin in transgenic Nicotianatabacum provides resistance to lepidopteran insects. Plant Physiol., 85: 1103-1109

Barwale, B. R. (2002). GM cotton approved in India. Hindustan Times, 28.3.2002.

Barwale, R. B., Gadwal, V. R., Zehr, U. and Zehr, B., (2004). Prospects for Bt cotton technology in India.AgBioForum, 7: 23-26.

Bennett, R., Kambhampati, U., Morse, S. and Ismail, Y., (2006). Farm-level economic performance of genetically modified cotton in Maharashtra, India. Rev Agricl Eco., 28: 50-71.

Bohnert HJ, Nelson DF, Jenson RG, (1995) Adaptation to environmental stresses. Plant Cell 7: 1099-1111

Bray EA, (1993) Molecular responses to water deficit. Plant Physiol., 103: 1035-1040

Bruce W. B. Edmeades GO, Barker TC, (2002). Molecular and physiological approaches to maize improvement for drought tolerance. J. Exp. Bot. 53: 13-25.

Carpenter, J., A. Felsot, T. Goode, M. Hammig, D. Onstad and S. Sankula, (2002). Comparative environmental impacts of biotechnology-derived and traditional soybean, corn and cotton crops. Council for Agricultural Science and Technology. Ames, IA: CAST

Carr, S. and Levidow, L., (1997). How biotechnology regulation separates ethics form risk. Outlook on Agric. 26: 145-150.

Chavez-Barcenas T., Neto-Jacobo M.F., Simpson J., Herrera-Estrella L., (2000). Molecular Genetic Approach to improve the nutritional quality of sample food crops, In: Impact of Agriculture on Human health and Nutrition, Vol II EOLOSS, UNESCO.

Chawla HS., (2000). Introduction to Plant Biotechnology. Enfield, NH, USA: Science Publishers Inc.

Choudhary B, Gaur K, (2008). The development and regulation of Bt brinjal in India (Eggplant/ Aubergine). ISAAA Briefs: 42-43.

Christou Paul, McCabe Dennis. (1992). Particle Gun Transformation of Crop Plants Using Electric Discharge (ACCELL™ Technology). Agracetus Inc., Middleton, WI

Colmer T., D Flowers TJ, Munns R., (2006). Use of wild relatives to improve salt tolerance in wheat. J. Exp. Bot. 57: 1059-1078.

Conway, G., (1999). The doubly green revolution: food for all in the twenty-first century. Cornell University Press.

Conway, G., and G. Toenniessen, (1999). Feeding the world in the twenty-first century. *Nature* 402(6761) Suppl: C55-58.

Cook, R. J. (2000). Science-based risk assessment for the approval and use of plants in agricultural and other environments. Agricultural biotechnology and the poor: proceedings of an international conference, Washington, D.C., 21-22 October 1999, Editors G. J. Persley, and M. M. Lantin. Washington, D.C.: Consultative Group on International Agricultural Research.

Crawley MJ, Brown SL, Hails RS, Kohn DD, Rees M., (2001). Transgenic crops in natural habitats. Nature 409: 682-683.

Crickmore, N., D.R. Ziegler, J. Fietelson, E. Schnepf, J. Van Rie, D. Lereclus, J. Baum and D.H. Dean, (1998). Revision of the nomenclature for Bacillus thuringiensispesticidal crystal proteins. Microbiol. Mol. Biol. Rev., 62: 807-813

CSSRI (2001). http://plantstress.com/files/salt_karnal.htm

Dahiya Neetu, (2009). Biotechnology: A new Track on Fungal Resistant Crops In: Advances in Fungal Biotechnology, Ed. RaiMahendra, I.K. International Publishing House Pvt. Ltd. New Delhi pp. 57

Daniell, H., (1999). GM crops: Public Perception and Scientific solutions. *Trends in Plant Science* 4(12): 467-69.

Davis Lawrence C., (2006). Genetic engineering, ecosystem change and agriculture: an update, Biotech. MolBiol Rev., 1(3): 87-102

de la Fuente, J. M., V. Ramirez-Rodriguez, J. L. Cabrera-Ponce, and L. Herrera-Estrella. (1997). Aluminum tolerance in transgenic plants by alteration of citrate synthesis. *Science* 276(5318): 1566-68.

de la Riva, G.A., González-Cabrera Joel, Vázquez-Padrón Roberto, Ayra-PardoCamilo, (1998). Agrobacteriumtumefaciens: a natural tool for plant transformation. Electronic Journal of Biotechnology, 1(3): p. 118-133.

Delannay, X., B.J. LaVallee, R.K. Proksch, R.L. Fuchs, S.K. Sims, J.T. Greenplate, P.G. Marrone, R.B. Dodson, J.J. Augustine, J.G. Layton and D.A. Fischhoff, (1989). Field performance of transgenic tomato plants expressing *Bacillus thuringiensis*varkurstaki insect control protein. *Bio. Technol.*, 7: 1265-1269

Domingo, J.L., (2000). Health risks of GM foods: many opinions but few data. Science 288: 1748-1749.

Dure L III, Crouch M, Harada J, Ho T-HD, Mundy J., (1989). Common amino acid sequence domains among the LEA proteins of higher plants. Plant Mol Biol.,12: 475-486

EPA, (1998). Pesticide fact sheet: Bacillus thuringiensis subsptolworthi Cry9 protein and the genetic material necessary for its production in corn. US Environmental Protection Agency.

Ewen, S.W.B. and Pusztai, A., (1999). Effects of diets containing genetically modified potatoes expressing Galanthusnivalislectin on rat small intestine. Lancet 354: 1353-1354.

Ferber D. (1999). Risks and benefits: GM crops in the cross hairs. Science 286: 1662-6.

Freifelder, D., (1986). Molecular Biology. Second ed. Boston: Jones and Bartlett Publishers Inc.

Gasser CS, Fraley RT., (1989). Genetically engineered plants for crop improvement. Science 244(4910): 1293-1299.

Gelvin S. B., (2003). Agrobacterium-mediated plant transformation: the Biology behind the Gene-Jockeying Tool. MicroMolBiol Rev., 67(1): 16-37.

Ghosh, S. K., Laskar, N., Senapati, S. K., (2003). Estimation of loss in yield of brinjal due to pest complex under Terai region of West Bengal. Environ. Ecol. 21: 764-769.

Giles, K.L., R.L. Hellmich, C.T. Iverson and L.C. Lewis, (2000). Effects of transgenic *Bacillus thuringiensis*corn grain on *B. thuringiensis*-susceptible *Plodiainterpunctella* (Lepidoptera: Pyralidae). J. Econ. Entomol., 93: 1011-1016

Glare, T.R. and M. O'Callaghan, (2000). *Bacillus thuringiensis*: Biology, Ecology and Safety. John Wiley and Sons Ltd., Chichester, New York.

Gonsalves, D. (1998). Control of papaya ringspot virus in papaya: a case study. Ann Rev Phytopatho. 415-37.

Goto F., Yoshihara R., Shigemoto N., Toki. S. and Takaiwa, F. (1999). Iron fortification of rice seed by the soybean ferritin gene. Nat Biotech., 17: 282-286.

Gould, F., (1998). Sustainability of transgenic insecticidal cultivars: integrating pest genetics and ecology. Ann. Rev. Entomol. 43: 701-726.

Griffiths Kate, PartisLina, Croan David, Wang Nicky, Emslie Kerry R. (2002). Review of Technologies for Detecting Genetically Modified Materials in Commodities and Food, available online at http://www.daff.gov.au/__data/assets/pdf_file/0008/196982/AGAL_report.pdf

Grover A, Kapoor A, Satya Lakshmi O, Agrawal S, Sahi C, Katiyar-Agarwal S, Agarwal M, Dubey H., (2001). Understanding molecular alphabets of the plant abiotic stress responses. Curr Sci., 80: 206-216

Gupta, AK., (2009). Transgenic Rice- Review towards Food Security. Available online at http://agropedia.iitk.ac.in/content/transgenic-rice-review-ak-gupta-phd Last updated on 02/02/2009

Hazra, C.R., K.C. Pandey, N.C. Sinha (1998). Integrated pest management in forage crops. In: Integrated pest and disease management (Eds.: R.K. Upadhyay, K.G. Mukherji, B.P. Chamola and O.P. Dubey). A.P.H. Publishing House, New Delhi. pp. 490-513

HeleniusElina, Boije Maria, Niklander-Teeri Viola, Palva E. Tapio, HeeriTeemu U. (2000). Gene Delivery into Intact Plants Using the Helios Gene Gun. *Plant Molecular Biology Reporter*; 18: 287a-2871.

Herdt, R. W. (1991). Rice Biotechnology, eds. Khush, G. S. &Toenniessen, G. H. (CAB International, Wallingford, U.K.), pp. 19-54.

Hofmann, C., P. Luthy, R. Hutter and V. Pliska (1988a). Binding of the δ-endotoxin from Bacillus thuringiensis to brush-border membrane vesicles of the cabbage butterfly (Pierisbrassicae). Eur. J. Biochem., 173: 85-91

Hofmann, C., H. Vanderbruggen, H. Hofte, J. Van Rie, S. Jansens and H. Van Mellaert (1988b). Specificity of Bacillusthuringiensis δ-endotoxins is correlated with the presence of high-affinity binding sites in the brush border membrane of target insect midguts. Proc. Natl. Acad. Sci. USA, 85: 7844-7848

Jain Kalpana, (2010). Genetic Engineering Panel to enforce Bt brinjal ban. Business Standard. 18 February 2010

James C., (1999). Global status of transgenic crops in 1999. ISAAA: Ithaca, New York.

James C., (2008). Global status of commercialized biotech/GM crops: 2008. ISAAA Brief 39. International Service for the Acquisition of Agri-biotech Applications: Ithaca, NY.

James, C. (2001). Global Review of Commercialized Transgenic Crops: 2001. ISAAA Brief No. 24. ISAAA: Ithaca, NY.

Karenlampi S., Schat H., Vangronsveld J., Verkleij J.A.C., van der Lelie D., Mergeay M., Tervahauta A.I., (2000). Genetic engineering in the improvement of plants for phytoremediation of metal polluted soils, Environmental Pollution 107: 225-231

Karihaloo J. L. and Kumar P. A., (2009). Bt Cotton In India: A Status Report, 2nd Edi. Asia-Pacific Consortium on Agricultural Biotechnology (APCoAB), New Delhi, India. p. 56.

Kasuga, M., Liu, Q., Miura, S., Yamaguchi-Shinozaki, K., Shinozaki, K., (1999). Improving plant drought, salt andfreezing tolerance by gene transfer of a single stress-inducible transcription factor. Nat. Biotechnol., 17: 287-291.

Katiyar-Agarwal S, Agarwal M, Grover A., (1999). Emerging trends in agricultural biotechnology research: use of abiotic stress induced promoter to drive expression of a stress resistance gene in the transgenic system leads to high level stress tolerance associated with minimal negative effects on growth. Curr Sci., 77: 1577-1579

Knowles, B.H. and J.A.T. Dow (1993). The crystal endotoxin of Bacillus thuringiensis: Models for their mechanism of action on the insect gut. Bioassays, 15: 469

Koziel, M. G., Beland, G. L., Bowman, C., Carozzi, N. B., Crenshaw, R., Crossland, L., Dawson, J., Desai, N., Hill, M., Kadwell, S., Launis, K., Lewis, K., Maddox, D., McPherson, K., Meghji, M. R., Merlin, E., Rhodes, R., Warren, G. W., Wright, M. & Evola S. V. (1993). Field Performance of elite transgenic maize plants expressing an insecticidal protein derived from Bacillus thuringiensis. Bio/Technology 11: 194-200.

Krattiger, A.F. (1997). Insect resistance in crops: A case study of Bacillus thuringiensis (Bt) and its transfer to developing countries. ISAAA Briefs No. 2. ISAAA: Ithaca, New York

Kumar Ashok and Srivastava Meena (2012). Molecular Biology, In: A text book on Molecular Cytogenetics, 1[st]Edn., NarendraPublishing House, New Delhi India

Kumar, P.A. (2006). Bt crops. In: Extended summaries, Natl. Seminar Transg. Crops Indian Agricul.: Status, Risks and Acceptance. Hisar, India. pp. 1-3

Kumar, Suresh (2001). Transgenics and their biosafety concerns. In: Abstracts, National symposium on plant physiology and biochemistry in transgenic Era and Beyond, Kolkata, India. p. 7

Kumar, Suresh (2002). GM Vs green revolution: Bt cotton raises new questions. Civil Services Chronicle, 12: 28-30

Lal, R., and S. Lal. (1993). Genetic engineering of plants for crop improvement. CRC Press, Boca Raton, FL.

Lemaux, Peggy G. (2008). Genetically Engineered Plants and Foods: A Scientists Analysis of the Issues (Part 1). Ann Rev Plant Bio. 59: 771-812

Lemaux, Peggy G. (2009). Genetically Engineered Plants and Foods: A Scientist's Analysis of the Issues (Part 11). Annual Review of Plant Biology, 60: 511-559.

Liu, Y., G. Wang, J. Liu, X Peng, Y. Xie, J. Dai, S. Guo, and F. Zhang. (1999). Transfer of E. coli gutD gene into maize and regeneration of salt-tolerant transgenic plants. *Science in China. Series C, Life Sciences* 42(1): 90-95.

Lopez-Bucio, J., Martinez de la Vega, O., Guevara-Garcia, A., Herera-Estrella, L. (2000). Enhanced phosphorous uptake in transgenic tobacco plants that overproduce citrate. Nat Biotech., 18: 450-453.

Losey JE, Rayor LS, Carter ME. (1999). Transgenic pollen harms monarch larvae. Nature, 399(6733): 214.

Lucca, P. (1999). Genetic engineering approaches to improve the bioavailability and the level of iron in rice grains. General Meeting of the International Program on Rice Biotechnology, Phuket, Thailand, September 20-24, 1999.

Mahyco (2009). M/s Maharashtra Hybrid Seeds Company Ltd. (Mahyco) Mumbai University of Agricultural Sciences (UAS), Dharwad and Tamil Nadu Agricultural University (TNAU), Report of the Expert Committee (EC-II) on Bt Brinjal Event EE-1

Manshardt R. (2004).Crop Improvement by conventional breeding or Genetic Engineering: How different are they? Biotechnology BIO-5. Published by Cooperative Extension Services, Department of Agriculture, US. pp 1-3

Meadows, M.P. (1993). *Bacillus thuringiensis*in the environment: ecology and risk assessment. In: *Bacillus thuringiensis*, New York. pp. 193-220

Metahelix: Germinating Hopes (2007). Metahelix Life Sciences Limited, Bangalore Karnataka. http://www.meta-helix.com/main/pbt_cotton.html

Milne, R. and H. Kaplan (1993). Purification and characterization of a trypsin like digestive enzyme from spruce budworm (*Christoneurafumiferana*) responsible for the activation of δ-endotoxin from Bacillus thuringiensis. Insect Biochem. Mol. Biol., 23: 663-673

Mundy J, Chua N-H (1988) Abscisic acid and water-stress induce the expression of a novel rice gene. EMBO J., 7: 2279-2286

Munkvold, G.P., R.L. Hellmich and L.G. Rice (1999). Comparison of fumonisin concentration in kernels of transgenic Bt corn hybrids and nontransgenic hybrids. Plant Dis., 83: 130-138

Nester, E.W., I. Altosaar and G. Stotzky (2002). 100 years of Bacillus thuringiensis: A critical scientific assessment. American Academy of Microbiology Colloquium Report. Based on Colloquium, Ithaca

Nordlee, J.A. S.L. Taylor, J.A. Townsend, L.A. Thomas and R.K. Bush. (1996). Identification of a Brazil-nut allergen in transgenic soybeans. New England Journal of Medicine 334: 688-692.

Oerke, E.C., H.W. Dehne, F. Schonbeck and A. Weber (1994). Crop production and crop protection: Estimated losses in major food and cash crops. Elsevier, Amsterdam, The Netherlands.

Pelletier David L. (2006). FDA's regulation of genetically engineered foods: Scientific, legal and political dimensions. Food Policy, 31: 570–591

Peng, J., D. E. Richards, N. M. Hartley, G. P. Murphy, K. M. Devos, J. E. Flintham, J. Beales, L. J. Fish, A. J. Worland, F. Pelica, D. Sudhakar, P. Christou, J. W. Snape, M. D. Gale, and N. P. Harberd. (1999). 'Green revolution' genes encode mutant gibberellin response modulators. *Nature,* 400(6741): 256-261.

Perlak F. J., Fuchs R. L., Dean D. A., McPherson S. L., Fischhoff D. A., (1991). Modification of the coding sequence enhances plant expression of insect control protein genes. Proc. Natl. Acad. Sci. 88(8): 3324–3328.

Pighin Jamie (2003). Transgenic Crops: How Genetics is providing New Ways to Envision Agriculture, The Science Creative Quarterly. Available online ta http://www.scq.ubc.ca/transgenic-crops-how-genetics-is-providing-new-ways-to-envision-agriculture/

Pinto, Y. M., R. A. Kok, and D. C. Baulcombe, (1999). Resistance to rice yellow mottle virus (RYMV) in cultivated African rice varieties containing RYMV transgenes. *Nature Biotechnology* 17(7): 702-707.

Purrington, C.B., and Bergelson, J. (1995). Assessing weediness of trangenic crops: industry plays plant ecologist. Trends Ecol. Evol., 10: 340-342.

Ribaut J-M., Poland D. (2004). http://www.cimmyt.org/english/docs/proceedings/molecApproaches/pdfs/MolecularApproaches.p

Rice, M.E. and C.D. Pilcher (1998). Potential benefits and limitations of transgenic Btcorn for management of the European corn borer (Lepidoptera: Crambidae). Am. Entomol., 44: 74-78

Robinson C. (2001). Genetic modification technology and food: Consumer health and safety. ILSI Europe Concise Monograph Series. ILSI Press Washington, DC.

Ross, M.A. and C.A. Lembi (1985). Applied weed science. Burgess Publishing Co., Minneapolis. p. 340

Sacchi V.F., Parenti P., Giordana B., Hanozet G.M., Luthy P. and Wolfersberger M.G. (1986). *Bacillusthuringiensis* inhibits K+ gradient dependent amino acid transport across the brush border membrane of Pierisbrassicaemidgut cells. FEBS Lett., 204: 213-218

Sahi C., Singh, A., Kumar, K., Bumwald, E. and Grover, A. (2006). Salt stress response in rice: genetics, molecular biology, and comparative genomics. Functional & Integrative Genomic, DOI 10.1007/s10142-006-0032-5

Shinozaki K, Yamaguchi-Shinozaki K (1999) Molecular responses to drought stress. In: Shinozaki K, Yamaguchi-Shinozaki K (eds) Molecular responses to cold, drought, heat and salt stress in higher plants. R.G. Landes Co., Austin, pp. 11-28

Shrivastava Umesh Prasad (2011). Transgenic Plants: Review on Risks and Benefits in the Environment of Developing Countries, Academic Voices: A Multidisciplinary Journal, 1(1): 84-93

Skriver K, Mundy J (1990). Gene expression in response to abscisic acid and osmotic stress. Plant Cell 5: 503-512

SoodJyotika, (2012). Whose germplasm is it? NBA struggles to frame case against Mahyco in first biopiracy case, Cover story, Down to Earth. Oct 15, 2012, Available on http://www.downtoearth.org.in/content/whose-germplasm-it

Souza, M. T., Jr. (1999). Analysis of the resistance in genetically engineered papaya against papaya ringspotpotyvirus, partial characterization of the PRSV.Brazil.Bahia isolate and development of transgenic papaya for Brazil. Thesis. Cornell University.

Stein, A.J. and Rodríguez-Cerezo E. (2009). The global pipeline of new GM crops: introduction to the database. JRC Technical Note EUR 23810 EN. Luxembourg: Office for Official Publications of the European Communities. http://ipts.jrc.ec.europa.eu/publications/pub.cfm?id=2199

Supli K, Gryson N, Messens K, De Loose M, Dewettinck K. (2001). Extraction and PCR analysis of soy DNA in chocolate. Mededelingen- FaculteitLandbouwkundige en ToegepasteBiologischeWetenschappen (Universiteit Gent), 66(3b): 631-634.

Sutton DW., Havstad, PK., Kemp JD. (1992). Synthetic cryIIIA gene from Bacillus thuringiensis improved for high expression in plants. Transgenic Res. 1: 228-236.

Tabien, R. (2000). Biotech for Agriculture. Presentation during the PhilRice Farmers Forum. July 17, 2000.

TadeleZerihun (ed.) (2009). New Approaches to Plant Breeding of Orphan Crops in Africa. Proceedings of an International Conference, 19-21 September 2007, Bern, Switzerland

The California Right to Know Genetically Engineered Food Act. (2012). Retrieved August 2012, from Label GMOs: http://d3n8a8pro7vhmx.cloudfront.net/labelgmos/pages/31/attachments/original/CA-Right-to-Know-Initiative12.pdf?1324916176

Tojo, A. and K. Aizawa (1983). Dissolution and degradation of δ-endotoxin by gut juice protease of silkworm, Bombyxmori. Appl. Environ. Microbiol., 45: 576-580

Torres, A. C., A. T. Ferreira, P. E. Melo, E. Romano, M. A. Campos, J. A. Peters, J. A. Buso, and D. de Castro Monte. (1999). Transgenic plants of Achat potato resistant to the mosaic virus (PVY). *BIOTECNOLOGIA* Ciência and Desenvolvimento, no. 7.

Trapmann, S., H. Schimmel and G.N. Kramer, (2002). Production of certified reference materials for the detection of genetically materials for the detection of genetically modified organisms. Journal of AOAC International, 85(3): 775-779.

Tribe, D. (2012). 450+ published safety assessments. Retrieved November 2012, from GMO Pundit a.k.a. David Tribe: http://gmopundit.blogspot.com/p/450-publishedsafety-assessments.html

USDA (2008). United States Department of Agriculture: controls over importation of transgenic plants and animals. Audit Report 50601-17-Te, 12 December. Washington, DC: USDA, Office of Inspector General. http://www.usda.gov/oig/rptsauditsmulti.htm

Vaeck, M., A. Reynaerts, H. Hoftey, S. Jansens, M. DeBeuckleer, C. Dean, M. Zabeau, M. Van Montagu and J. Leemans (1987). Transgenic plants protected from insect attack. Nature, 327: 33-37

Vinocur B, Altman A (2005). Recent advances in engineering plant tolerance to abiotic stress: achievements and limitations. CurrOpinBiotechnol 16: 123-132

Vogel KE. (2009). Backcross breeding. Methods Mol Biol. 526: 161-169.

Vranova E, Inze D, Van Breusegem F., (2002). Signal transduction during oxidative stress. J Exp Bot., 53: 1227-1236

Winicov I, Bastola DR (1997). Salt tolerance in crop plants: new approaches through tissue culture and gene regulation. ActaPhysiol Plant., 19: 435-449

Witkowski J.F., J.L. Wedberg, K.L. Steffey, P.E. Sloderbeck, B.D. Siegfried, M.E. Rice, C.D. Pilcher, D.W. Onstad, C.E. Mason, L.C. Lewis, D.A. Landis, A.J. Keaster, F. Huang, R.A. Higgins, M.J. Haas, M.E. Gray, K.L. Giles, J.E. Foster, P.M. Davis, D.D. Calvin, L.L. Buschman, P.C. Bolin, B.D. Barry, D.A. Andow and D.N. Alstad. (2002). Bt Corn & European Corn Borer: Long-Term Success Through Resistance Management, Ed K.R. Ostlie, W.D. Hutchison and R. L. Hellmich. Available at http://www.extension.umn.edu/distribution/cropsystems/dc7055.html

Wolfersberger, M.G., (1989). Neither barium nor calcium prevent the inhibition by *Bacillus thuringiensis* – Endotoxin of sodium or potassium gradient dependent amino acid accumulation by tobacco hornworm midgut brush border membrane. Arch. Insect Biochem. Biophys., 12: 267-277

Wunn, J., Kloti, A., Burkhardt, P. K., Ghosh-Biswas, G. C., Launis, K., Iglesias, V. A., Potrykus, I. (1996). Transgenic Indica rice breeding line IR58 expressing a synthetic cryIA(b) gene from *Bacillus thuringiensis* provides effective insect pest control. Bio/Technology 14 (2): 171-176.

Yamaguchi-Shinozaki K, Urao T, Iwasaki T, Kiyosue T, Shinozaki K (1994). Function and regulation of genes that are induced by dehydration stress in *Arabidopsisthaliana*. JIRCAS J., 1: 69-79

Ye X., Al-babili S., Kloti A., Zhang J., Lucca P., Beyer P and Potrykus I. (2000). Engineering the Provitamin A (beta-carotene) biosynthetic pathway into (carotenoid-free) rice endosperm. Science. 287: 303-305.

Zhai, W., X. Li, W. Tian, Y. Zhou, X. Pan, S. Cao, X. Zhao, B. Zhao, Q. Zhang, and L. Zhu. (2000). Introduction of a blight resistance gene, Xa21, into five Chinese rice varieties through an Agrobacterium mediated system. *Science in China (Series C)* 43(3): 361-368.

15

Tissue Culture Technology for Crop Improvement in the Challenging Environment

Maya Kumari[1*] and Ramesh Namdeo Pudake[2]

[1]*Defence Institute of Bio-energy Research (DRDO), Goraparao, Haldwani Nainital-263139, Uttarakhand*

[2]*Department of Agronomy, School of Agriculture Lovely Professional University, Jalandhar-Delhi G.T. Road (NH-1) Phagwara-144411, Punjab, India*

ABSTRACT

Demographic pressures, land degradation, loss of agro-biodiversity, increased water scarcity and over-exploitation of natural resources, has ignited the fire of global climate change. The impact of climate change can be envisaged on crops, cropping systems, crop rotations, flora and fauna transformation. It has been shown that past climatic change has been a major driver of the processes of speciation and extinction. Plants can only function physiologically and successfully complete their life cycles under specific environmental conditions, changes to climate are likely to have significant impacts on plants and ecosystem or biome as a whole. The negative effects of climate change on agricultural productivity and food security are extreme temperature, drought, salinity, pests and diseases. In the context of climate change, biotechnology can respond positively towards reducing vulnerability of natural systems to climate change effects. This chapter highlights different tissue culture techniques that can be employed to address the negative effects of climate change for crop improvement and conservation. Establishment of cell, tissue and organ culture and regeneration of plantlets under *in vitro* conditions has opened up new avenues in the area of plant biotechnology. Different tissue cultures techniques serve as a tool for crop improvement and also assist conventional plant breeding. An *in vitro* study for development of stress tolerant variants has the potential for phytoremediation of degraded lands. Micropropagation is widely used for propagation of horticultural, ornamental and medicinal crops. It also helps in conservation of rare endangered plant species which might get extinct with the changing climate. Meristem culture enables rapid production of disease-free high priced plants. Wide hybridization has become a success due to embryo/ovule rescue technique. Protoplast fusion allows wide hybridization. Anther culture followed by colchicine treatment is important for the development of doubled haploids in polyploidy research. Plant regeneration is one of the major prerequisites for plant genetic engineering. Development of synthetic seeds and its cryopreservation

*Corresponding author's E-mail: madhulogin@gmail.com

Roychowdhury R. (Ed.), Crop Improvement in the Era of Climate Change.

can help long-term conservation of plant species. The recent advancements in tissue culture techniques will significantly contribute and form the basis of crop improvement in agriculture and horticulture in lieu of the changing climatic conditions.

Keywords: Climate change, crop improvement, tissue culture techniques.

1. CLIMATE CHANGE AND ITS IMPACT ON PLANTS

The negative impact of climatic changes can be seen on world food production and rise in food and fuel prices. Environmental variables do not act in isolation, but in combination with one other and with other pressures like ever-increasing human population, massive industrialization, air pollution, atmospheric ozone depletion, increase in UV-B radiation, increase in atmospheric CO_2 concentration, highly unpredicted rainfall pattern, its timing and distribution, shift in cropping pattern due to change in photoperiod and changing food habit, intermittent dry spell, temperature extremes, degradation of water and soil resources are gradually placing a pressure on the existing natural resources like water, arable land, environment, and germplasm (Engler *et al.*, 2011). Temperature and water supply are critical drivers for seed dormancy, germination, seed longevity, and soil pathogen activity (Walck *et al.*, 2011). The formidable task for the planners to overcome the weather uncertainties are complex due to the inability to make accurate region specific predictions particularly of rainfall patterns, damage to ecosystems, greenhouse gas emissions and climate effects. Climatic change has been a major driver of the processes of speciation, extinction and biodiversity patterns. More/less equitable climate will alter geographic distribution for species. With the changing climatic conditions, some of the species are on the verge of extinction and some new species are being evolved. There are strong evidences that plant species are shifting their ranges in altitude and latitude, resulting in a change in phenology of some species as a response to changing regional climates. The result of these changes is likely to increase the rate of extinction risk and invasion of one species in the habitat of another species. Plant genetic composition may change in response to the selection pressure of climate change. Some plant communities or species associations may be lost as species move and adapt at different rates. Many species are projected to become vulnerable to climate changes (Thomas *et al.*, 2004; Thuiller *et al.*, 2005), with consequent effects on the life. Climate change could lead to a disproportionate loss of evolutionary history (Nee and May, 1997; Heard and Mooers, 2000; Mace *et al.*, 2003). Thuiller *et al.* (2011) estimated the consequences of climate change on the phylogenetic diversities of plants, birds and animals across Europe. Even though reductions in diversity in one region and gains in regions of high latitude or altitude are expected, losses will not be offset by gains owing to loss of diversity.

Crop improvement in agriculture and horticulture is totally dependent on natural resources and climate. Water scarcity is the most important challenge, and the increase in temperature will add new complexity to drought research, especially in the semi-arid tropic areas, where the evaporative demand is high and increased temperature may further strain plant-water relations. Temperature increase will variably accelerate the onset of flowering. Recent research has shown an interdependence of C and N nutrition in the N performance of legumes, a balance that may be altered under climate change (Vadez *et al.*, 2012). An increased frequency of heat waves will require the development of varieties capable of setting and filling seeds at high temperature.

Finally, increase in temperature and CO_2 will affect the geographical distribution of pests, diseases, and weeds, presenting new challenges to crop management and breeding programmes (Vadez *et al.*, 2012). The increase in atmospheric CO_2 levels has direct impact on photosynthesis, can alter growth rates, offset the competitiveness between crop plants and weeds, distribution of weeds and insect pests, and their impact on agricultural productivity, for example, cotton is highly responsive to elevated CO_2 whereas wheat is much less responsive (Jain, 2010). Increased CO_2 can also lead to increased carbon: nitrogen ratios in the leaves of plants. Increased level of UV-B radiation as a result of depletion in protective ozone shield could lead to negative effects on the flora and fauna wealth of earth. The ambient ozone concentration alone or in combination with other pollutants also decrease yield of several important crops. Extreme temperatures can be harmful when beyond the physiological limits of a plant. Increases in temperature raise the rate of many physiological processes such as photosynthesis in plants. Changes in precipitation may lead to flood and drought like conditions due to uneven distribution. Climate change may become disastrous to agriculture and horticulture crop production especially by appearance of new insects, pests and diseases. Pests and diseases of minor nature may become major, threatening crop production while, farmer friendly predators may disappear in the course of evolution. It is, therefore, important to develop strategies to cope with effects of climate change on insect-pests and diseases of important crops (Chattopadhyay *et al.*, 2011).

Indian agriculture is extremely vulnerable to weather and climate. Significant variations in the weather and climatic parameters are expected to have substantial impact on crop production in the country in future. In recent past there was substantial loss of crop in the country due to extreme and unusual weather conditions (Chattopadhyay, 2011). Climate changes were large enough in some countries to offset a significant portion of the increases in average yields that arose from technology, carbon dioxide fertilization, and other factors (Lobel *et al.*, 2011). Many studies warn that climate change may undermine global food security. Results show that rice, wheat and maize production in middle income countries especially were vulnerable to drought compared to harvests in countries with higher investments in agriculture (Simelton *et al.*, 2012). Due to climate change, global maize and wheat production declined by 3.8 and 5.5 %, respectively. For soybeans and rice, gain and loss largely balanced out (Lobel *et al.*, 2011). Recent research indicates that monsoon rainfall became less frequent but more intense in India, thus increasing the risk of drought and flood damage to the country's kharif rice crop. Drought and extreme rainfall negatively affected rice yield in predominantly rainfed areas during 1966–2002, with drought having a much greater impact than extreme rainfall (Auffhammer *et al.*, 2012). Climate change has evidently already negatively affected India's hundreds of millions of rice producers and consumers (Auffhammer *et al.*, 2012).

Models that couple habitat suitability with demographic processes offer a potentially improved approach for estimating spatial distributional shifts and extinction risk under climate change in five species of Australian plants with contrasting demographic traits. Climate-driven changes in range area are sensitive to the underlying habitat model, regardless of whether demographic traits and their interaction with habitat patch configuration are modeled explicitly. Dispersal dynamics and habitat patch structure have the strongest influence on the amount of movement of the trailing and leading edge of the range margin, explaining roughly 40% of modeled structural deviance (Fordham *et al.*, 2012). Direct estimation of extinction risk incorporates important demographic and ecosystem processes, which potentially influence species vulnerability to extinction due to climate change. Scherrer and Körner (2011) aimed to explore thermal habitat preferences in Swiss central alpine vegetation (2200–2800 m.a.s.l) across mosaics of topographically controlled micro-habitats and quantify the shift in micro-habitat conditions under the influence of climate warming. High-resolution infra-red thermometry and small data loggers were used to assess the spatial and temporal variation of plant-surface, ground temperatures and snow-melt patterns across the alpine slopes of contrasting exposure. They observed a substantial variation between micro-habitats in seasonal mean soil temperature, surface temperature and season length. Plant species with low indicator values for temperature grew in significantly colder micro-habitats than plants with higher indicator values found on the same slope. Their results demonstrate that the topographically induced mosaics of micro-climatic conditions in an alpine landscape are associated with local plant species distribution. Engler *et al.* (2011) assessed the impact of climate change on 2632 plant species across all major European mountain ranges, using high-resolution species samples and data expressing four future climate scenarios. Their analyses also reveal unequal distribution of this threat across the various mountain ranges. Change in precipitation, not only warming, plays an important role in determining the potential impacts of climate change on vegetation. Climate change impacts on vegetation are mediated by soil processes that regulate rhizosphere water balance, nutrient dynamics, and ground-level temperatures. Estimates of species turnover across the soil depth gradient suggest that communities of shallow microsites are responding directly to micro environmental changes induced by climate manipulation, while those of the deepest microsites are shifting in response to changes in competitive interference from more nutrient-demanding species (Fridley *et al.*, 2011). Moreover, for several species in summer drought and winter heated treatments, climate response in deep microsites was opposite that of shallow microsites, suggesting microsite variation is contributing to community stability at the whole-plot level. This study demonstrates a strong link between community dynamics and substrate properties, and suggests ecosystems typified by fine-scale substrate heterogeneity may possess a natural buffering capacity in the face of climate change.

Chaturvedi *et al.* (2011) assessed the impact of projected climate change on forest ecosystems in India, based on climate projections of the Regional Climate Model of the Hadley Centre (HadRM3) and the dynamic global vegetation model IBIS for A2 and B2 scenarios. Based on the dynamic global vegetation modeling, the vulnerability index suggests that upper Himalayas, northern and central parts of Western Ghats and parts of central India are most vulnerable to projected impacts of climate change, while Northeastern forests are more resilient. Thus, their study points on the need for developing and implementing adaptation strategies to reduce vulnerability of forests to projected climate change. Behera *et al.* (2012) collected data on microclimate between 2010 and 2011 in five forest communities in a tropical moist deciduous forest in Uttar Pradesh, India to compare how vegetation structure affects microclimate. Diurnal variations in microclimatic variables (photosynthetically active radiation (PAR), air and soil surface temperature, ambient CO_2, air absolute humidity) were measured with LI-COR (LI-COR, USA) in each forest community. The diurnal trend in microclimatic parameters showed wide variations among communities. Among the five communities, the highest PAR value was observed in savannah and lowest in sal mixed forest. In savannah ecosystem, diurnal trend of ambient CO_2 was totally different from rest four communities. According to Canonical correspondence analysis, PAR and ambient CO_2 are most important in establishment of forest community, among microclimatic variables.

2. TISSUE CULTURE TECHNOLOGY TO COMBAT CLIMATE CHANGE

The major challenge by 2050 is of increasing the food production by about 50% to cater for an additional three billion people, in the scenario of shrinking arable land, soil degradation, nutrient deficiencies, increased water scarcity, and uncertainty due to climatic changes (Vadez *et al.*, 2012). The new challenges posed by climatic changes have left the breeders and agronomists on task to cope up with the pressure to sustain food production in order to feed the growing population. The food and fuel prices are soaring worldwide and the heat can be felt in both developed and developing countries, facing economic crisis, however, the developing and underdeveloped countries are the worst hit. To solve the world food problem and sustain food production, there is a need to identify the most appropriate cost effective techniques for crop improvement. Genetic variation is the raw material for any breeding programme. Variation can be created either through conventional breeding like hybridization followed by selection, mutagenesis, etc. or by resorting to biotechnological tools. Conventional breeding has its own limitations like genetic barrier to crossing, season bound, time duration required is minimum 7-10 yrs. Under the climate change regime, conventional breeding in combination with modern tissue culture techniques such as *in vitro* mutagenesis, plant biotechnology and genetic engineering may prove to be helpful for the improvement of crops of great economic importance and value for sustainable utilization. The extinction of plant diversity can be minimized through germplasm conservation using tissue culture techniques. *In vitro* cell and tissue-based systems have tremendous potential in fundamental research and for commercial applications such as clonal propagation, genetic engineering and production of valuable secondary metabolites. The different tissue culture strategies along with their case studies are discussed in the following section.

3. CROP IMPROVEMENT THROUGH THE APPLICATION OF TISSUE CULTURE TECHNIQUES

The recent advancements in plant tissue culture technology form the basis of crop improvement in economically and ecologically important crop species in agriculture and horticulture across the world. Plant cell, tissue and organ culture technique is effective because almost all plant cells are 'totipotent' means that each cell possesses the genetic information and cellular machinery necessary to generate an entire organism. This is the miracle of nature gifted to plants only, since this concept of 'totipotency' to do hold true for other higher organisms like animal and human cells. Establishment of cell, tissue and organ culture and regeneration of plantlets has opened up new avenues in the area of plant biotechnology serving as a tool for crop improvement and also assist conventional plant breeding. (The history of different tissue culture techniques is presented in Table 1). Plant tissue culture involves culturing of plant cells, tissues or organs on synthetic media under aseptic environment and controlled conditions of light, temperature, and humidity. The micropropagation technique allows mass propagation of true to type elite lines. Micropropagation is widely used for propagation of horticultural, ornamental and medicinal crops. It also helps in conservation of rare endangered plant species which might get extinct

Table 1 History of different tissue culture techniques

Year of discovery	*Technique*	*Crop*	*Pioneer worker*
1902	Cell culture	*Laminum purpureum*	Gottlieb Haberlandt
1909	Protoplast fusion	–	Kuster
1929	Culture of embryos to overcome the incompatibility	Linum spp.	Laibach & Hered
1934	Root culture	Tomato	White
1941	Embryo rescue	*Datura*	Van Overbeek and coworkers
1946	Micropropagation	Lupinus	Ernest Ball
1952	Meristem culture	Dahlia	Morel and Martin
1953	Pollen culture	*Ginkgo biloba*	Tuleke
	Ovule culture		San Noeum
1954	Cell suspension culture for production of sec metabolites	*Tagetes erecta* and *Nicotiana tabacum*	Muir
1958	Somatic embryogenesis	Citrus nuselus	Maheswari & Rangaswamy
1960	Enzymatic protoplast isolation	Plant tissue	Cocking
1962	Role of tissue culture in plant genetic engineering	–	Kanta and Maheshwari
1964	Anther culture	Datura	Guha and Maheshwari
1966	Somatic embryos from cells in suspension	Carrot Daucus carota	Reinert and Steward
1970	Synthetic seeds	Carrot	Backs-Husemann and Reinert
1972	Somatic hybridization	First hybrid between *N. Glauca* and *N. langsdorffii*	Carlson and coworkers
1973-74	Microspore culture	*Nicotiana and datura*	Nitsch
1977	Use of bacteria for plant transformation	–	Chilton and coworkers
1978	Intergeneric hybridization	Tomato and potato	Melcher
1981	Cryopreservation for germplasm conservation	Any plant propagule	Kartha
1981	Somaclonal variation	–	Larkin and scrowfort
1983	*In vitro* selection	For disease resistance	Miller and maxwell
1984	Gametoclonal variation	–	Evans and coworkers
1984	First transgenic	Tobacco	Horsch and coworkers
1983-87	*In vitro* mutagenesis	Herbicide tolerance in tobacco Methyl tryptophan tolerance in Datura innoxia	Jacob; Huges; Ranch and coworkers
1892	Protoplast isolation	Onion	Klercker

with the changing climate. Meristem culture enables rapid production of disease-free high priced plants. Wide hybridization has become a success due to embryo/ovule rescue technique. Protoplast fusion allows wide hybridization, change in ploidy level and genetic transformation in plants. Tissue culture coupled with somatic hybridization helps in evolving new cultivar in a short time. Anther/pollen/microspore culture is important for development of haploids, fixation of hetrogygosity, genetic stability, fixing apomixes, expression of recessive alleles, shorts breeding time for the development of homozygous lines, etc. Development of somatic embryos through somatic embryogenesis for making synthetic/artificial seeds and its cryopreservation can help conserve plant species. The application, advantages and limitations of different tissue culture techniques are presented in Table 2.

Plant tissue cultures can be initiated from almost any part of a plant called 'explant'. The parent plant must be healthy, vigorous and free from disease or decay. The plants themselves must be actively growing, and not about to enter a period of dormancy. Younger tissue is considered to be more responsive to a callus initiation and regeneration under *in vitro* culture conditions, since they contain a higher proportion of actively dividing cells. The regeneration potential of an explant depends largely on the genotype and is different for each plant species (Kumari *et al.*, 2012a). Each genotype often has a particular set of cultural requirements. Hence, the culture protocol for regeneration has to be optimized for each variety by designing experiments. Plant tissue culture has direct commercial applications as well as plays a vital role in carrying out basic research into cell biology, cell development, genetics and cytogenetics, physiology and biochemistry. The various applications of tissue culture include:

- Micropropagation using meristem and shoot tip/axillary node culture to produce large numbers of genetically identical individuals.
- *In vitro* mutagenesis and screening for variants for abiotic stress like cold, heat, salinity, heavy metal tolerance and disease, pesticide and herbicide tolerance.
- *In vitro* screening can be done for finding the potential of a plant species for phytoremediation of the affected land.
- Cell suspension culture technique can be utilized for large-scale growth of plant cells in liquid culture for the production of secondary products of commercial value.
- Makes possible crossing of distantly related species through protoplast fusion and regeneration of the novel hybrid.
- Haploid and dihaploid plants from haploid cultures can be produced through anther culture to achieve homozygous lines more rapidly than conventional breeding strategies.
- *In vitro* regeneration is one of the prerequisites for *Agrobacterium* mediated gene transfer for transgenic development.
- Meristem culture is popular for propagation of disease-free planting material that also fulfills the new WTO regime (issue of phytosanitary certificate) which stresses on the production of disease free planting material.
- Somatic embryos can be used as synthetic seeds for true to type propagation and storage for long time.
- Germplasm conservation through cryopreservation of *in vitro* propagules.

Table 2 Tissue culture techniques: their applications, advantages and limitations

S.N.	*Technique*	*Applications*	*Advantages*	*Limitations*
1.	Micropropagation	True to type multiplication of elite lines and disease free lines	Genetically identical individuals can be propagated in less space and time.	Costly as it needs a well established lab. Contamination loss
2.	Somaclonal variation	Callus culture for creation of variation	Epigenetic changes. New variants at genetic or epigenetic level might emerge.	It is a nuisance when true to typeness is needed. The variation may be selected against and get lost subsequently. Genetic stability is the question. Not all changes are desirable; in fact most are deleterious or of no agronomic value.
3.	*In vitro* mutagenesis	Mutation induction for creation of genetic variation. Useful for horticulture and floriculture species	More cells exposed to mutagen in less space and the frequency of mutants increases. Involves less time, labour and land. Can be screened *in vitro*.	Mutagens are hazardous to handle. The stability of mutants is a question.
4.	*In vitro* selection	Induce desirable, heritable changes in regenerated plants by subjecting a population of cells to a selection pressure, e.g., for disease and abiotic stress screening *in vitro*. Phytore mediation to salts and heavy metals.	Selection pressure can be applied at the cellular level to screen more cells in less space and time.	The selection agent may affect the plant regeneration potential and retard growth. The performance needs to be confirmed after field growing of tissue culture screened plants.
5.	Meristem culture	Disease-free propagation of plants	Disease-free, e.g., virus-free material can be generated through meristem culture	It is difficult to keep the plant disease free once it is exposed to the field environment. Quarantine and virus indexing are still recommended to verify that the material is disease-free.
6.	Cryopreservation	Propagule storage for long term	This allows storage of plant propagules in the form of seeds, explants, somatic embryos etc for long term without damage. Preserve rare endangered species from extinction	Vitrification Loss of regeneration potential with storage time. Continuous power supply required which increases the cost of maintenance.

Contd...

Contd...

7.	Cell suspension culture	Production of secondary metabolites, enzymes and hormones	Decrease time to obtain plants. Constant supply of embryogenic-competent cells; and scaling-up. Used in industries	Somaclonal variation and slow cell growth rate.
8.	Protoplast fusion	Somatic hybridization	Creation of variation	Regeneration of protoplast is often the major limitation to this technique
9.	Anther/ovule culture	Embryo rescue after wide hybridization.	Fixation of homozygosity. Creation of variation. Creation of haploids and doubled haploids in less time.	Confirmation of regeneration through gametes only is difficult. Regeneration may occur from somatic tissues of anther wall.
10.	Germplasm conservation	Cryopreservation and gene/ seed bank	Rare, threatened and endangered species can be conserved for future use.	Needs huge resource and testing time to time to check for viability
11.	Somatic embryogenesis	Creation of somatic seeds	Can be seeded from air and germinate to give complete plants.	Dehydration, loose viability. Somaclonal variations
12.	Ploidy manipulation	Somatic fusion, doubling of chromosomes through cholchicine treatment	Creation of variation.	May lead to unwanted polyploids.
13.	Genetic manipulation	Transgenic development	Creation of genetic variation for the desired trait. No barrier to gene transfer is involved.	Copy number

One by one each tissue culture technique along with their applications, advantages, limitations and case studies will be discussed under separate subheadings.

3.1 Micropropagation

Micropropagation (tissue culture or *in-vitro* culture) refers to the multiplication of plants, in aseptic condition and in artificial growth medium, from very small plant parts. The German plant physiologist Hebertland (1902) first described the principles of tissue and organ culture, after which rapid advancements have been made in plant tissue culture technique. Micropropagation technology is applied commercially for large-scale plant multiplication, germplasm conservation, pathogen elimination, genetic manipulations and continuous supply of selected plants. Plant cell and tissue culture techniques provide an alternative to mass propagate the species which are difficult to cultivate, perennial in nature, are seed propagated and vegetative propagation is season bound or where conventional methods are inadequate to meet the demand of planting material. Selection of plant material to be propagated is the first prerequisite. Clean, disease free healthy explant from the mother plant is selected to obtain healthy plants. The mother explant can be maintained for longer period through continuous subculturing, without losing its regeneration potential. Regeneration under *in vitro* conditions is largely dependent on explant type, genotype, nutrients, growth hormones and culture conditions (Kumari *et al.*, 2012a). For micropropagation the explant material is surface sterilized, cut into small pieces, placed on a growth medium containing sucrose as an energy source and one or more plant growth hormones. For efficient plant regeneration, all these parameters need to be optimized by varying the concentrations and combinations of hormones and additives. For true to type multiplication of plants, meristematic tissues like shoot tip, axillary node and cotyledonary node explants are used which give regeneration without any intervening callus phase. Callus is not desired for true to type propagation, since it leads to somaclonal variations. The regenerated plants after elongation and rooting is removed from the media and transferred in potting mix for hardening under culture room conditions for a month, thereby transferring in field for further growth and development. Micropropagation is assisting conventional crop improvement. This also protects the natural resources from being cut for propagation in nature. The regeneration protocol developed for mass multiplication can also be used for genetic transformation for crop improvement by transferring agriculturally desired traits (Li *et al.*, 2007; Divakara *et al.*, 2010).

Advantages over traditional plant propagation techniques

Micropropagation offers several advantages not possible with conventional propagation techniques. Micropropagation proves useful for propagation of sexually sterile species like triploids, aneuploids which cannot be perpetuated by seeds, seedless plants like banana and sugarcane, cross-bred perennials where heterozygosity is to be maintained, mutant lines like auxotrophs which cannot be propagated *in vivo* and disease free planting material of fruit and ornamentals. Once the regeneration protocol is established, actively dividing cultures are a continuous source of propagules ensuring round the year supply in huge number, without seasonal interruption.

- Genotype constitution is maintained due to propagation of genetically identical progenies.
- Micropropagation facilitates long distance transport of propagation material and long term storage of clonal materials.
- Round the year supply irrespective of season in less space can be ensured.
- Reduce growth cycle and allow rapid multiplication of shoots through subculturing.
- Selective multiplication can be done for, e.g., papaya, seabuckthron females can be selected for propagation.
- Micropropagation allows production of pathogen-free material from plants raised and maintained through meristem culture, e.g., horticultural fruit and ornamental crops.
- Traits of economic importance identified as somaclonal variations can be further maintained in its true form by micropropagation.
- It is the only viable method of regenerating genetically modified cells after protoplast fusion.
- It is useful in multiplying plants which produce little or no seeds or produce no viable seeds or when the seed cannot be stored (recalcitrant seeds).

- Micropropagated plants are observed to establish more quickly, grow more vigorously and taller, have a shorter and more uniform production cycle, and produce higher yields than conventional propagules.
- Micropropagation often produces more robust plants, compared to similar plants produced by conventional methods — like seeds or cuttings.
- Some plants with very small seeds like orchids are most reliably grown from seed in sterile culture.
- It is the most reliable technique for the propagation and conservation of germplasm of rare, endangered and threatened (RET) species.

Limitations

The major limitation in the use of micropropagation for many plants is the cost of production.

- It needs skilled labour, is very expensive, and requires special setup.
- Slight infection may damage the entire lot of plants.
- The disease-free material may break down due to monoculture of the clones rendering them more vulnerable to the same infections for which it was propagated.
- It is often hard to get rid of inborn infections without the use of antibiotics in the medium.
- The regeneration protocols developed by one lab may not work in another so it needs to be standardized for every lab conditions.

3.2 Somatic Embryogenesis

Somatic embryogenesis is a process involving redifferentiation of meristematic cells into non-zygotic somatic embryo which are capable of germinating into complete plant. Somatic embryos are bipolar structures with radical and plumule is contrast to monopolar shoot bud with only plumular end in organogenesis. While developing into somatic embryo, the mersitematic cells break any cytoplasmic or vascular connections with other cells around it and become isolated. Therefore, unlike shoot bud, the somatic embryos are easily separable from explants. Somatic embryogenesis involves three distinct steps which are absent in organogenesis. First is induction phase where callus cells divide and differentiate into groups of meristematic cells called embryogenic clumps that develop into initial stages of somatic embryo, i.e., globular stage. Maturation phase where somatic embryos develop into mature embryos by differentiating from globular to heart shaped, torpedo to cotyledonary stages (Fig. 1). The mature embryo here undergoes biochemical changes to acquire hardiness. Third is conversion phase where embryos germinate to produce seedlings.

Somatic embryogenesis is strongly depends on the plant genotype, physiological state of the explants, the composition of the culture medium and the conditions of cultivation. The process could be direct when the embryogenic structures develop directly from the initial explant, or indirect through a callus stage. Whole plants develop from the somatic embryo grow like normal seeds upon sowing. Somatic embryos can be produced from cells growing in suspension, thereby making possible batch culture techniques which can be scaled-up easily. Cell suspension

showed anthocyanin pigmentation. Plant species where no obvious difference can be observed between parental lines, vital stains such as fluorescein isothiocyanate (FITC, stains green) and rhodamine isothiocyanate (RITC, stains red) can be used for selection of hybrid cells as the fusion product gives yellow colour. Fused cells with fluorescent dyes can be isolated using flow cytometer. Besides these, two stains do not affect the viability of cells.

Application of protoplast fusion for somatic hybridization

This section describes the protoplast fusion technology and its biotechnological applications.

Somatic hybridization minimizes the time taken for cytoplasm transfer to one year from 6-7 years required in back cross method. Cybrid is produced when cytoplasm of two species or genera is fused with nuclear genome of only one cell (nuclear-cytoplasmic combination). Also, this method allows cytoplasm transfer between sexually incompatible species. Cybrids have cytoplasm from both parents but nucleus of only one. Nucleus of other parent is irradiated. This approach has been potentially used to transfer cytoplasmic male sterility and resistance to atrazine herbicide, both coded by cytoplasmic genes in Brassica to different crops like tobacco, rice, etc. Somatic hybridization could facilitate the introduction of the desirable traits like tolerance to biotic and abiotic stresses from cultivars and wild species to cultivated species (Litz, 2004). Sigareva and Earle (1997) transferred cold tolerant Ogura cytoplasmic male-sterile (CMS) cytoplasm from broccoli to fertile but cold sensitive cabbage via leaf protoplast fusion. The nuclear DNA of the broccoli from the fusion protoplast was eliminated. Diploid cabbage plants were identified by flow cytometry and morphological characters. PCR using primers for an Ogura CMS-specific mitochondrial DNA sequence were used for the identification of CMS cybrids. Plants transferred into soil 8 months after fusion produced male-sterile flowers. Somatic hybrids were produced by PEG-induced symmetric and asymmetric protoplast fusions in order to transfer resistance to *Alternaria brassicicola, A. brassicae, Phoma lingam, Plasmodiophora brassicae* and *Turnip mosaic virus* (TuMV) into *Brassica oleracea var. capitata* (cv. 'Toskama') and *botrytis* (cv. 'Korso') (Scholze *et al.*, 2010). As resistance donors, ten species belonging to several genera of the family Brassicaceae including wild relatives were used. 73.8% of the somatic hybrids were resistant against at least one of the pathogens, indicating a successful transfer of resistance trait. Five hundred and twenty-two hybrids showed multiple resistances to two, three and, in a single case, to four pathogens. Irrespective of the donor parents used in the fusion, a broad variability in symptom ranging from 0 (without symptoms) to 9 (highly susceptible) could be observed. With regard to the *Alternaria* pathogens, *Sinapis alba*, *B. nigra* and *B. juncea* were the most effective resistance donors, whereas fusions with *Raphanus sativus* resulted in the most hybrids with resistance to clubroot and TuMV. Some combinations in which both parents were highly susceptible, e.g., *R. sativus* (+) *B. oleracea* var. *capitata*, yielded hybrids that exhibited strong resistance, e.g., to *A. brassicicola*, revealing the development of a new type of resistance. With regard to the *Alternaria* pathogens, resistance expression was very unstable. Many hybrids with resistance from donors, such as *B. vulgaris*, *S. alba* and *B. carinata*, became highly susceptible. For reliable characterization of the resistance response, hybrids should be subjected to several resistance tests during growth period of the host, at least until flowering. Most of the classic experiments in somatic hybridization have been carried out in petunia and tobacco. Somatic hybrids have been produced in Citrus spp. (Fu *et al.*, 2003; Khan and Grosser,

2004), grapevine (Matt and Blaich, 2000), banana (Matsumoto *et al.*, 2002), persimmon (Tamura *et al.*, 1995) and Actinida spp. (Xiao *et al.*, 2004). Rezazadeh *et al.* (2011) attempted somatic hybridization of mango via protoplast fusion through enzymatic isolation of protoplasts from leaves of greenhouse-grown seedlings of cvs. 'Tommy Atkins', 'Keitt' and 'Haden' and from proembryonic masses (PEMs) of cv. 'Kensington Pride'. Flow cytometry analysis revealed 41 tetraploid lines of which DNA fingerprinting of the regenerated embryos from the tetraploid lines showed that only four lines were somatic hybrids, all resulting from 'Haden' + 'Kensington Pride' protoplast fusions. The tetraploid lines from 'Keitt' + 'Kensington Pride' and 'Tommy Atkins' + 'Kensington Pride' were autotetraploids.

Uses of protoplast fusion

- To combine two complete genomes creating polyploids
- Partial transfer of genome (for few traits) between species
- Genetic engineering
- Transfer of cytoplasmic organelles like mitochondria and chloroplast between species

Advantages

- Protoplast fusion offers the possibility of one-step transfer of organelles between plant species, replacing the tedious procedure involving repeated back-crosses.
- Protoplast fusion has overcome the impediment of reproduction in haploid, triploid and aneuploid plants. Also, genomes of asexually reproducing plants can be recombined using this approach.
- The incompatibility barriers in sexual recombination at interspecific or intergeneric levels are also overcome by somatic hybridisation.
- Organelle transfer by protoplast fusion is based on the segregation of nuclei in the primary fusion products, and subsequent segregation of organelles resulting in new combinations of nuclei and organelles.
- Transfer of chloroplasts has been achieved using this method.
- Protoplast fusion has been used as an important tool to transfer useful genes such as disease resistance, protein quality, frost hardiness, drought resistance, herbicide resistance, heat and cold resistance from one species to another.

Protoplast fusion would continue to be an existing area of research in modern biotechnology. This technique in the future will be one of the most frequently used research tools for tissue culturists, molecular biologists, biochemical engineers and biotechnologists. To date, cell suspension cultures have been usually used as a protoplast donor material in carrot studies, especially in research on intra- and inter-specific as well intergeneric somatic hybridization (Han *et al.,* 2009). Despite the fact, that suspension cultures are an excellent source of protoplasts due to their high embryogenic ability, the establishment and maintenance of suspension culture is laborious and time consuming, usually requiring several weeks in addition to the time required for callus induction. In contrast, leaf tissue can serve as an easy and direct source of protoplasts. Several types of plant materials like leaves, cotyledons or hypocotyls are used as protoplast sources in many species.

Commercial application of protoplast fusion

Protoplast fusion technique has a great potential for industrially useful microorganisms (Murlidhar and Panda 2000). The direct bioconversion of cellulosic materials to ethanol by the intergeneric fusants between *T. reesei* and *Saccharomyces cerevesie* appears to be the best technique for ethanol production. Protoplast fusion of *T. reesei* and *A. niger* resulted in the production of cellulases (Ahmed and Berkley, 2006). Rygielska (2004) obtained interspecific fusants of starch fermenting yeasts *Schwanniomyces occidentalis* ATCC 48086 and *Saccharomyces diaastiacus* ATCC 13007. Pasha *et al.* (2007) through protoplast fusion of thermotolerant *Saccharomyces cerevisiae* and mesophilic, xylose-utilizing *Candida shehatae* followed by sequential mutations obtained a stable and well performing fusant with maximum utilization of reducing sugars in the media. Somatic hybrids for transfer of useful genes such as disease resistance, abiotic stress resistance or genes of industrial use, for example, Datura hybrids (*D. innoxia* + *D discolor, D. innoxia* + *D stramonium*) show heterosis for scopolamine (alkaloid) content which is 20-25% higher than in parent species and therefore has industrial application.

3.3.2 Protoplast Culture and Somaclonal Variations

Somaclonal variants via protoplast culture have been obtained in various species for improved disease resistance and other traits. Protoclones (protoplast derived plants) with desired traits have been selected with improved resistance to early blight and late blight in potato cv Russet Burbank and Bintje (Compton *et al.*, 2000). In tomato protoclones with improved resistance to fusarium wilt has been obtained in cv UC82 and protoclones with elevated lanatoside C content were regenerated from *Digitalis lanata* protoplasts (Compton *et al.*, 2000).

3.3.3 Protoplast Fusion for Genetic Transformation

Protoplast culture can also be used for genetic transformation studies. Gentic transformation has been achieved by co-cultivating protoplasts with *Agrobacterium tumefacians* or by direct DNA uptake using polycationic chemicals, electroporation, sonication, liposomes or microinjection (Sawahel and Cove, 1992; Fisk and Dandekar, 1993). Protoplasts are mixed with agrobacterium culture and incubated for 24-30h. after co-cultivation, selection of transformed cells are done by adding antibiotic (kanamycin, hygromycin, etc) to the culture medium, that inhibit growth of non-transformed cells. Transformation frequency ranging from 0.26 to 15.3% has been reported. Another method of genetic transformation via protoplast is polybrene (Hexademethrine bromide) method used to trasnfrom maize cv Black Mexican Sweet protoplast (Antonelli and Stadler, 1990). This method has low cell toxicity, promotes transformation frequencies and in not limited by DNA size. Electroporation method has been used to transfer PEP gene and GUS A reported gene in *Sorghum bicolor* (Compton *et al.*, 2000).

3.3.4 Protoplast Fusion and Polyploidy Induction

Very often protoplast fusion leads to polyploidization of the fusion product. *Jatropha curcas* L. has been considered one of the most promising alternatives for biofuel production (Kumari *et al.*, 2012b). *In vitro* tissue culture techniques such as organogenesis and embryogenesis have been conducted for mass clonal propagation of elite *J. curcas* lines. Oliveria *et al.* (2012) attempted to induce *in vitro* polyploidy in plantlets generated from shoot tips of *J. curcas* 'Gonçalo' ($2n = 2\times = 22$). Tetraploid and mixoploid plantlets were obtained from different treatments, with 0.5 mM colchicine for 96 h. The plantlets were recovered and clonally propagated in tissue culture medium supplemented with indole-3-acetic acid (IAA) and 6-benzylaminopurine (BAP). DNA ploidy level screening by flow cytometry was used as a practical and rapid strategy for selection of diploid, mixoploid and tetraploid plantlets.

3.4 Somaclonal Variation and Cell Line Selection

Since the invention of plant cell and tissue culture techniques more than half a century ago, scientists have been trying to understand the morphological, physiological, biochemical and molecular changes associated with tissue culture responses (Anjanashree and Wang, 2012). Establishment of de novo developmental cell fate *in vitro* is governed by factors such as genetic make-up, stress and plant growth regulators. The genetic variability present in somatic cell progenies derived from cells/tissue cultured *in vitro* is called somaclonal variation. Larkin and Scrowcroft coined this term in 1981 for all plant variants derived from any form of cell or tissue culture. Chromosomal rearrangements are an important source of this variation. The variations can be genotypic or phenotypic, which in the latter case can be either genetic or epigenetic Somaclonal variation is particularly common in plants regenerated from callus (Miguel and Marum, 2011). Typical genetic alterations are: changes in chromosome numbers, chromosome structure and DNA sequence (base mutations).

Occurrence of natural somaclonal variations is very common in tissue culture-derived plants but most of them go undetected. For selection to occur, the frequency of somaclonal variation should be high for the desirable variants, and the selected lines should perform well under open environmental conditions (Duncan, 1997). Selection under field conditions requires huge amount of money, space, labour and time, while *in vitro* selection shortens the time considerably for the selection of a desirable trait under the selection pressure. Field evaluation of somaclonal variants includes genotype x environmental interactions as well (Duncan, 1997). Somaclonal variation is desirable when variation in plant for its improvement is required but is a nusiance when true to type progenies are required. Somaclonal variation leads to the creation of additional genetic variability. Somaclonal mutants can be enriched during *in vitro* culture for resistance to disease pathotoxins, herbicides and tolerance to chemical stress, as well as for increased production of secondary metabolites. Heath-Pagliuso *et al.* (1988) selected *Fusarium oxysporum* f.sp.*apii* resistant celery somaclones superior to the parental type. A serious disadvantage of somaclonal variation occurs where clonal uniformity is required, as in the horticulture, floriculture and forestry industries where tissue culture is employed for rapid propagation of elite genotypes. However, high production cost has limited the application of this technique to more valuable ornamental crops and fruit trees.

Somaclonal variation due to genetic and epigenetic changes

In this section, we discuss the genetic and epigenetic changes that take place under *in vitro* conditions. *In vitro* culture is believed to destabilize the genetic and epigenetic programme of intact plant tissue and can lead to chromosomal and DNA sequence variations, methylation changes, transposon activation, and generation of somaclonal variants (Neelakandan and Wang, 2012). Epigenetic mechanisms are highly dynamic events that modulate gene expression. As more accurate and powerful tools for epigenetic analysis become available for application in a broader range of plant species, analysis of the epigenetic landscape of plant cell cultures may turn out to be crucial for understanding variant phenotypes. Genetic variation due to DNA methylation seems to be much more frequent and has been directly implicated in phenotypic variation. Recent advances in the field of epigenetics have uncovered highly dynamic mechanisms of chromatin remodelling occurring during cell dedifferentiation. It is hoped that a precise and comprehensive knowledge of the molecular basis of these variations would help to devise strategies to improve the embryogenic capability in recalcitrant species and genotypes, for clonal propagation.

Major causes of somaclonal variation

(i) **Physiological and Epigenetic:** Variations induced by physiological factors in culture medium, for example, exposure to PGRs. Often such variations are epigenetic and hence do not follow Mendelian inheritance.

(ii) **Genetic:** Changes at chromosomal level are grouped under genetic cause of variation observed in regenerants.

- Chromosomal rearrangements such as deletion, duplication, translocation, inversion polyploidy, aneuploidy, have been reported to be the chief source of somaclonal variation.
- Meiotic crossing over involving symmetric and asymmetric recombination could also be responsible for variation observed among somaclones.
- Transposable elements like Ac-Ds in maize have been shown to get activated in *in vitro* culture. For example, in maize (*Zea mays* L.) and broad beans (*Vicia faba* L), late replicating heterochromatin is the main cause of somaclonal variation.
- Single gene mutations in cultures also give rise to variations which are not detected in plants regenerated *in vitro* from any cell or tissue (R0 plants) but express in R1 plants (produced by selfing R0 plants).

(iii) **Biochemical:** The most common biochemical variation is the change in metabolism leading to failure of photosynthesis, viz., albinos is rice, change in other cell processes like starch biosynthesis, pigment pathway, antibiotic resistance etc also lead to somaclonal variation.

Application for crop improvement

Somaclonal variation represent useful source for creation of variations, occurs at much higher frequencies than induced mutants, allow plant breeder to have greater control on selection process as here they have the option to select from large amount of genetically uniform material. Therefore, this is the only approach for genetic improvement in perennial species limited by narrow germplasm and long regeneration cycle, asexually propagated plants like bananas, for isolation of biochemical mutants like auxotrophs. Somaclonal variants have been isolated for variety of valuable traits like disease resistance, stress (salt, low temperature) resistance, improved yield and efficient nutrient uptake etc. Bio-13 is a somaclonal variant of *Citronella java,* a medicinal plant which yields 37% more oil and 39% more citronellol than the control varieties. Indian Agricultural Research Institute (IARI) developed Brassica cv Pusa Jai Kisan, with bolder seeds and higher yield developed through somaclonal variation of *Brassica juncea* variety 'Varuna'. IARI also developed somaclonal variants in Lathyrus with low toxin level.

Gametoclonal variation

The variation observed among haploid plants having gametic chromosome number developing from anther culture is called gametoclonal variation. Such variations resulting in desirable traits are subjected to selection at haploid level followed by diploidisation to get homozygous plants which can be released as new varieties. Even recessive mutations are expressed in haploids unlike diploids where they express only in segregating generation. Therefore, in several crops desirable mutants including traits like resistance to diseases, antibiotics, salts etc have been isolated from haploids derived from anther culture. For example, tobacco mutants resistant to black shank disease and wheat lines resistant to scab (*Fusarium graminearum*) have been selected and used as improved cultivars.

3.5 *In vitro* Mutagenesis for Creation of Variation

Genetic variability is the basic ingredient of any plant breeding programme in developing plant varieties for sustainable food production. Plant breeders, in the past have successfully combined the desired genes from the existing gene pool and the related plant species by sexual hybridization, and successfully developed new cultivars with desirable traits such as high yield, quality, abiotic and biotic stress resistance (Jain *et al.,* 2006; Jain 2010). Mutation techniques have been widely applied to improve crop yield, quality, disease and pest resistance (Liu *et al.*, 2005). About 2335 varieties were released through mutagenesis in the world, in which ornamental crops and decorative crops are 5528. Its efficiency has been proved in several food crops and horticultural crops. In International Atomic Energy Agency (IAEA) mutant database, over 3000 officially released mutant varieties have been released worldwide in cereals, ornamental plants, fruits, vegetables, and oil crops. However this convention method of mutant generation has several limitations. The frequency of occurrence of spontaneous mutation in nature is very low and difficult to detect. Such mutations even though it occurs may get lost in the selection process. It requires large population for screening. This involves huge cost, time and labour. In perennial crops it is very difficult to find the mutant expression. The purpose of *in vitro* mutation induction is to enhance mutation rate in a short time in small working area. By *in vitro* selection, desirable mutants with useful agronomical traits can be isolated in a short period of time. Combining mutation technique with *in vitro* selection, allow the selection of traits controlled by their recessive or dominant genes, can improve recovery of the desired mutant. The genetic fidelity of the regenerated plants is highly desirable for developing new improved plant varieties for feeding the ever-growing human population, especially under climate change and shrinking arable land.

In vitro mutation and selection has several advantages over traditional mutation breeding being more efficient and cost effective. Induced mutations have over the past 50 years played a major role in the development of superior crop varieties translating into a tremendous economic impact on agriculture and food production. The purpose of induced mutations is to enhance the mutation frequency rate in order to select appropriate variants for plant breeding. Mutations are induced by physical (e.g., gamma radiation) and chemical (e.g., ethyl methane sulfonate) mutagen treatment of both seed and tissue explant. The mutagen treatment results in chromosomal mutations and that enable plant breeders to select useful mutants. *In vitro* induced mutations are passed through several generations of selfing or clonal propagation and leads to the fixation of the mutation events. Thus, mutation induction is a flexible, workable, unregulated, non-hazardous and low-cost alternative to genetically modified organisms (GMOs). In order to screen target characters in a large number of mutagenised cells, it is essential to have efficient selection agents (Liu *et al.*, 2005). Toxic culture filtrate and purified toxins have been used in *in vitro* selection of disease resistant plants (Thakur *et al.*, 2002; Svabova and Lebeda 2005; Kumar *et al.,* 2008). It has been attempted at the plant level where a large population of plants, raised through *in vitro* callus cultures, have been screened for resistance in the field capacity or via a more targeted approach of regeneration of disease resistant plants through resistant callus culture selected against fungal toxins (Saxena *et al.*, 2008). *In vitro* mutagenesis has given successful new ornamental cultivars of *Saintpaulia* and *Rhododendron.* Banerjee and Kallo (1989) obtained disease and pest resistant lines in cultivated tomato through *in vitro* selection. Jain (1997) irradiated strawberry plants axillary buds with gamma rays, giving 5% mutants which were selected in the presence of *Phytophthora cactorum* crude extract. These plants were also able to withstand both drought and disease tolerance. While studying somaclonal and *in vitro* mutagen-induced variability in grapevine, Kuksova *et al.* (1997) achieved induced polyploidy in somatic seedlings after gamma irradiation and the variability was seen among regenerated plants after field testing.

Advantages

- This technique involves creation of variation in the plant at cellular level using mutagens under *in vitro* conditions.
- Mutagen treatment can be given to a larger cell mass or tissues, fast multiplication of mutant cells via callus phase, *in vitro* selection is simple, and it requires less space selection of mutants and also allows its fast multiplication.
- After regeneration, the plantlets can be subjected to selection and multiplication of the desired genotypes.
- Since the frequency of somaclonal variation is low, it becomes empirical to increase the genetic variability by combining mutagenesis and tissue culture for crop improvement.

3.6 *In vitro* Selection for Crop Improvement

Tissue culture generates a wide range of genetic variability in plants through somaclonal variation (Larkin and Scowcroft, 1981) and can be effectively used by plant breeders for crop improvement. *In vitro* selection usually involves subjecting a population of cells to a suitable selection pressure and recovering any variant lines which have developed resistance or tolerance to the stress treatment. Tissue culture as a possible technology for obtaining the desired variants can lead the variation toward the expected outcomes while the probability to accomplish an *in vitro* selection depends on the accessibility of an effective regeneration system associated with an efficient selective agent (Rai *et al.*, 2011). If the tolerance has a genetic basis then tolerant lines can be selected for abiotic stresses such as salt, freezing, herbicide, and various chemicals such as heavy metals and through breeding programme the trait can be transferred to other plants as well. *In vitro* tissue culture based screening of tolerant cell lines is based on spontaneous or induced mutation *in vitro* (Rai *et al.*, 2011). Somaclonal variations could be the result of mutation caused by genetic and epigenetic changes during tissue culture processes (Ehsanpour *et al.*, 2007).

3.6.1 *In vitro* Screening for Abiotic Stress Tolerance

Biotic and abiotic stresses impose a major threat to agriculture. Abiotic stresses like soil salinity, alkalinity, metal toxicity, heat, cold and drought are expected to be more frequent in future due to disturbances in global climate, pose a serious challenge for scientists and researchers worldwide to ensure growing demand of food supply for the growing world population. Therefore, the efforts to develop stress-tolerant plants are of immense importance to increase crop productivity. In recent years, tissue culture based *in vitro* selection has emerged as a feasible and cost-effective tool for developing stress-tolerant plants. Plants tolerant to both the biotic and the abiotic stresses can be acquired by applying the selecting agents such as NaCl (for salt tolerance), PEG or mannitol (for drought tolerance) and pathogen culture filtrate, phytotoxin or pathogen itself (for disease resistance) in the culture media. Only the explants capable of sustaining such environments survive in the long run and are selected. *In vitro* selection is based on the induction of genetic variation among cells, tissues and/or organs in cultured and regenerated plants. The selection of somaclonal variations appearing in the regenerated plants may be genetically stable and useful in crop improvement. Here we present some case studies focusing on the progress made towards the development of stress-tolerant lines through tissue culture based *in vitro* selection.

Drought is the most serious threat to agriculture, limiting plant growth, development and productivity (Azami *et al.*, 2010). The situation demands crop breeding for drought-prone areas which does not seem feasible with classical plant breeding strategies, due to slow process, restricted gene pool availability, species barrier and other biological limitations (Razzaq *et al.*, 2004). Drought tolerant plants can be obtained by supplementing polyethylene glycol (PEG) or mannitol as selection agents in culture media containing explants. Only explants competent to tolerate selective agent survive in the long run and are selected. Cells resistant to selective agent are selected and subsequently regenerated into plantlets. Screening for drought tolerance and the traits that enhance drought tolerance is not straightforward in large mapping populations. Kumari *et al.* (2011) investigated the possibility of screening a mapping population *in vitro* for PEG-induced water deficit stress and recovery potential. In order to identify potato QTLs for drought tolerance and recovery potential an SNP marker-rich integrated linkage map was used. A total of 23 QTLs were detected under control, stress and recovery treatments explaining 10.3–22.4% of the variance for each phenotypic trait. Successful *in vitro* selection for drought tolerance using polyethylene glycol (PEG), sorbitol, mannitol and agar as selection agents has been also applied to several crops such as coconut, banana, rice, sugarcane, potato, sweet potato, and alfalfa (Bidabadi *et al.*, 2012). Significant improvement in drought tolerance has been reported in crops like maize (El-Aref, 2002), sugarcane, rice (Lutts *et al.*, 2001), wheat (Mahmood *et al.*, 2012) and mung bean (Gulati and Jaiwal, 1994) by exploiting somaclonal variation. PEG was used to simulate *in vitro* osmotic stress in the culture media and drought tolerant somaclones were regenerated.

Soil salinity is another widely recognized major threat to global food security. Several approaches have been advocated to address the salinity problem, but the most logical solution to maintain crop productivity in salt-affected areas is to enhance salt tolerance of crop plants (Karan and Subudhi, 2012). Due to the genetic and physiological complexity associated with salt tolerance, efforts to breed salt-tolerant plants have met with limited success. Although progress has been made in deciphering the genetic basis of salt tolerance, sustained efforts are needed to screen and utilize the natural variability in the available germplasm for improving crop adaptation in saline environments. With an enhanced understanding of molecular mechanisms and the associated genes for salt tolerance, it would be possible to breed salt-tolerant plants using an integrated approach involving conventional breeding and tissue culture technology. *In vitro* selection against salt tolerance is commonly achieved as a result of cells ability to compartment the excessive salts into the vacuoles, and survive by adjusting the osmotic pressure, but this adaptation causes reduction in cell division and expansion (Bressan *et al.*, 1985). The mechanism of this relationship is not yet known. However, some plants become so well adapted to high salts that they are unable to grow in the absence of salt selection pressure (Remotti, 1998). Bressan *et al.* (1985) found stable salt tolerant lines in tobacco indicating that high salt concentration acted as a selective as well as mutagenic agents leading to heritable salt tolerance. Glycinebetaine is known as osmoprotectant which is accumulated in certain plant species under salt and drought conditions. Sabry *et al.* (1995) found accumulation of glycinebetaine and sucrose in salt stressed wheat plants. Though there is a long list of *in vitro* selection against salt stress in several crop species, long-term salt adaptation requires more studies so that salt tolerant selected plants grow more like halophytic plants (Remotti, 1998). Luan *et al.* (2007) succeeded to obtain salt tolerant cultivars of sweet potato from induced mutation caused by EMS.

Cold stress is one of the abiotic stresses limiting crop production in temperate regions. In order to screening for chilling tolerance, Bertin *et al.* (1996) subjected embryo-derived calli of four rice varieties to different periods of chilling stress at 4°C, with a total length varying from 2 to 6 weeks, along with control calli cultivated without cold stress. Regenerated plants (R0) and their descendants in R1, R2 and R3 generations were cold-screened together with control plants. In all varieties, significantly higher chilling survival rates were obtained in R3 with *in vitro* grown plants than with control plants.

3.6.2 *In vitro* Screening for Phytoremediation

Heavy metals like cadmium, lead, mercury, being a highly toxic metal pollutant of soils, inhibits root and shoot growth, yield production, affects nutrient uptake and homeostasis, inhibits chlorophyll synthesis and photosynthesis, is frequently accumulated by agriculturally important crops and then enters the food chain with a significant potential to impair animal and human health (Liu *et al.*., 2003). Excessive amount of heavy metals may cause decreased uptake of nutrient elements, inhibition of various enzyme activities, induction of oxidative stress including alterations in enzymes of the antioxidant defence system (Sandalio *et al.*, 2001). At increased concentrations, both essential (copper, iron, manganese, molybdenum, zinc) and non-essential metals (e.g., cadmium, chromium, lead, and mercury) are toxic. Mercury and selenium can also be converted by plants into a volatile form to release into the atmosphere. Lead (Pb), a potentially toxic heavy metal results in soil fertility deterioration resulting in yield decline (Majer *et al.*, 2002). Phytoremediation is the ability of plants to remove pollutants from the environment. It is currently used as a simple and low-cost cleaning technology. The sensitivity of plants to heavy metals depends on an interrelated network of physiological and molecular mechanisms such as uptake and accumulation of metals, adsorption, transport and translocation, hyperaccumulation, volatilization, sequestration into vacuoles, efflux of heavy metals from cytoplasm to extranuclear compartments like vacuoles, complex formation of heavy metal ions inside the cell by various substances, for example, organic acids, amino acids, phytochelatins, and metallothionein, accumulation of osmolytes and osmoprotectants and induction of antioxidative enzymes, activation of plant metabolism and rapid repair of damaged cell structures (Cho *et al.*, 2003).

Little is known about the ability of certain plants to tolerate, detoxify, and store high concentrations of heavy metals. Plant cell and tissue culture is considered an important tool for that offer the advantage of testing the effects of contaminants under controlled conditions, allows fundamental studies that provide information about the plant-contaminant relationships, help to predict plant responses to environmental contaminants, improve the design of plants with enhanced characteristics for phytoremediation and enables study of the complex system of plant, soil, and microbial interaction in order to evaluate the effect of stress factors on metabolism, specific enzymes and metabolites involved in plant response to the pollutant (Couselo *et al.*, 2012). Plant tissue cultures such as callus, cell suspensions, hairy roots, and shoot multiplication cultures are applied frequently in phytoremediation research to study the interactions between plants and pollutants under aseptic conditions. Its usefulness has been demonstrated in many sites and this technology is now used by several environmental companies (Giachetti and Sebastiani, 2006). The hairy root culture constitutes a valuable tool for phytoremediation research and development (Flocco and Giulietti, 2007). Hairy roots of plants are among the several experimental systems which have been employed to improve the efficiency of phytoremediation. Hairy root disease, is caused by *Agrobacterium rhizogenes*, a Gram-negative soil bacterium causes a massive production of adventitious hairy roots at the wound site. Hairy roots possess tremendous potential for phytoremediation through accumulation and biodegradation of hazardous compounds. *In vitro* cultures offer a range of experimental advantages in studies aimed at examining the intrinsic metabolic capabilities of plant cells and their capacity for toxicity tolerance. The ability to identify the contributions of plant cells to pollutant uptake and detoxification without interference from microorganisms is of particular significance in the search for fundamental knowledge about plants. The ability of plants to remove pollutants from the environment is currently used in a simple and low-cost cleaning technology known as phytoremediation. Unfortunately, little is known about the metabolic pathways involved in the transformation of xenobiotic compounds and the ability of certain plants to tolerate, detoxify, and store high concentrations of heavy metals. *In vitro* phytoremediation involves identification of plant species which could tolerate toxic levels of heavy metals in soil. Many studies have demonstrated that plant tissue cultures are an extremely valuable tool in phytoremediation research. The results derived from tissue cultures can be used to predict the responses of plants to environmental contaminants (Doran, 2009). Many plant species have an inherent ability to accumulate/metabolize a variety of pollutants, but they normally produce little biomass. However, fast-growing trees are excellent candidates for phytoremediation because of their rapid growth, extensive root system, and high water uptake (Couselo *et al.*, 2012). Phytoremediation on a large scale through metal uptake by trees can be more effective, mainly because of deeper root system and a greater biomass (Fernández *et al.*, 2008). Poplars and willows are particularly suitable for remediation purposes (Di Lonardo *et al.*, 2011).

3.7 Meristem Culture for Production of Disease-Free Planting Material

Meristem is a zone of actively dividing cells, about 0.1 mm to 0.3 mm in diameter, situated in the top of buds, and extremities of roots. Endogenous contaminants do not easily invade in the meristem due to fast multiplication of meristem cells, and also because the phloem and xylem vessels are not present in meristem (Audet and Charest, 2008). Meristem culture was pioneered by Morel (1958-60) who cultured orchids and dahlias freed from viral disease. Conventional method of breeding has so far been used in plant breeding for developing disease resistant or tolerant lines. However, it has its own limitations like, it requires a donor which often carries many undesirable genetic loads. Also this needs more space, time and labour. Producing and maintaining a virus-free plant material is the key in terms of control of viral diseases (Mink *et al.*, 1998). These limitations can be overcome to some extent through meristem-tip culture either alone or in combination with heat therapy, chemotherapeutic agents or chemical treatment for generating disease-free planting material *in vitro*. The excised meristem tip is typically small and is removed by sterile dissection under the microscope. The explant comprises the apical dome with few leaf primordla, and excludes any differentiated provascular or vascular tissues. The pathogen cannot cope up with the fast multiplication of meristem cells giving rise to disease-free cells which regenerate into disease-free plantlets. Such cultures, once screened and certified, can form the basis of a guaranteed disease-free stock for further propagation Shoot development directly from the meristem avoids callus formation and adventitious organogenesis, ensuring genetic stability and minimize somaclonal variation. The axillary buds of *in vitro* plantlets derived from meristem-tip culture may also be used as a secondary propagule. This technique adds a high micropropagation rate to the original meristem-tip culture technique, which is commercially important to the horticulture industry. A further advantage of meristem culture is that the technique preserves the arrangement of cell layers necessary if a chimeral genetic structure is to be maintained. In a typical chimera, the surface layers of the developing meristem are of differing genetic background, producing the desired characteristics, e.g., variegated flower colour. Meristem culture combined with micropropagation technique to produce large number of disease-free plants has been used successfully in the removal of viruses from many plants (potato, sugarcane and strawberry) and is now used routinely for the eradication of many viral diseases from plant material. Three approaches, apical meristem culture, thermotherapy and thermotherapy combined with apical meristem culture, were used to obtain the virus-free potato plantlets (Faccioli, 2001).

Advantages of meristem culture

- It allows clonal propagation of disease-free material with maximal genetic stability.
- The meristem tip can be cryopreserved and stored as a propagule for future use.
- Easy to transport and easily meets quarantine requirement for export.

Limitations

- Meristems are difficult to work with, hence require skill.
- Confirmation at flowering and further generation is required.
- It still remains highly susceptible to infection in the presence of inoculums.

3.8 Cryopreservation for Germplasm Conservation

In vitro techniques are very useful for conserving plant biodiversity, including (a) genetic resources of recalcitrant seed and vegetatively propagated species, (b) rare, endangered and threatened plant species, (c) medicinal plant species, (d) disease-free plant propagules, and (e) elite genotypes and genetically engineered plant material (Engelmann, 2011). It ensures production and rapid multiplication of disease-free material. Germplasm can be stored *in vitro* in as isolated protoplasts, cells from callus or suspension cultures, meristem tips, shoot tips, somatic embryos, gametic embryos, or any other propagules. Short and medium term conservation is achieved by subculturing at intervals in to fresh media. Slow growth storage protocols are routinely employed for a large number of species, including numerous endangered plants, from temperate and tropical origin. The methods involved for *in vitro* germplasm conservation are slow growth culture and cryopreservation. Slow growth culture is a simple, effective and economic method and can be used in all species where shoot tip/ nodal explant are available. In this technique, growth is suspended by either cold storage or lowering oxygen concentration. Periodic renewal of cultures through serial subculturing is required. However, subculturing invites the risk of contamination by pathogen and genetic changes may also occur.

Cryopreservation (liquid nitrogen −196°C) also called freeze preservation is the most popular and effective technique for indefinite storage, protected from contaminations and with limited maintenance. At ultra low temperature (–196°C, using liquid nitrogen) metabolic processes and growth activity of plant cell and tissue propagule is brought to a halt retaining the viability of cryopreserved cells (Benson, 2008). Theoretically, the plant materials are in a stable state for an infinite period of time, provided ice formation inside the cells is low enough to allow plant recovery. Cryopreservation has become a very important tool for the long-term storage of plant germplasm with minimal space and maintenance (Sakai and Engelmann, 2007). Cryopreservation for germplasm purposes utilizes shoot tips and buds only but protoplasts, cells, tissues and somatic embryos can also be cryopreserved for other tissue culture processes. Cryopreservation requires the cells in the late lag or exponential phase ideal for freeze preservation. After thawing, these cytoplasm rich cells are able to retain their viability and grow again in culture medium. For tissues like meristems, shoot tips, embryos, etc., may easily get damaged by freezing due to the presence of highly vacuolated cells in large number challenging its viability for further regeneration. Prefreezing treatment given to cells before freezing results in their hardening and increased survival rates. Such hardening treatments include growing culture in presence of cryoprotectant or at low temperature or in the presence of osmotic agents like sucrose. These treatments function by either changing the cell water content, metabolite content or membrane permeabilites. Cryoprotectant such as Dimethyl sulphoxide (DMSO) glycerol, polyvinyl pyrollidone, polyethylene glycol (PEG) are added gradually to prevent plasmolysis of the cells for imparting protection to withstand low temperature. The sensitivity of cells to low temperature varies with the species. Rapid freezing (@ 200°C/minute) by dipping the vials with plant materials directly in liquid nitrogen has been successful for germplasm conservation of large number of species where plant material with small size and low water content has been chosen. For the controlled freezing method, the plant material is cooled stepwise from room temperature to intermediate temperature (–20°C), maintained at that temperature for thirty minutes followed with rapid freezing by dipping into liquid nitrogen. This method is applicable to wide range of plant materials including shoot apices, buds and suspension cultures. For better survival of preserved samples, thawing rapidly (from –196°C to about 22°C), minimizes the damaging effects of cellular ice crystal formation.

Application

Cryopreservation is an ideal method for long-term preservation of germplasm resources. It can also ensure the genetic stability of the materials, and it requires less storage space with the low cost for maintenance (Fang *et al.*, 2008). Cryopreservation is well advanced for vegetatively propagated species, and techniques are ready for large-scale experimentation in an increasing number of cases. Research is much less advanced for recalcitrant species due to their seed characteristics, viz., very high sensitivity to desiccation, structural complexity and heterogeneity in terms of developmental stage and water content at maturity. In comparison with crop species, limited research has been performed on cryopreservation of rare and endangered species. In addition, cryopreservation is receiving attention as a possible method for removing viruses (Helliot *et al.*, 2002; Wang *et al.*, 2003).

Advantages of Cryopreservation:

- Allows long-term preservation as metabolism comes to halt
- Low maintenance is required as liquid nitrogen only needs to be replenished
- There are no chances of contamination
- It is applicable to all species responding to *in vitro* culture conditions

Limitations

There are a few limitations on the use of this very useful technique. Loss of seed viability, loss of regeneration potential of explant, vitrification and albinism and power failure for long time are the major constraints which can be taken care of time to time.

- Sophisticated equipment and facilities required
- Skill and expertise is needed for its operation
- Cells/tissues get damaged due to ice crystal formation or high solute concentration during dessication.

Even though routine use of cryopreservation is still limited, an increasing number of examples where cryopreservation is used on a large scale can be found both in genebank for crops and in botanical gardens for endangered species.

3.9 Synthetic/Artificial Seed Production

Synthetic seed is a bead of gel containing somatic embryo or shoot bud with growth regulator, nutrients, fungicides, pesticides, etc., needed for development of complete plantlet. They do not need hardening and can be sown directly in field. Synthetic seeds from somatic embryos make a promising technique for the propagation of non-seed producing plants (e.g., orchids, grapes, watermelon, banana, etc.), transgenic plants, polyploids, apomicts, elite breeding lines with highly desirable traits, and for recalcitrant species where seed cannot be stored and offer high degree of genetic variations (e.g., mango, jack fruit, coconut, cocoa, etc.). Synthetic seeds can be adopted for commercially important agronomic and horticultural crops. Somatic embryos being bipolar structure, just like zygotic embryos, but somatic in origin, can germinate as normal seeds upon seeding. In contrast to zygotic embryos, somatic embryos are genetically identical as the process of their development does not involve sexual recombination. This shortens the generation time as no recombination and segregation is involved, speeding up the breeding process. The basic requirement for the application of synthetic seed technology in micropropagation is the production of high quality, vigorous somatic embryos that are capable to produce plants comparable to natural seeds. It requires development of large number of somatic embryos with synchronous maturation. The somatic embryos are encapsulated using gelling agents like agar, gelrite, alginate, carboxy methyl cellulose, carrageenan, guargum, sodium pectate, etc. Alginate hydrogel is the most widely used encapsulating agent because of its moderate viscosity and low spinnability of solution, low toxicity, quick gellation, low cost and biocompatibility characteristics. Alginate beads also provide better protection against mechanical injury. Since somatic embryos lack endosperm, nutrients and growth regulators are also added to the encapsulation matrix which serves as artificial endosperm and helps to increase the germination percentage of synthetic seeds upon sowing. The synthetic seeds can be stored (4°C) for long time say six months without losing viability. In addition to nutrients and growth regulators, fungicides, pesticides, antibiotics, activated charcoal and useful microbes can be integrated in the encapsulation matrix along with somatic embryos. The artificial seeds produced *in vitro* are free of pathogens. This offers the advantage of propagation of disease free propagules even across international borders, passing quarantine. The artificial seeds can also be cryopreserved for long-term storage and also for germplasm conservation.

Advantages

- Can be produced in large numbers
- Maintains genetic uniformity in plant population.
- It can be seeded directly in the field, with no transportation cost, any time, anywhere required.
- Low production cost per explant due to rapid multiplication rate.
- It has the potential for long-term storage without losing viability.

Limitations

- Lack of somatic embryos with synchronous maturation.
- Production of large number of high quality vigorous somatic embryos is difficult.

3.10 Embryo Rescue (Ovule Culture) After Wide Hybridization

Viable seeds are produced by the plants after intraspecific crosses, i.e., cross between two similar species. On the other hand, if inter-specific (cross between two different species) or intergenic cross (cross between two genera) occurs; no viable seed develops in the ovary, due to abnormal development of endosperm which results in death of embryo before maturity and sterile seed formation (Miyajuma, 2006). Besides, there are other barriers also which operate during the different stages of fertilization. However, in some cases in spite of pollination and fertilization, embryo does not develop due to incompatibility between the developing embryo and endosperm (Sandra, 2005). From such seeds, immature embryo can be taken out and grown on artificial medium. This technique of culturing immature embryo on artificial nutrient medium is called embryo rescue. This technique nurtures the immature or weak embryo, thus allowing it the chance to survive. In plant breeding, wide hybridization crosses can result in small shrunken seeds which indicate that fertilization has occurred, however the seed fails to develop, thus embryo rescue can assist in circumventing this problem (Bridgen, 1994). The first successful *in vitro* embryo culture was performed by Hanning in 1904 (Mehetre and Aher, 2004). Embryo rescue technology is very useful nowadays for regenerating plants from the embryos wide hybridization, playing an important role in modern plant breeding, allowing the development of many interspecific and intergeneric food and ornamental crop hybrids (Sage *et al.*, 2010). Recently, fruit breeding programmes have greatly increased the interest in exploiting interploid hybridization to combine desirable genetic traits of complementary parents at the triploid level through immature embryo rescue for the purpose of developing improved seedless fruits (Shen *et al.*, 2011). Seedlessness is one of the most important characteristics for mandarins for the fresh-fruit market and mandarin triploid hybrids have this trait. Triploid embryos are found in small seeds that do not germinate. Aleza *et al.* (2010) gave an effective methodology to recover triploid hybrids from 2x × 2x hybridizations based on *in vitro* embryo rescue and ploidy determination in citrus. Geerts *et al.* (2011) made wide crosses in *Phaseolus vulgaris* to transfer resistant genes from its wild relatives. However, due to post-zygotic incompatibility barriers, early hybrid embryo aborted limiting the success of F1 crosses. This problem was overcome by rescuing 2-day-old *P. vulgaris* embryos using a pod culture technique. Interspecific crosses allow for the introgression of important alleles of agricultural interest from wild species, such as the resistance or tolerance to abiotic and biotic stresses (Fratini and Ruiz, 2011). Fratini and Ruiz, (2011) rescued interspecific hybrids of Lens *in ovulo* and recovered interspecific hybrid plantlets.

Embryo rescue refers to the development of an immature or weak embryo into a viable plant in the artificial nutrient culture medium (Fig. 3). Embryo rescue has been widely used for producing plants from hybridizations in which embryo abortion occurs due to improperly developed endosperm. Embryo rescue techniques are among the oldest and most successful *in vitro* procedures. One of the primary uses of embryo rescue has been to produce interspecific and intergeneric hybrids. While interspecific incompatibility can occur for a wide variety of reasons, such as embryo abortion, production of small, shrunken seed post-fertilization. Embryo rescue have been very successful in overcoming these barriers to wide hybridization in several plant species (Clarke *et al.*, 2006). In addition, embryo rescue has been used to recover maternal haploids that have developed as a result of chromosome elimination following interspecific hybridization, to obtain progeny from intraspecific hybridizations that do not normally produce viable seed for, e.g., triploids have been recovered from crosses between diploid and tetraploid members of the same species, for overcoming seed dormancy and studying seed development and germination. Depending on the organ cultured, embryo rescue is sometimes preceded by ovule or ovary culture. One advantage of this technique is that embryo excision is delayed until the embryo becomes large enough to remove without damage. Also, for those species affected by dormancy, removing the embryo may overcome any inhibitory effects imposed by the surrounding ovular tissues. Embryos are difficult to excise when very young or from small-seeded species. To prevent damaging embryos during the excision process, ovules are cultured, referred to as ovule culture or *in ovolo* embryo culture. Ovaries are collected prior to embryo abortion, surface sterilized and the ovules removed and placed into culture (Rangan, 1984). Nurse cultures have been used for rescuing embryos (Williams *et al.*, 1982). This technique involves inserting the embryo from an incompatible cross into endosperm removed from a related compatible cross. In ovary or pod culture, the entire ovary after surface sterilization is cultured in the medium, without damaging the ovary. Ovary culture has been used for rescuing *Tulipa* interspecific hybrid embryos (Van Creij *et al.*, 1999), peach (*Prunus persica* L. Batsch) (Pinto *et al.*, 1994).

Embryo rescue technique involves emasculation of female flower bud (unopened) and hand pollination of stigma with fresh pollen from the wild species. Plant growth regulators may be applied to the petiole of the crossed flower to delay abscission (Mallikarjuna, 1999) and other open flowers and selfed pods should be removed to divert photo-assimilates for the development of crossed pods. Immature pods, 14 to 22 days (optimized for chickpea) after cross-pollination, are to be removed from mother plants before they abort. Embryos are rescued *in ovulo* (embryos excised when they broke through the seed coat 2 to 6 weeks later) and hybrid plantlets regenerated (Clarke *et al.*, 2006). Ovule culture following hybridization in some species has been done to observe the processes of fertilization and embryo development.

Applications

- Breeding of incompatible interspecific and intergeneric species
- Overcoming seed dormancy
- Determination of seed viability
- Recovery of maternal haploids that develop as a result of chromosome elimination following interspecific hybridization
- To study the physiology and biochemistry of seed germination and development

Limitations

This phenomenon is widely reported across many plant species and is a major bottleneck in wide hybridization and doubled haploid culture (Kumari *et al.,* 2009). Complete to partial albino plants were observed following embryo rescue *in vitro* between chickpea cultivars (Clarke *et al.*, 2006). Varying degrees of albinism, from partial to complete pigment loss, occur following interspecific hybridization. In interspecific hybrids, incompatibility between the plastome (plastid genome) and nuclear genome has long been recognized as a major factor inhibiting chloroplast development and chlorophyll or pigment formation (Hagemann 1992; Kirk and Tilney-Basset 1978).

3.11 Anther/Pollen/Microspore Culture for Haploid Production

The conventional breeding techniques are time consuming and self-incompatibility act as barrier for distant hybridization and fertilization. An anther, pollen or microspore culture technique offers great opportunities for accelerating breeding progress. The technique of haploid induction has great value in vegetable breeding as it has the ability to shorten the breeding period. The production of homozygotes is important both for genetic studies and hybrid seed production in highly cross-pollinated crops. It involves the culture of excised anthers/pollen/microspore on sterile nutrient medium. The ploidy level of haploid plantlets derived from pollen grains can be doubled using colchicine to obtain doubled haploid plant, completely homozygous in nature. Culture with pollen of different species may enable certain crosses to be made that are normally impossible due to incompatibility factors in the stigma.

Anther culture

Culturing anther on a suitable media to regenerate into haploid plants is called anther culture (Fig. 4). Haploid plants were first discovered in *Datura stramonium* by A.D. Bergner in 1921. Guha and Maheshwari in 1964 pioneered the formation of embryos from anthers of *Datura innoxia*. After this, haploid plants have been produced via anther culture in more than 170 species. The anther culture technique is useful in haploid production. Haploid plant is defined as a sporophyte with gametophytic chromosome number. Haploid plants produced from anther culture cannot form viable gametes and hence no seed setting takes place for further perpetuation. Therefore, it is necessary to double the chromosome number of haploids to obtain homozygous diploids or dihaploid plants followed by their transfer to culture medium for further growth. Culture medium and genotype are the main factors affecting androgenesis (Çömlekçioğlu *et al.*, 2001). For the success of anther/pollen/microspore culture, stage of pollen development is the most critical which corresponded with the bud size for selection using stereozoom microscope and heamocytometer for cell count. Cold or heat pretreatment is also one of the important factors affecting anther culture, apart from genotype, media, nutrients, hormones, additives and growth conditions. The advantage of this method is that there exist thousands of microspores in each anther and numerous haploid plants can be obtained from a single anther. The main principle of anther culture is the prevention of the pollen cells development that forms gamete, and instead the induction of immature pollen cells to directly form embryos similar to somatic cells. After that anther culture for haploid plant production was studied by a lot of researchers in many plant species to develop hybrids. Taskin *et al.* (2011) cultured anthers of five pepper genotypes in order to optimize the frequency of embryo production. The most significant application of haploid plant production is the rapid production of homozygous doubled haploid lines even with self-incompatible species. Dihaploid plants undergo normal meiotic segregation, they can be used to determine and isolate recessive mutants.

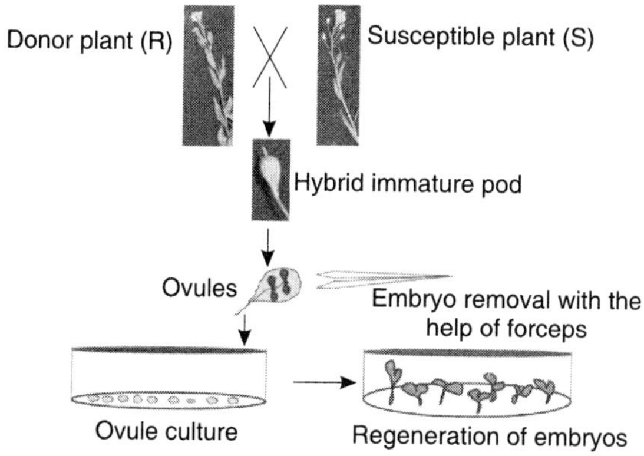

Fig. 3 Embryo rescue following wide hybridization for transfer of disease resistance gene.

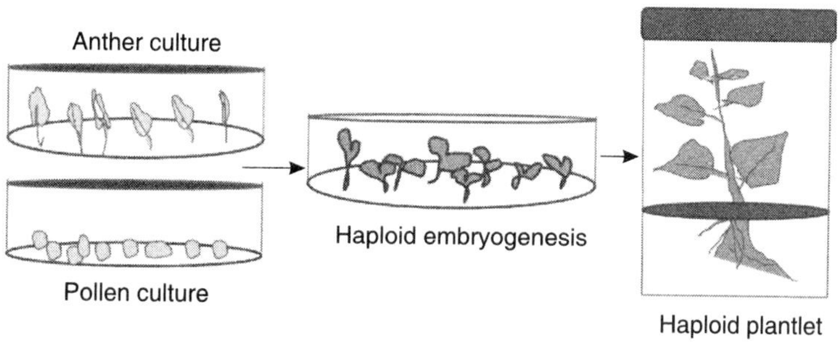

Fig. 4 Anther culture and haploid plant regeneration.

Microspore culture

Microspore embryogenesis is an important tool in plant breeding for the generation of haploids, isogenic lines and new varieties through double-haploid plants (Guha and Mukherjee, 1999). Microspores can switch their normal gametophytic development *in vitro* towards an embryogenic pathway. Microspore culture is highly dependent on the species, genotype and physiological state of the donor plants, induction conditions, culture media composition and state of development of the explants. Stage of development of anthers need to be optimized by selecting appropriate bud size and division stage of the pollen grains for haploid plant development. In some species it is possible to culture isolated pollen grains. This precludes the possibility that any resulting plantlets might be derived from somatic tissue of anther rather than pollen. However, in culturing isolated pollen grains, anther tissue may have to be used as nurse tissue. The most widely used method to induce microspore embryogenesis in several species that are particularly recalcitrant for plant breeding purposes is anther culture. Basic studies on the process have been mainly pursued in model species, such as rape seed, tobacco and barley (Chupeau *et al.*, 1998), but reports on other economically interesting crops are very limited. It has been applied to many species, including crops like potato (Chani *et al.*, 2000), tomato (Zagorska *et al.*, 2004), pepper (Gyulai *et al.*, 2000), capsicum (Barany *et al.*, 2005).

Application

Diploidisation of haploid plants result in rapid achievement of homozygous traits in doubled haploids. Doubled haploid homozygous lines can be used directly as cultivars or may be used in breeding programme, e.g., doubled haploids have been used for rapid development of inbred lines in hybrid maize programme. The anthers from F1 hybrids of desirable cross are excellent breeding material for raising anther derived homozygous doubled haploids in which complementary parental characteristics are combined in one generation. The doubled haploid plants are subjected to selection for superior plants. This approach is described as hybrid sorting where recombinant superior gametes are virtually being selected since the heterozygous gene combination in the F1 hybrid is transformed into homozygous combinations. Hybrid sorting reduces the time required for haploid breeding by 4-5 years as in conventional breeding by pedigree/ bulk method, the same requires ten years. Also, selection among DH lines reduces the size of breeding population.

Homozygous lines can also be obtained via conventional methods, although this process takes 10 to 12 years in open pollinated plants, and 6 to 7 years in self-pollinated plants (Dunwell, 2010). However, this period can be shortened for 1 to 2 years with the anther culture method. Anther culture has been used for crop improvement in several species. In fruit crops, high heterozygosity, long generation cycle with long juvenile period, large size, and often self-incompatibility do not allow the haploidization easily through the conventional methods (Germanà, 2006). The production of haploid plants through anther culture is widely used for breeding purposes, as an alternative to the numerous cycles of inbreeding or backcrossing usually needed to obtain pure lines in conventional breeding (Debnath *et al.*, 2007).

Advantages

- One of advantages of tissue culture techniques is that the protocols may permit the recovery of variants not easily obtained by conventional breeding.
- Production of double haploids through anther culture is a rapid approach to homozygosity that shortens the time required for the development of new cultivars as compared to conventional methods, which require at least 6-7 generations (Xa and Lang, 2011).
- Dominant gene effect disappears and additional gene effects double.
- Haploid and doubled haploid plants are valuable materials for genetic and cytogenetic studies as well as for plant breeding purposes such as production of pure line, hybrid seeds, gene mapping, resistance to many fungal and viral diseases that affect this crop.
- Anther culture shortens the breeding cycle of new varieties and allows early expression of recessive genes.
- This technology can aid plant breeding programmes by accelerating production of homozygous lines and its fixation.

Limitations

Although, anther culture is an important method in homozygous line production, the yield of haploid plants is low in most species (Kristiansen and Andersen, 1993). Beside the advantages, anther culture has certain constraints, like, callus production, low plant regeneration, and high proportion of albino plants produced.

- Many species are not yet amenable for haploid production.
- Deleterious mutations may be induced during *in vitro* phase.
- Occurrence of gametoclonal variation limits the use of anther derived embryos for genetic transformation.

3.12 Establishment of Regeneration for Genetic Transformation

Biotechnology has enormous potential in increasing crop productivity in a sustainable manner by developing transgenic crops that are resistant to viral, bacterial and fungal diseases, pest attacks, high in nutritional value and tolerant to several biotic and abiotic stresses (Kumari *et al.*, 2010). Crop productivity and quality are seriously affected by several biotic and abiotic stresses, which destabilize rural economies in many countries. With the advent of molecular genetics and biotechnology, it is now possible to genetically engineer plants to be more tolerant to many environmental adversities like cold, drought, heat and salinity, which cause major economic loss to agriculture worldwide. Despite continued efforts, traditional breeding gave limited success in imparting crop plants with better stress tolerance. Various traits such as biotic stress resistance, quality and storage life have been successfully engineered into vegetable crops and some of them have been commercialized. Significant progress has been made to manipulate vegetable crops for abiotic stress tolerance, quality improvement, pharmaceutical and industrial applications. Genetic engineering and tissue culture technology has been used effectively to isolate, clone and characterize useful genes for their further use to develop transgenics (Kumari *et al.*, 2010).

Crop improvement through genetic engineering involves direct and indirect methods for gene transfer in target crop. Effective regeneration and transformation systems are the prerequisite for successful genetic transformation (Kumari *et al.*, 2010). Transgenic development through tissue culture mode mainly comprises four major steps: establishment of regeneration protocol, preparation of gene construct, transformation system, stable gene integration and finally expression and segregation analysis in ensuing generations. *Agrobacterium* and biolistics are the most widely adopted method. Still today the most widely and cost effective method of transformation is *Agrobacterium* mediated marker-free method. One of the major prerequisite of genetic transformation is regeneration protocol for that particular species. The protocol developed for mass multiplication can also be used for genetic transformation. Antibiotic sensitivity assay becomes a must to check for selection after transformation.

Application

The accumulated hectarage planted with biotech crops, during the 15 years (1996 to 2010) exceeded 1 billion hectares. It took 10 years to reach the first 500 million hectares in 2005, but only half that time, 5 years, to plant the second 500 million hectares to reach a total of 1 billion hectares in 2010 (James, 2010). A record 87-fold increase in hectarage between 1996 and 2010, making biotech crops the fastest adopted crop technology in the history of modern agriculture. Insect pest attack is one of the major biotic stress affecting crop yield, productivity and quality. Chemical control measures are environmentally hazardous and non-sustainable. Bt-transgenics have proved to be highly effective in management of lepidopteran pests of several crops. The most successful example of transgenic global adoption is of Bt cotton developed by Monsanto,

USA, for insect resistant. This Bt gene variants have been successfully transferred in canola, soybean, rice, brinjal, etc. Other genes such as lectins, protease inhibitors, etc. have also shown promise. Genetic engineering offers elegant ways to fortify food with vitamins and minerals, as demonstrated with rice, canola and tomato. "Amflora", a potato with high quality starch is approved by EU for commercialization for the first time. In potato, a gene conferring broad spectrum resistance to late blight disease has been transferred from the wild species, *Solanum bulbocastanum* to the cultivated species through somatic hybridization (Hagelson *et al.*, 1998). This gene can now be incorporated into any potato variety through conventional breeding and would not attract biosafety regulation. In 2010, 15.4 million farmers in developing countries grew biotech crops. Herbicide tolerant soybean continued to be the principal biotech crop in 2010, followed by biotech maize, biotech cotton and biotech canola. Herbicide tolerance has consistently been the dominant trait (soybean, maize, canola, cotton, sugarbeet and alfalfa). In 2010, the stacked double and triple traits occupied a larger area (22% of global biotech crop area) than insect-resistant varieties. Biotech maize with eight genes, named Smartstax^{TM}, was released in the USA and Canada in 2010 with eight different genes coding for several pest resistant and herbicide tolerant traits. Future stacked crop products will comprise both agronomic input traits for pest resistance, tolerance to herbicides and drought plus output traits such as high omega-3 oil in soybean or enhanced pro-Vitamin A in Golden Rice. Global value of the biotech seed market alone was valued at US$11.2 billion in 2010 with commercial biotech maize, soybean grain and cotton valued at ~US$150 billion for 2010 (for more details visit: http://www.isaaa.org).

Benefits from biotech crops

Diversity conservation

Biotech crops are a land-saving technology, capable of higher productivity, and thereby can help preclude deforestation and protect biodiversity in forests and in other *in situ* biodiversity sanctuaries.

Enhanced food and feed production and poverty alleviation

If the 229 million tonnes of additional food, feed and fibre produced by biotech crops during the period 1996 to 2009 had not been produced by biotech crops, an additional 75 million hectares of conventional crops would have been required to produce the same tonnage (James, 2010).

Helping mitigate climate change and reducing greenhouse gases.

Biotech crops contribute to a reduction of greenhouse gases and help mitigate climate change in two principal ways: reduced use of pesticides and herbicides, secondly, additional savings from conservation tillage for biotech food, feed and fibre crops, led to soil carbon sequestration.

Even though genetic engineering has a lot of potential in genetic improvement of crops, this technology may not readily be operational because of regulatory embargoes, environmental concerns, biosafety concerns and consumer acceptance of genetically modified (GM) food, etc. Keeping in view the issues relating to biosafety aspects and risk assessment as per the regulatory bodies set guidelines, the transgenics can either be released directly after testing as essentially derived variety or can be used as non-recurrent parent in further breeding programme to transfer the gene in any desirable background without inviting biosafety concerns. Presently commercialization of transgenics is under hammer in lot of countries owing to its unpredicted effect on environment and health in long term. Also because of IPR protection of the technology and the gene construct, it cannot be freely used for crop improvement as its commercialization will invite royalty to the IPR holder. Although the progress in commercialization of transgenic crops has been relatively slow, transgenic crops will contribute significantly to agriculture in near future.

3.13 Cell Suspension Culture for Production of Secondary Metabolites

The compounds/biochemicals which are not directly involved in primary metabolic processes like respiration, photosynthesis, etc., are known as secondary metabolites. Plant cell culture can also be used to produce a variety of compounds like alkaloids, terpenoids, phenyl propanoids, etc., with various biological activities like antimicrobial, antibiotic, insecticidal, valuable pharmacological and pharmaceutical activities. Plants are the source of a large variety of biochemicals which are produced as both primary and secondary metabolites. Primary metabolites include nucleic acids, proteins, carbohydrates and fats which along with their intermediates function for survival of cell and organism. Compounds like alkaloids, non-protein amino acids, terpenoids and phenolics are grouped under secondary metabolites which do not participate in vital metabolic function of cell. Primary metabolites essentially provide the basis for growth and reproduction, while secondary metabolites for adaptation and interaction with the environment. As secondary metabolites provide industrially important natural products like colour, insecticides, antimicrobials and fragrances, therapeutics, etc., they are of great economic importance. Therefore, plant tissue culture is being potentially used as an alternative to plants for production of secondary metabolites. This technology is now available on industrial scale, for example, the commercial production of compounds "shikonin" from *Lithospermum erythrorhizon* cell culture. Secondary metabolites can be obtained from plant cell suspension and hairy root cultures and their exudates. The efficiency for product recovery of metabolites can be increased by various methods, based on the principle of continuous product release into the cultivation medium. The most common methods for metabolite recovery are elicitation, influencing membrane permeability, and *in situ* product removal (Zhenzhen *et al.*, 2012). The biosynthetic pathways can be influenced by cultivation conditions, transformation, or application of elicitors. The membrane permeability can be altered through the application of chemical or physical treatments. Product removal can be greatly increased through a two-phase system and the introduction of absorbents into the cultivation medium.

The capability to cultivate plant cells, tissues and organs in liquid media has made an important contribution to modern plant biotechnology with respect to the production of commercially valuable compounds like secondary metabolites and bioactive compounds such as hormones, enzymes, proteins, antigens, food additives and natural pesticides from plants, since 1950's for various industrial uses like in pharmaceuticals, cosmetics, food industry, etc. (Su and Lee, 2007). For many plant species of commercial interest, mass cultivation may not be possible under natural conditions due to environmental, ecological or climatic conditions and/or insufficient lands may limit the cultivation processes. Under such circumstances plant cell and tissue culture techniques under controlled and reproducible conditions, independent of geographic and climatic conditions can be useful. The main advantages associated with *in vitro* plant systems include the manipulation of environmental conditions, the rapidity of production, and cheaper downstream processing for product recovery from the culture medium (Doran, 2000).

Cell culture studies begin with callus initiation, screening for genetic stability, evaluation of each cell line so that the best performing lines can be selected for cell suspension culture scale up. The final critical step is the bioreactor studies leading to a possible commercial production of secondary metabolites. Bioreactors act as a biological factory for the production of high-quality products and provide many advantages like controlled supply of homogeneous culture in large volumes, better control of cultural and physical environment for easy optimization of growth parameters for achieving optimum yield of metabolites and bioactive compounds, simpler and faster harvest of cells, offer the opportunity to perform biosynthetic and/or biotransformation experiments related to metabolite production with enzyme availability (Fulzele, 2000; Su, 2006). The first patent for the cultivation of plant tissue was received in 1956. Shortly afterwards, the National Aeronautics and Space Administration (NASA) started to support research in the field of plant cell cultures for regenerative life support systems (Sajc *et al.*, 2000). Industrial companies in Japan, The Japan Tobacco Inc.'s applied this technology for the commercial production of tobacco cells as raw materials for cigarettes. Culture of hairy roots produced by transformation with *Agrobacterium rhizogenes* has been shown to be a more efficient system than cell culture for the production of compounds which are normally synthesized in roots of intact plants. Due to scale-up and automation advantages bioreactors have been used for the cultivation of hairy roots for secondary metabolite production (Ziv, 2000). Banasiuk *et al.* (2012) demonstrated the application of *in vitro* culture for the production of pharmacologically important compounds (e.g., naphthoquinones, flavonoids, anthocyanins, phenolic compounds from carnivorous plant Drosera belonging to endangered species. *Drosera herba* extracts have been reported to possess bacteriostatic and antitumour activity. In spite of all these advantages, cell culture techniques have not been effectively exploited on industrial scale bioreactors due to the slow growth rates of plant cell cultures and the high capital costs involved and have been limited to a handful of applications such as the production of shikonin from *Lithospermum erythrorhizon*, berberine from *Coptis japonica* and ginsenosides from *Panax ginseng* (Celiktas *et al.*, 2010). Phyton Biotech located in Germany is another good example of industrial exploitation of plant cell culture for the production of the Taxanes on large scale (Phyton Biotech, 2006). Callus cells obtained from the transgenic plants can also be grown in liquid media to establish transgenic cell suspension cultures for recombinant protein production (Su, 2006). Whole plant system can also be used for the synthesis of recombinant proteins after genetic transformation with target gene, e.g., genetically transformed banana with edible vaccines. Cell culture technique is being used as an efficient system for production of high-value natural plant products at industrial level. Also, cultured cells of many plant species produce novel biochemicals which have otherwise not been detected in whole plants. In the 1950s and 1960s, great efforts were made by the Pfizer Company to culture plant cells in liquid medium (suspension culture) similar to culture of microbes for production of natural plant products as an alternative to whole plants. Different kinds of bioreactors have been designed for large-scale cultivation of plant cells. Mitsui Petrochemical Co. of Japan for the first time in 1983 commercially produced shikonin, a natural colouring substance used in dyes, from the cell cultures of *Lithospermum erythrorhizon* through suspension culture. It is used as antiseptic and as dye for cosmetics. Since then many valuable secondary metabolites like taxol, berberine, etc., has been obtained using tissue culture. However, genetic instability associated with callus cells due to somaclonal variation is a potential disadvantage in using cultured plant cells for recombinant protein production (Su, 2006).

Application of cell suspension culture

Cell suspension cultures are particularly suitable for physiological, biochemical and molecular studies of the process of somatic embryogenesis and its different stages – induction development, maturation and conversion. The knowledge of genome and development of protocol of cell suspension will provide the opportunity to identify and characterize genes involved in the whole process of somatic embryogenesis. The use of fine suspension culture offers the opportunity to confirm single-cell origin of somatic embryos and the asymmetry of the first cell division, which starts the process. The other advantage of cell suspension culture is the use of single embryogenic cells and somatic embryos as a source for direct gene transfer via particle bombardment, transit gene expression and confocal microscopy observation. Induced mutagens produce more frequency of mutants than spontaneous ones and screening them at cellular level also inhibits chimeric formation which is a drawback in mutation breeding. Cell cultures are effectively utilized for production of bio-chemicals and secondary metabolites on a commercial scale.

Advantages of cell culture systems over the conventional cultivation of whole plants are:

- Quicker and higher yield of product from very small amount of initial culture of cells, for example, yield of shikonin produced from cell culture is 20% more than from plants.
- In case of plant materials available in limited numbers like *Lerythrorhizon*, cordyceps, gynoderma, *in vitro* culture of mycelium cells will conserve such germplasm.
- Cell culture ensures continuous supply of metabolites. In conventional system, source plant may be seasonal, location specific and also subject to environmental degradation.
- Low cost precursors are supplied as substrates to cell cultures for conversion to the high cost final product, thus minimizing labour, cost and time. Also, specific substrates can be converted to more valuable products through bioconversion.
- Mutant cell lines can be utilized to produce novel compounds not found naturally in plants, e.g., cell suspension cultures of *Rauwolfia serpentina* produce novel glucosides of ajmaline (alkaloid).

Limitations

- High production cost is involved.
- Lack of knowledge of biosynthetic pathways of many compounds is a major bottleneck.
- Cultured plant cells are often unable to produce high value compounds.

4. PROBLEMS ASSOCIATED WITH TISSUE CULTURE

The development of tissue culture protocols is a rigorous procedure that involves optimizing the various chemical, physical and environmental growth parameters (Bairu and Kane, 2011). The artificial growth conditions are not the perfect substitutes for the natural ones. Despite the meticulous efforts involved in growing plants *in vitro* and the advances made in plant tissue culture, the application of this technique is still hampered by various physiological and developmental problems. Furthermore, the natural variation in physiology and growth requirement among plant species warrants the need to optimize protocols for individual species. The problems faced are abnormal growth, genetic instability, shoot-tip necrosis, fasciation, tissue proliferation, epigenetic changes, somaclonal variation, effect of phenolics on rooting and vitrification (Bairu and Kane, 2011).

5. STATUS OF TISSUE CULTURE TECHNOLOGY APPLICATION IN INDIA

Looking into the application of plant tissue culture technology, several plant tissue culture laboratories (govt., govt. supported and private owned, private support and owned) have come up worldwide, and more so in India. In India, there are about 100 commercial plant tissue culture units with a minimum production capacity of about 1 million plants per year from each of the units. Among these, at least 20 of the units have larger production capacities, with 5 to 10 million plants/year. In addition, there are more than a dozen smaller units with 0.2 to 0.5 million plant production capacities where single crops are being produced. The Government of India has identified micropropagation industry as a priority area for further research, development and commercialization (Singh and Shetty, 2011). Over the past 20 years, the Ministry of Science and Technology, DBT, DST, CSIR and other funding agencies has supported huge number of projects for research and development in this field for micropropagation and crop improvement. The major consumers of tissue culture plants are the State Agriculture and Horticulture Departments, Agri Export Zones, sugar and paper industries, private farmers and floriculturists and herbal medicines. It is important to note that the demand for some crops high value crops and recalcitrant crops like banana, grapes, pineapple, strawberry, sugarcane, potato, turmeric, ginger, cardamom, vanilla and ornamentals like anthuriums, orchids, chrysanthemums, rose, lily, and gerberas are on the rise (Singh and Shetty, 2011). A concerted effort is being made by the government and the industry to ensure that plant tissue culture technology with enormous commercial potential, emerges as an important industrial activity. For more details on the commercialization of tissue culture refer to the article by Singh and Shetty, 2011. Table 3 shows some of the varieties released through the application of different tissue culture techniques (the list for transgenic crops is not included here).

Table 3 Varieties developed through the application of plant tissue culture

Crop	*Technique involved*	*Advantage over wild type*
Banana, sugarcane, jatropha, ornamental plants	Micropropagation	True to type propagation of elite and healthy propagules
Sugarcane	Somaclonal variation	Increase in cane and sugar yield, and resistance to eye-spot disease
Potato	Somaclonal variation	Improvement of tuber shape, colour and uniformity, and late blight resistance
Tomato	Somaclonal variation	Increased solids, resistance to *Fusarium* race2

Contd...

Contd...

Banana	Somaclonal variation	CIEN BTA-03, a variant of Williams variety is resistant to yellow Sigatoka disease
Sugarcane	Meristem culture	AT626 & BT 627 of sugarcane variant resistant to sugarcane mosaic virus A and B
Sugarcane	Meristem culture	'Ono', a sugarcane variant from variety, Pindar resistant to Fiji disease
Tobacco	Somaclonal variation	ATCC 40463, a tobacco variety with enhanced flavour
Corn	Somaclonal variation	DK 671, corn variety with higher yield, with lasting green colour and higher seedling vigour
Citronella	Somaclonal variation	Bio-13 a variety of citronella
Brassica	Anther culture and doubled haploid	Bio-902, Bio-YSR variants of *Brassica* parent 'Varuna' with enhanced seed yields
Tomato	Transgenic	Flavr-savr for slow ripening and increased shelf life
Cotton, canola, soybean, rice etc.	Transgenic (Bt gene)	Insect pest resistance

6. CONCLUSION

The rapid increase in population and economic development has led to severe environmental degradation acting as driving forces for the change in climate. Recent analyses of the possible adverse effects of climate change on agriculture in developing countries have raised food security concerns, due to fall in crop productivity. The projected climate change under various scenarios is likely to have implications on food production, water supply, biodiversity and livelihoods. Agriculture typically plays a larger role in developing economies than in the developed world. In India nearly 55% of the population depends directly on the climate sensitive sectors such as agriculture, fisheries and forests for their livelihood (Ahmad *et al.*, 2011). Any change in countries rainfall and temperature pattern impacts agriculture and hence the country's economy and food security. It has become increasingly clear in recent years that for any poverty reducing growth strategy to be sustainable, it must address environmental concerns and sustainable utilization of generally limited natural resources. Improvements in the economic well-being of the population can only be sustained in the long run if the natural resources are utilized in a sustainable manner. In the changing climate scenario, tissue culture technologies can contribute immensely for increasing crop productivity, assist conventional breeding, and generation of novel crops and germplasm conservation for future use. Plant tissue culture techniques have an important role to play in the production of agricultural, horticultural and ornamental plants and in the manipulation of plant for improved agronomic performance. Plant tissue culture research is a multidimensional science that offers exciting prospects to future improvements in crop productivity while most nursery have introduced the techniques of micropropagation, other dimensions of tissue culture research have been less well publicized for example: the potential for selecting pathogen or stress resistant plant clones, the creation of novel genetic combinations through somatic hybridization are some of the novel techniques that have not been frequently transposed to the nursery industry. Micropropagation can be used to produce a large number of plants that are genetically identical to parent, as well as one another. Virus-free plants can be raised using apical meristem of virus-infected plants. Homozygous plants can be obtained in a single generation by diploidization of the haploid cells through anther or pollen culture. Protoplast technology has made it possible to develop somatic hybrids and cybrids of distantly related plant species and genus. Protoplasts are also a suitable material for genetic engineering of plants in a manner similar to gene transfer into bacteria. Cell culture may be an important source of induction and selection of cell variants for production of new varieties of economically important plants.

References

Ahmad, J., Alam, D., and Haseen, S. (2011). Impact of climate change on agriculture and food security in India. *Int. Jr. of Agril., Env. and Biotech.* 4 (2): 129-137.

Ahmed, M., and Barkly, E. (2006). Gene Transfer between Different *Trichoderma* Species and *Aspergillus niger* through intergeneric protoplast fusion to convert ground rice straw to citric acid and cellulose. *Appl. Biochem. Biotechnol.* 135: 117-132.

Aleza, P., Juárez, J., Cuenca, J., Ollitrault, P., and Navarro, L. (2010). Recovery of citrus triploid hybrids by embryo rescue and flow cytometry from 2x x 2x sexual hybridisation and its application to extensive breeding programmes. *Plant Cell Rep.* 29(9): 1023-34.

Antonelli, N.M., and Stadler, J. (1990). Genomic DNA can be used with cationic methods for highly efficient transformation of maize protoplasts. *Theor. Appl. Genet.* 80: 395-401.

Audet, P., and Charest, C. (2008). Allocation plasticity and plant-metal partitioning: Meta-analytical perspectives in phytoremediation. *Environmental Pollution* 156 (2): 290-296.

Auffhammer, M., Ramanathan, V., and Vincent, J.R. (2012). Climate change, the monsoon, and rice yield in India. *Climatic Change* 111(2): 411-424.

Bairu, M.W. and Kane, M.E. (2011). Physiological and developmental problems encountered by *in vitro* cultured plants. *Plant Growth Regul.* 63: 101-103.

Ball, E. (1946). Development in sterile culture of stems tips and subjacent regions of *Tropaeolum malus* L. and of *Lupinus albus* L. *Am. J. Bot.* 33: 301-318.

Banasiuk, R., Kawiak, A., and Królicka, A. (2012). *In vitro* cultures of carnivorous plants from the Drosera and Dionaea genus for the production of biologically active secondary metabolites. *BioTechnologia.* 93(2): 87-96.

Banerjee, M.K., and Kalloo. (1989). Role of phenols in resistance to tomato leaf curl virus, *Fusarium* wilt and fruit borer in *Lycopersicon. Curr. Sci.* 58: 575-576.

Barany, I., Melendi, P.G., Fadon, B., Mityko, J., Risueno, M.C., and Testillano, P.S. (2005). Microspore-derived embryogenesis in pepper (*Capsicum annuum* L.): subcellular rearrangements through development. *Biol. Cell* 97: 709-722.

Behera, S.K., Mishra, A.K., Sahu, N., *et. al.* (2012) The study of microclimate in response to different plant community association in tropical moist deciduous forest from northern India. *Biodiversity and Conservation* 21 (5): 1159-1176.

Benson, E.E. (2008). Cryopreservation Theory. In: Plant Cryopreservation: A Practical Guide, B.M. Reed, (ed.). Springer Press, Germany.

Bertin, P., Kinet, J.M., and Bouharmont, J. (1996). Heritable chilling tolerance improvement in rice through somaclonal variation and cell line selection. *Australian Journal of Botany* 44(1): 91-105.

Bidabadi, S.S., Meon, S., Wahab, Z., Subramaniam, S., Mahmood, M. (2012). *In vitro* selection and characterization of water stress tolerant lines among ethyl methanesulphonate (EMS) induced variants of banana. *AJCS* 6(3): 567-575.

Bressan, R.A., Singh, N.K., Handa, A.K., Kononowicz, A.K., and Hasegawa, P.M., (1985) In: *Plant Genetics*, M. Freeling, (ed.). New York: Liss, pp 755-769.

Bridgen, MP. (1994). A review of plant embryo culture. *Hortsci.* 29: 1243-1246.

Carlson, P.S., Smith, H.H., and Dearing, R.D. (1972). Parasexual interspecific plant hybridization. *Proc. Natl. Acad. Sci.* USA, 69: 2292-2294.

Celiktas, O.Y., Gurel, A., and Sukan, F.V. (2010). Large scale cultivation of plant cell and tissue culture in bioreactors, *Transworld Research Network* 37/661 (2), Fort P.O. Trivandrum, Kerala, India: 1- 54 ISBN: 978-81-7895-474-5.

Chani, E., Veilleux, R.E., and Boluarte-Medina, T. (2000) Improved androgenesis of interspecific potato and efficiency of SSR markers to identify homozygous regenerants. *Plant Cell Tissue Organ Cult.* 60: 101-112.

Chattopadhyay, C., Bhattacharya, B.K., Kumar, V., Kumar, A., and Meena, P.D. (2011). Impact of climate change on pests and diseases of oilseeds Brassica – the scenario unfolding in India. *Journal of Oilseed Brassica* 2(2): 48-55.

Chattopadhyay, N. (2011). Climate Change and Food Security in India. Climate Change and Food Security in South Asia, pp. 229-250.

Chaturvedi, R.K., Gopalakrishnan, R., Jayaraman, M., Bala, G., Joshi, N.V., Sukumar, R., and Ravindranath, N.H. (2011). Impact of climate change on Indian forests: a dynamic vegetation modeling approach. *Mitigation and Adaptation Strategies for Global Change* 16 (2): 119-142.

Cho, M., Chardonnens, A.N., and Dietz, K.J. (2003). Differential heavy metal tolerance of *Arabidopsis halleri* and *Arabidopsis thaliana*, a leaf slice test. *New Phytologist* 158: 287-293.

Chupeau, Y., Caboche, M., and Henry, Y. (1998). Androgenesis and haploid plants, Springer–Verlag, Berlin, Heidelberg.

Clarke, H.J., Wilson, J.G., Kuo, I., Lülsdorf, M.M., Mallikarjuna, N., Kuo, J., and Siddique, K.H.M. (2006) Embryo rescue and plant regeneration *in vitro* of selfed chickpea (*Cicer arietinum* L.) and its wild annual relatives. *Plant Cell Tissue Organ Cult.* 85: 197-204.

Cocking, E.C. (1960). A method for the isolation of plant protoplasts and vacuoles. *Nature* 187: 927-929.

Çömlekçioğlu, N., Büyükalaca, S., and Abak, K. (2001). Effect of silver nitrate on haploid embryo induction by anther culture in pepper (*Capsicum annuum*). XIth EUCARPIA Meeeting on Genetics and Breeding of Capsicum & Eggplant. Antalya-Turkey. pp: 133-136.

Compton, M.E., Saunders, J.A., Veilleux, R.E. (2000). Use of protoplasts for plant improvement. In: Plant tissue culture concepts and laboratory exercises, 2nd edition, CRC Press LLC, pp: 249-261.

Couselo, J.L., Corredoira, E., Vieitez, A.M., and Ballester, A. (2012). Plant tissue culture of fast-growing trees for phytoremediation research. *Methods Mol Biol.* 877: 247-63.

Debnath, S.C., Jaime, A., and Silva, T.D. (2007). Strawberry culture in vitro: Applications in genetic transformation and biotechnology. *Fruit, Vegetable and Cereal Science and Biotechnology* 1: 1-12.

Di Lonardo, S., Capuana, M., Arnetoli, M., Gabbrielli, R., and Gonnelli, C. (2011). Exploring the metal phytoremediation potential of three *Populus alba* L. clones using an *in vitro* screening. *Environmental Science & Pollution Research* 18: 82-90.

Divakara, B.N., Upadhyaya H.D., Wani S.P., and Gowda C.L. (2010). Biology and genetic improvement of *Jatropha curcas* L.: A review. *Appl. Ener.* 87: 732-742.

Doran, P.M. (2000). Foreign protein production in plant tissue cultures, *Current Opinion in Biotechnology* 11: 199-204.

Doran, P.M. (2009). Application of plant tissue cultures in phytoremediation research: incentives and limitations. *Biotechnol Bioeng.* 103(1): 60-76.

Duncan, R.R. (1997). Tissue culture-induced variation and crop improvement. *Adv. Agron.* 58: 201-240.

Dunwell, J.M. (2010). Haploids in flowering plants: origins and exploitation. *Plant Biotech. J.* 8: 377-424.

Ehsanpour, A.A., Madani S., and Hoseini, M. (2007). Detection of somaclonal variation in potato callus induced by UV-C radiation using RAPD-PCR. *Gen. Appl. Plant Physiol.* 33(1-2): 3-11.

El-Aref, H.M. (2002). Employment of maize immature embryo culture for improving drought tolerance. Proc. 3rd Scientific Conf. Agric. Sci., Fac. Agric. Assiut Univ., Assiut, Egypt. October 20-22. pp. 463-477.

Engelmann, F. (2011). Use of biotechnologies for the conservation of plant biodiversity. *In vitro Cellular & Developmental Biology – Plant* 47 (1): 5-16.

Engler, R., Randin, C.F., Thuiller, W., et. Al. (2011). 21st century climate change threatens mountain flora unequally across Europe. *Global Change Biology* 17(7): 2330-2341.

Faccioli, G. (2001). Control of potato viruses using meirstem culture and stem-cutting cultures, thermotherapy and chemotherapy, In: Loebenstein, G., Berger, P.H., Brunt, A., and Lawson, R.H. (eds.). Virus and virus-like diseases of potatoes and production of seed-potatoes, Kluwer, Dordrecht, pp. 365-390.

Fang, D., Shuang, J., Ni, H., Yun, Z., Qing, C., Gan, Y., and Guo, W. (2008). Vitrification-cryopreservation, an efficient method for eliminating *Candidatus Liberobacter* asiaticus, the citrus Huanglongbing pathogen, from *in vitro* adult shoot tips. *Plant Cell Rep.* 27: 241-250.

Fernández, R., Bertrand, A., Casares, A., García, R., González, A., and Tamés, R.S. (2008). Cadmium accumulation and its effect on the in vitro growth of woody fleabane and mycorrhized white birch. *Environmental Pollution* 152 (3): 522-529.

Fisk, II.J., and Dandekar, A.M. (1993). The introduction and expression of transgene in plants. *Scientia Hort.* 55: 5-36.

Flocco, C.G., and Giulietti, A.M. (2007). *In vitro* hairy root cultures as a tool for phytoremediation research. *Methods in Biotechnology* 23: 161-173.

Fordham, D.A., Akçakaya, H.R., Araújo, M.B., *et. al.* (2012). Plant extinction risk under climate change: are forecast range shifts alone a good indicator of species vulnerability to global warming? *Global Change Biology* 18 (4): 1357-1371.

Fratini, R., and Ruiz, M.L. (2011). Wide crossing in lentil through embryo rescue. *Methods Mol. Biol.* 710: 131-9.

Fridley, J.D., Grime, J.P., Askew, A.P., Moser, B., and Stevens, C.J. (2011). Soil heterogeneity buffers community response to climate change in species-rich grassland. *Global Change Biology* 17(5): 2002-2011.

Fu, C.H., Guo, W.W., Liu, J.H., and Deng, X.X. (2003). Regeneration of *Citrus sinensis* (+) *Clausena lansium* intergeneric triploid and tetraploid somatic hybrids and their identification by molecular markers. *In vitro Cell Dev. Biol. Plant* 39: 360-364.

Fulzele, D.P. (2000). Bioreactor technology for large scale cultivation of plant cell suspension cultures and production of bioactive compounds. BARC Newsletter, (http://www.barc.ernet.in/webpages/letter/2000/200010-04.pdf).

Geerts, P., Toussaint, A., Mergeai, G., and Baudoin, J.P. (2011). Phaseolus immature embryo rescue technology. *Methods Mol. Biol.* 710: 117-29.

Germanà, M. (2006). Doubled haploid production in fruit crops. *Plant Cell, Tissue and Organ Cult.* 86 (2): 131-146.

Giachetti, G., and Sebastiani, L. (2006). Metal accumulation in poplar plant grown with industrial wastes. *Chemosphere* 64: 446-454.

Guha, S., and Maheshwari, S.C. (1964). *In vitro* production of embryos from anthers of Datura. *Nature* 204: 497.

Guha, S., and Mukherjee, S.C. (1999). The discovery of haploid production by anther culture. *In vitro Cell Dev. Biol.-Plant* 35: 357-360.

Gulati, A., and Jaiwal, P.K. (1994). *In vitro* selection and characterization of *Vigna radiata* cellline resistant to PEG-induced drought stress. *Acta Physiol. Plant.* 16 (1): 53-60.

Gyulai, G., Ge´mesne´, J.A., Sa´, gi, Z., Venczel, G., et. Al. (2000). Doubled haploid development and PCR-analysis of F-1 hybrid derived DH-R-2 paprika (*Capsicum annuum* L.) lines. *J. Plant Physiol.* 156: 168-174.

Haberlandt, G. (1902). Kulturversuche mit isolierten Pflanzenzellen. Sitzungsber. Akad. Wiss. Wien. Math.-Naturwiss. Kl., *Abt. J.* 111: 69-92.

Hagelson, J.P. *et al.* (1998) Somatic hybrids between *Solanum bulbocastanum* and potato: A new source of resistance to late blight. *Theor. Appl. Genet.* 96: 738-742.

Hagemann, R. (1992). Plastid genetics in higher plants. In: R.G. Herrmann (ed.). Cell organelles, Springer-Verlag, Wien, Austria, pp. 65-96.

Han, L., Zhou, C.H., Shi, J., Zhi, D., and Xia, G. (2009). Ginsenoside Rb1 in asymmetric somatic hybrid calli of *Daucus carota* with *Panax quinquefolius*. *Plant Cell Rep.* 28: 627-638.

Heard, S.B., and Mooers, A.O. (2000). Phylogenetically patterned speciation rates and extinction risks change the loss of evolutionary history during extinctions. *Proc. R. Soc. Lond.* B 267: 613-620.

Heath-Pagliuso, S., Pullman, J., and Rappaport, L. (1988). Somaclonal variation in celery: screening for resistance to *Fusarium oxysporum* f.sp.*apii*. *Theor. Appl. Genet.* 75: 446-451.

Helliot, B., Panis, B., Poumay, Y., Swenen, R., Lepoivre, P., and Frison, E. (2002). Cryopreservation for the elimination of cucumber mosaic and banana streak viruses from banana. *Plant Cell Rep.* 20: 1117-1122.

Jain, S.M. (1997). Creation of variability by mutation and tissue culture in improving plants. *Acta Hort.* 447: 69-78.

Jain, S.M. (2010). Mutagenesis in crop improvement under the climate change. *Rumanian Biotechnological Letters*. 15: 88-106.

Jain, S.M., and Spencer, M.M. (2006). Biotechnology and mutagenesis in improving ornamental plants. In: Floriculture, ornamental and plant biotechnology: advances and topical issues. J.A. Teixeira da Silva, (ed.). Global Science Books, Isleworth, UK. pp. 589-600.

James, C. (2010). Global Status of Commercialized Biotech/GM Crops. ISAAA Brief No. 42. ISAAA: Ithaca, NY.

Karan, R., and Subudhi, P.K. (2012). Approaches to increasing salt tolerance in crop plants. *Abiotic Stress Responses in Plants*, 63-88.

Khan, I.A., and Grosser, J.W. (2004). Regenartion and characterization of somatic hybrid plants of *Citrus sinensis* (sweet orange) and *Citrus micrantha*, a progenitor species of lime. *Euphytica* 137: 271-278.

Kirk, J.T.O., and Tilney-Bassett, R.E.A. (1978). The plastids. Their chemistry, structure, growth and inheritance, 2nd ed. Elsevier/North Holland, Amsterdam.

Kuksova, V.B., Piven, N.M., Gleba, Y.Y. (1997). Somaclonal variation and *in vitro* induced mutagenesis in grapevine. *Plant Cell Tiss. Org. Cult.* 49: 17-27.

Kumar, A., and Cocking, E.C. (1987). Protoplast fusion: a novel approach to organelle genetics in higher plants. *American J. Botany* 78(8): 1289-1303.

Kumar, S., Sunil K., Negi, S.P., and Kanwar, J.K. (2008). *In vitro* selection and regeneration of chrysanthemum (*Dendranthema grandiflorum* Tzelev) plants resistant to culture filtrate of *Septoria obesa* Syd. *In vitro Cellular and Developmental Biology-Plant* 44: 474-479.

Kumari, A.M., Dolstra, O., Vosman, B., Visser, R.G.F., and Linden, C.G. (2011). *In vitro* screening and QTL analysis for drought tolerance in diploid potato. *Euphytica* 181(3): 357-369.

Kumari, M., Clarke, H.J., Small, I., and Siddique, K.H.M. (2009). Albinism in plants: A major bottleneck in wide hybridization, androgenesis and doubled haploid culture. *Crit Rev Plant Sci.* 28: 393-409.

Kumari, M., Patade, V.Y., and Ahmad, Z. (2010). Transgenic approach towards development of cold stress tolerant vegetables for high altitude areas. *The Scientific Temper* 1: 1-14.

Kumari, M., Patade, V.Y., and Ahmed, Z. (2012a). Establishment of high regeneration potential in sweet pepper (*Capsicum annuum* L) cultivars Yolo Wonder and California Wonder. *World Journal of Science and Technology* 2(7): 26-35.

Kumari, M., Patade, V.Y., Arif M., and Ahmed Z. (2012b). An improved simple protocol for high frequency adventitious regeneration from leaf, cotyledon and petiole explant of *Jatropha curcas:* an energy plant. *Biochem. Cell. Arch.* 11 (2): 281-288.

Larkin, P.J., and Scowcroft, W.R. (1981). Somaclonal variation–a novel source of variability from cell cultures for plant improvement. *Theor. Appl. Genet.* 60: 197-214.

Li, M., Li, H., Jiang, H., Pan, X., and Wu, G. (2007). Establishment of an Agrobacteriuim-mediated cotyledon disc transformation method for *Jatropha curcas*. *Plant Cell Tiss. Org. Cult.* 92: 173-181.

Litz, R.E. (2004). Biotechnology and mango improvement. *Acta Hort.* 645: 85-92.

Liu, J.G., Li, K.Q., Xu, J.K., Zhang, Z.J., Ma, T.B., Lu, X.L., Yang, J.H., and Zhu, Q.S. (2003). Lead toxicity, uptake, and translocation in different rice cultivars. *Plant Sci.* 165: 793-802.

Liu, S., Wang, H., Zhang, J., Fitt, B.D.L., Xu, Z., Evans, N., Liu, Y., Yang, W. and Guo, X. (2005). *In vitro* mutation and selection of doubled-haploid Brassica napus lines with improved resistance to *Sclerotinia sclerotiorum*. *Plant Cell Rep.* 24: 133-144.

Lobel, D.B., Schlenker, W., and Roberts, J.C. (2011). Climate trends and global crop production since 1980. *Science* 333 (6042): 616-620.

Luan, Y.S., Zhang, J., and Gao, X.R. (2007). Mutation induced by ethylmethanesulphonate (EMS), *in vitro* screening for salt tolerance and plant regeneration of sweet potato (*Ipomoae batatas* L.). *Plant Cell Tiss. Org. Cult.* 88: 77-81.

Lutts, S., Kinet, J.M., and Bouharmont, J. (2001). Somaclonal variation in rice after two successive cycles of mature embryo derived callus culture in the presence of NaCl. *Biol. Plant.* 44 (4): 489-495.

Mace, G.M., Gittleman, J.L., and Purvis, A. (2003). Preserving the tree of life. *Science* 300: 1707-1709.

Mahmood, I., Razzaq, A., Ashraf, M., Hafiz, I.A., Kaleem, S., Qayyum, A., and Ahmad, M. (2012). *In vitro* selection of tissue culture induced somaclonal variants of wheat for drought Tolerance. *J. Agric. Res.* 50(2): 177-189.

Majer, B.J., Tscherko, D., and Paschke, A. (2002). Effects of heavy metal contamination of soils on micronucleus induction in Tradescantia and on microbial enzyme activities: a comparative investigation. *Mut. Res.* 515: 111-124.

Mallikarjuna, N. (1999). Ovule and embryo culture to obtain hybrids from interspecific incompatible pollinations in chickpea. *Euphytica* 110: 1-6.

Matsumoto, K., Vilarinhos, A.D., and Oka, S. (2002). Somatic hybridization by electrofusion of banana protoplasts. *Euphytica* 125: 317-324.

Matt, A., and Blaich, R. (2000). Somatic hybridization of grapevine protoplasts. *Acta Hort.* 528: 411-414.

Mehetre, S.S., and Aher, A.R. (2004). Embryo rescue: A tool to overcome incompatible interspecific hybridization in Gossypium Linn. -A review. *Indian J. of Biotechnol.* 3: 29-36.

Miguel, C., and Marum, L. (2011). An epigenetic view of plant cells cultured *in vitro*: somaclonal variation and beyond. *J. Exp. Bot.* 62 (11): 3713-3725.

Mink, G.I., Wample, R., and Howell, W.E. (1998). Heat treatment of perennial plants to eliminate phytoplasms, viruses, and viroids while maintaining plant survival, *In*: Plant Virus Disease Control. Hadid, A., Khetarpal, R.K., and Koganezawa, H. (eds.). MN, APS Press, St. Paul, USA, pp. 332-345.

Miyajuma, D. (2006). Ovules that failed to form seeds in zinnia (*Zinnia violacec* Cav). *Sci. Hortic.* 107 (2): 176-182.

Morel, G. (1960). Producing virus-free cymbidium. *Am. Orchid Soc. Bull.* 29: 495-497.

Muir, W.H., Hildebrandt, A.C., and Riker, A.J. (1954). Plant tissue cultures produced from single isolated plant cells. *Science* 119: 877-878.

Murlidhar, R.V., and Panda, T. (2000). Fungal protoplast fusion: A revisit. *Bioprocess Biosyst. Engg.* 22: 429-431.

Nee, S., and May, R.M. (1997). Extinction and the loss of evolutionary history. *Science* 288: 328-330.

Neelakandan, A.K., and Wang, K. (2012). Recent progress in the understanding of tissue culture-induced genome level changes in plants and potential applications. *Plant Cell Rep.* 31(4): 597-620.

Pasha, C., Kuhad, R.C., and Rao, C.V. (2007). Strain improvement of thermotolerant *Saccharomyces cerevessie* VS3 strain for better utilization of lignocellulosic substrates. *J. Appl. Microbiol.* 103: 1480-1489.

Phyton Biotech (2006). http://www.phytonbiotech.com/about.htm (240709).

Pinto, A.C.Q, Rogers, S.M.D., and Byrne, D.H. (1994). Growth of immature peach embryos in response to media, ovule support method, and ovule perforation. *Hort Science*, 29: 1081-1083.

Rai, M.K., Kalia, R.K., Singh, R., Gangola, M.P., and Dhawan, A.K. (2011). Developing stress tolerant plants through *in vitro* selection – An overview of the recent progress. *Environ. Exp. Bot.* 71(1): 89-98.

Rangan, T.S. (1984). Culture of ovules. In: *Cell Culture and Somatic Cell Genetics of Plants, Vol. 1: Laboratory Procedures and Their Applications*. Vasil, I.K. (ed.). Academic Press, New York. pp. 227-231.

Razzaq, A., Zhiying, M., and Haibo, W. (2004). Genetic transformation of wheat (*Triticum aestivum* L): A Review. *Mol. Pl. Breed.* 2(4): 457-464.

Remotti, P.C. (1998). Somaclonal variation and induced mutations in crop improvement, In: Jain, S.M., Brar, D.S., and Ahloowalia, B.S. (eds.). Kluwer Academic Publishers, Dordrecht, pp. 169-201.

Rezazadeh, R., Harrison, D.K., and Williams, R.R. (2011). Intraspecific somatic hybridization of mango (*Mangifera indica* L.) through protoplast fusion. *J. Applied Hort.* 13(2): 101-107.

Rygielska, J.K. (2004). Obtaining hybrids of distialliary yeasts characterized by the ability of fermenting starch. *Elec. J. Pol. Agri. Univ. Series* Biotechnol 7(2): p://www.ejpau.media.pl/volume7/issue2/biotechnology/abs-02.html.

Sabry, S.R.S., Smith, L.T., and Smith, G.M. (1995). Osmoregulation in Spring wheat under drought and salinity stress. *J. Genet. Breed.* 49: 55-60.

Sage, T.L., Strumas, F., Cole, W.W., and Barret, S. (2010). Embryo rescue and plant regeneration following interspecific crosses in the genus Hylocereus (Cactaceae). *Euphytica* 174: 73-82.

Sajc, L., Grubisic, D., and Vunjak-Novakovic, G. (2000). Bioreactors for plant engineering: an Outlook for further research. *Biochemical Engineering Journal* 4: 89-99.

Sakai, A., and Engelmann, F. (2007). Vitrification, encapsulation-vitrification and droplet-vitrification: a review. *CryoLett.* 28: 151-172.

Sandalio, L.M., Dalurzo, H.C., Gomez, M., Romero-Puertas, M.C., and del Rio, L.A. (2001). Cadmium-induced changes in the growth and oxidative metabolism of pea plants. *J. Exp. Bot.* 52: 2115-2126.

Sandra, R. (2005). Plant development and biotechnology. R.N. Trigiano and D.J. Gray (eds.). CRC Press. pp. 235-239.

Sawahel, W.A., and Cove, D.J. (1992). Gene trasnfer strategies in plants. *Biotech. Adv.* 10: 393-412.

Saxena, G., Verma, P.C., Rahman, L., Banerjee, S., Shukla, R.S., and Kumar, S. (2008). Selection of leaf blight-resistant *Pelargonium graveolens* plants regenerated from callus resistant to culture filtrate of *Alternaria alternate*. *Crop Protection* 27: 558-565.

Scherrer, D., and Körner, C. (2011). Topographically controlled thermal-habitat differentiation buffers alpine plant diversity against climate warming. *Journal of Biogeography* 38 (2): 406-416.

Scholze, P., Krämer, R., Ryschka, U., Klocke, E., and Schumann, G. (2010). Somatic hybrids of vegetable brassicas as source for new resistances to fungal and virus diseases. *Euphytica.* 176 (1): 1-14.

Shen, X., Gmitter, F.G. Jr, and Grosser, J.W. (2011). Immature embryo rescue and culture. *Methods Mol. Biol.*710: 75-92.

Sigareva, M.A., and Earle, E.D. (1997). Direct transfer of a cold-tolerant Ogura male-sterile cytoplasm into cabbage (*Brassica oleracea* ssp. capitata) via protoplast fusion. *Theor. Appl. Genet.* 94 (2): 213-220.

Simelton, E., Fraser, E.D.G., Termansen, M., *et al.* (2012). The socioeconomics of food crop production and climate change vulnerability: a global scale quantitative analysis of how grain crops are sensitive to drought. *Food Security.* 4(2): 163-179.

Singh, G., and Shetty, S. (2011). Impact of tissue culture on agriculture in India. *Biotechnol. Bioinf. Bioeng.* 1(2): 147-158.

Su, W., (2006). Bioreactor engineering for recombinant protein production using plant cell suspension culture. In: Plant Tissue Culture Engineering, Gupta, S.D., and Ibaraki, Y. (eds.),. Springer. Printed in the Netherlands. pp. 135-159.

Su, W., and Lee, K.T. (2007). Plant cell and hairy root cultures – Process Characteristics. In: Products, and Applications Bioprocessing for Value-Added Products from Renewable Resources New Technologies and Applications, pp. 263-292.

Tamura, M., Tao, R., and Sugiura, A. (1995). Regeneration of Somatic hybrids from electrofused protoplasts of Japanese persimmon (*Diospyros kaki* L.). *Plant Sci.* 108: 101-107.

Taskin, H., Büyükalaca, S., Keles, D., and Ekbiç, E. (2011). Induction of microspore-derived embryos by anther culture in selected pepper genotypes. *Afr. J. of Biotechnol.* 10(75): 17116-17121.

Thakur, M., Sharma, D.R., and Sharma, S.K. (2002). *In vitro* selection and regeneration of carnation (*Dianthus caryophyllus* L.) plants resistant to culture filtrate of *Fusarium oxysporum* f. sp. *dianthi. Plant Cell Rep.* 29: 825-828.

Thomas, C. D., Cameron, A., Green, R. E., *et. al.* (2004). Extinction risk from climate change. *Nature* 427: 145-148.

Thuiller, W., Lavergne, S., Roquet, C., Boulangeat, I., Lafourcade, B., and Araujo, M.B. (2011). Consequences of climate change on the tree of life in Europe. *Nature* 470: 531-534.

Thuiller, W., Lavorel, S., Arau′ jo, M. B., Sykes, M.T., and Prentice, I.C. (2005).Climate change threats to plant diversity in Europe. *Proc. Natl Acad. Sci.* USA 102, pp. 8245-8250.

Vadez, V., Berger, J.D., Warkentin, T., *et. al.* (2012). Adaptation of grain legumes to climate change: a review. *Agronomy for Sustainable Development* 32 (1): 31-44.

Van Creij, M.G.M., Kerckhoffs, D.M.F.J., and Van Tuyl J.M. (1999). The effect of ovule age on ovary-slice culture and ovule culture in intraspecific and interspecific crosses with *Tulipa gesneriana* L. *Euphytica* 108: 21-28.

Walck, J.L., Hidayati, S.N., Dixon, K.W., Thompson, K., and Poschlod, P. (2011). Climate change and plant regeneration from seed. *Global Change Biology* 17(6): 2145-2161.

Wang, Q.C., Munir, M., Li, P., Gafny, R., Sela, I., and Tanne, E. (2003) Elimination of grapevine virus A (GVA) by cryopreservation of *in vitro*-grown shoot tips of *Vitis vinifera* L, *Plant Sci.* 165: 321-327.

Williams, E.G., Verry, I.M., and Williams, W.M. (1982). Use of embryo culture in interspecific hybridization. In: *Plant Improvement and Somatic Cell Genetics.* Vasil, I.K., Scowcroft, W.R., and Frey, K.J. (eds.). Academic Press, New York. pp. 119-128.

Xa, T.T., and Lang, N.T. (2011). Rice breeding for high grain quality through anther culture. *Omonrice* 18: 68-72.

Xiao, Z.A., Wan, L., and Han, B.W. (2004). An interspecifi c somatic hybrid between *Actinida chinensis* and *Actinida kolomikta* and its chilling tolerance. *Plant Cell Tiss. Organ Cult.* 79: 299-306.

Zagorska, N.A., Shtereva, L.A., Kruleva, M.M., Sotirova, V.G., Baralieva, D.L., and Dimitrov, B.D. (2004). Induced androgenesis in tomato (*Lycopersicon esculentum* Mill.). III. Characterization of the regenerants. *Plant Cell Rep.* 22: 449-456.

Zhenzhen, C., Anja, K., Dietrich, K., and Smetanska I. (2012). Exudation: an expanding technique for continuous production and release of secondary metabolites from plant cell suspension and hairy root cultures. *Plant Cell Rep.* 31(3): 461-477.

Ziv, M., (2000). Bioreactor technology for plant micropropagation. In *Horticultural Reviews*, Volume 24, Jules Janick (ed.). John Wiley & Sons, Inc. ISBN 0-471-33374-3.

16

Tailoring of Abiotic Stress Adaptive Traits to Diminish the Effect of Changing Climate on Crop Productivity

Paramasivam Arunachalam[1*], Mahalingam Govindaraj[2,3], and Kannan Pandian[1], Natarajan Sritharan[4]

[1]Dryland Agricultural Research Station, Tamil Nadu Agricultural University Chettinad, Tamil Nadu, India

[2]Centre for Plant Breeding and Genetics, Tamil Nadu Agricultural University Coimbatore, Tamil Nadu, India

[3]Millet Breeding, ICRISAT, Patancheru, Andhra Pradesh, India

[4]Agricultural Research Station, Tamil Nadu Agricultural University, Kovilpatti Tamil Nadu, India

ABSTRACT

The world population is likely to exceed 10 billion by 2050, thereby increasing food, feed and fuel production demand. On the other hand, global climate change (drought, heat, salinity, elevated CO_2 and extreme cold) hostile the global agricultural productions. The changes in climatic factors perhaps influence the crop distribution, affect the crop growth and yield, and increase the risks of farming and human health consequences in developing countries. Crop breeding is one of the approaches to fight environmental challenges in agriculture. Available literatures imply that genotypes of different crop species are expressing greater phenotypic variability to tolerate abiotic stresses by inherent constitutional. Hence, there is an opportunity for utilizing the existing variability in abiotic stress tolerance traits. The gene sources for abiotic tolerance are available in germplasm collection, landraces, or wild relatives, if not, with less frequency it can be created as transgenes, somoclones or mutants. However, to make significant advancement in abiotic stress breeding requires accurate and reproducible phenotyping under well-imposed stress environment. The targeted trait for abiotic stress tolerance should have high positive correlation with yield attributes and be amenable for scoring in given environment.

The traits introgressed for abiotic stress tolerance vary with stress scenario, timing and intensity of stress encountered by the crop species. Most of the traits that confer abiotic stress tolerance are quantitative in nature. The conventional crop improvement strategy followed to transfer abiotic stress tolerance is by recurrent selection and backcross breeding, which delivered limited success. The recent advancement through rapid and high-throughput phenotyping and genotyping have given much hope for tailoring desirable traits to evolve climate-resilient cultivars. The gene pyramiding strategy

Corresponding author's E-mail: arunachalp@gmail.com

Roychowdhury R. (Ed.), Crop Improvement in the Era of Climate Change.

is useful to accumulate desirable abiotic tolerant traits into a commercially preferred cultivar. Further, transgenic and double haploid approaches will help in accelerating the trait pyramiding strategy. The climate-resilient cultivars with climate-smart farming will offer sustainable and cost-effective solution to the changing agro-climatic situations.

Keywords: Abiotic stress, breeding, phenotyping, tolerance, trait pyramiding.

1. EXTENT AND IMPLICATIONS OF ABIOTIC STRESS IN AGRICULTURE

The global temperature over the last 100 years has increased by 0.74°C and is expected to increase by about 0.2°C per decade over next two decades. Global sea level has risen by 17 cm during the 20th century, in part because of the melting of snow and ice from many mountains and polar regions (Green Facts, 2007). Changes in temperature, rainfall pattern and increase in CO_2 levels will have drastic impact on global agriculture, especially on dryland ecosystems which supports 2.5 billion inhabitants (Anderson and Morton, 2008). In coming years, direct impact of climate changes in agriculture mainly on decreasing the water availability and soil moisture, increasing the ambient air temperature thereby increase land degradation and desertification. By the 2050s, 50% of agricultural lands are very likely to be subjected to desertification and salinization in some areas of Latin America. Yields from rain-fed crops could be halved by 2020 in some countries in Africa (UNFCCC, 2007). The climate changes influence the crop distribution, decline in crop growth and yield, and increase the risks that are associated with agriculture. At least 22% of the cultivated area under the world's most important crops is projected to face negative impacts from climate change by 2050 (Campbell *et al.*, 2011). Abiotic stresses are the principal cause of crops failure worldwide which reduce the yield by more than 50%. A projected global impact of climate change by 2030 in agriculture for five major crops in each region is depicted in Fig. 1. The major abiotic stresses that are expected to increase in response to climate change are heat, drought, salinity, waterlogging and inundation (Bray *et al.*, 2000). These stresses are location specific, occurrence is unpredictable, degree of stress varies with crop season or year, one stress may increase or decrease the level of another stress for instance, in saline soil the moisture stress would enhance salinity stress (Singh, 2001). Experts say human population might exceed to 10 billion by 2050, in order to nourish the teeming millions, FAO estimated that the food production should be increased to 70% over the next four decades which is nearly impossible with current progress in densely populated developing countries like India. This implies that there is a pressure on agricultural scientists to evolve climatic resilient varieties and develop suitable management technologies to cope with abiotic anxieties. These hassles vary with timing, duration and intensity, thus sufficient knowledge of the target environment is essential in abiotic stress tolerance breeding. Hence, it is time to look for suitable crop hybridization strategies to mitigate various kinds of stresses and to ensure food security for teeming millions.

2. DROUGHT STRESS

Agricultural drought results in reduction of crop yield due to shortage of soil moisture availability in the root zone (especially on rain-fed or dryland conditions) and it is highly variable in its timing, duration and severity, which results in high environmental variation and genotype × environment interaction. The impact of drought stress in rice is depicted in Figure 2 using the data from Boonjung and Fukai (1996), which shows 20% yield loss at maturity. Similarly, groundnut germplasm were screened during rainy and post-rainy seasons, post-rainy crop suffered with soil moisture stress to an extent of 50 per cent during flowering and pod development stages. The soil moisture stress in peanut delayed the flowering and maturity (Figure 3), and showed negative influence on plant height, number of pods per plant and pod yield per plant (Arunachalam *et al.*, 2012).

The drought resistance cultivar has the ability to withstand or minimize the yield loss due to moisture stress by various mechanisms, viz., drought escape, dehydration avoidance and dehydration tolerance. The dehydration avoidance genotypes can be further viewed as 'water savers' by reduced transpiration and 'water spender' by increased water uptake, these were well described in many literatures. Developing of crop varieties with low-water requirements and better water-use efficiency is essential to minimize the total production loss.

2.1 Managing Drought Environment for Screening

The plant traits associated with drought avoidance and tolerance can be *constitutive* that differs between genotypes, or *adaptive* which vary with the stage of the life cycle. In plants, constitutive traits are inherent to the genotypes like early flowering, rooting patterns, water use efficiency, epicuticular wax, stomatal index and phenology — which are not under control of stress responsive genes and do not require drought conditions for selection. Acquired traits/mechanism are upregulated under stress such as osmotic adjustment, changes in abscisic acid (ABA), tolerance to oxidative stress and reserve mobilization to the grain (partially stress responsive). Such drought responsive and adaptive traits requires specific stress environment to express in plants (Blum, 2011). Stress environment is apparent to identify the stress tolerant genotype or to score the desired trait. Hence, drought-screening fields require a careful management to avoid epigenetic variation among plots. To identify sources of drought resistance the screening methods should be simple and reproducible under the target environment.

2.1.1 Moisture Stress Environment in Field Conditions

The genotypes can be exposed to varied moisture stress levels in the field condition by the following ways.

(i) *Artificially create stress in the "normal" growing season:* Conducting experiment in normal growing season helps to avoid genotype x season problems. The varied moisture regime during particular crop growth stage can be achieved by (a) providing controlled irrigation in the field; (b) line-source-sprinkler irrigation method used to create a gradient of drought stress and help to evaluate large numbers of genotypes at varying intensity of drought (Serraj *et al.*, 2003), the statistical parameter to assess the sensitivity of genotype to drought was dealt by Singh *et al.* (1991); and (c) drip irrigation system also can be designed to deliver the desired volume of water and manage the stress levels for wide spaced and row crops. It is important to eliminate the undesirable rainfall on the experimental field by rain-out shelter and is a costly affair.

(ii) *Evaluating in 'desert' environment:* It is ideal to carry out the phenotyping for drought tolerance in the 'desert' environment, where no unexpected rainfall can be interfere with the managed stress. Breeder can have total control of the moisture regime through irrigation (Blum, 2011). But, the phenotyping field should be surrounded by sufficiently wide crop-border to avoid the 'fetch' or dry wind effect in desert.

(iii) *Managing water-deficits in a non-growing season:* To avoid unpredictable seasonal rainfall during phenotyping, the drought screening experiments are planned in offseason or dry environment. But it needs application of uniform moisture stress environment. In this method, there is a possibility to maximize genetic component of the observed variation. But the extrapolation of results to 'natural' target environment or normal growing season can be difficult.

(iv) *Delayed Planting:* The delayed planting was done such a way that crop exposed to terminal drought. This makes amenable for phenotyping the flowering traits and grain development under moisture stress and selecting the genotypes for terminal stress tolerance.

2.1.2 Managed Moisture Stress Environment in Pot Culture

Imposition of uniform drought stress under field conditions at every plot is difficult. On the other hand, pot culture studies serves better managed stress environments. Different methodologies have been used to stimulate the moisture stress to known quantum at specific stages at whole plant level in pot experiments. An ideal technique used for the simulation of drought stress in pot studies should (a) maintain uniform moisture content within the root zone, (b) allow the rate at which stress develops to be precisely controlled, (c) permit a range of physiologically distinct stress levels to be imposed and maintained indefinitely, and (d) provide a mean of accurately quantifying stress level. The stress can be imposed for pot studies by gravimetric method or computer automated null-balance lysimeter as described by Earl (2003).

2.2 Drought Tolerant Traits and their Phenotyping

Phenotyping is the comprehensive measurement of individual quantitative parameter that forms the basis for the complex traits such as plant growth and development, plant architecture, yield, and physiological mechanism under stress environment. The traits underlying the different mechanism by which the plant mitigates the drought and ways to phenotype those traits are discussed below.

2.2.1 Drought Escape

Genotypes having rapid growth, early flowering and maturation are preferred to avoid the terminal drought. This can be easily scored by assessing day to flowering and maturity.

2.2.2 Dehydration Avoidance

In case of dehydration avoidance, the changes in the anatomical and morphological traits help the plant to increase water uptake and reduce water spending. The water uptake is increased by extensive root system with large active surface area and shoot/root ratio shifted in favour of the roots. The water loss is reduced by reducing transpiration through timely stomatal closure, thick cuticle, epicuticular waxes, white hairs on leaves, leaf angle, leaf rolling and plant senescence. The screening traits used to assess dehydration avoidance response of genotypes and their relevance, and ways to measure the trait of interest is discussed below:

(i) *Dehydration symptom in leaves:* It can be measured by scoring of leaf rolling or per cent reduction in leaf width resulting from rolling. Based on wilting symptoms the slow witling or delayed leaf rolling genotypes under water deficit conditions can be identified for the drought tolerance.

(ii) *Plant water status:* The total plant water potential is measured by the pressure chamber technique and thermocouple psychrometer method. The relative leaf water content (RLWC) can be measured at flowering and grain or pod development stage. The canopy temperature (CT) variation among the genotypes can be measured by infrared thermometer (IRT). It is an efficient method for rapid, non-destructive monitoring of whole-plant response to water stress. The genotypes relatively lower canopy temperature should be given importance in selecting to drought tolerance.

(iii) *Osmotic adjustment:* As a plant detects a water-deficit, it may accumulate a variety of osmotically active compounds such as amino acids, sugars and ions inside its cells, resulting in a lowering of the cell osmotic potential. Then the water present in inter-cellular spaces flows towards the inside of those cells. This process is called "osmotic adjustment" (OA). The amino acid, proline accumulation was observed significantly more in tolerant rice genotype under aerobic condition (Sritharan and Vijayalakshmi, 2007). Osmometer is used to measure osmotic potential. The leaf water potential is measured by pressure bomb apparatus.

(iv) *Stomatal activity and transpiration:* Leaf transpiration partitioned into stomatal and non-stomatal (cuticular) transpiration. Stomatal transpiration or conductance can be assessed by porometer. It is measured on a fully expanded young leaf. The cuticular transpiration can be phenotyped by two approaches (a) quantifying epicuticular wax load by colorimetric assay (Ebercon *et al.*, 1977), and (b) assessing leaf water loss by diffusion porometer and weight of detached leaves over specific time interval under standard conditions. Genotypes with a thick cuticle layer retain their leaf turgor for longer periods of time after the onset of a water-stress. The drought tolerant rice genotype exhibited better epicuticular wax content under aerobic condition (Sritharan and Vijayalakshmi, 2007).

(v) *Non-senescence (stay-green):* The non-senescence or the chlorophyll retention parameter under stress condition is an important indicator to identify the drought tolerant types. Stay green trait can be assessed by visual scoring, measuring the per cent green leaf area and SPAD meter reading. The stay-green type is also judged by (a) estimation of normalized difference vegetation index (NDVI) at physiological maturity; and (b) the rate of senescence (RS). Lopes and Reynolds (2012) suggested a simple and integrated way to measure stay-green in large sets of germplasm using a "GreenSeeker" sensor to measure NDVI during the grain-filling stage in wheat. The precision of estimation will increase with the number of NDVI measurements taken and, probably after mid-grain filling; two weekly measurements should be taken under stressed environments.

(vi) *Measuring roots:* In abiotic stress breeding the root phenotyping is play an important role. In spite of number of constrains, the out-of-field methods are widely used for root phenotyping. The various versions of the containers like upright (split) PVC tubes, large polycarbonate boxes, big pots or containers and temporary rhizo-structures to mimic field conditions are used to grow the plants. The plants are removed without damaging roots, and the root traits, namely, root length, root length density, root volume is studied. The root parameters can also be measured by minirhizotron image analysis by non-destructive means in the field. The hydroponics also used to study the root architecture. The root function can be indirectly phenotyped by assessing plant water status in a drought managed conditions (Blum, 2011). Root penetration capacity is an important factor in avoiding drought where soil hindrances exist. To assess the root penetration capacity the root penetration of a wax layer method is widely used.

(vii) *Leaf anatomy:* The traits like waxiness, pubescence, rolling, thickness, leaf angle and leaf movement help to decrease radiation load on the leaf surface when the plant is in moisture stress. Which lower evapotranspiration rate and reduced risk of irreversible photo-inhibition. However, they may also reduce radiation use efficiency, in turn, reduce yield under favorable conditions.

2.2.3 Dehydration Tolerance

It is the ability of the plant as a whole or in any of its components to function under low plant water status (Blum, 2011). The dehydration tolerance ability of the genotypes can be assessed by following parameters.

(i) *Survival and phenology:* Seedling survival or recovery is a useful index of dehydration tolerance. This can be assayed by the 'box' test or seedling wilting test (Blum, 2011). Delayed heading or flowering, short anthesis-to-silking interval (ASI) in maize, scoring fertility and seed set can be used as selection index. Photosynthetic rate of genotypes can be measured using portable photosynthesis system under stress and non-stress conditions.

(ii) *Cell membrane stability (CMS):* It is the ability of the cell to sustain cell activity and survive under dehydration or reduced water content or high temperature. Otherwise it is the capacity of the cell membrane to prevent electrolyte leakage at decreasing water content. The tolerant rice genotype showed better CMS under aerobic condition than other tested genotypes (Sritharan and Vijayalakshmi, 2008). Measurements of CMS reported to be correlated with yields under drought/temperature stress in different crops. CMS is estimated by subjecting leaf disc to stress and measuring the leakage of solutes into deionised water conductometrically.

(iii) *Partitioning and stem reserve mobilization for grain filling:* As photosynthesis becomes inhibited by moisture or heat stress, the grain filling process becomes increasingly dependent on stem reserve utilization. So the grain number maintenance, grain filling duration and grain filling index can be accounted to asses this trait under stress and non-stress conditions. Blum *et al.* (1983) proposed the use of chemical desiccation of the canopy after flowering as means for inhibiting current plant photosynthesis. Chemical desiccation of plant canopies at the onset of grain filling was developed as a tool for revealing genotypic differences in grain filling from stem reserves in the absence of current photosynthesis (Blum *et al.,* 1983a). Potassium iodide (KI) is a chemical contact-desiccant was used for assessing genotypic diversity in grain filling under drought stress (Tyagi *et al.*, 2000).

(iv) *Plant growth and productivity*:

(a) **Plant growth attributes:** The key indices assessed to understand growth responses are relative growth rate (RGR), net assimilation rate (NAR), leaf area ratio (LAR), specific leaf area (SLA) and leaf weight ratio (LWR).

(b) **Yield parameters:** The yield response of genotypes to moisture stress can be assessed by the different drought tolerance indices such as mean productivity (MP) under stress and non-stress conditions (Rosielle and Hamblin, 1981), drought susceptibility index (DSI) (Fischer and Maurer, 1978), drought tolerance efficiency (DTE%) (Fischer and Wood, 1981) and stress tolerance index (STI) (Fernandez 1992) and harvest index (HI). The genotypes with high value of MP, DTE, STI and value below 1 of DSI were considered drought tolerant genotypes.

(c) **Water use efficiency:** Water use efficiency (WUE) can be determined gravimetrically as per Udayakumar *et al.* (1998) or by using a computer automated gravimetirc lysimeter in association with physiological parameters. The carbon isotope discrimination (CID) technique is also used to determine the WUE; but it needs $\delta^{13}C$ analysis by isotope ratio mass spectrometer (IRMS). Alternate strategies may be adopted to measure the WUE using the surrogate traits such as SLA, SCMR and stomatal index. SLA is the ratio of leaf area to leaf dry weight. It is an indirect measure of leaf expansion. Higher SLA represents larger surface area for transpiration and hence SLA and WUE would be inversely correlated (Karaba *et al.*, 2011). Sritharan and Vijayalakshmi (2010) reported that δ^{13} C is negatively correlated with WUE in rice genotypes under aerobic condition. However, it had positive correlation with total drymatter production, leaf rolling, chlorophyll stability index, relative water content, total chlorophyll content, soluble protein and nitrate reductase activity (Sritharan and Vijayalakshmi, 2012). Stomatal index (SI) can be computed by impression method to count the stomata under microscope. SPAD chlorophyll meter reading (SCMR) is a non-destructive method of measuring chlorophyll content in the leaves by hand-held chlorophyll meter. It determines the relative amount of chlorophyll present in the leaf. The SPAD value is measured on a fully expanded young leaf of the plant.

v. Hormonal regulation: The abscisic acid (ABA) is known as *stress hormone* and its concentration increases under moisture stress in plants. The ABA-mediated stomatal closure is a known mechanism under drought to reduce the water loss by leaf transpiration, reduction in leaf expansion and promotion of root growth. On the other hand, ABA has negative effect on plant during flowering by decreasing the pollen viability and seed set.

3. SALT STRESS

The climate change causes melting of glaciers can threaten mountain settlements, water resources and increase the damage associated with coastal flooding. Global average sea level is expected to rise by 18 to 59 cm by the end of the 21st century (GreenFacts, 2007). This leads increased salinization of arable land is expected to have devastating global effects, resulting in 30% land loss within next 25 years and up to 50% by the middle of 21st century (Wang *et al.*, 2003). Agricultural land and thus food security affected by sea level rise, inundation, soil salinization, seawater intrusion into freshwater lenses (UNFCCC, 2007). To cope with this situation, it is essential to evolve saline tolerant cultivars to sustain crop yield in these areas.

3.1 Salt Stress Effects on Plant

Salts taken up by plant influence the plant growth by affecting turgor, photosynthesis or enzymes. Buildup of salt in old leaves accelerates their death, and loss of these leaves affects the supply of assimilates or hormones to their growing regions and thereby affect plant growth (Munns, 1993). The deleterious effect of salt stress on plants is physiologically lowering of osmotic pressure by decrease osmoprotectants and lowering antioxidants. Salinity affects the germination, vigour, root and shoot length; morphologically stunted plant growth, chlorosis, wilted and senescence, and arrest of plant growth leads to plant death. Plants are classified as glycophytes or halophytes according to their ability to grow on high salt environment. Most plants are glycophytes and cannot tolerate salt stress due to the following reasons:

- Soil salinity decreases the osmotic potential of soil solution creating water stress in plants and affects plant growth and development. High concentrations of Na^+ disturb osmotic balance and results in "physiological drought", preventing plant water uptake. Whereas, halophytic plants tolerate sodium toxicity, osmotic stress might be the main reason of growth inhibition (Türkana and Demiral, 2009). Many glycophytes respond to relatively low salt concentrations (< 6,000 mg/L, or roughly 100 mM) by "salt exclusion," particularly through low rates of net transport of sodium or chloride, or both, from root to shoot. Most of these salt-excluding glycophytes cannot adjust osmotically to the low external water potential by increased synthesis of organic solutes and, therefore, suffer from a decrease in turgor. Hence, salinity may induce an osmotic stress in glycophyte (Lauchi and Epstein, 1984).
- Infux of Na^+ dissipates the membrane potential and facilitates the uptake of Cl^- down the chemical gradient, which disrupts cell ionic equilibrium (Mahajan and Tureja, 2005). This leads to severe ion toxicity to cell metabolism and cause deleterious effect on some enzymes, since Na^+ is not readily sequestered into vacuoles in glycophytes as in halophytes. High Na^+ levels also lead to reduction in photosynthesis and production of reactive oxygen species (Flowers *et al.*, 1977; Greenway and Munns, 1980; Yeo, 1998).
- The interactions of salts with mineral nutrition may result in nutrient imbalances and deficiencies in crop plants.

3.2 Salt Stress Tolerance Mechanism

The salt-resistant glycophyte can adjust osmotically to a saline medium. Salinity stress is first sensed in the root, but osmotic adjustment as well as growth inhibition and ion toxicity are most apparent in the shoot. Thus, in addition to cellular processes, root-shoot interactions and the coordination of the whole plant are an integral part of the responses to salinity (Lauchi and Epstein, 1984). Mechanisms for salt tolerance are of two main types: those minimizing the entry of salt into the plant, and those minimizing the concentration of salt in the cytoplasm (Munns, 2002).

To overcome the detrimental effects of salt stress, plants have evolved many biochemical strategies like (i) selective build-up or exclusion of salt ions, (ii) control of ion uptake by roots and transport into leaves, (iii) ion compartmentalization, (iv) synthesis of compatible osmolytes, (v) alteration in photosynthetic pathway, (vi) changes in membrane structure, (vii) induction of antioxidative enzymes and (viii) stimulation of phytohormones (Parida and Das, 2005).

In order to prevent water loss from the cell and to protect the cellular proteins, plants accumulate many metabolites with an osmolyte function such as sugars, mainly fructose and sucrose, sugar alcohols and complex sugars like trehalose and fructans. Since water moves from high water potential to low water potential, the accumulation of osmolytes make the water potential low inside the cell and prevent the intracellular water loss (Mahajan and Tureja, 2005). Tavakkoli and coworkers (2012) suggested that salt exclusion coupled with a synthesis of organic solutes is an important component of salt tolerance in the tolerant genotypes. The halophyte *Mesembryanthemum crystallinum* has emerged as a model system for understanding the molecular response to salt stress. This plant switches from C3 photosynthesis to crassulacean acid metabolism (CAM) in response to salt or drought stress (Sairam and Tyagi, 2004).

Salinity stress response is multigenic, as a number of processes involved in the tolerance mechanism, such as various compatible solutes/osmolytes, polyamines, reactive oxygen species and antioxidant defence mechanism, ion transport and compartmentalization of injurious ions (Sairam and Tyagi, 2004). Salinity tolerance was defined as the genotypic differences in biomass production in saline versus non-saline conditions over prolonged periods, of 3–4 weeks (Munns and James, 2003).The mechanisms of salt tolerance are categorized as (a) tolerance to osmotic stress, (b) Na^+ exclusion from leaf blades, and (c) tissue tolerance and the detailed are discussed by Munns and Tester (2008).

3.3 Salt Stress Imposition in Plants

The screening for salt tolerance requires reliable and controlled methods to maintain the salinity stress levels; and precise quick techniques to measure the salinity tolerance of plants. The saline environment can be created by adding salt (NaCl) solution to the growing media or soil. The nutrient contents should be sufficiently maintained to avoid the interference of nutrient deficiency during the salt stress screening either in field or in pots. The *in vitro* and *in vivo* techniques used to create varied salt stress environment for the screening programme is described below:

- ***In vitro* methods:** The techniques like filter papers with saline solution, salt-containing agar plants are used to assess tolerance potential of germination and seedling traits. In pot grown plants, the salt concentration is increased gradually by adding saline water in short time intervals for plant to adjust. Then periodically pots to be watered with saline solution to maintain the different levels of salt stress. The *in vitro* protocol for salinity screening in rice using seedling float for seedling, vegetative and reproductive stages is described in detail by Gregorio *et al.* (1997). Hydropnic cultures can also be used for salt stress evaluation purpose.
- **Field screening:** To create and maintain the salt stress level in field condition, non-saline field is irrigated with known quantum and concentration of saline solution. This strategy is used to assess the response of the genotypes to salt stress under field environment.

- **Screening in micro-plots:** It is a series of dug-out cavity structures made of brick-mortar-concrete materials and filled with artificially prepared or transported problem soil. It helps to mimic field environments and to maintain the varying levels of controlled salinity and sodicity. This technique is useful to avoid the soil heterogeneity and spatial variability in field that affect the reliability of genotypes response.

3.4 Phenotyping for Salinity Tolerance

Salt tolerant plants differ from salt-sensitive ones in having a low rate of Na^+ and Cl^- transport to leaves, and the ability to compartmentalize these ions in vacuoles to prevent their build-up in cytoplasm or cell walls and thus avoid salt toxicity (Munns, 2002). The genotypes with salt exclusion property need not to be a high yielder. If the osmotic effect of the salinity is dominated in salt tolerant lines it may affect the yield potential. Measurements of germination, survival, growth, and leaf gas exchange will indicate the osmotic effect of salinity. If the experiments are conducted for lengthy periods of time, measurements of growth, survival and gas exchange will also reflect toxic effects of salts in the leaves. An ideal high yielding salt tolerant variety should have good initial vigour, low Cl^- uptake, low Na^+/K^+ ratio, good Na^+ excluder either in shoot or leaf, high tissue tolerance and agronomically superior characteristics with high yield potential. The selection criteria are to assess salt tolerance in plants is narrated below:

(i) Measuring cell survival, germination and survival of the plant under salt stress.

(ii) Leaf injury/death or senescence: Measuring the electrolyte leakage from leaf discs, chlorophyll content by SPAD meter and chlorophyll fluorescence. Decrease in chlorophyll content with increase in salinity level was observed in rice genotypes. The tolerant genotypes recorded higher chlorophyll stability index (CSI) than susceptible cultivars (Sritharan and Mallika Vanangamudi, 2006).

(iii) The parameters like Na/K ratio, Na and K uptake can be estimated to assess the salt tolerance of the genotypes. It was reported that the tolerance of a genotypes in rice have ability to restrict toxic ion uptake like Na^+ and have preferential uptake of the balancing ion like K^+. Lower Na/K ratio is considered desirable trait as it maintains the ion balance.

(iv) Tissue tolerance is measured in terms of LC_{50}. It is the concentration of sodium in the leaf tissue which causes a 50% loss of chlorophyll (Yeo and Flowers, 1983).

(v) The damage to chlorophyll apparatus can be assessed by measuring chlorophyll fluorescence by chlorophyll fluorometer.

(vi) The effect on salt stress on plant growth and yield can be measured by root elongation, leaf elongation, dry matter accumulation or biomass production and yield in saline environment over non-stress condition.

While screening for the salt tolerance, it is important to consider the concept of a two-phase growth response to salinity as suggested by Munns (1993).

- **Phase 1 (osmotic stress):** In this phase, visible quick reduction in growth is due to the salts outside the roots. It is essentially a water stress or osmotic phase, for which little genotypic variation among the genotypes. The growth reduction is presumably regulated by hormonal signals coming from the roots.

- **Phase 2 (salt-specific effect):** There is a second phase of growth reduction, which takes time to develop, and results from internal injury. It is due to salts accumulating in transpiring leaves to excessive levels, exceeding the ability of the cells to compartmentalize salts in the vacuole. This will inhibit growth of the younger leaves by reducing the supply of carbohydrates to the growing cells.

4. HEAT STRESS

The earth's climate is predicted to warm by an increase in mean ambient temperatures between 1.8 and 5.8°C by the end of the 21st century mainly due to both anthropogenic and natural factors. Future climates will also be affected by greater variability in temperature and increased frequency of hot days (IPCC 2007). Heat stress is often defined as the rise in temperature beyond a threshold level for a period of time sufficient to cause irreversible damage to plant growth and development. In general, a transient elevation in temperature, usually 10 to15°C above ambient, is considered heat shock or heat stress. Heat stress, singly or in combination with drought, is a common constraint during anthesis and grain filling stages in many cereal crops of temperate regions (Wahid *et al.*, 2007).

4.1 Response of Plants to Heat Stress

In response to high temperature the plants manifest different mechanisms, including long-term evolutionary phonological and morphological adaptations and short-term avoidance or acclimation mechanisms. The crop plants tend to reduce heat-induced damage by phonological and adaptation changes. The adaptation mechanism varies with the crop stage at which exposed to heat stress. The leaf rolling, leaf shedding, reducing leaf size, thickening leaves, reducing growth duration, transpirational cooling or alteration of membrane lipid compositions and other adjustments in morphology and ontogeny mechanism help plants to overcome the heat stress (Wahid *et al.*, 2007).

Leaf senescence starts early in response to heat stress, particularly when stress occurs during post-flowering stages of grain filling stages. Poor fruit/grain set at high temperature cannot be attributed to a single factor; decreases in pollen germination and/or pollen tube growth are among the most commonly reported factors. Therefore, pollen viability has been suggested as an additional indirect selection criterion for heat tolerance (Wahid *et al.*, 2007). Grain development is impacted by heat stress because assimilate translocation and grain-filling duration and rate are influenced directly by changes in ambient temperature. The extent of heat-driven damage is dependent on the level of heat stress (Farooq *et al.*, 2011). The maintenance of leaf chlorophyll and photosynthetic capacity at maturity, called 'stay-green', is considered an indicator of heat tolerance (Fokar *et al.*, 1998).

The heat-shock response is a reaction caused by exposure of an organism's tissue or cells to sudden high temperature stress, and it is characterized by a transient expression of heat-shock proteins (HSPs). In plants following temperature elevation, the heat signal is probably transduced by several pathways that will, however, coalesce into the final activation of heat-shock transcription factors (HSFs), the expression of HSPs and the onset of cellular thermo-tolerance (Saidi *et al.*, 2011).

Heat resistance is defined as the ability of a cultivar to yield more than other cultivars when subjected to heat. The plant overcomes the high temperature by heat avoidance and heat tolerance mechanisms. *Heat avoidance* is the ability of the genotype to dissipate the radiation energy to avoid rise in plant temperature. The plant reduces the heat by the process of transpiration cooling, leaf reflectance to reduce heat by pubescence and glaucousness. *Heat tolerance* is the ability grow and produce economic yield under high temperatures by improving membrane stability, reduced heat sensitivity to photosystem II, photosynthate translocation, stem-reserve mobilization and osmoregulation.

4.2 Selection Environment to Heat Stress

The selection environments for heat stress can be of *in vitro* or field environment (Singh, 2001), and moisture stress to plants should be avoided during screening for heat stress.

(i) *Screening under 'abnormal' field environment:* The screening at the location having high temperature or during off-season (summer).

(ii) *Managed environments:* The plant growth chambers can be used to manage the desirable level of heat stress during the specific period of plant growth.

(iii) *In vitro methods:* Screening of genotypes for high temperature stress in natural conditions, which are highly variable and is very difficult. The best alternative is to follow suitable laboratory procedures for screening. Temperature induction response (TIR) is a screening protocol developed based on the principle of "acquired tolerance" in which exposure of seedlings to a sub-lethal level of specific stress is used to induce tolerance to a subsequent lethal level of stress. By adapting this technique the genotypes were screened for thermo-tolerance (Kheir *et al.*, 2012). TIR techniques wherein the genetic variability for temperature stress response can be examined in crop plants by assessing the survival, recovery and growth rates after exposing the germinated seedlings to lethal temperature stress. TIR is a robust and powerful technique and can be used to screen breeding lines or germplasms to identify thermo-tolerant lines. The usefulness of TIR technique was demonstrated in groundnut and cotton to assess the response of genetic lines to heat stress (Gangappa *et al.*, 2006; Kheir *et al.*, 2012).

4.3 Phenotyping for Heat Stress

Based on the adaptation mechanism expressed by the crop plants at different stages to heat stress, the selection of suitable traits for screening programme is vital to breeding for heat tolerance. The physiological and morphological traits have been evaluated for heat tolerance in plants are membrane thermo-stability, canopy temperature, leaf chlorophyll, leaf conductance, higher photosynthetic rates, stay-green, spike number, biomass, and flowering. Dias and Lidon (2009) proposed that high grain-filling rate and high potential grain weight as useful selection criteria for improving heat tolerance. The indices have been developed to measure the thermo-tolerance in crops based on the comparison of midday foliage, air temperature and their differential (Singh *et al.*, 2007). They are canopy temperature depression (CTD), crop water stress index (CWSI) and thermal stress index (TSI). To screen for leaf tolerance to heat in breeding programme the canopy temperature depression (CTD) and leaf conductance can be effectively used for early generation selection (Reynolds *et al.*, 2001). The screening traits used to assess heat tolerance and ways to quantify the trait of interest is discussed below:

(i) Germination, seedling mortality and recovery growth, canopy temperature depression and cellular membrane thermo-stability (CMT).

(ii) *Leaf conductance:* Leaf diffusive resistance was linearly related to leaf water potential (O'Toole and Moya, 1978). The stomatal conductance/resistance can be measured by steady state porometer or portable photosynthesis system.

(iii) *Chlorophyll content and fluorescence:* Heat stress reduces photosynthesis through disruptions in the structure and function of chloroplasts, and reductions in chlorophyll content (Xu *et al.,*1995). Photosystem II and chloroplast are sensitive to heat leads to damage to PSII in turn reduce the photosynthesis. Chlorophyll content meter (CCM) or SPAD meter can be used to assess the leaf chlorophyll content. The use of chlorophyll fluorescence as a screening trait, given that association between plant tolerance and lower fluorescence signals have been reported in a number of crops. Chlorophyll fluorescence can be measured through chlorophyll fluorometer.

(iv) *Pollen viability, fruit set and grain filling:* The heat sensitivity of genotypes in reproductive phase can be assessed by flowering, pollen sterility and seed set. The high grain-filling rate and high grain weight are useful selection criteria for improving heat tolerance.

(v) *Growth and yield:* The biomass, yield and harvest index under heat stress are used as selection criteria for heat tolerance.

5. SCREENING AND BREEDING FOR CLIMATE-RESILIENT VARIETIES

The crop improvement strategies, followed to evolve abiotic stress resistant varieties, are direct and indirect. In direct breeding, the genetically diverse materials are raised under stress environment and selection is done based on survival, yield and traits contributing to stress resistance. Whereas, in indirect approach the breeding materials are not deliberately developed for stress tolerance, but they are evaluated for stress resistance under specified environmental condition after it has been developed (Singh, 2001).

Importance of screening and phenotyping: The abiotic stresses are rarely singular, more likely they impart on plant populations as a set of interacting multiple stresses present in the given environment (Jarvis *et al.*, 2005). The plants immobility limits the range of their behavioural responses to environmental signals and places a strong emphasis on cellular and physiological mechanisms of adaptation and protection to abiotic stresses (Wahid *et al.*, 2007). The genotypes of different crop species express greater phenotypic variability in their ability to tolerate/resist abiotic stresses or in different environment.

Hence, it is impossible to make significant advancement in crop improvement without accurate and reproducible phenotyping under mutable stress environment (McClean *et al.*, 2011). The rapid advancement can be achieved in abiotic stress tolerance by screening and selecting breeding materials in the target environment rather than doing in controlled conditions. The stress tolerant traits are the plant characters or a biochemical mechanism that facilitates cell survival and plant growth under stress condition. The screening trait for the abiotic stress should have desirable characteristics (Serraj *et al.*, 2003) of strong link with higher or more stable grain yield in target environment, high level of heritability and easily measurable expression of tolerance is highly preferred.

A major challenge in traditional breeding for abiotic stress tolerance is the identification of reliable screening methods and effective selection criteria to facilitate detection of abiotic tolerant traits in plants under controlled stress environment. Controlled stress environment means imposing stress to appropriate duration and severity of stress at the desired plant growth stage. The phenotyping of abiotic stress tolerance for drought, heat and salt stress is described in the respective sections.

Care to be taken for managing stress environment in field conditions: The site homogeneity and control over the water regime or stress level are two important factors that determine the success or failure of managing stress environment in field conditions (Blum, 2011).

(i) *Site homogeneity:* It is most crucial in phenotyping of genotypes under field conditions. The important generators of site variability are (a) topography, (b) soil variability such as soil pH, and mineral deficiency, (c) bordering trees and their root extensions affect the edges of sites and (d) poor distribution of agricultural inputs and water affect the site homogeneity. Poor homogeneity in especially soil moisture supply can destroy field screening. Hence, the experimental field selected for drought phenotyping should be assured for maximum possible homogeneity. The field heterogeneity can be assessed by micro-level soil profiling and mapping.

(ii) *Experimental station fault:* The crop growing conditions are often different from those in the farmer's field. The experimental stations are characterized by high soil fertility, which is not typical of dryland farming. Influence of non-conventional crop grown in the previous season can influence the experiment.

(iii) *Controlling moisture regime:* Control over the moisture or salt regime is achieved by eliminating undesirable rainfall during experimental period using rain-out shelters.

(iv) *Replicating the experiments:* Repeating the screening/evaluation trails in the desired environments and follow suitable analysis methodology will increase the precision of screening.

5.1 Conventional Breeding Approach

The crop resistance to stress may be defined as the mechanism(s) causing minimum yield loss in a stress situation relative to the maximum yield in a stress-free environment. The conventional or empirical breeding approach for abiotic stress tolerance is based on the selection of yield and its components under a given environment (Serraj *et al.*, 2003). A commonly followed conventional method of selecting plants for heat-stress tolerance is to grow breeding materials in a hot target production environment and identify individuals/lines with greater yield potential (Ehlers and Hall, 1998). Native varietal halo-tolerance was exploited to characterize differences between salt-sensitive and salt-tolerant barley varieties (Hurkman *et al.*, 1989). In India, the salt tolerant lines developed in wheat are *Kharchia* and *Rata*. Salt-tolerant varieties developed in rice by pure line selections from local traditional cultivars are *Damodar* (CSR 1), *Dasal* (CSR 2), *Getu* (CSR 3), *Hamilton* and by pedigree method, namely, CSR 10, CSR13 CSR23, CSR27 and CSR36. In case of rice, the salinity-resistant varieties have been developed by crossing with *Pokkali*, a salinity-resistant local landrace (Sairam and Tyagi, 2004). Through conventional breeding, number of wheat and maize drought tolerant varieties was developed by CIMMYT and adapted in African countries.

Ideotype or trait based approach: The trait based breeding programme can help to improve the crop adaptation and to enhance productivity under abiotic stress conditions. In breeding for stress tolerant types combining empherical selection assisted with physiological traits will be of more advantages (Reynolds *et al.*, 2000; Araus *et al.*, 2002) in the following respects.

- Helps to eliminate inferior agronomic phenotypes at an early generation itself.
- Selecting superior physiological phenotypes using rapid detection techniques in intermediate generations.
- Selecting for higher performance types in yield trials in advanced generations.

In ideotype approach, plant types preferred for different abiotic situation is decided in advance. Accordingly, the suitable donor source is identified and transferred through elite genetic background. Plant types preferred for different drought stress scenario and timing of stress are extracted from Blum (2011) and the concise details are presented in Table 1.

Table 1 The different plant types preferred for drought tolerance

Stress situations	*Ideotype preferred*
1. Drought situations	
i. Soil moisture is always available at depth (sub-soil moisture)	Deep rooted genotype[1]
ii. Water is available at depth, but soil is hard and offers large resistant to root penetration	Deep rooted genotypes[2]
iii. Stored soil moisture with limited or no seasonal rainfall	Genotypes with reduced plant size and early flowering
iv. Water is not available in depth	Genotypes with large root length density and small root (hydraulic) resistance[3]
2. Timing of drought stress	
i. Seedling development stage	Selecting better seedling survival types[4]
ii. Vegetative (pre-flowering) growth stage	(a) Genotypes able to continue growing during stress and/or be capable of recover upon rehydration. (b) Genotypes with dehydration avoidance[5]
iii. Flowering stage (most sensitive stage)	Low plant ABA status[6]. Delayed heading or flowering, short anthesis-to-silking interval (ASI) in maize
iv. Fruit or grain development stage	Genotypes with non-senescence (stay-green) or stem reserve utilization for grain filling

[1]In case of cereals, limited tillering at the vegetative stage can enhance deeper root development especially when top soil dries out.

[2]Hard soil resistance can constrain root growth into deep soil despite genotype has ability to grow long roots. Such situations management practice (deep tillage) offer solution to this problem.

[3]For greater soil moisture extraction by roots under lower soil water potential.

[4]By looking into constitutive accumulation of osmotically active solutes and other protective agents in seedlings.

[5]By way of maintaining plant water potential or by osmotic adjustment as reflected by RWC.

[6]Observing phenotypic expression of fertility dysfunction, because the ABA accumulation promotes stomatal closure but it causes distinct sterility symptom during flowering stage stress.

A conceptual model for drought tolerant ideotype in wheat was proposed by Reynolds *et.al.*, (2001a) with high expression of the following traits, but not all of which would be useful in all drought environments:

- Seed size and coleoptile length to improve early crop establishment.
- Leaf anatomical traits, e.g., waxiness, pubescence, rolling, thickness (reduce risk of photo-inhibition), high tiller survival and stay-green.
- Early ground cover and pre-anthesis biomass to reduce evaporation of soil moisture.
- Stem reserves/remobilization and spike photosynthesis that help grain filling during severe post-anthesis stress.
- High stomatal conductance as an indicative of roots which are able to extract soil water at depth.
- High osmotic adjustment to maintain cell functions at low water potential.
- Accumulation of abscisic acid for pre-adapts cells to stress.
- Heat tolerance (heat stress may be caused by low leaf transpiration rates under drought).

Difficulties encountered in conventional breeding approach for abiotic stress tolerance are (a) the abiotic stress tolerant traits are mostly governed by polygenes and are difficult to transfer through simple breeding programmes, (b) linkage drag of negative alleles/trait from donor parent to recurrent parent is another major cause of concern, and (c) handling large number of segregating population and identifying the desired genotype by simple empirical selection is difficult.

5.2 Tailoring Abiotic Stress Tolerant Traits by Gene or Traits Pyramiding

In climate changing era, the job of plant breeder is to device the suitable strategy to incorporate the tolerant traits in suitable genetic background in a reasonably shorter time. The choice of trait to be transferred for abiotic tolerance varies with drought or stress situation and time of stress. Both marker-assisted breeding and genetic engineering strategies can be efficiently used for trait pyramiding. Molecular marker technologies can be used to dissect defined quantitative traits by quantitative trait locus (QTL), so that individual loci can be targeted in marker-assisted selection (Witcombe *et al.*, 2008).

Marker-assisted trait/QTL pyramiding: It is easy to pyramid genes controlling qualitative traits like disease and pest resistance having major effect on phenotype and easy to measure. Whereas, most of the traits conferring abiotic stress tolerance are quantitative in nature. Hence, molecular markers and QTL help in screening and introgression programmes. Large QTLs were identified and reported for the abiotic stress tolerant traits (ASTTs) such as root traits in rice and maize, osmotic adjustment and leaf rolling in rice, leaf ABA concentration and anthesis silking index in maize, stay-green type in sorghum, rice and wheat. The QTLs for submerged tolerance in rice, cold and salt tolerance in rice and wheat were also identified. The list of QTLs identified for abiotic stress tolerant traits and stress responsive genes in different crops is available in biotech issues at 'www.plantstress.com'. QTL(s) identified for the stay-green character (Kumar *et al.,* 2010), senescence-related traits (Vijayalakshmi *et al.,* 2010) and canopy temperature under heat (Pinto *et al.,* 2010) in wheat.

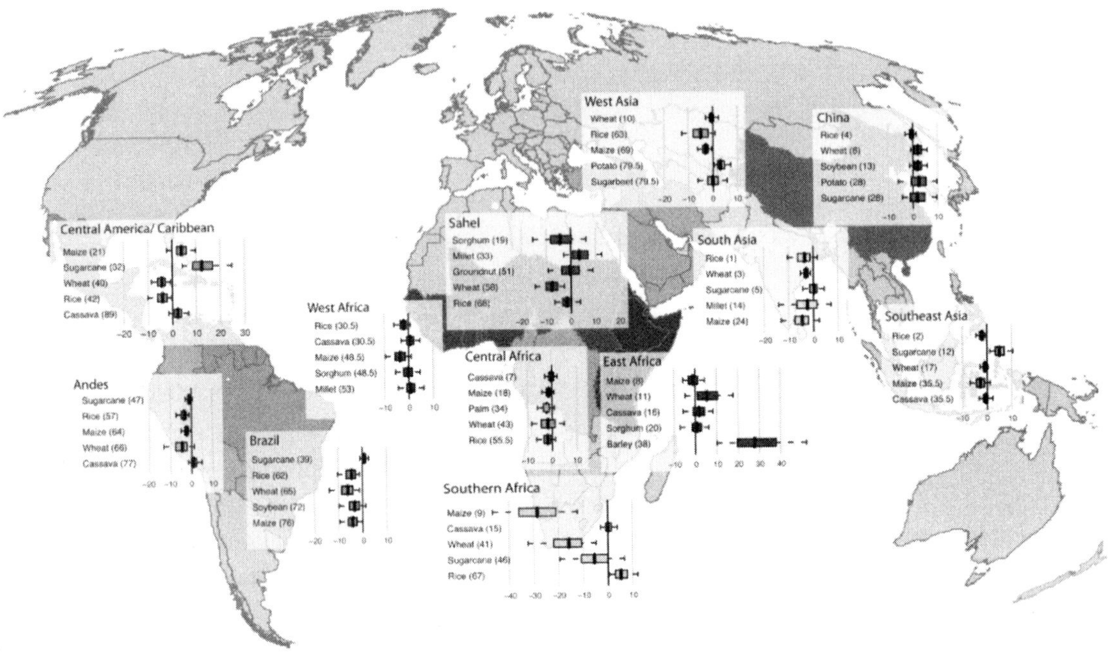

Fig. 1 Projected impacts of climate change by 2030 for five major crops in each region. For each crop, the dark vertical line indicates the middle value out of 100 separate model projections, boxes extend from 25th to 75th percentiles, and horizontal lines extend from 5th to 95th percentiles. Number in parentheses is the overall rank of the crop in terms of importance to food security, calculated by multiplying the number of malnourished in the region by the per cent of calories derived from that crop (adapted with permission from Lobell *et al.*).

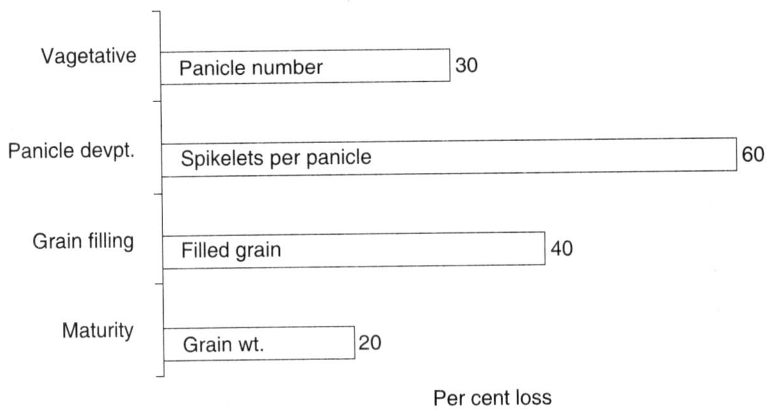

Fig. 2 Impact of water stress at different growth phases in rice.

The marker-assisted selection (MAS) comes in handy, if the marker is linked to the target genes. MAS is useful for (a) simply inherited traits, but difficult to phenotype or do not have consistent phenotypic expression under certain specific selection environment; (b) if the major genes affecting the quantitative trait are identified, it is easy to pyramid those genes by marker-assisted breeding; (c) the genes have similar phenotypic effect can be pyramid through MAS, otherwise difficult in conventional phenotypic selection, and MAS helps to control the linkage drag and accelerated the recovery of recurrent genome (Ye and Smith, 2008). Different marker assisted breeding methods are used to accumulate abiotic stress tolerant trait QTLs into the desirable genetic background is describe below:

(i) *Marker-assisted backcross breeding (MABC):* This method is used to transfer desirable ASTT governed by major genes/QTLs from donor to agronomically superior recurrent parent. It needs genetic markers that are associated with genes or QTLs affecting whole plant stress tolerance or individual components contributing to it. The steps involved (Kulwal *et al.*, 2011) in MABC are: (a) identification of molecular markers associated with ASTTs, (b) validation of identified markers in the genetic background of targeted genotype (recurrent parent) to be improved, and (c) MABC to transfer the QTL/gene from donor into the recurrent parent.

MABC combines 'foreground' selection of donor alleles linked to QTLs and 'background' selection of recurrent parent alleles in the BC_2 and later generations. Foreground selection involves the use of two flanking markers for selecting a particular gene or a QTL. Background selection uses markers that are not associated with the desirable QTL/gene and selects backcross progeny with the highest proportion of genome of the recurrent parent (Witcombe *et al.*, 2008; Gaur *et al.*, 2012). The stepwise and simultaneous-stepwise transfer of abiotic stress tolerant trait(s) to recurrent parent is schematically represented in Figs. 4 and 5, respectively. The improved cultivars of submerged tolerance was developed using *Sub 1* gene/QTL in rice (Septiningsih *et al.*, 2009) and for pre-harvest sprouting tolerance in wheat (Kumar *et al.*, 2010a) by marker-assisted introgression.

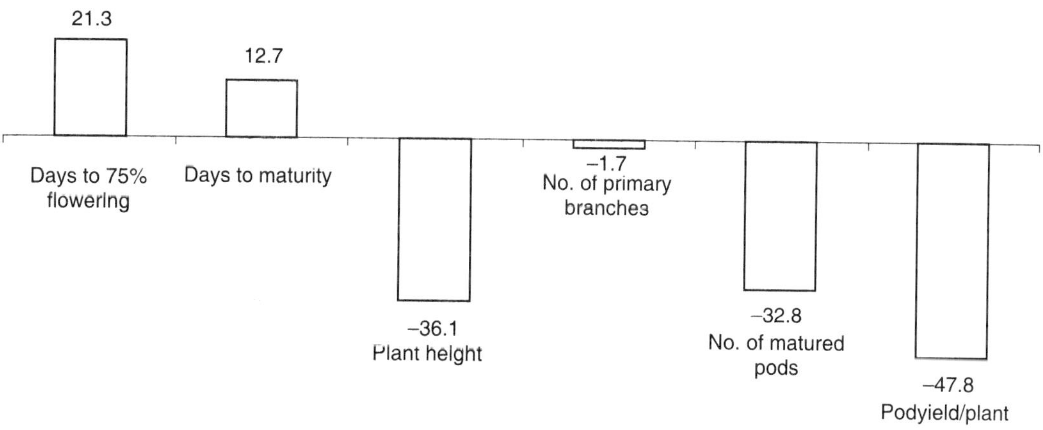

Fig. 3 Per cent influence of soil moisture stress during flowering and pod development stages in peanut traits.

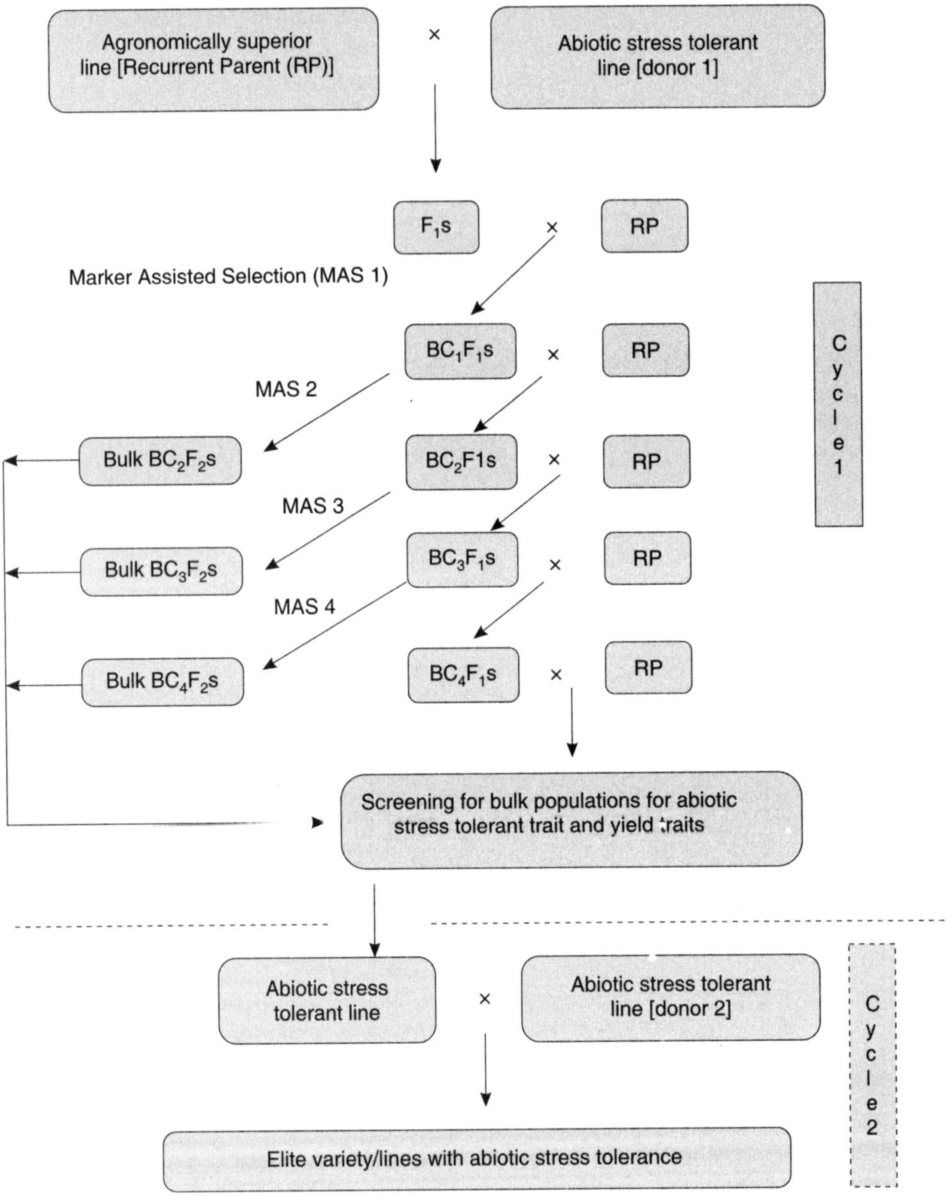

Fig. 4 Stepwise transfer of abiotic stress tolerant trait(s) by marker assisted backcross.

(ii) *Marker-assisted gene pyramiding by multiple cross:* In this method, genes of desirable trait from multiple parents are converged into a single genotype to enhance desirable trait performance. Evolving of abiotic tolerant lines through this method in a single breeding cycle involves creation of root genotype by two-or three-way crosses. This brings desirable genes into single genotype as heterozygous from different parents. Then fix these genes

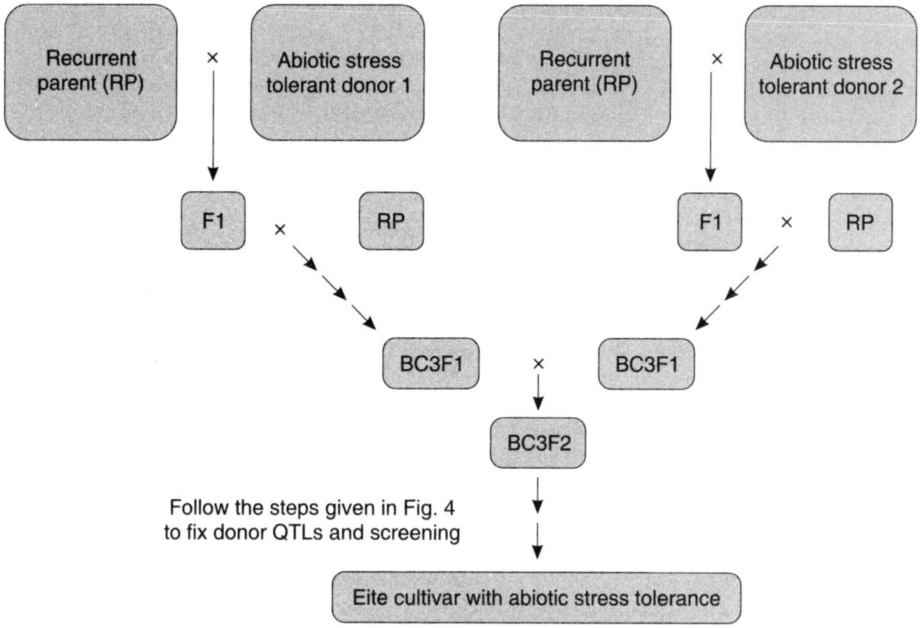

Fig. 5 Simultaneous and stepwise transfer of abiotic stress tolerant trait(s) by marker assisted selection.

as homozygous state in inbred lines by selfing or double haploid strategies. To ensure the presence of genes of target traits, it is essential to genotype by MAS in all the steps of breeding cycle. Bringing four favourable genes of an abiotic stress tolerant trait from different parents to develop elite inbred line(s) is represented in Fig. 6.

(iii) *Marker-assisted recurrent selection (MARS):* Most of the ASTTs are quantitative in nature, and are governed by polygenes with minor effects. The transfer of such minor genes/ QTLs through MABC is much difficult as a large number of progenies are needed to select desirable ones. In such situation, the MARS will be of very useful in pyramiding desirable alleles in single genotype. Conventionally the recurrent selection is used to accumulate the desirable alleles for yield and other ploygenic traits in cross-pollinated crops. In MARS, the individuals for intercrossing are selected using selection index-constructed, based on QTL associated markers. The genetic gain using such selection index is higher than the conventional phenotypic selection (Gaur *et al.*, 2012).

(iv) *Advanced backcross QTL analysis (AB-QTL):* This method was proposed by Tankley and Nelson (1996), and is useful for simultaneously to identify and introgress favourable alleles from unexplored donors into elite genetic background. This method readily applied to annual crops and is difficult to apply to crops for which inbred lines do not exist. It would be difficult to apply this method to highly heterozygous, cross-pollinated and clonally propagated crops. The advantages of AB-QTL analysis for introgression of valuable QTLs as compared with conventional strategies for QTL detection is discussed by Tanksley and Nelson (1996). If the identified markers are available for abiotic stress tolerant traits,

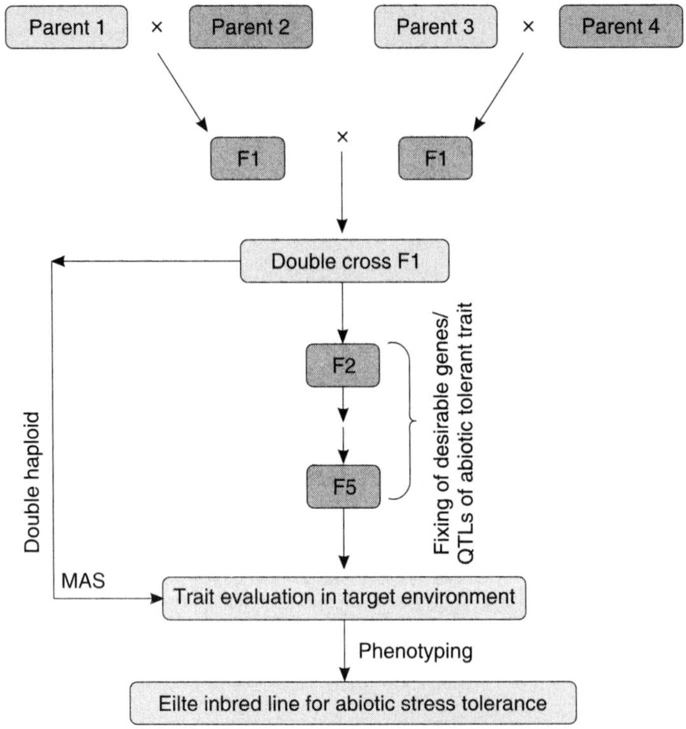

Fig. 6 Pyramiding of four genes/QTLs of a quantitative trait for abiotic stress by double cross and marker-assisted selection.

the AB-QTL strategy can be utilized to transfer such traits into agronomically superior varieties. The steps involved in this method are narrated below:

- The commercially preferred cultivar/recurrent parent is crossed with wild species or donor parent.
- Backcrossing is effected with recurrent parent up to BC_2 or BC_3 generations. This method adopts strategy of delay the QTL and marker analysis to BC_2 or BC_3, generations beyond the BC_3 are likely to have low statistical power to detect most QTLs. The molecular marker/QTLs are used to conduct negative genotype and phenotype to eliminate the undesirable alleles from donor parent (Grandillo and Tanksley, 2003). This allows to reduce the frequency of the donor-parent genome in each of the advanced backcross lines and to avoid the masking effect from deleterious wild recessive alleles or donor parent.
- The BC_2 or BC_3 are backcrossed with donor parent to generate BC_2F_1 or BC_3F_1 population in hybrid crops (BC_2S_1 or BC_3S_1 for open-pollinated crops).
- Evaluation of BC_2F_1 or BC_3F_1 families to be carried out for stress tolerant and yield traits under monitoring of QTLs. Once favourable QTLs are identified not more than two generations required to isolate targeted QTL-NILs.
- Evaluation of NILs is done for stress tolerant and yield traits under target environment to identify the elite lines and for varietal development.

(v) *Genomic selection (GS):* The advantage of genome-wide or genomic selection is that it does not need to workout the marker-trait association as required in MABC or MARS. It calculates the marker effects (genomic estimated breeding values) of lines across the entire genome which explains the entire phenotypic variations of particular trait (Kulwal *et al.*, 2011). The superior lines can be selected based on breeding values to advance the progenies. Mayor and Bernardo (2009) have shown that double haploid populations are very useful in genome-wide selection compared to F_2 populations, especially for complex traits that are controlled by many QTL. The in-built advantages of this method can be effectively exploited for the abiotic stress breeding to evolve tolerant cultivars in a quicker time.

(vi) *MAPS breeding scheme:* Zong and coworkers proposed a novel QTL pyramid breeding scheme based on marker-assisted and phenotype selection (MAPS) that allowed pyramiding of as many as 24 QTLs at a single hybridization without massive crossing work. The MAPS QTL pyramiding scheme, not only overcomes the shortcomings of crossing NILs, but also efficiently pyramid the desired QTLs without any backcrossing. Combining the genome-wide association studies (GWAS) results with MAPS QTL pyramiding scheme, it is possible to breed stress tolerant lines in a novel way by assembling more number of QTLs as detailed by Zong *et al.*, (2012) in rice.

5.3 Trait Pyramiding by Transgenic Approach

The recent advances in plant molecular biology have thrown out a number of genes associated with abiotic stress mechanisms/adaptive traits. Some of these genes were validated for their function. In case of drought stress, the tolerance mechanism is amenable to genetic engineering of specific genes. However, whole-plant avoidance strategies depend more on conventional breeding schemes and QTL analysis (Vinocur and Altman, 2005). The major requirement for developing drought/abiotic stress resistance cultivar through transgenic approach are (a) the genes identified should improve drought resistance significantly; (b) no phenotypic changes for other traits; (c) no yield penalty under non-stress conditions; and (d) ensure over-expression of a single copy transgene rather than multiple copies of a gene that can leads to instability of expression and inheritance or even gene silencing (Xiao *et al.*, 2007; Blum, 2011). The transgenic lines were produced for over-expression of ion transport proteins; osmoprotectants mechanism for improving salt tolerance is reviewed in detail by Ashraf and Akram (2009). The stress adaptive genes can be exploited through genetic engineering of 'designer' traits as narrated below:

(i) *Use of abiotic stress tolerant genotypes for transgenics:* Identify the genotype(s) with superior water relation (drought avoidance) and use them as recipient genotype to develop transgenic. Transfer and express validated genes regulating other abiotic stress adaptive mechanism (cellular tolerance) in recipient background (Karaba *et al.*, 2011), and follow stringent evaluation for stress tolerance to identify elite genotypes (Fig. 7).

(ii) *Co-expression of validated abiotic stress genes:* Through currently available transformation technology, it is possible to transfer multiple genes, which may act synergistically and additively to improve plant stress tolerance (Wahid *et al.*, 2007; Karaba *et al.*, 2011). Co-express validated genes regulating diverse mechanisms using multi-gene expression cassettes. Co-expression of multi-genes into plant genome can be achieved by the various

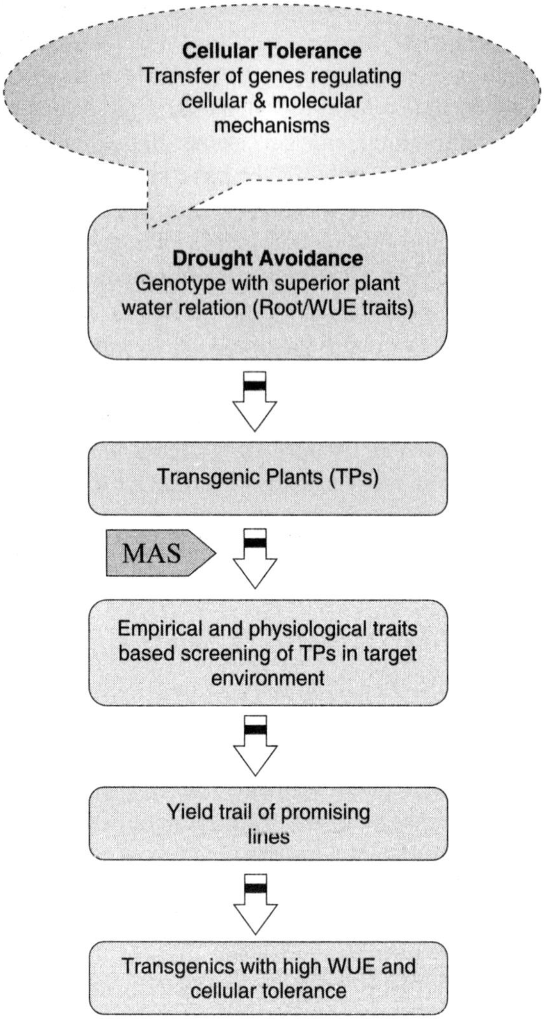

Fig. 7 Pyramiding of drought tolerant trait(s) by transgenic approach.

strategies such as (a) crossing between two transgenic plants; (b) co-transformation with multiple genes; (c) transformation of single plasmid carrying several linked transgenes or multi-genes; and (d) polycistron or polyprotein strategy.

6. IMPACT OF CLIMATE CHANGE ON PLANT GENETIC DIVERSITY

Genetic diversity in crop plants leads agriculture to withstand moderate climate changes experienced over past year (Lane and Jarvis, 2007). The erosion of crop biodiversity (so called genetic erosion) by the human intervention and modern agricultural practices threatened the changes for naturally adapted varieties/land races. To exploit the wealth of gene sequence information provided by the 'genomics revolution' in major crops and mine agricultural germplasm for genetic diversity, high resolution, high throughput technologies in plant physiology are required for bridging the gap between genotype and phenotype (Furbank, 2009).

Crop production systems that rely on highly selected genetic resources might be increasingly vulnerable to climate change impacts such as epidemic disease spread. If the production trend declines, there will be pressure on marginal lands to cultivate or implement unsustainable practices that over the long term, degrade lands and resources, and adversely impact biodiversity on and near agricultural areas. In this regard, already food insecure people in the developing countries will be most adversely affected by climate change. These changes have been seen to cause a decrease in the variability of those loci controlling physical responses to climate (Jump *et al.*, 2005). Therefore genetic variation holds the key to the ability of populations and species to persist over evolutionary time through changing environments (Shaffer *et al.*, 1985). No organism can predict the future (and evolutionary theory does not require them to), nor can any organism be optimally adapted for all environmental conditions. Nonetheless, the current genetic composition of a species influences how well its members will adapt to future physical and biotic environments.

The population can "migrate" across the landscape over generations. By contrast, populations that have a narrower range of genotypes (and are more phenotypically uniform) may simply fail to survive and reproduce at all as conditions become less locally favourable. Such populations are more likely to become extirpated (locally extinct), and in extreme cases the entire species may end up at risk of extinction. For example, the Florida Yew (*Torreya taxifolia*) is currently one of the rarest conifer species in North America. But in the early Holocene (10,000 years ago), when conditions in southeastern North America were cooler and wetter than today, the species was probably widespread. For reasons that are not completely understood, *T. taxifolia* failed to migrate northward as climate changed during the Holocene. Today, it is restricted to a few locations in the Apalachicola River Basin in southern Georgia and the Florida panhandle. As the *T. taxifolia* story illustrates, once species are pushed into marginal habitat at the limitations of their physiological tolerance, they may enter an "extinction vortex," a downward cycle of small populations, and so on (Shaffer *et al.*, 1985; Gilpin *et al.*, 1986). Reduced genetic variability is a key step in the extinction vortex. Gene-banks must be better to respond to novel and increased demands on germplasm for adapting agriculture to climate change. Researchers informed that 16-22% of wild species might go extinct (Jarvis *et al.*, 2005). Some species can adapt to projected future environments, some cannot this referred to as phenotypic plasticity. Therefore monitoring changes in crop (any) biodiversity is desirable. It is urged to have collaborative effort of public and private sector researchers to broaden and enhance the cultivated crop germplasm base. Gene-banks need to include different characteristics in their screening processes and their collections need to be comprehensive, including minor crops. Thus breeding efforts will require the continued collection, evaluation, deployment and conservation of diverse crop genetic material to sustain the crop production.

7. CONCLUSION

Abiotic stress breeding demands an integrated effort of plant breeder, crop physiologist, molecular biologist and soil scientist to evolve climate flexible cultivars. The success of abiotic stress breeding is lies in (a) adapting suitable methodologies to impose the stress at required intensity, desired stage of crop growth and duration; (b) choosing appropriate screening tools to assess the response of abiotic tolerant trait/mechanism under the given environment; (c) identifying genotypes with abiotic stress tolerant traits suited for stress scenario as a donor by stringent screening; (d) converge different traits that confers abiotic stress tolerance in a agronomically superior genetic background through modern crop improvement tools for rapid advancement; and (e) validating the elite genotypes with 'designer' traits under target stress environments before releasing as a variety. The available high-throughput techniques can be effectively used for genotyping and phenotyping to screening large number of genotypes and to identify small differences in targeted traits. It is better to select parents for breeding programme, only after the results of controlled condition are validated in field conditions.

The breeders' business does not rest with converging different abiotic stress tolerant traits in evolving climate-resilient varieties to mitigate the climate change, but also ensure the produce quality requirement of different communities. The crop breeding only can offer limited solution to climate change in short term; hence further emphasis should be given for the climate-smart agricultural practices such as conservation tillage, residue management, building soil organic matter, agro-forestry, efficient use of water can further help to manage stress like drought and salinity. Identifying and executing cost-effective and rapid genotyping and phenotyping methods in near future will help to evolve climate adapted cultivars, which hopefully address the climate change challenges in agriculture in short term. Plant biodiversity is yet another reservoir of variable adaptive traits for long-term solution and need to be conserved with well-database management in every locality to save from natural disasters including climate change.

References

Anderson, S. and Morton, J. (2008). Drylands and climate change. Social dimensions of Climate change workshop. International Institute for Environmental development.

Araus, J.L., Slafer, G.A., Rcynodls, M.P. and Royo, C. (2002). Plant breeding and drought in C_3 cereals: what should we breed for? *Ann. Bot.* 89: 925-940.

Arunachalam, P., Balasubramaniyan, P. and Kannan, P. (2012). Screening for drought tolerant bunch type groundnut (*Arachis hypogaea* L.) lines suitable for rainfed alfisol. National Seminar on Breeding for abiotic stress in crop plants, 22 March 2012, AC&RI, Madurai, India.

Ashraf, M. and Akram, N.A. (2009). Improving salinity tolerance of plants through conventional breeding and genetic engineering. *An analyt.comp.Biotech. Adv.* 27: 744-752.

Blum, A. (2011). *Plant Breeding of Water-limited Environments*, Springer-limited+Bussiness media.

Blum, A., Poyarkova, H., Golan, G. and Mayer, J. (1983). Chemical desiccation of wheat plants as a simulator of post-anthesis stress. I. Effects on translocation and kernel growth. *Field Crops Res.* 6: 51-58.

Blum, A., Mayer, J. and Gozlan, G. (1983a). Chemical desiccation of wheat plants as a simulator of post-anthesis stress II. Relations to drought stress. *Field Crop Res.* 6: 149-155.

Boonjung, H. and Fukai, S. (1996). Effect of soil water deficit at different growth stages on rice growth and yield under upland conditions 2. Phenology, biomass production and yield. *Field Crops Res.* 48: 47-55.

Bray, E.A., Bailey-Serres, J. and Weretilnyk, E. (2000). Responses to abiotic stresses, In: W. Gruissem, B. Buchannan, R. Jones (Eds.), Biochemistry and Molecular biology of plants, American Society of plant Biologists, Rockville, MD, pp. 158-1249.

Campbell, B., Mann, W., Melendez-ortiz, R., strack C. and Tennigkeit T. (2011). Agriculture and climate change; A scoping report. Washington, DC; Meridian Institute; 2011.

Dias, A. S., and Lidon, F. C. (2009). Evaluation of grain filling rate and duration in bread and durum wheat, under heat stress after anthesis. *J. Agron. Crop Sci.* 195: 137-147.

Earl, H.J. (2003). A precise gravimetric method for simulating drought stress in pot experiments. *Crop Sci.* 43: 1868-1873.

Ebercon, A., Blum, A. and Jordan, W.R. (1977). A rapid colorimetric method for epicuticular wax content of sorghum leaves, *Crop Sci.* 17: 179-180.

Ehlers, J.D. and Hall, A.E., (1998). Heat tolerance of contrasting cowpea lines in short and long days. *Field Crops Res.* 55: 11-21.

Farooq, M., Bramley, H., Palta, J.A. and Siddique, K.H.M. (2011). Heat stress in wheat during reproductive and grain-filling phases, *Critical Rev. Plant Sci.* 30: 1-17.

Fernandez, G.C.J. (1992). Effective selection criteria for assessing stress tolerance. In: C.G. Kuo, ed. *Proceedings of the international symposium on adaptation of vegetables and other food crops in temperature and water stress*, Tainan, Taiwan.

Fischer, K. S. and Wood, G. (1981). Breeding and selection for drought tolerance in tropical maize. In *Proc. symp. on principles and methods in crop improvement for drought resistance with emphasis on rice*, IRRI, Philippines, 23-25th May 1981.

Fischer, R. A. and Maurer, R. (1978). Drought resistance in spring wheat cultivars. I. Grain yield response. *Aust. J. Agric. Res.* 29: 897-912.

Flowers, T.J., Troke, P.F. and Yeo, A.R. (1977). The mechanisms of salt tolerance in halophytes. *Annu. Rev. Plant Physiol.* 28: 89-121.

Fokar, M., Blum, A., and Nguyen, H. T. (1998). Heat tolerance in spring wheat. II. Grain filling. *Euphytica* 104: 9-15.

Furbank, R.T. (2009). Plant phenomics: from gene to form and function. *Funct. Plant Biol.* 36, v-vi.

Gangappa, E., Ravi, K. and Veera Kumar, G.N. (2006). Evaluation of groundnut *(Arachis hypogaea* L.) genotypes for temperature tolerance based on Temperature Induction Response (TIR) technique. *Indian* J. *Genet.,* 66(2): 127-130.

Gaur, P.M., Jukanti, A.K., and Varshney, K.A., (2012). Impact of Genomic Technologies on Chickpea Breeding Strategies, *Agronomy*, 2: 199-221.

Gilpin, Michael E., and Michael E. Soulé. (1986). Minimum viable populations: the processes of species extinctions. In *Conservation biology. The science of scarcity and diversity*, Soulé, M.E. (ed.), Sunderland, MA: Sinauer Associates.

Grandillo, S. and Tanksley, S.D. (2003). Advanced backcross QTL analysis: Results and perspectives. In *Proceedings of the International Congress "In the Wake of the Double Helix: From the Green Revolution to the Gene Revolution,* Tuberosa R., Phillips R.L., Gale M. (eds.), , Bologna, Italy, 115-132.

GreenFacts (2007). Facts on climate change: A summary of the 2007 assessment report of the IPCC.

Greenway, H. and Munns, R. (1980). Mechanisms of salt tolerance in non halophytes. *Annu. Rev. Plant Physiol.* 31: 149-190.

Gregorio, G.B., Senadhira, D. and Mendoza, R.D. (1997). Screening rice for salinity tolerance, IRRI discussion paper series no. 22, IRRI, Phillippines.

Hurkman, W.J., Fornari, C.S. and Tanaka, C.K. (1989). A comparison of the effect of salt on polypeptides and translatable mRNAs in roots of a salt-tolerant and a salt-sensitive cultivar of barley. *Plant Physiol.* 90: 1444-1456.

IPCC (Intergovernmental Panel on Climate Change). 2007. *Intergovernmental Panel on Climate Change fourth assessment report: Climate change 2007 Synthesis Report.* World Meteorological Organization, Geneva, Switzerland.

Jarvis, D., Mar, I. and Sears, L. (eds.). (2005). Enhancing the use of crop genetic diversity to manage abiotic stress in agricultural production systems, Budapest, Hungary.

Jump, A.S. and Peñuelas, J. (2005). Running to stand still: adaptation and the response of plants to rapid climate change. Ecol. Lett. 8: 1010-1020.

Karaba, N.N., Rama, N. Rohini Sreevathsa and Kumaraswamy (eds.). (2011). Characterisation of drought adaptive traits and molecular approaches to introgress them for crop improvement, Manual series: DCP04-2011, University of Agricultural Sciences, Bengaluru, India.

Kheir, E.A., Sheshshayee, M.S., Prasad, T.G. and Udayakumar, M. (2012). Temperature induction response as a screening technique for selecting high temperature-tolerant cotton lines. *J Cotton Sci.* 16: 190-199.

Kulwal, P.L., Thudi, M. and Varshney, R.K. (2011). Genomics interventions in crop breeding for sustainable agriculture. In *Encyclopedia of sustainability science and technology*, Springer.

Kumar, U., Joshi, A. K., Kumari, M., Paliwal, R., Kumar, S., and R¨oder, M. S. (2010). Identification of QTLs for stay green trait in wheat (*Triticum aestivum* L.) in the 'Chirya 3' x 'Sonalika' population. *Euphytica* 174: 437-445.

Kumar, J., Mir, R.R., Kumar, N., Kumar, A., Mohan, A., Prabhu, K.V., Balyan, H.S. and Gupta, P.K. (2010a). Marker assisted selection for pre-harvest sprouting tolerance and leaf rust resistance in bread wheat. *Plant Breed.* 29(6): 617-621.

Lane, A. and Jarvis, A. (2007). Changes in Climate will modify the Geography of Crop Suitability: Agricultural Biodiversity can help with Adaptation. *SAT eJournal* 4(1): 1-12

Lauchi, A and Epstein, E. (1984). Mechanisms of salt tolerance in plants. *California Agriculture*, October 1984, pp. 18-20.

Lobell, D., Burke, M., Tebaldi, C., Mastrandrea, M., Falcon, W. and Naylor, R., Prioritizing climate change adaptation needs for food Security to 2030. *Program on food security and the environment - policy* brief. http://iis-db.stanford.edu/pubs/22099/science_policy_brief.pdf.

Lopes, M.S. and Reynolds, M.P. (2012). Stay-green in spring wheat can be determined by spectral reflectance measurements (normalized difference vegetation index) independently from phenology. *J. Experi. Bot.* 63(10): 3789-3798.

Mahajan, S. and Tuteja, N. (2005). Cold, salinity and drought stresses: An overview. *Arch. Biochem. Biophy.* 444: 139-158.

Mayor, P.J.; Bernardo, R. (2009). Genome wide selection and marker-assisted recurrent selection in doubled haploid versus F_2 populations. *Crop Sci. 49*: 1719-1725.

McClean, P.E., Burridge, J., Beebe, S., Rao, I.M. and Porch, T.G. (2011). Crop improvement in the era of climate change: an integrated, multi-disciplinary approach for common bean (*Phaseolus vulgaris*). *Funct. Plant Biol.* 38, 927-933.

Munns R. (1993). Physiological processes limiting plant growth in saline soil: some dogmas and hypotheses. *Plant Cell Environ.*16: 15-24.

Munns, R. (2002). Comparative physiology of salt and water stress. *Plant Cell Environ.*25: 239-250.

Munns, R. and James, R.A. 2003. Screening methods for salinity tolerance: a case study with tetraploid wheat, *Plant and Soil* 253: 201-218.

Munns, R. and Tester, M. (2008). Mechanisms of salinity tolerance. *Annu. Rev. Plant Biol.* 59: 651-81.

O'Toole, J.C. and T.B., Moya. (1978). Genotypic variation in maintenance of leaf water potential in rice. *Crop Sci.* 18: 873-876.

Parida, A.K. and Das, A.B. (2005). Salt tolerance and salinity effects on plants: a review. *Ecotoxicol. Environ. Saf.* 60: 324-349.

Pinto, R. S., Reynolds,M. P., Mathews, K. L.,McIntyre, C. L.,Olivares-Villegas, J.J., and Chapman, S. C. (2010). Heat and drought adaptive QTL in a wheat population designed to minimize confounding agronomic effects. *Theor. Appl. Genet.* 121: 1001-1021.

Reynolds, M.P., Van Ginkel, M. and Ribaut, J.M. (2000). Avenues for genetic modification of radiation use efficiency in wheat. *J. Exp. Bot.* 51: 459-473.

Reynolds, M.P., Nagarajan, S., Razzaque,M.A. and Ageeb, O.A.A. (2001). Heat tolerance, In *Application of Physiology in Wheat Breeding,* Reynolds, M.P., Ortiz-Monasterio, J.I. and McNab, A. (eds.), CIMMYT, Mexico.

Reynolds, M.P., Trethowan, R.M., Van Ginkel, M. and Rajaram, S. (2001a). Application of Physiology in Wheat Breeding. In *Application of Physiology in Wheat Breeding,* Reynolds, M.P., Ortiz-Monasterio, J.I. and McNab, A. (eds.), CIMMYT, Mexico.

Rosielle, A.A. and Hamblin, J. (1981). Theoretical aspects of selection for yield in stress and non-stress environments. *Crop Sci.* J. 21, 943-946.

Saidi, Y., Finka, A. and Goloubinoff, P. (2011). Heat perception and signalling in plants: a tortuous path to thermotolerance. *New Phytologist.* 190: 556-565.

Sairam, R.K. and Tyagi, A. (2004). Physiology and molecular biology of salinity stress tolerance in plants, *Current Sci.*, 86(3): 407-421.

Septiningsih, E,M., Pamplona, A.M., Sanchez, D.L., Neeraja, C.N., Vergara. G.V., Heuer, S., Ismail, A.M. and Mackill, D.J. (2009). Development of submergence-tolerant rice cultivars: the *Sub1* locus and beyond. *Ann. Bot.* 103: 151-160.

Serraj, R., Bidinger, F.R., Chauhan, Y.S., Seetharama, N., Nigam, S.N. and Saxena, N.P. (2003). Management of drought in ICRISAT cereal and legume mandate crops. In *Water productivity in agriculture: limits and opportunities for improvement,* Kinje, J.W., Barker, R. and Molden, D. (eds.) CAB International, p. 127-144.

Shaffer, Mark L., and Fred B. Samson. (1985). Population size and extinction: A note on determining critical population sizes. American Naturalist 125 (1).

Singh, B.D. (2001). *Plant breeding: Principles and methods*, Kalyani Publishers, Ludhiana.

Singh, M., Rao., R.C.H. and Williams, J.H. (1991). A statistical assessment fo genotype sensitivity of groundnut (*Arachis hypogaea* L.) to drought in line source sprinkler experiments. *Euphytica* 57: 19-25.

Singh, R.P., Vara Prasad, P.V., Sunita, K., Giri, S.N. and Reddy, K.R. (2007). Influence of high temperature and breeding for heat tolerance in cotton: a review. *Ad. Agron.* 93: 313-385.

Sritharan, N, Vijayalakshmi, C. and Selvaraj, P.K. (2010). Effect of Micro-irrigation technique on physiological and yield traits in aerobic rice. *International Journal of Agriculture Environment & Biotechnology* 3(1): 39-41.

Sritharan, N. and Mallika Vanangamudi. (2006). Screening of rice genotypes for salt tolerance. *Plant Archives* 6(2): 815-818.

Sritharan, N. and Vijayalakshmi, C. (2007). Relative water content, chlorophyll stability index, proline, epicuticular wax content and yield attributes of rice genotypes under aerobic condition. *Plant Archieves,* 7(1): 301-304.

Sritharan, N. and Vijayalakshmi, C. (2008). Chlorophyll pigments, gas exchange parameters, membrane integrity and yield potential of rice genotypes under aerobic condition. *Advances in Plant Sci.,* 21(2): 545-548.

Sritharan, N., Vijayalakshmi, C. (2012). Physiological and biochemical basis of aerobic rice. Lambert Academic Publishing, Germany.

Tanksley, S.D. and Nelson, J.C. (1996). Advanced backcross QTL analysis: a method for simultaneous discovery and transfer of valuable QTL from unadapted germplasm into elite breeding lines. *Theor. Appl. Genet.* 92: 191-203.

Tavakkoli, E., Fatehi, F., Rengasamy, P. and McDonald, G.K. (2012). A comparison of hydroponic and soil-based screening methods to identify salt tolerance in the field in barley. *J. Experi. Bot.* 63(10): 3853-3868.

Türkana, I. and Demiral, T. (2009). Recent developments in understanding salinity tolerance, *Environ. Exper. Bot.* 67: 2-9.

Tyagi, P.K., Singh, D.P. and Pannu, R.K. (2000). Effect of post anthesis desiccation on plant-water relation, canopy temperature, photosynthesis and grain yield in wheat genotypes. *Annals of Biol.* 16: 111-119.

Udayakumar, M., Rao, R.C.N., Wright, G.C. Ramaswamy, G.C., Gangadhar, G.C.R. and Aftab Hussain. I.S. (1998). Measurement of transpiration efficiency in field condition. *J. Plant. Physiol. Biochem.* 1: 69-75.

UNFCCC. (2007). Climate change: Impacts, vulnerabilities and adaptation in developing countries, p. 64.

Vijayalakshmi, K., Fritz, A. K., Paulsen, G. M., Bai, G., Pandravada, S., and Gill, B. S. (2010). Modeling and mapping QTL for senescence-related traits in winter wheat under high temperature. *Mol. Breed.* 26: 163-175.

Vinocur, B. and Altman, A. (2005). Recent advances in engineering plant tolerance to abiotic stress: achievements and limitations. *Current Opinion in Biotech.* 16: 123-132.

Wahid, A., Gelani, S., Ashraf, M., and Foolad, M. R. (2007). Heat tolerance in plants: an overview. *Environ. Exp. Bot.* 61: 199-223.

Wang, W., Vinocur, B. and Altman, A. 2003. Plant responses to drought, salinity and extreme temperatures: towards genetic engineering for stress tolerance, *Planta* 218: 1-14.

Witcombe, J.A., Hollington1, P.A., Howarth, C.J., Reader, S. and Steele, K.A. (2008). Breeding for abiotic stresses for sustainable agriculture, *Phil. Trans. R. Soc. B* 363: 703-716.

Xiao, B., Huang. Y., Tang, N. and Xiong, L. (2007). Over-expression of a LEA gene in rice improves drought resistance under the field conditions. *Theor. Appl. Genet.* 115: 35-46.

Xu, Q., Paulsen, A. Q., Guikema, J. A., and Paulsen, G.M. (1995). Functional and ultrastructural injury to photosynthesis in wheat by high temperature during maturation. *Environ. Exp. Bot.* 35: 43-54.

Ye, G. and Smith, K.F. (2008). Marker-assisted gene pyramiding for inbred line development: basic principles and practical guidelines. *Int. J. Plant Breed.* 2(1): 1-10

Yeo, A. R. and Flowers, T. J. (1983). Varietal differences in the toxicity of sodium ions in rice leaves. *Physiologia Plantarum* 59: 189-195.

Yeo, A.R. (1998). Molecular biology of salt tolerance in the context of whole plant physiology, *J. Exp. Bot.* 49: 915-929.

Zong , G., Wang, A., Wang, L., Liang G., Gu, M., Sang, T. and Han, B. (2012). A pyramid breeding of eight grain-yield related quantitative trait loci based on marker-assistant and phenotype selection in rice (Oryza sativa L.). *J. Genet.Genomics*, 39: 335-350.

PART D

Implications of Advanced Technologies to Improve Demanding Crops in Changing Climate

17

Improvement of Melon Fruits Using Biotechnological Strategies for the Changing Climatic Conditions of the 21st Century

Rubaiyat Sharmin Sultana[1*] and Md. Mahabubur Rahman[2]

[1]*Department of Botany, University of Rajshahi, Rajshahi 6205, Bangladesh*
[2]*Research Institute for Sustainable Humanosphere, Kyoto University, Uji, Kyoto 611-0011, Japan*

ABSTRACT

Two major melon genera are widely produced and consumed around the world, including the watermelon [*Citrullus lanatus* (Thunb.) Matsum. & Nakai] and muskmelon (*Cucumis melo* L.) as fleshy fruits. Many different cultivars of them are grown according to consumer and grower preferences, and market conditions (i.e., seedless varieties, good shelf life, etc.). China is the world's largest producer of melons, accounting for over 52% production by weight. In the past, melons considered to be a 'seasonal delight' because of their limited availability throughout the year, however, for production flexibility, imports availability and popularity, melons are now available year-round. Till recently, the melon production was 20% of the total fruits production worldwide. Over the next three decades, a forty per cent increase in the production of melons is required for the growing population. For the decade to 2009, the production of melons approximately doubled by 2005 and after which it was stable. The inherent complexities in achieving this unparalleled increase in the demand of melon fruits is worsened by the yield-depressing consequences of climate alterations or dissimilarities. With declining agricultural land and water resources, and assailing microbes and pests, the expected increases in the melons production cannot be reached at an efficient level. Elite varieties of melons such as, tolerance to drought condition and inferior lands, or disease resistant that yield more with fewer inputs will be pivotal to success. Moreover, varieties of melons can improve with the traits such as, seedless, good shelf life, excellent flesh colour and good shapes that will make them more attractive to consumers. Plant breeding with respect to conventional or molecular breeding methods must be reoriented in order to generate these elite varieties in melon fruits to meet the global demand for next decades. This paper highlights some of the scientific and technological tools that ought to be necessary for all improvement programmes of the melon production.

Keywords: Biotechnology, climatic change, muskmelon, reorientation, variety improvement, watermelon.

*Corresponding author's E-mail: sultanaru@yahoo.com

Roychowdhury R. (Ed.), Crop Improvement in the Era of Climate Change.

1. INTRODUCTION

Melons belong to the family Curcubitaceae. Two genera of melons widely consumed in the human diet, *Citrullus* in which *C. lanatus* (Thunb.) Matsum. & Nakai, is commonly known as watermelon, and another *Cucumis* in which *C. melo* L., is commonly known as muskmelon or by some varietal names. Watermelon differs in producing berries with a harder rind when compared with the other melons. In many regions of the world, there are more than 1200 varieties of watermelon grown, which are varied with shape and colour (Compton *et al.,* 2004). There are a large number of commercial cultivars in the United States, Japan and elsewhere. Those recommended as Dixic Queen, Big Top, Sweet Marvel, Kleckly Sweet, Sweet-Bally, Sweet Baby, Atx-2, Yamato, Sugar-Baby, etc. Among the *C. melo* varieties, there are also notable differences in the rind with the smooth skinned Honeydew, Crenshaw and Casaba muskmelons, the netted skinned cultivars such as Cantaloupe, Santa Claus or Christmas muskmelon and some partly netted varieties such as the Persian muskmelon, Chinese Hami muskmelons and Charentais (Goldman 2002; Zheng and Wolff, 2000). Moreover, different cultivars of both melons had been crossbred to improve suitability for commercial production (e.g., robustness for transport, disease resistance) and consumer appeal (e.g., seedless and sweeter cultivars). According to FAOSTAT (2011), total 609 million tonnes production of all fruits have been achieved in the year 2009, where only melons production was approximately 126 million tonnes. For the decade to 2009, the production of total melons approximately doubled by 2005 and after which it was stable although forty per cent increase in the production of melons is required for the growing population over the next three decades.

Global climate has changed ever since industrial revolution and by now it is revealed that major greenhouse gases, especially CO_2 increased by 30%. Tropospheric ozone has increased two- to fivefolds since the last century. By latest report of United Nation's Intergovernmental Panel on Climate Change (IPCC), global mean temperature would rise between 0.9 and 3.5°C by the year 2100 (Chakraborty *et al.,* 2000). Therefore, the frequent occurrences of drought and floods always, result in acute foods shortages. Evans (2009) reported that an additional 40 to 170 million more people will be underfed in poor developing countries as a direct consequence of climate change according to the recent data of IPCC. Since a huge number of peoples will be undernourished, it should be given more attention to production of major crops, e.g., cereals, vegetables and tubers. So, it is a little chance to cultivate fruit crops in poor developing countries, results in fruits production rate will go down from the current level. Indeed, the great forecast is that extreme weather events such as heavy precipitation, heat waves, and rising sea levels will occur in many parts of the world during the 21st century (IPCC 2012) resulting in floods, drought, and salinity as the most critical consequences. There are no more redundant water resources and arable lands to deploy for enlargement the already over-stretched ones in many parts of the world. This generational challenge of producing enough fruit crops for a rapidly growing population under extreme.

The rising global population results in land progressively used for non-agricultural activities. Besides, due to change in climate, areas of inferior land like drought and saline lands are increased. As a result, productive land resources shrink and worries of climate change only magnify the challenge of increasing fruits production to feed the expanding population (Secretariat of the Pacific Regional Environment Programme, 2009). Other notable challenge for fruits production is the competition for fertile lands within fruits, and other major food crops, feedstock of bio-energy and livestock feeds cultivation. Economic and environmental costs of the operation of further agricultural chemicals as means for boosting yields is another challenge. For success in producing more fruits under decline climatic conditions, it is necessary to provide elite variety and low cost cultivation techniques for achieving high yields per unit of land area with low input.

There are many experiments on how climate change may affect plant diseases and on the host and pathogen interaction. However, interactions are more complex in the actual situation, where multiple climatological and biological factors vary simultaneously. Climate change has the potential to modify host's physiology and resistance, as well as alter the stages and rates of development of pathogens (Prestidge and Pottinger, 1990). The most likely impacts would be shifted in the geographical distribution of the host and pathogen, change in the physiology of host-pathogen interactions and change in crop losses. As under climate change, plants may potentially be unable to migrate or adapt as readily as environmental conditions change. But most pathogens have advantage over plants because of their shorter generation time and in many cases the ability to move readily through wind dispersal. Because of these characteristics, rate of evolution will be highest in pathogens compared with hosts (Sutherst *et al.*, 1996). However, it is summarized that environmental factors are important for disease in host plants. In Mexico, it was illustrated that environmental changes decreased muskmelons (Cantaloupe) production. With soil-borne disease damage, Cantaloupe production declined by 24% and exports decreased in North America by 92% in the year between 1999 and 2005. These trade gaps are then readily filled by other countries and the original position could be recovered only after a long time for Mexico (Avendano *et al.*, 2009).

The melon fruits production is potential part of the fruits production worldwide, which is important for the human health care by supplying vitamins, minerals and fibres. By the climate changing, amplify the challenge for rising fruit production to feed the population for the next decades. Introduction of varied elite melon fruit varieties with more yield and will be critically important for achieving sustainable fruits production security. This also requires reorientation of cultivation strategies for high yield, and less input in cultivation of melons.

In this chapter, breeding technologies are discussed for improvement of the melon fruits. The current status of scientific and technological tools for improvement of melon fruits are also being described elaborately.

2. NUTRITIVE VALUE, PRODUCTION AND TRADE OF MELONS

Fruits and vegetables come in a rainbow of colours. To consume everyday a variety of colorful fruits and vegetables (red, yellow/orange, white, green and blue/purple) is good for health. These may lower the risk of some cancers. Melons are found in the red, yellow/orange and green colour groups. Red colour group of melon: red watermelon, it is good for heart, memory function and urinary tract. Yellow/orange colour group of melon: Cantaloupe, Casaba, Piel De Sapo, Sugar Melon and Yellow Watermelon, is good for heart, vision and healthy immune system; and green colour group of melon: Honeydew, Crenshaw, Horned Melon and Galia, maintain vision and strong bones and teeth.

Most melons are similar in their calorie and carbohydrate content, but the vitamin and mineral contents vary between each type of melon. Watermelon is lower in vitamins and minerals than most other melons, but contains antioxidants that provide it red colour. On the other hand, muskmelon does not have many antioxidants, but supplies more vitamins than watermelon. One cup of raw, diced watermelon contains 46 calories, nearly all of which come from 11.5 g of carbohydrates. One cup of watermelon contains less than 1 g of protein and fat. Although many fruits are high in fibre, watermelon is not. One cup contains only 0.6 g, which is less than 5% of the recommended dietary allowance (RDA), of fibre for most adults. One cup of watermelon supplies less than 10% of the RDA of all vitamins and minerals except vitamin C. One cup provides 16% of the RDA of vitamin C for women, and 14% for men. Watermelon also contains lycopene, an antioxidant that may help prevent cell damage and boost the immune system. One cup of raw, diced cantaloupe contains 53 calories, which come from approximately 85% carbohydrates, 10% protein and 5% fat. One cup contains 0.3 g of fat, 1.3 g of protein and 12.7 g of carbohydrates. The cantaloupe contains 1.4 g of fibre, which is the same amount as one cup of honeydew melon. One cup of cantaloupe provides 21% of the EMR of potassium and more than 60% of the RDA of vitamin C. Cantaloupe is also high in vitamin A, with one cup supplying more than 25% of the RDA. Vitamin A is important for proper vision and a healthy immune system.

China, Iran, Turkey, Spain and USA are the top melon-producing countries. Production of watermelon and muskmelon globally was estimated 126 million tonnes in 2009 (FAOSTAT, 2011). Watermelons accounted for 80% of the share while muskmelons for 20%. China is by far the main producing country accounting for 52% of global production, while other countries such as Iran, Turkey, Spain and USA, etc. follow with far lower shares (Table 1). According to the latest statistics, the growth rate of the melon production is increasing in Afghanistan among of melon-producing countries in the world.

Table 1 Melon producing, importing and exporting countries

Producing countries	*Importing countries*	*Exporting countries*
China (over 52%)	Germany	Spain
Turkey (6.1%)	Canada	Japan
Iran (4.4%)	United States	Mexico
USA (4.2%)	France	Guatemala
Spain (3.9%)	United Kingdom	Costa Rica
Others (29.4%)		

Exports of melons (watermelon and muskmelon) were valued at 1.36 million USD in 2010. Export values of these commodities reached a peak in 2007 with 2.2 million USD, and then re-bounced in 2010. Exports have experienced a growth trend overall, during this decade, having increased by 15 times, compared to 2000. Today, not only is the United States a net importer of melons (imports minus exports) but it is the largest importer of cantaloupes and other melons worldwide. France and the United Kingdom are also large import markets for cantaloupes and other melons on the world market. Germany is the largest importer of watermelons followed by the United States and then Canada. The total value of melons imported into the United States was 281.8 million USD. The majority of melon imports are in the winter season from December through May. These imports predominately originate in Latin American countries such as Mexico, Guatemala, and Costa Rica. Mexico has been and remains the largest supplier of USA melon imports. Since 2002, Guatemala has been the second largest importer to the United States, followed by Costa Rica. About $100 million, or nearly one third, of the total value of all melon imports is attributed to watermelon imports. Mexico is the dominant source, accounting for 91.2% of watermelon imports. Guatemala and Costa Rica are the next largest sources of watermelon imports, accounting for 3.5% and 2.4% respectively. Total U.S. imports of cantaloupes were valued at 117.3 million USD. Although historically Mexico has been the largest supplier of cantaloupes to the United States, by 2002, both Guatemala and Costa Rica had become the main suppliers of USA cantaloupes. One reason for this shift was the decline in USA imports of Mexican cantaloupes following a series of salmonella outbreaks in the United States that were traced back to Mexican cantaloupes. In 2000, Mexico accounted for 27.9% of USA cantaloupe imports. By 2004 Mexico accounted for just 4% of cantaloupe imports while Guatemala accounted for 41% and Costa Rica 32%. However, Mexico has been and still remains the dominant supplier of USA imports of watermelon and other melons with the exception of cantaloupe. In 2004, Mexico accounted for 91% of USA watermelon imports and 54% of other melon imports.

The world's largest exporter of cantaloupes and other melons is Spain, followed by the United States and then Costa Rica (FAO). Although the United States is a net importer of melons, in 2004 USA melon exports were valued at 98.1 million USD, up from 68.4 million USD in 1990. Canada has been, and remains, the major USA export market for melons. USA exports to Canada increased from 56.1 million USD in 1990 to 85.2 million USD in 2004. Exports to Canada account for nearly 87% of all USA melon exports, 99% of all watermelon exports, and nearly 80% of cantaloupe and honeydew exports. Japan is the second largest market for USA melon exports. In 2004, Japan accounted for 7% of total USA melon exports, and 13% of cantaloupe and other melon exports.

3. GENETIC IMPROVEMENT OF WATERMELON

3.1 Watermelon Improvement by Conventional Selection Breeding

For watermelon breeding, two things immediately focus the attention, one, that most of our knowledge on the inherent characters in watermelon has been obtained by plant breeders interested in the development of watermelons and second, peoples interested in the production and consumption of developed watermelons. In breeding for disease resistant, a variety not only must be resistant to disease but also must possess high quality. In a few cases, the horticulturists, geneticists or plant breeders have studied the watermelon (McKay, 1936). Investigators have striven to develop varieties resistant to wilt, caused by *Fusarium oxysporum* f. *niveum* Synder and Hansen, and to anthracnose, due to *Colletotrichum lagenarium* (Pass.) Ell. & Hals. Other diseases that would appear susceptible to control through breeding are downy mildew, caused by *Peronoplasmopara cubensis* Rost., and gummy-stem blight, due to *Mycosphaerella citrullina* (Smith) Gros. 1, for differences in susceptibility does exist. It was thought that resistance to *Fusarium* wilt is inherited as a recessive character, since F1 plants become diseased when grown on wilt-infested soil. Segregation of resistance and susceptibility takes place in the F2 generation, but no one has obtained ratios (Porter and Melhus, 1932). In these instances, wilt resistance was obtained from a citron (Conqueror) or from the progeny of a citron × watermelon (Iowa Belle or Iowa King). First clue that wilt resistance might be inherited differently was found by Welch and Melhus (1942), who crossed resistant × susceptible watermelons and discovered that the F1 progeny were not always susceptible to wilt. In one cross, Dixie Queen × (Iowa Belle × Jugoslavia 7 × Iowa Belle), the F1 were 70 to 85 per cent (Porter and Melhus, 1932) found that edible watermelon varieties are highly resistant, if not immune, to mosaic, but that the red seeded citron is susceptible. They crossed the inedible Kafir, White Seeded, Majorta, Mammoth White, Stock and Preserving Citron with five watermelon varieties and found that the F1 were highly susceptible to the disease. When White Seeded and Kafir were used as parents, the F2 were less susceptible than when the others were the parents. Walker (1933) recorded the first natural occurrence of mosaic. Walker (1945) found galls on the roots of citron plants and also on the F1 of crosses of watermelon on citron. No causal agent was associated with the enlargements, and their nature is imperfectly understood.

Inbreeding does not seem to affect vigour, earliness and uniformity characteristics. The advantages of watermelon hybrids extensively progressed in Japan by triploid (seedless) hybrids production, as first reported in Japan (Kihara and Nishiyama, 1947; Kihara, 1951). Seedless watermelons have gradually found their way onto the market. Triploid watermelon exhibit many desirable characteristics besides the novelty of seedlessness (Kihara 1951).

3.2. Watermelon Improvement by Recombinant Technology

The *in vitro* plant regeneration systems are one of the fundamental resources for the breeding based on recombinant technology. Several reports have been published on regeneration systems of watermelon. The use of shoot tip explants has potential application for the propagation of elite triploid cultivars and tetraploid watermelon breeding lines. Several attempts have been made to adapt tissue culture procedures for the propagation of triploid genotypes (Adelberg and Rhodes, 1989). Ahad *et al.* (1994) reported that establishment of an efficient protocol for plant regeneration from its immature and mature embryo axis explants of diploid watermelon. As shown in the Fig. 1 (A), shoots regenerated from shoot tip of diploid watermelon through micropropagation. Regeneration of adventitious shoots organogenesis has been reported from a wide range of diploid and tetraploid watermelon cultivars (Dong and Jia, 1991, Compton and Gray, 1993). In all reports, cotyledons of *in vitro*-germinated seedlings were the best source of explants. Two- to five-day-old seedlings with their cotyledons in close contact have displayed the greatest organogenic competence (Compton and Gray, 1993; Dong and Jia, 1991). Our research group (Sultana *et al.*, 2004) reported that leaf-derived callus showed an excellent performance for shoot regeneration through adventitious organogenesis. A plant regeneration system from leaf-derived callus of diploid watermelon was presented in Fig. 1 (B, C and D) in terms of shoot formation (Fig. 1B), root induction (Fig. 1 C) and plantlet acclimatization (Fig. 1D). Somatic embryogenesis method is practicable for plant propagation and a resource of genetic transformation. A limited study has been done on somatic embryogenesis of watermelon (e.g., Kahar *et al.,* 2009). Pro-embryos formation in the cell suspension culture from leaf-derived callus has been reported in diploid watermelon (Sultana and Rahman, 2012). The successful tissue culture techniques further used in the transgenic technology for the genetic improvement of watermelon, is described below:

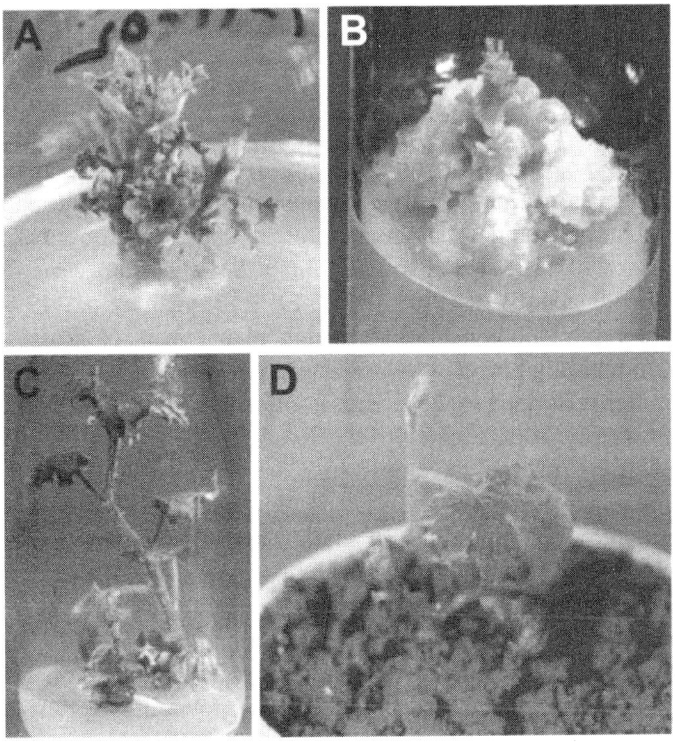

Fig. 1 Shoot regeneration and plant production of diploid water through micropropagation and organogenesis. (A) Shoots regeneration from a shoot tip explant *in vitro.* (B) Developmental shoot buds and a shoot from leaf-derived callus using indirect organogenesis. (C) A rooted shoot derived from leaf derived callus. (D) Acclimatized plantlets.

As a tool of genetic engineering, tissue culture methods are used as the last step, i.e., transformant regeneration in transgenic technology. Insertion of bacterial, fungal and virus resistance genes through recombinant DNA technology would facilitate the development of new disease-resistant genotypes without significantly altering the genetic composition and desirable phenotypic qualities of accepted cultivars with high fruit quality. Many stable genetic transformations has been reported for watermelon. Recovery of transgenic watermelon shoots and plantlets *in vitro* during transformation has generally been used by the reporter GUS and *npt*II genes for selection

ethylene. Clendennen *et al.* (1999) claimed that by expressing *SAMase* in a regulated manner by a fruit-specific promoter, transgenic fruits produced less ethylene than non-transgenic fruit resulting in a modified ripening and post-harvest phenotype. Silva *et al.* (2004) obtained transgenic muskmelon plants cv. Védrantais by inserting and over expressing ACC oxidase from apple and not from muskmelon as Ayub *et al.*, (1996) protocol. All fruit quality oriented transgenic muskmelon plants have been obtained using just three muskmelon cultivars, such as Védrantais and lines 'A' and 'B', belonging to one single variety, i.e., *cantaloupensis* (Ayub *et al.*, 1996; Clendennen *et al.*, 1999; Silva *et al.*, 2004). Considering that there are seven commercial and horticultural important muskmelon varieties (*cantaloupensis, reticulatus, saccharinus, inodorus, flexuosus, conomon* and *dudaim*) (Guis *et al.*, 1998), and hundreds of muskmelon cultivars, much more work would be needed for introducing more improved cultivars since no transgenic lines are released to growers till now.

4.3. Elite Muskmelon by Polyploidization and Somaclonal Variation in Tissue Culture

When plant tissue culture was applied to *Cucumis melo* in order to obtain reliable regeneration protocols, somaclonal variation was a common observable fact, therefore tetraploid, octaploid, mixoploid, and aneuploid plants were easily recovered from *in vitro* cultures (Ezura *et al.*, 1992; Debeaujon and Branchard, 1992; Ezura and Oosawa, 1994). According to Ezura *et al.*, (1995); somaclonal variation could be used to obtain variants lines with low-temperature germinability in muskmelon. Changes in fatty acid patterns have been found in muskmelon callus tissue (Halder and Gadgil, 1984), as well changes in a repetitive DNA sequences during callus culture have been detected (Grisvard *et al.*, 1990). However, somaclonal variation has to be avoided in research, where genetic transformation is involved because genomic stability in transgenic plants has to be maintained in order to express the inserted transgene. The production of tetraploid regenerated muskmelon plants has been observed from somatic embryogenesis and organogenesis (Ezura *et al.*, 1992), and protoplast regeneration (Debeaujon and Branchard, 1992). Nevertheless, each morphogenetic pathway has a different effect on the frequency of recovered tetraploid plants, i.e., somatic embryogenesis (31%), adventitious shoots (30%), shoot primordia (4%), and axillary buds (0%) (Ezura *et al.*, 1992). Therefore, when a callus stage is involved in the regeneration process, the chance to augment the ploidy level in the regenerated plant is increased as well. In addition, explant origin affects the frequency of tetraploid plants from muskmelon tissue cultures (Adelberg *et al.*, 1994). Immature cotyledons produced more tetraploid regenerants than mature cotyledons, while explants from apical meristems produced fewer or no tetraploid plants (Adelberg *et al.*, 1994). Ezura and Oosawa (1994) reported that the capacity of diploid muskmelon cells to generate *in vitro* shoots was greater than tetraploid cells. The ability of tetraploid cells to differentiate into somatic embryos was greater than diploid cells. These same authors reported that the ability of somatic embryos to develop into plantlets decreased in the following order: diploid > tetraploid > octaploid. Ezura and Oosawa (1994) reported the longer muskmelon cells are kept under *in vitro* conditions, the greater the possibility to increase the ploidy levels in those cells. The frequency of chromosomal variation leading to aneuploid (hyperploid and hypoploid) plants at diploid, tetraploid and octaploid levels also increases.

The tetraploid line of muskmelon production using somaclonal variation as well as colchicine treatment is important for production of a triploid muskmelon by hybridization a tetraploid and diploid although nowadays the triploid is not good hybrid cultivar for commercial production.

5. ELITE MELON VARIETIES FOR THE 21ST CENTURY

According to FAO (2011), 'a genetically diverse portfolio of improved agronomic plant varieties, suited to a range of agroecosystems and farming practices, and resilient to climate change' is key to sustainable production intensification. The new elite varieties of melons predicted to address the bourgeoning drivers for production insecurity must be adapted to extreme weather conditions and the attendant continually evolving new strains and biotypes of pests and diseases. Extreme and changing patterns of drought and salinity are probably the most critical consequences of climate change and variations for which plant breeding of melons must develop well-adapted varieties. Additionally, the 21st century plant breeding must cater to different prevailing farming systems and conditions including rain-fed agriculture that accounts for a significant proportion of global melon production in places where erratic rainfall patterns are expected. The new elite varieties of melons must make more efficient use of inputs, and have improved nutritional qualities that meet the myriad dietary preferences of an increasingly more affluent, health-conscious, and generally more discerning consumer. Breeding objectives and strategies for melons must also lead to these crop varieties that fit into ecosystem-based approaches such as conservation agriculture that emphasizes zero tillage. The breeding of melon varieties with nutrients are severally suited for use as good feed.

6. NEW TECHNIQUES FOR GENETIC IMPROVEMENT OF MELONS

The traditional selection breeding is a tedious and very old system nowadays. Using this system, watermelon has improved by a few characteristics, especially with seedless or low seeded watermelon. On the other hand, a hybrid muskmelon 'Galia' is introduced with the hybridization between *cantaloupensis* and *reticulatus* varieties, which is successfully cultivated in several countries. Using breeding based on the transgenic technology, no variety or cultivar for both melon species is released to growers although many researches has been carried out as well as reported, especially on disease resistant traits. All transgenic lines of melon are till now under the field trial conditions. The transgenic technology, especially co-cultivation technique for plant improvement is a very dynamic field but breeders find more efficient and convenient systems. The genetic improvement techniques in plants are changing continually. The status of upcoming technologies is provided by Lusser *et al.* (2011) in response to a request by the European Commission 'to provide information on the state of adoption and possible economic impact of new plant breeding techniques'. The authors identified eight new such techniques and concluded that the new varieties from these techniques might be released within a few years. These new techniques and their features are: (i) Zinc finger nuclease (ZFN): Single mutations are generated or new genes are introduced into predetermined target sites of the genome. (ii) Oligonucleotide directed mutagenesis (ODM): Targeted mutations of one or a few nucleotides are induced. (iii) Cisgenesis and intragenesis: GMOs are produced by the insertion of hereditary materials derived from the species itself or from a cross-compatible species and are contiguous and unchanged (cisgenesis) or the inserted DNA may be a new combination of DNA fragments but must still be from the species itself or from a cross-compatible species. (iv) RNA-dependent DNA methylation (RdDM): Still being refined, modified gene expressions are epigenetic with the new phenotypes inherited only over a few generations. (v) Grafting (on GM rootstock): Desired improvements are achieved by the grafting of non-transgenic scions onto GM rootstock. (vi) Reverse breeding: A combination of recombinant DNA techniques and cell biology procedures is used to generate suitable transgene-free homozygous parental lines rapidly for reconstituting elite heterozygous genotypes. (vii) Agro-infiltration: Used mostly in research settings, for example, to study plant-pathogen interaction in living tissues, to select parental lines or to evaluate the efficacy of transgenes, a liquid suspension of *Agrobacterium* sp. containing the desired gene(s) is used to infiltrate plant tissues, mostly leaves, so that the genes are locally and transiently expressed at high levels. (viii) Synthetic genomics: Large functional DNA molecules that are synthesized without any natural templates are used for constructing viable minimal genomes which can serve as platforms for the biochemical production of chemicals such as biofuels and pharmaceuticals.

Among new techniques, ODM, cisgenesis/intragenesis, and agro-infiltration were the most commonly used for developing crops and brought them at commercial development phase (Lusser *et al.* 2011). In plant breeding, the ZFN technology, RdDM, grafting on GM rootstocks, and reverse breeding were less used. Lusser *et al.* (2011) further projected that the first commercial products derived from these technologies that will be released for production would be herbicide resistant oilseed rape and maize using ODM, and fungal resistant potatoes, drought tolerant maize, scab resistant apples, and potatoes with reduced amylose content developed using cisgenesis and/or intragenesis. The clearly identified needs for the further modification of technical obstacles to the routine adoptions and use of these new techniques notwithstanding, it would appear that policy regulations that are expensive to comply with and public perceptions, rather than the ability to innovate, are holding back the unleashing of the incredible advances of science and technology in plant improvement. However, advanced techniques may impact positively melon improvement. Although some techniques among them are utilized in the crops improvement, such methods are not attended in case of melon improvement. The worsenl ing threats to global melon production security might be possible to rescue with application of advanced breeding techniques.

7. CONCLUSION

Production of melon is on topmost level among fruits production by weight worldwide. In fruits including apple, strawberry, orange, grape, etc., the advanced technologies are applied for their improvement. Therefore, it is more urgent now to improve melon fruits production due to its yield at the highest level. A small scale improvement in the melon production has been achieved with the classical selection breeding technique. Although the advanced transgenic breeding system has been applied to improve melons production, it is still in infant stage due to fundamental tools for genetic improvement programmes, i.e., genetic information, transformation and regeneration do not establish as general methods/information. Changing climate increases drought and saline lands, which magnify the challenge of increasing fruits production to feed the expanding population. A forty per cent increase in the production of melons is required for a growing population over the next three decades but the rate of total melons is stable. Herein, we illustrated the current status in the improvement of melon fruits using scientific and technological tools that would be the necessary for the further steps on research to improve melons production.

References

Adelberg, J., Rhodes, B., and Skorupska, H. (1993). Generating tetraploid melon in tissue culture. *Acta Hort.* 336: 373-380.

Adelberg, J., and Rhodes, B.B. (1989). Micropropagation from zygotic tissues of watermelon. In: *Proceedings of Cucurbitaceae 89: Evaluation and Enhancement of Cucurbit Germplasm*, C.E. Thomas (ed.). USDA/ARS, pp. 110-112.

Adelberg, J.W., Rhodes, B.B., Skorupska, H.T., and Bridges, W.C.. (1994). Explant origin affects the frequency of tetraploid plants from tissue-cultures of melon. *HortSci.* 29: 689-692.

Ahad, A., Islam, R., Hossian, M., Khalekuzzaman, M., and Joarder, O.I. (1994). Plant regeneration from immature and mature embryo axes of watermelon. *Plant Tissue Cult.* 2: 39-44.

Akasaka-Kennedy, Y. Tomita, K.O., and Ezura, H. (2004). Efficient plant regeneration and *Agrobacterium*-mediated transformation via somatic embryogenesis in melon (*Cucumis melo* L.). *Plant Science.* 166: 763-769.

Akashi, K., Morikawa, K., and Yokota, A. (2005). *Agrobacterium*-mediated transformation system for the drought and excess light stress-tolerant wild watermelon (*Citrullus lanatus*). *Plant Biotechnol.* 22(1): 13-18.

Anonymous. (2002). Information Systems for Biotechnology. Available at http://www.isb.vt.edu/cfdocs/fieldtests1.cfm.

Avendano, B., Narrod, C., and Tiongco, M. (2009). Food safety requirements for cantaloupe exports from Mexico and their impact on small farmers' access to export markets. Internet Fd. Pol. Res. Institute. IFPRI Discussion Paper.

Ayub, R., Guis, M., BenAmor, M., Gillot, L., Roustan, J.P., Latche, A., Bouzayen, M., and Pech, J.C. (1996). Expression of ACC oxidase antisense gene inhibits ripening of cantaloupe melon fruits. *Nature Biotechnology.* 14: 862-866.

Bordas, M., Montesinos, C., Dabauza, M., Salvador, A., Roig, L.A., Serrano, R., and Moreno, V. (1997). Transfer of the yeast salt tolerance gene *HAL1* to *Cucumis melo* L. cultivars and *in vitro* evaluation of salt tolerance. *Transgenic Research.* 6: 41-50.

Boyhan, G.J., Gudauskas, R.T., Norton, J.D., and Abrahams, B.R. (1994). Evaluation of watermelon and related germplasm for resistance to the Egyptian strain of zucchini yellow mosaic virus. *Plant Dis.* 78: 100.

Chakraborty, S., Tiedemann, A.V., and Teng, P.S. (2000). Climate change: potential impact on plant diseases. *Environ. Pollution.* 108: 317-326.

Chen, W.S., Chiu, C.C., Liu, H.Y., Lee, T.L., Cheng, J.T., Lin, C.C., Wu, Y.J., and Chang, H.Y. (1998). Gene transfer via pollen-tube pathway for anti-fusarium wilt in watermelon. *Biochem. Mol. Biol. Int.* 46: 1201-1209.

Cho, M.A., Moon, C.Y., Liu, J.R., and Choi, P.S. (2008). *Agrobacterium*-mediated transformation in *Citullus lanatus*. *Biol. Plant.* 52: 365-369.

Choi, P.S., Soh, W.Y., Kim, Y.S., Yoo, O.J., and Liu, J.R. (1994). Genetic transformation and plant regeneration of watermelon using *Agrobacterium tumefaciens. Plant Cell Rep.* 13: 344-348.

Clendennen, S., Kellogg, J.A., Wolf, K.A., Matsumura, W., Peters, S., Vanwinkle, J.E., Copes, B., Pieper, M., and Kramer, M.G. (1999). Genetic engineering of cantaloupe to reduce ethylene biosynthesis and control ripening. In *Biology and Biotechnology of the Plant Hormone Ethylene II*, A. Kanellis, C. Chang, H. Klee, A.B. Bleecker, J.C. Pech, and D. Grierson (eds.). Kluwer Academic Publishers, Netherlands, pp. 371-379.

Clough, G.H., and Hamm, P.B. (1995). Coat protein transgenic resistance to watermelon mosaic and zucchini yellows mosaic-virus in squash and cantaloupe. *Plant Dis.* 79: 1107-1109.

Compton, M.E., Gray, D.J., Hiebert, E., and Lin, C.M. (1993). Expression of the β-glucuronidase gene in watermelon cotyledon explants following particle bombardment or infection with *Agrobacterium tumefaciens. HortSci.* 28: 138.

Compton, M.E., Gray, D.J., Hiebert, E., and Lin, C.M. (1994). Microprojectile bombardment prior to co-cultivation with *Agrobacterium* improves GUS expression in watermelon cotyledons. *In Vitro Cell. Dev. Biol.* 30A: 62.

Compton, M.E., Gray, D.J., and Elmstrom, G.W. (1996). Identification of tetraploid regenerants from cotyledons of diploid watermelon cultured *in vitro. Euphytica* 87: 165-172.

Compton, M.E., Gray, D.J., and Gaba, V.P. (2004). Use of tissue culture and biotechnology for the genetic improvement of watermelon. *Plant Cell, Tiss. Org. Cult.* 77: 231-243.

Compton, M.E., and Gray, D.J. (1993). Shoot organogenesis and plant regeneration from cotyledons of diploid, triploid and tetraploid watermelon. *J. Am. Soc. Hort. Sci.* 118: 151-157.

Dane, F. (1991). Cytogenetics of the genus *Cucumis*. In *Chromosome Engineering in Plants: Genetics, Breeding, Evolution,* vol. part B., P. Gupta, and T. Tsuchiya (eds.). Elsevier, Amsterdam, pp. 201-214.

Debeaujon, I., and Branchard, M. (1992). Induction of somatic embryogenesis and caulogenesis from cotyledon and leaf protoplast-derived colonies of melon (*Cucumis melo* L). *Plant Cell Rep.* 12: 37-40.

Dong, J.Z., Yang, M.Z., Jia, S.R., and Chua, N.H. (1991). Transformation of melon (*Cucumis melo* L.) and expression from the cauliflower mosaic virus-35s promoter in transgenic melon plants. *Bio-Technology.* 9: 858-863.

Dong, J.Z., and Jia, S.R. (1991). High efficiency plant regeneration from cotyledons of watermelon (*Citrullus vulgaris* Schrad.). *Plant Cell Rep.* 9: 559-562.

Ellul, P., Rios, G., Atare, A., Roig, L.A., Serrano, R., and Moreno, V. (2003). The expression of Saccharomyces cerevisiae *HAL1* gene increases salt tolerance in transgenic watermelon [*Citrullus lanatus* (Thunb.) Matsum. & Nakai.]. *Theor. Appl. Genet.* 107: 462-469.

Evans, A. (2009). The Feeding of the Nine Billion: Global Food Security for the 21st Century. London: Chatham House.

Ezura, H., Amagai, H., Kikuta, I., Kubota, M., and Oosawa, K. (1995). Selection of Somaclonal Variants with Low-Temperature Germinability in Melon (*Cucumis melo* L). *Plant Cell Rep.* 14: 684-688.

Ezura, H., Amagai, H., Yoshioka, K., and Oosawa, K. (1992). Efficient production of tetraploid melon (*Cucumis melo* L.) by somatic embryogenesis. *Jap. J. Breed.* 42: 137-144.

Ezura, H., Amagai, H., and Oosawa, K. (1993). Efficient production of triploid melon plants by *in vitro* culture of abnormal embryos excised from dried seeds of diploid X tetraploid crosses and their characteristics. *Jap. J. Breed.* 43: 193-199.

Ezura, H., Yuhashi, K.I., Yasuta, T., and Minamisawa, K. (2000). Effect of ethylene on *Agrobacterium tumefaciens*-mediated gene transfer to melon. *Plant Breed.* 119: 75-79.

Ezura, H., and Oosawa, K. (1994). Ploidy of somatic embryos and the ability to regenerate plantlets in melon (*Cucumis melo* L). *Plant Cell Rep.* 14: 107-111.

FAOSTAT, (2011). FAO statistical databases data sets. Available at http://faostat.fao.org/.

Fallik, E., Aharoni, Y., Copel, A., Rodov, V., Tuvia-Alkalai, S., Horev, B., Yekutieli, O.,Wiseblum, A., and Regev, R. (2000). Reduction of postharvest losses of Galia melon by a short hot-water rinse. *Plant Pathol.* 49: 333-338.

Fang, G.W., and Grumet, R. (1990). *Agrobacterium tumefaciens*-mediated transformation and regeneration of muskmelon plants. *Plant Cell Rep.* 9: 160-164.

Food and Agriculture Organization of the United Nations (FAO). (2011). Save and grow - a policy maker's guide to the sustainable intensification of smallholder crop production. Rome.

Fuchs, M., Klas, F.E., McFerson, J.R., and Gonsalves, D. (1998). Transgenic melon and squash expressing coat protein genes of aphid-borne viruses do not assist the spread of an aphid non-transmissible strain of cucumber mosaic virus in the field. *Transgenic Res.* 7: 449-462.

Fuchs, M., McFerson, J.R., Tricoli, D.M., McMaster, J.R., Deng, R.Z., Boeshore, M.L., Reynolds, J.F., Russell, P.F., Quemada, H.D., and Gonsalves, D. (1997). Cantaloupe line CZW-30 containing coat protein genes of cucumber mosaic virus, zucchini yellow mosaic virus, and watermelon mosaic virus-2 is resistant to these three viruses in the field. *Mol. Breed.* 3: 279-290.

Gaba, V., Kless, H., and Antignus, R. (1992). Transformation of melon by particle acceleration. *Suppl. Plant Physiol.* 99: 137-137.

Gaba, V., Zelcer, A., and Gal-On, A. (2004). Cucurbit biotechnology - the importance of virus resistance. *In Vitro Cell. Dev. Biol. Plant.* 40: 346-358.

Garcia, E., Jamilena, M., Alvarez, J.I., Arnedo, T., Oliver, J.L., and Lozano R. (1998).Genetic relationships among melon breeding lines revealed by RAPD markers and agronomic traits. *Theor. Appl. Gene.* 96: 878-885.

George, E.F. (1993). Plant Propagation by Tissue Culture. Part 1 The Technology, 2nd edn. Exegetics Ltd., Edington, UK.

Goldman, A. (2002). Melons for the passionate grower. Artisan, New York, NY. pp. 176.

Gonsalves, C., Xue, B., Yepes, M., Fuchs, M., Ling, K.S., Namba, S., Chee, P., Slightom,J.L., and Gonsalves, D. (1994). Transferring cucumber mosaic virus-white leaf strain coat protein gene into *Cucumis melo* L. and evaluating transgenic plants for protection against infections. *J. Am. Soc. Hort. Sci.* 119: 345-355.

Gray, D., Hiebert, E., Kelley, K.T., Compton, M.E., and Gaba, V.P. (1995). Comparison of methods to transform embryogenic cotyledons of melon. *HortSci.* 30: 788-788.

Gray, D.J., McColley, D.W., and Compton, M.E. (1993). High-frequency somatic embryogenesis from quiescent seed cotyledons of *Cucumis melo* cotyledons. *J. Am. Soc. Hort. Sci.* 118: 425-432.

Grisvard, J., Sevignac, M., Chateau, M., and Branchard, M. (1990). Changes in a repetitive DNA-sequence during callus-culture of *Cucumis melo*. *Plant Sci.* 72: 81-91.

Guis, M., Botondi, R., BenAmor, M., Ayub, R., Bouzayen, M., Pech, J.C., and Latche, A. (1997). Ripening-associated biochemical traits of cantaloupe charentais melons expressing an antisense ACC oxidase transgene. *J. Am. Soc. Hort. Sci.* 122: 748-751.

Guis, M., Roustan, J.P., Dogimont, C., Pitrat, M., and Pech, J.C. (1998). Melon biotechnology. Biotechnol. *Gen. Eng. Rev.* 15: 289-311.

Halder, T., and Gadgil, V.N. (1984). Comparison of fatty-acid patterns in plant-parts and respective callus-cultures of *Cucumis melo*. *Phytochem.* 23: 1790-1791.

Intergovernmental Panel on Climate Change (IPCC), (2012) Summary for policymakers. In *Managing the Risks of Extreme Events and Disasters to Advance Climate Change Adaptation*, C.B. Field, V. Barros, T.F. Stocker, D. Qin, D.J. Dokken, K.L. Ebi, M.D. Mastrandrea, K.J. Mach, G-K. Plattner, S.K. Allen, M. Tignor, and P.M. Midgley (eds.). Cambridge, Cambridge University Press.

Jaynes, J.M., Xanthopoulos, K.G., Destefano-Beltran, L., and Dodds, J.H. (1987). Increasing bacterial disease resistance in plants utilizing antibacterial genes from insects. *Bio-Essays.* 6: 263-270.

Kahar, L.K., Karande, S.S., and Belhekar, B.M. (2009). Study of biochemical mechanism of embryogenesis in watermelon by using molecular marker. *Intl. J. Plant Sci.* 4(1): 333-337.

Kajikawa, M., Morikawa, K., Abe, Y., Yokota, A., and Akashi, K. (2010). Establishment of a transgenic hairy root system in wild and domesticated watermelon (*Citrullus lanatus*) for studying root vigour under drought. *Plant Cell Rep.* 29: 771-778.

Karchi, Z. (2000). Development of melon culture and breeding in Israel. *Acta Hort.* 510: 13-17.

Kathal, R., Bhatnagar, S.P., and Bhojwani, S.S. (1994). Plant regeneration from the callus derived from root explants of *Cucumis melo* L. cv. Pusa sharbati. *Plant Sci.* 96: 137-142.

Kihara, H. (1951). Triploid watermelons. *Amer. Soc. Hort. Sci. Proc.* 58: 217-231.

Kihara, H., and Nishiyama, I. (1947). An application of sterility of autotriploids to the breeding of seedless watermelons. *Seiken Ziho.* 3 (III): 5-15.

Li, R., Sun, Y., Zhang, L., and Li, X. (1990). Plant regeneration from cotyledon protoplasts of Xinjiang muskmelon. *Plant Cell Rep.* 9: 199-203.

Lin. C-Y., Ku, H-M., Chiang, Y-H., Ho, H-Y., Yu, T-A., and Jan, F-J. (2012). Development of transgenic watermelon resistant to Cucumber mosaic virus and Watermelon mosaic virus by using a single chimeric transgene construct. *Transgenic Res.* 21: 983-993.

Lopez-Sese, A.I., Staub, J.E., and Gomez-Guillamon. M.L. (2003). Genetic analysis of spanish melon (*Cucumis melo* L.) germplasm using a standardized molecular-marker array and geographically diverse reference accessions. *Theor. App. Gen.* 108: 41-52.

Lusser, M., Parisi, C., Plan, D., Rodríguez-Cerezo, E. (2011). New plant breeding techniques. In *State-of-the-art and prospects for commercial development*, M. Lusser, C. Parisi, D. Plan, E. Rodríguez-Cerezo (eds.). Luxembourg, Publications Office of the European Union. pp. 104-110.

McKay, F.W. (1936). Factor interaction in *Citrullus.* Seed coat color, fruit shape and markings show evidence of Mendelian inheritance in watermelon crosses. *J. Hered.* 27: 110-112.

Molina, R.V., and Nuez, F. (1995). Correlated response of *in vitro* regeneration capacity from different source explants in Cucumis melo. *Plant Cell Rep.* 15: 129-132.

Nuñez-Palenius, H., Cantliffe, D.J., and Klee, H.J. (2002). Effect of explant source on regeneration and transformation efficiency in Galia melon (*Cucumis melo* L.). In *10th IAPTC&B Congress Plant Biotechnology 2002 and Beyond*, I. Vasil (ed.). The International Association for Plant Tissue Culture & Biotechnology, Orlando, FL, pp. 100-A.

Ohki, S., Nasuda, K., Mori, Y., and Katsuta, H. (1991). *In vitro* nursery system for vegetable crops – tomato and melon. In *Biotechnology in Agriculture and Forestry*, vol. 17, Y.P.S. Bajaj (ed.). Berlin, Springer, pp. 344-358.

Park, S.M., Lee, J.S., Jenal, S., Jeon, S.L., Shin, Y.S., Her, N.H., Lee, J.H., Lee, M., Ryu, K.H., Yang, S.G., and Harn, C.H. (2005). Transgenic watermelon rootstock resistant to CGMMV (*Cucumber green mottle mosaic virus*) infection. *Plant Cell Rep.* 24: 350-356.

Pitrat, M., Chauvet, M., and Foury, C. (1999). Diversity, History and Productivity of Cultivated Cucurbits. In *First International Symposium on Cucurbits*, K. Abak, and Büyükalaca (eds), Adana, Turkey, pp. 21-28.

Porter, D.R., and Melhus, I.E. (1932). The pathogenicity of *Fusarium niveum* (E.F.S.) and the development of wilt resistant strains of *Citrulus vulgaris* (Schrad.). *In. Agr. Exp. Sta, Res. Bull.* 149: 123-184.

Prestidge, R.A., and Pottinger, R.P. (1990). The Impact of Climate Change on Pests, Diseases, Weeds and Beneficial Organisms Present in New Zealand Agricultural and Horticultural Systems. MAF Technology, Ruakura Agricultural Centre, Hamilton, NZ.

Provvidenti, R. (1991). Inheritance of resistance to the Florida strain of zucchini yellow mosaic virus in watermelon. *HortSci.* 26: 407-408.

Rane, K.K., and Latin, R.X. (1992). Bacterial fruit blotch of watermelon: association of the pathogen with seed. *Plant Dis.* 76: 509-512.

Reed, J., Privalle, L., Powell, M.L., Meghji, M., Dawson, J., Dunder, E., Suttie, J., Wenck, A., Launis, K., Kramer, C., Chang, Y.F., Hansen, G., Wright, M., and Chang, Y.F. (2001). Phosphomannose isomerase: an efficient selectable marker for plant transformation. *In Vitro Cell. Dev. Biol. Plant.* 37 :127–132.

Robinson, R.W., and Decker-Walters, D.S. (1999). Cucurbits. CAB International, Wallingford, New York, N.Y., pp. 226.

Secretariat of the Pacific Regional Environment Programme (SPREP). (2009). Climate Change, Variability and Sea- Level Change. Available at http://www.sprep.org/topic/climate.htm.

Shetty, K., Shetty, G.A., Nakazaki, Y., Yoshioka, K., Asano, Y., and Oosawa, K. (1992) Stimulation of benzyladenine-induced *in vitro* shoot organogenesis in *Cucumis melo* L. by proline, salicylic acid and aspirin. *Plant Sci.* 84: 193-199.

Silva, J.A., da Costa, T.S., Lucchetta, L., Marini, L.J., Zanuzo, M.R., Nora, L., Nora, F.R., Twyman, R.M., and Rombaldi, C.V. (2004). Characterization of ripening behavior in transgenic melons expressing an antisense 1-aminocyclopropane-1-carboxylate (ACC) oxidase gene from apple. *Postharvest Biol. Tech.* 32: 263-268.

Sultana, R.S., Bari, M.A., Rahman, M.H., Rahman, M.M., Siddique, N.A., and Khatun, N. (2004). *In vitro* regeneration of plantlets from leaf explant of watermelon (*Citrullus lanatus* Thunb.). *Biotech.* 3: 131-135.

Sultana, R.S., and Rahman, M.M. (2012). Ontogeny of embryogenic aggregates in suspension culture of diploid watermelon [*Citrullus lanatus* (Thunb.)]. *Intl. J. Agron. Agri. Res.* 2: 40-46.

Sutherst, R.W., Yonow, T., Chakraborty, S., O'Donnell, C., and White, N. (1996). A generic approach to dening impacts of climate change on pests, weeds and diseases in Australasia. In *Greenhouse, Coping with Climate Change.* W.J. Bouma, G.I. Pearman, and M.R. Manning, (Eds.). CSIRO, Australia, pp. 281-307.

Tabei, Y., Kanno, T., and Nishio, T. (1991). Regulation of organogenesis and somatic embryogenesis by auxin in melon, *Cucumis melo* L. *Plant Cell Rep.* 10: 225-229.

Tabei, Y., Oosawa, K., Nishimura, S., Hanada, K., Yoshioka, K., Fujisawa, I., and Nakajima, K. (1994). Environmental risk-evaluation of the transgenic melon with coat protein gene of cucumber mosaic-virus in a closed and a semiclosed greenhouses. 1. *Breed. Sci.* 44: 101-105.

Tricoli, D.M., Carney, K.J., Russell, P.F., Quemada, H.D., McMaster, R.J., Reynolds, J.F. and Deng, R.Z. (2002). Transgenic plants expressing DNA constructs containing a plurality of genes to impart virus resistance. *United States Patent* No. 6: 337-431.

Walker, M.N.A. (1933). Occurrence of watermelon mosaic. *Phytopath.* 23: 741-744.

Walker, M.N.A. (1945) Galls on the roots of citron watermelon hybrids. *Phytopath.* 35: 480-482.

Welch, A., and Melhus, I.E. (1942). Wilt resistance in F1 hybrid watermelons. *Phytopath.* 32: 181-182.

Yadav, R.C., Salah, M. T., and Grumet, R. (1996). High frequency shoot regeneration from leaf explants of muskmelon. *Plant Cell Tiss. Organ Cult.* 45: 207-214.

Yoshioka, K., Hanada, K., Harada, T., Minobe, Y., and Oosawa, K. (1993). Virus-resistance in transgenic melon plants that express the cucumber mosaic-virus coat protein gene and in their progeny. *Jap. J. Breed.* 43: 629-634.

Yoshioka, K., Hanada, K., Nakazaki, Y., Minobe, Y., Yakuwa, T., and Oosawa, K. (1992). Successful transfer of the cucumber mosaic-virus coat protein gene to *Cucumis melo* L. *Jap. J. Breed.* 42: 277-285.

Zhang, X.P., Rhodes, B.B., and Adelberg, J.W. (1994). Shoot regeneration from immature cotyledons of watermelon. *Cucurbit Genet. Coop. Rep.* 17: 111-115.

Zheng, X.Y., and Wolff, D.W. (2000). Ethylene production, shelf-life and evidence of RFLP polymorphisms linked to ethylene genes in melon (*Cucumis melo* L.). *Theor. Appl. Gene.* 101: 613-624.

Zitter, T.A., Hopkins, D.L., and Thomas C.E. (1998). Compendium of Cucurbit Diseases. APS Press, St. Paul, Minn. pp. 87.

18

Conventional Breeding and Biotechnological Approaches in Biodiesel Crops for Crop Improvement in the Era of Climate Change

Dharman Dhakshanamoorthy* **and Radhakrishnan Selvaraj**

Division of Molecular Biology and Biotechnology, Department of Botany, Annamalai University Annamalainagar-608002, Tamilnadu, India

ABSTRACT

Climate is the average weather in a place over a long period. Climate change is the altering of weather and climate patterns that previously have been relatively stable. The changes in climate may be due to natural forces or from human activities. There is growing concern among the scientific community regarding climate change and scientists have proved its occurrence through ample evidence and tested models. According to the latest assessment report of the United Nations Intergovernmental Panel on climate change (IPCC), there is now unequivocal evidence that the earth's climate system is warming, very likely due to anthropogenic activities, greenhouse gas (GHG) and emission from automobiles. Due to this climate change, oil crisis of the 1970s and recognition of the limitation of world oil resources due to the overexploitation, most of the oil importing countries including India has been motivated to develop alternative, renewable sources of energy. Although researchers developed the technology of biodiesel as the best alternative to petroleum based conventional diesel and is defined as the mono-alkyl esters of vegetable oils and animal fats some two or three decades ago, its use is not widespread, mainly due to the higher production cost involved. Intensive production and commercialization of biodiesel from edible-oil grade sources would raise some critical environmental concerns and create a supply chain shortage and a rise in food prices. In this context, we are searching for crop species that have high agricultural and technological advantages and chiefly produce non-edible oil, and do not compete with human food needs that include some tree species especially *Ricinus communis*, *Pongamia pinnata* and *Jatropha curcas*. Studies of crop improvement and biotechnological intervention in tree species are very limited and more research needs to be carried out to bring the genetic changes in these species. Therefore, this chapter gives an overview on application of conventional breeding programmes and biotechnological tools in crop improvement in tree species in respect of climate change.

Keywords: Biodiesel crops, biotechnological approaches, climate change, crop improvement, plant breeding.

**Corresponding author's E-mail: biofueldd2009@yahoo.com*

Roychowdhury R. (Ed.), Crop Improvement in the Era of Climate Change.

1. CLIMATE CHANGE

Climate is the average weather in a place over a long period. Climate change is the altering of weather and climate patterns that previously have been relatively stable. The climate change is a never ending process. It is dynamic and keeps changing every minute because of the interaction of biotic and abiotic components of the ecosystem. Natural forces ensure that the earth has experienced a changing climate since the beginning of time. However, since the last century, anthropogenic (human) activity has threatened significant climate change over a relatively short time period. The term 'global warming' is well documented and refers to the measured increase in the earth's average temperature. This is caused by the build-up of key greenhouse gases in the atmosphere accumulated from continual combustion of fossil fuels and land use changes over the 20th century (Weubles and Jain, 2001). The anthropogenic signal has now become increasingly evident in the climate record where the rate and magnitude of warming due to greenhouse gases are directly comparable to actual observed increases of temperature (Watson, 2001). Any change to the composition of the atmosphere requires a new equilibrium to be maintained; a balance ultimately achieved by changes to the global climate.

1.1 Major Cause of Climate Change

Oil is the dominant fuel source for transportation with road transport accounting for 81% of total energy use by the transport sector. This dependence on fossil fuels makes transport sector a major contributor of greenhouse gases and is one of few industrial sectors where emissions are still growing. The impact of transport on the global climate is not limited to vehicle emissions as the production and distribution of fuel from oil, a 'wells to wheels' approach, produces significant amounts of greenhouse gases itself. For example, consideration of total CO_2 emissions from an average car showed that 76% are from fuel usage whereas 9% from manufacturing of the vehicle and a further 15% from emissions and losses in the fuel supply system. Buses, taxis and inter-cities coaches all play a significant role, but the major contributor is road freight which typically accounts for just under half of the road transport total. Away from road transport, the biggest contributor to climate change is aviation. This is due to its fumes being released directly into the upper atmosphere, where the localized effects can be more damaging then the effects of CO_2 alone. Besides these, anthropogenic activities is one of the major causes of climate change. Transport was one of the key sectors highlighted to be tackled by the 1997 Kyoto Protocol. The aim was to reduce worldwide greenhouse gas emissions by 5.2% of 1990 levels by 2012. Therefore, since 1997, transport has featured heavily in the political agendas of the 38 developed countries who signed the agreement. Technological change will play an increasing role in the reduction of fumes/ greenhouse gases from transport. A car manufactured today emits less CO_2 and pollutants than a car manufactured a decade ago. However, there is not enough evidence to say these new engines are zero risk, but they are certainly of lower risk than before. At the same time, experts say new diesel engines spew out fewer fumes but further studies are needed to assess any potential dangers.

1.2 Diesel Fume/Exhaust Causes Lung Cancer in Humans

According to the World Health Organization's cancer agency recently (June 12, 2012), diesel exhaust causes cancer and it makes exhaust as important a public health threat as second-hand smoke. The last time the agency considered the status of diesel exhaust was in 1989, when it was labelled a "probable" carcinogen. Reclassifying diesel exhaust as carcinogenic puts it into the same category as other known hazards such as asbestos, alcohol and UV radiation. The US government, however, still classifies diesel exhaust as a likely carcinogen. The risk of getting cancer from diesel fumes is small, but since so many people breathe in the fumes in some way, the science panel says raising the status of diesel exhaust to carcinogen from "probable carcinogen" is an important shift. Since so many people are exposed to exhaust, there can be many cases of lung cancer connected to the contaminant. Experts in Lyon had analysed published studies, evidence from animals and limited research in humans. They found that one of the biggest studies was published in March, 2012 by the US National Cancer Institute. That paper analysed 12,300 miners for several decades starting in 1947. Researchers found that miners heavily exposed to diesel exhaust had a higher risk of dying from lung cancer.

1.3 The Energy–Climate Challenge

Energy is one of the most important resources for mankind and its sustainable development. Today, the energy crisis becomes one of the global issues confronting us. Fuels are of great importance because they can be burned to produce significant amounts of energy. Many aspects of everyday life rely on fuels, in particular the transport of goods and people. Main energy resources come from fossil fuels such as petrol oil, coal and natural gas. Fossil fuel contributes 80% of the world's energy needs. Most industries use diesel machines for the production process. In the transportation sector, private vehicles, buses, trucks, and ships also consume significant amounts of diesel and gasoline. This situation leads to a strong dependence of everyday life on fossil fuels. However, the growth of the population is not covered by domestic crude oil production. Fossil oils are fuels which come from ancient animals and microorganisms. Fossil fuel formation requires millions of years. Thus, fossil oils belong to non-renewable energy sources. An increase of the oil price often leads to economic recessions, as well as global and international conflicts. Especially in some developing countries, the great development in the economy in fossil fuel resources will be consumed in only 65 more years. As previously discussed, most countries experience more and more international pressure on climate change and global warming issues. Hence, renewable, green and clean alternative fuels have received increasing attention for current and future utilization.

The challenges turn out to be heavily technological in character. If we were assessing the challenges to optimally using the world's plentiful conventional and non-conventional fossil fuel resources, economics would be paramount. Most energy economists would emphasize the role of prices in conserving energy use and in bringing forth new supplies. While conventional oil reserves may peak in the next decade or two (even this is not certain, at $50+ a barrel) and its production may begin to fall, there are huge non-conventional oil reserves to tap (e.g., the tarsands in Canada). Economists would acknowledge that at $10–20 barrel oil shortages will appear. At $50 or more, it is not at all clear that looming "shortages" is a good description of our present situation. In any event, long before severe shortages appear, plentiful coal will be liquefied. China is already getting ready to do so, in order to reduce its dependence on oil imports (Aldhous, 2005). What really is threatened by these developments is any hope of stabilizing the atmospheric concentration of CO_2 at levels which would not "dangerously interfere" with climate.

1.4 Why Should Globe Care About Climate Change?

Previous studies show that some of the most adverse effects of climate change will be globally, where populations are most vulnerable and least likely to easily adapt to climate change. Changes in temperature, water supply and quality will impact on agricultural production, human settlement and health, biodiversity and animal migratory patterns (IPCC, 2001a,b). Understandably, vulnerability to the adverse impacts of climate change is one of the most crucial concerns of developing countries engaged in climate policy discussions. It is also a critical element in planning any long term climate and development strategy. A survey of studies undertaken since the IPCC Second Assessment Report (SAR) showed aggregate monetized impact estimates based on current economic conditions and populations, for a 1.5-2.5°C temperature increase (Tol *et al.*, 2000). These studies indicate a greater economic vulnerability of developing countries to climate change. But, at higher levels of climate change (more than 2–3° C), net damages occur in almost all regions of the world (IPCC, 2001a, b). Developing nations face greater vulnerability because of their reliance on agriculture, their lower tolerance to coastal and water resource changes, and lower financial, technical, and institutional capacity to adapt (e.g., causing higher health impacts). While sustainable development might reduce this vulnerability, uncertainties about the rate of climate change and pattern of economic development in poorer countries raise questions about whether development could occur fast enough to make a difference. Few studies have considered dynamic responses to steadily increasing greenhouse gas (GHG) concentrations, and the implications of multiple stress factors. This issue was highlighted recently by IPCC as critical for further research (IPCC, 2001a, b). In the absence of dynamic analyses, it is hard to determine whether certain impacts are in fact best mitigated or avoided through GHG emissions reduction or through other policies, such as improved health care or infrastructure development (OECD, 2001). South Asia is extremely vulnerable to climate change impacts, given its high levels of poverty, low human development indices, inadequate legal and governance mechanisms, and vulnerability to extreme weather events (Rahman *et al.,* in press). An Asian Development Bank (ADB, 1994) study analysed the direct impacts on the Sri Lankan economy, using estimates for loss of agricultural production. Estimates for losses were based on available research findings, expert opinion and subjective judgement and resulted from changes in the production of rice, rubber, coconut, coarse grain, vegetables, and tea. The result was an adverse impact on the economy estimated to be in the range of 1.5–2.0% of GDP in 2010.

With a temperature increase of 2°C and an accompanying precipitation increase of 7%, farm level total net revenue is estimated to fall by 9%, whereas, with a temperature increase of 3.5°C and precipitation increase of 15%, the fall in farm level total net revenue would be nearly 25%. Beyond the direct economic impacts, crop failure due to climate change could also increase unemployment, destabilise food security, further increase competition for scarce resources, and increase social inequity. In addition, there are significant risks associated with sea-level rise in some of the world's poorest regions.

1.5 Renewable Bioenergy

Since the oil crisis of the 1970s and recognition of the limitations of world oil resources, most of the oil importing countries including India has been motivated to develop alternative sources of energy. The use of biomass to provide energy has been fundamental to the development of civilization. Biomass contributes a significant share of global primary energy consumption and its importance is likely to increase in future world energy scenario.

In view of growing interest for renewable energy resources, liquid bioenergy production from vegetable oils is one of the possible options to reduce greenhouse gas emissions. Biodiesel production from vegetable oils during 2004-2005 was estimated to be 2.36 million tonnes globally. Of this, European countries (1.93 million tonnes) with expectation of 30% annual increase and the USA (0.14 million tonnes) together accounted for 88% and rest of the world (0.29 million tonnes) for the remaining 12%. Global biodiesel production is set to grow at slightly higher rate than bioethanol and will reach 24 billion litres by being the largest share by 2017 (OECD-FAO, 2008).

1.6 Advantages of Emissions Produced by Biodiesel

Biodiesel can replace fossil fuel as a "clean energy source". It can protect the environment by reducing carbon dioxide (CO_2), sulphur dioxide (SO_2), carbon monoxide (CO) and hydrocarbon (HC). Plants absorb CO_2, which is more than those discharged by the biodiesel combustion process. Thus, using biodiesel can more effectively reduce the emission of CO_2, protect the natural environment and maintain the ecological balance, compared to the use of fossil fuel. The emission of SO_2 in the combustion process of biodiesel is much lower than normal diesel oil because of the low sulphur content in it. Thus, the use of biodiesel instead of normal diesel oil will effectively reduce acid rain, which represents a serious threat to the environment and human infrastructure in forms of acidification of soil, surface and groundwater, and increased corrosion of buildings and historical monuments made from calcium containing stones. Furthermore, CO, HC and particulate matters will be less discharged, because ester compounds in biodiesel contains oxygen promoting clean burning. Using biodiesel can also reduce air pollution. The use of biodiesel in a conventional diesel engine results in a substantial reduction of hydrocarbons, aromatic hydrocarbons, carbon monoxide, alkenes, aldehydes, ketones, and particulate matter. Nitrogen oxide emissions are slightly increased if the engine management remains unchanged. However, this can be optimized using special software and biodiesel sensors. Using biodiesel decreases solid carbon fraction particulate matter and eliminates the sulphate fraction. Increasing the percentage of biodiesel blended with petroleum diesel fuel progressively eliminates sulphates. Biodiesel works well with new technologies such as catalysts, particulate traps, and exhaust gas recirculation. In addition, diesel engine exhaust from biodiesel was found to have a lower mutagenic potential than that from conventional diesel fuel. This effect is believed to result from a lower content of polycyclic aromatic hydrocarbons in the particle emission of biodiesel. Biodiesel is the first alternative fuel that has fully completed the health effects testing requirements of the Clean Air Act.34.

1.7 Biodiesel Status and Crops in India

Biodiesel is expanding very fast because of demand, necessary policy support and technological availability. India consumes approximately 40% million tonnes of diesel and ranked fifth in the world after the US, China, Russia and Japan in terms of fossil fuel consumption. In the recent past, Government of India launched "National Mission on Bio-diesel" with a view to find a cheap, better and renewable liquid fuel based on vegetable oil. However, shortage of raw material to produce biodiesel is a major constraint. The total number of oil-bearing species range from 100 to 300, and of them 63 belonging to 30 plant families hold promise for biodiesel production. Many developed countries are using edible oil-seed crops such as canola, soybean, sunflower, rapeseed, groundnut and sesame for production of biodiesel. However, developing countries like India having dearth of huge quantity of edible oil (6.31 million tonnes) for commercial consumption, cannot afford to use edible oils for production of biodiesel. A vast area of land (about 42%) in India is represented by arid and semi-arid conditions that unsuitable for plantation of food and other commercial crops. Thus, the use of non-edible oil seeds such as *Jatropha* (*Jatropha curcas*), Karanja (*Pongamia pinnata*), Kokum (*Garcinia indica*), Mahua (*Madhuca indica*), *Simarouba* (*Simarouba glauca*), Wild apricot (*Prunus armeniaca*), Jojoba (*Simmondsia chinesis*), Kusum (*Schleichera trijuga*), Ball nut (*Calophyllum inophyllum*), Neem (*Azadirachta indica*), English walnut (*Juglans regia*), Salvadora (*Salvadora oleoides*), Sal (*Shorea robusta*), Rubber (*hevea brasiliensis*), castor (*Racinus communis*) and cotton (*Gassypium hirusutum*) are explored for biodiesel production along with meeting additional criteria of greening the waste land without compromising the food and fodder security, and improve livelihoods. However, a group of oilseed bearing shrubs, from castor bean (*Ricinus communis*), to karanja (*Pongamia pinnata*), ball nut (*Calophyllum inophyllum*), to black vomit nut (*Jatropha curcas*) have been proposed, especially on websites of government panels, various NGOs, and commercial groups promoting their favourite species.

2. CONVENTIONAL BREEDING APPROACHES FOR CROP IMPROVEMENT

Conventional plant breeding is better suited for improving many traits simultaneously, or improving traits controlled by many genes, or traits for which the controlling gene has not been identified. It is also relatively inexpensive, technically simple, and free of government regulation. It involves crossing carefully chosen parent plants, then selecting the best plants from the resulting offspring to be grown on for further selection.

For plant species especially cereals, hundreds of individual crosses are carried out by hand to create seed for the first filial (F1) generation. The resulting F1 plants are uniform, but in the following generation several hundred thousand different plants are produced. Because of the way gene works, the new combinations produced from each cross are not revealed until the second (F2) generation. It is this enormous diversity of new gene combinations which may hold the key to a successful new variety. The plant breeder's task is to select the plants most likely to meet his breeding objectives. Seed from the best of these F2 plants is grown on in small rows or plots and the best plants again selected. This process is repeated year after year until only the very best plants remain. As promising new lines emerge, tests are conducted on each plot to assess factors such as yield, disease resistance and endure quality. Once the best lines are purified to ensure that every plant has the same characteristics, the process of multiplying seed begins. These 'inbred' lines are then ready for entry into official trials, some six to ten years after the initial cross.

Since 1900, Mendel's laws of genetics provided the scientific basis for plant breeding. As all traits of a plant are controlled by genes located on chromosomes, conventional plant breeding can be considered the manipulation of the combination of chromosomes. In general, there are three main procedures to manipulate plant chromosome combination. First, plants of a given population which show desired traits can be selected and used for further breeding and cultivation, a process called (pure line-) selection. Second, desired traits found in different plant lines can be combined together to obtain plants which exhibit both traits simultaneously, a method termed hybridization. Heterosis, a phenomenon of increased vigour, is obtained by hybridization of inbred lines. Third, polyploidy can contribute to crop improvement. Finally, new genetic variability can be introduced through spontaneous or artificially induced mutations.

2.1 Selection

Selection in plant breeding means, selection of individual plants having desirable characteristics from a group of plants (populations, breeding materials, etc.). The critical factor, however, is that only the selected plants are taken to the next generations, that is, allowed to reproduce. This is necessary because if not done, the selection will not be effective. Selection thus brings changes in gene frequencies of different combinations of genes and hence the genetic value of the population in the desired direction. However, it may be noted that selection can act effectively only on heritable differences and that it cannot create variability but acts on that which already exists.

2.1.1 Natural Selection

(a) **Stabilizing selection:** The process of selection is continuously ongoing in nature. Under natural conditions, directional selection favours plants adapting to new environmental conditions, namely, new geographic areas and changing environmental conditions in the original habitat. For characters such as viability, fertility, etc., which directly affect fitness once nature has produced the optimal phenotype for a given habitat, further selection will operate to perpetuate this phenotype as long as the habitat itself remains stable. Under these circumstances extreme expression, whether positive or negative, is at a disadvantage as nature selects for a population mean. For example, at flowering time, natural selection in such circumstances will favour neither the very early nor the late types in the population. This type of natural selection is known as stabilizing selection. Characters which have been subjected to stabilizing selection show quite a different architecture compared to those subjected to directional selection. Here, dominance will be low or absent, and if present may be ambidirectional in nature, whereas epistasis will not generally occur. Stabilizing selection promotes largely additive variation.

(b) **Disruptive selection:** Natural habitats encompass a number of distinct 'ecological niches' among these, the differences may be spatial (microniches) or temporal (seasonal or long-term cycles). Each of these ecological conditions favours selection of different optima in form of function which often involves several related processes, such as density-dependent and/or frequency-dependent interactions between genes or genotypes. Inevitably, the process favours diversity (polymorphism in genetically controlled traits) rather than uniformity within the populations. This is known as disruptive selection. The effect of disruptive selection on population structure depends on the following two factors.

1. Whether the different optimal phenotypes are independent of each other or dependent on each other for maintenance or functioning. For example, coexisting differences such as male and female forms in dioecious species are interdependent in reproduction and so gene exchange is 100 per cent ensured. Thus they share a common genetic background which has to ensure that both channels of development are equally adapted.
2. The rate at which genes are exchanged between the differentially selected genotypes.

(c) **Directional selection:** In plant breeding the aim of selection is to change existing populations, varieties, etc., to achieve specific objectives. Selection is therefore 'directional' towards maximal expression of the targeted trait(s), or on optimal expression. This requires reassociation of genes affecting a number of connected characters in producing a fully balanced phenotype. Directional selection for a character leads to the establishment of dominance and epistasis. Several selection methods have been devised to help the plant breeder make accurate decisions. These methods are based on genetic concepts, namely, progeny test or family evaluation and types of gene action, and experimental procedures that attempt to minimize the effect of environments on the expression of genotypes.

2.1.2 Selection Procedures

(a) **Pure line selection:** The progeny of a single homozygous self fertilized plant forms a pure line. Such lines remain homozygous as long as self pollination takes place. The variation in pure line may be due to mutation but its frequency is very low. The success of this method of breeding depends upon the existence of variability in the original population. This method is used for the improvement of autogamous vegetables like peas, beans and sweet pepper and for creation of parental lines for producing F1 hybrids. Inbreeding in cross-pollinated vegetables lead to breeding depression and here selfing is done to attain uniformity in plant characters, to improve yield by individual plant selection as in cucurbits in which there is no inbreeding depression and to recombine suitable inbred in F1 hybrids and synthetics.

(b) **Mass selection:** Mass selection is widely used in cross-pollinated vegetable crops for the improvement as well as maintenance of varieties. As the selection is based on the phenotype, heritability of the character determines the success of mass selection. This method is effective in improving qualitative or highly heritable characters such as colour and shape of roots and head, ascorbic acid and sugar contents in vegetables and oil yielding plants. The selection intensity in each generation depends on the size of the population from which the selection is to be made. The decision on selection intensity also takes in account the crops, the character under selection, genetic make up of the variety, heritability and linkage relationship of the character.

(c) **Recurrent selection:** Recurrent selection is effective in improving yield and other characters by accumulation of additive genetic variance. This method is highly efficient in cross-pollinated plant for creating genetically superior germplasm. In simple recurrent selection, the best individuals are selected and their selfed progenies intercrossed and further selections are made in the intercross progenies. When a population plateau is off its potential recurrent selection may be useful. Involvement of a large number of parents may be helpful for accelerating the desirable gene frequency in the population. The last intermated population is used as the base population for further improvement through different breeding methods.

2.2 Hybridization or Sexual Hybridization

Crop improvement entails genetically manipulating plants in a predetermined way, which often involves the transfer of genes from one source or genetic background to another. When a plant breeder has determined the direction in which a crop is to be improved, the next crucial step is to find a source of the appropriate gene(s) for making the desired change(s). Once an appropriate plant source has been found, the next step is to transfer the gene(s) to the parent to be improved. In flowering plant, the conventional breeding method of gene transfer is crossing or sexual hybridization. This procedure causes genes from the two parents to be assembled into a new genetic matrix. The end product of sexual hybridization is called a hybrid. Sexual hybridization can occur naturally through agents of pollination. Even though self-pollinating species may be causally viewed as "self-hybridizing", the term hybridization is reserved for crossing between unidentical parents (the degree of divergence is highly variable). Artificial sexual hybridization is the most common conventional breeding method of generating a segregating population for selection in breeding flowering species. In some breeding programmes, the hybrid (F_1) is the final product of plant breeding. However, in most situations, the F_1 is selfed to give and F_2 generation, to generate recombinants (as a result of recombination of the parental genomes) or a segregating population, in which selection is practised.

2.2.1 Natural Hybridization/Cross-Pollinating Species

Several mechanisms occur in nature by which cross-pollination is ensured, the most effective being dioecy. Hermaphrodites have both male and female sexual organs and hence may be capable of self-fertilization. On the other hand, unisexuals, having one kind of sexual organ, are compelled to cross-fertilize. Each mode of reproduction has genetic consequences, hermaphrodity, promoting a reduction in genetic variability, whereas unsexuality, through cross-fertilization, promotes genetic variability. The dioecious species are those in which a plant is either female or male but not a hermaphrodite (e.g., hemp, dates, palm). When such species are cultivated from grain or fruit, it is critical that the producer provides pollinator rows. Less stringent mechanism is monoecy (monoecious plants) as monoecious species can receive pollen from their own male flowers. Dichogamy occurs in hermaphroditic flowers, whereby cross-pollination may be enforced when the stamens mature before the pistil is mature and receptive (a condition called protandry) or the reverse (a condition called protogyny). Sometimes, the pollen from a flower is not tolerated by its own stigma, a condition known as self-incompatibility. Male sterility, the condition whereby the pollen of the male is sterile, compels the plant to receive pollen from different flowers. Similarly, a conditional called heterostyle is one in which significant difference in the lengths of the stamen and pistil makes it less likely for self-pollination to occur, and thereby promotes cross-pollination. Cross-pollinated species depend on agents of pollination, especially wind and insects.

(a) **Genetic and breeding implications of cross-pollination:** The genotype of the sporophytic generation is heterozygous while the genotypes of gametes of a single plant are all different. The genetic structure of a cross-pollinated species is characterized by heterozygosity. Self-incompatibility occurs in such species. Unlike self-pollinated species in which new gene combinations are prohibited, cross-pollinated species share a wide gene pool from which new combinations are created to form the next generation. Furthermore, when cross-pollinated species are selfed, they suffer inbreeding depression. Deleterious recessive alleles that were suppressed because of heterozygous advantage have opportunities to be homozygous and therefore become expressed. However, such depression is reversed upon cross-pollination. In addition to hybrid breeding, population-based improvement methods such as mass selection, recurrent selection, and synthetic cultivars are common methods of breeding cross-pollinated species.

2.2.2 Artificial Hybridization

Artificial hybridization is the deliberate crossing of selected parents. There are specific methods for crossing different species, which differ according to factors including floral morphology, floral biology, genetic barriers, and environmental factors. However, there are certain basic factors to consider in preparation for hybridization.

(a) The parents should be unidentical but reproductively compatible. Generally, corn is crossed with corn, and tomato with tomato. Further, parents to be crossed are usually obtained from the same species.

(b) The parents together should supply the critical genes needed to accomplish the breeding objective.

(c) Once parent is usually designated as female. Whereas some breeding methods may not require this designation, breeders usually select one parent to be a female and the other a male. This is especially so when hybridizing self-pollinated species. In some cases, selected parents of cross-pollinated species may be isolated and allowed to randomly cross-pollinate each other.

(d) The female parent usually needs some special preparation. In complete flowers (with both male and female organs), the flowers of the parent selected to be female are prepared for hybridization by removing the anthers, a tedious procedure called emasculation. Emasculation is eliminated in some crossing programmes by taking advantage of male sterility when it occurs in the species.

Pollen is often physically or manually transferred. Artificial hybridization often includes artificial pollination, whereby the breeder physically deposits pollen from the male parent onto the female stigma. However, when hybridization is conducted on a large scale (commercial hybrid seed development), hand pollination is not feasible option in nearly all cases.

(a) **Emasculation:** The process of making a bisexual flower female by removing the male parts or incapacitating them is called emasculation. It should be pointed out right away that emasculation is not a universal requirement for artificial crossing of plants. Species with infertility-regulating mechanisms (male sterility, self-incompatibility, protandry, protogyny, monoecy, dioecy) may be crossed without the often tedious and time-consuming process of emasculation.

(b) **Methods of emasculation:** There are several techniques of emasculation used by plant breeders that include the use of instruments or chemicals. A pair of forceps or tweezers is one of the most widely used instruments in the emasculation of flowers. Different shapes and sizes are used according to the size and structure of the flower. The methods of emasculation may be classified as direct or indirect.

(c) **Direct emasculation:** The technique of removing anthers from selected flowers is the most common procedure for the emasculation of flowers (by using a pair of forceps). When handling plants with inflorescence, it is important to first thin out the bunch by removing immature flowers as well as old flowers. This will improve the survival of the emasculated flowers. Breeders of various crops have developed their own convenient ways of removing the anthers. Sometimes, the sepals are first removed, followed by the petals, before access is gained to the anthers. In soybean and sesame, a skilled person may be able to remove the petals and anthers in one attempt. In flowers such as soybean, the pedicel is easily broken as a result of physical handling of the delicate flower during emasculation. In wheat and barley, the florets are clipped with scissors.

(d) **Indirect emasculation:** In these methods, the anthers are incapacitated without being removed from the flower. Incapacitation is achieved in various ways.

(e) **Thermal inactivation:** The inflorescence is first thinned out to leave only flowers at the proper stage for emasculation. It is then immersed in hot water (e.g., held in a thermos bottle) to kill the pollen without injuring the pistil. The temperature and time of emasculation is variable (e.g., 43°C for 5 minutes in rice; 47-48°C for 10 minutes in sorghum). The inflorescence is allowed to dry before pollinating in about 30-60 minutes time.

(f) **Alcohol emasculation:** In species such as alfalfa, the raceme is immersed in 57% ethanol for 10 seconds and then rinsed in water for a few seconds.

(g) **Commercial gametocides:** These are chemicals designed to kill the anthers (e.g., sodium methyl arsenate).

Note: If pollination will not immediately follow emasculation, the flowers should be covered and tagged with an appropriate label, to exclude contaminants.

2.2.3 Pollination

(a) **Collection and storage:** In some species pollination immediately follows emasculation. In this case, there is no need for storage. Fresh pollen gives the best success of crossing. Good pollen flowers may be picked and placed in a Petri dish or some suitable container for use. In some species, mechanical vibrations may be used to collect pollen. Pollen is most copious at peak anthesis. Generally, pollen loses viability quickly. However, in some species, pollen may be stored at a cool temperature and appropriate humidity for the species for an extended period of time.

(b) **Application of pollen:** Commonly, pollen is applied directly to the stigma by using a fine brush or dusting off the pollen onto the stigma directly from the flower of the pollen source. Sometimes, a tooth pick or pointed object is used to deposit pollen on the stigma. In some flowers, pollen deposition is made without direct contact with the stigma. Instead, pollen may be injected into a sack covering the emasculated inflorescence and agitated to distribute the pollen over the inflorescence. A key precaution against contamination during pollination is for the operators to disinfect their hands and tools between pollinations, when different varieties involved. It is critical to tag the pollinated flower for identification at the time of harvesting.

(c) **F1 progeny and its crosses for hybrid production:** As result of hybridization, F_1 hybrid is produced. Generally, a few hundred cross combinations per crop per year would be adequate for most purposes for species in which F_1 is not the commercial product. Some breeders make a small number of carefully planned crosses, while others may make thousands of cross combinations. More crosses may be needed for species in which hybrids are commonly produced, in order to discover heterotic combinations. Breeding programmes that go beyond the F_1 usually require very large F_2 populations.

2.2.4 Types of Populations Generated through Hybridization

A breeding programme starts with an initial population that is obtained from previous programmes, existing variable populations or is created though a planned cross. Hybridization may be used to generate a wide variety of populations in plant breeding, ranging from the very basic two-parent cross (single cross) to very complex populations in which hundreds of parents could be involved. Simple crosses are widely used in breeding. Commercial hybrids are mostly produced by single crosses. Complex crosses are important in breeding programmes where the goal is population improvement. Hybridization may be used to introgress new alleles from wild relatives into breeding lines. Because the initial population is critical to the success of the breeding programme, it cannot be emphasized enough that it be generated with much planning and thoughtfulness.

Various mating designs and arrangements are used by breeders and geneticists to generate plant populations. These designs require some type of cross to be made. Factors that affect the choice of a mating design, as outlined by C. Stuber include: (i) the predominate type of pollination (self- or crossed-pollinated), (ii) the type of crossing used (artificial or natural), (iii) the type of pollen dissemination (wind or insect), (iv) the presence of a male-sterility system, (v) the purpose of the project (breeding or genetic studies) and (vi) the size of the population required. In addition, the breeder should be familiar with how to analyze and interpret or use the data to be generated from the mating.

The primary purpose of crossing is to expand genetic variability by combining genes from the parents involved in the cross to produce offspring that contain genes they never had before. Sometimes, multiple crosses are conducted to generate the variability in the base population to begin the selection process in the programme. Based on how the crosses are made and their effects on the genetic structure of the plants or the population, methods of crossing may be described as either divergent or convergent.

(a) **Divergent crossing:** Genetically, divergent parents are crossed for recombination of their desirable genes. To optimize results, parents should be carefully selected to have the maximum number of positive traits and a minimum number of negative traits. This way, recombinant that possesses both sets of desirable traits will occur in significant numbers in the F2. The F_1 contains the maximum number of desirable genes from both parents. There are several methods to conduct divergent crosses.

(i) **Single cross:** If two elite lines are available that together possess adequate traits, one cross [single cross = (A parent × B parent)] may be all that is needed in the breeding programme.

(ii) **Three-ways cross:** Sometimes, desirable traits occur in several cultivars or elite germplasm. In this case, multiple crosses may be required in order to have the opportunity of obtaining recombinants that consist of all the desirable traits. The method of three-way crosses [(A parent × B parent) × C parent] may be used. If a three-way cross product will be the cultivar, it is important that the third parent (C) be adapted to the region of intended use.

(iii) **Double cross:** A double cross is a cross of two single crosses [(A parent × B parent) × (C parent × D parent)]. The method of successive crosses is time-consuming. Further, complex crosses such as double crosses have a low frequency of yielding recombinants in the F_2 that possesses a significant number of desirable parental genes. When this method is selected, the targeted desirable traits should be small (about 10). The double-cross hybrid is more genetically broad-based than the single cross hybrid but is more time-consuming to make the end product.

(iv) **Diallel cross:** A diallel cross is one in which each parent is crossed with every other parent in the set (complete diallel), yielding $n - (n\text{-}1)/2$ different combinations (where n is the number of entries). This method entails making a large number of crosses. Sometimes, a partial diallel is used in which only certain parent combinations are made. The method is tedious to apply to self-pollinated species. Generally, it is a crossing method for genetic studies.

(b) **Convergent crossing:** These are conservative methods of crossing plants. The primary goal of convergent crossing is to incorporate a specific trait into an existing cultivar without losing any of the existing desirable traits. Hence, one or several parents serve as a donor of specific genes and are usually involved in the cross only once. Subsequent crosses entail crossing the desirable parent (recurrent parent) repeatedly to the F_1, in order to retrieve all the desirable traits. A commonly used convergent cross is the backcross.

2.2.5 Application of Crossing in Plant Breeding

Sometimes, crossing is done for specific purposes, within the general framework of generating variability. Hybridization precedes certain methods of selection in plant breeding to generate variability.

(a) **Gene transfer:** Sometimes, only a specific gene (or a few) needs to be incorporated into an adapted cultivar. Crossing is used for the gene transfer process, followed by additional strategic crossing to retrieve the desirable genes of the adapted cultivar.

(b) **Recombination:** Genetically diverse parents may be crossed in order to recombine their desirable traits. The goal of recombination, which is a key basis of plant breeding, is to forge desirable linkage blocks.

(c) **Break undesirable linkages:** Whereas forging desirable linkage blocks is a primary goal of plant breeding, sometimes crossing is applied to provide opportunities for undesirable linkages to be broken.

(d) **For heterosis:** Hybrid vigour is the basis of hybrid seed development. Specially developed parents are crossed in a predetermined fashion to capitalize on the phenomenon of heterosis for cultivar development.

(e) **For genetic analysis:** Geneticists make planned crosses to study the inheritance and genetic behaviour of traits of interest.

(f) **For evaluation of parental line:** Inbred lines for hybrid seed development are evaluated by conducting planned crosses to estimate combining abilities, in order to select appropriate parents for using in hybrid seed development.

(g) **For maintenance of parental lines:** In hybrid seed development programmes, crossing is needed to maintain special parents used in the breeding programme such as cytoplasmic male-sterile (CMS) lines and maintainer lines.

(h) **For maintenance of diversity in a gene pool:** Plant breeders may use a strategy of introgression and incorporation to develop dynamic gene pools from which they can draw materials for crop improvement.

(i) **Number of oil yielding crops improved:** Soybeans, groundnut, rape-seed, mustard, sesame, linseed and sunflower are improved through crossing.

2.3 Mutation Breeding

Mutation is defined as sudden heritable changes in the karyotype of an organism constituting the raw material of species transformation. Mutations created by natural forces are called spontaneous, as they occur spontaneously and those directed by forces under the guidance of man are induced mutations. The knowledge of mutation first originated from the observations of the Dutch Botanist, Hugo de Vries in 1880 on variations in the plants of evening primrose (*Oenothera lamarkiana* L.) growing in Holland. Soon after the fundamental physical discoveries of X-rays and radium elements in 1898, Hugo de Vries heralded the remarkable statement in his work "Die mutation theories". Knowledge of the principles of mutation will certainly some time in the future enable a fully planned artificial induction of mutations, i.e., the creation of new properties in plants and animals. The naturally occurring spontaneous mutation frequency is very less, hence induced mutagenesis through physical and chemical mutagens gained importance over the years. Muller (1927) demonstrated for the first time that the artificial mutation can be induced in *Drosophilla* by X-rays. Use of certain chemicals for the same consequences came to light with the discoveries of Auerbach (1943), Oehlkers (1943), and Rapoport (1946). Thus, two types of mutagens could be recognized; physical and chemical. Currently, arrays of both physical and chemical mutagens are under use for crop improvement in plant species.

In most situations, mutation induction should be looked upon as a tool that in combination with other tools such as cross-breeding can be used to achieve the desired plant breeding goals. Mutation techniques are the best methods to enlarge the genetically conditioned variability of a species within a short time and have played significant role in the development of many crop varieties. The first successful induction of mutation was made in tobacco (*Nicotiana tobaccum*) by Tolleanaar (1934). These mutants were utilized directly or indirectly via crosses with other gene sources in the development of plant cultivars. Upto 1960, there were only few reports of improved cultivars developed by mutagenesis and those few cases were more or less byproducts of radiobiology and fundamental mutagenesis research. In 1964 an international symposium organized by FAO and IAEA received the results and prospects of practical application of mutagenesis for developing better crop cultivars (FAO/IAEA, 1965). In this symposium, 77 mutant cultivars were reported. In 1990, in another FAO/IAEA symposium, records were presented with 1365 mutant cultivars, which derived valuable traits from mutagenesis. Literature revealed that more than 2200 mutant varieties of different crops with improved agronomic traits have been developed and released to the farmers for general cultivation all over the world.

Gene mutations occur naturally as errors in deoxyribonucleic acid (DNA) replication. Most of these errors are repaired, but some may pass the next cell division to become established in the plant offspring as spontaneous mutations. This means that every plant may carry one or more spontaneous mutations into the next generation. Gene mutations without phenotypic (visible) expressions are usually not recognized. Consequently, genetic variation appears rather limited, and scientists have to resort to mutation induction. There are no other economic ways of altering genes, except to wait a long time for spontaneous mutations to occur. Artificial induction of mutations by ionizing radiation dates back to the beginning of the 20th century. But it took about 30 years to prove that such changes could be used in plant breeding. In past one decade (2000-2010) the work of mutation breeding in crop plants is more and achieved good results for the betterment of persistent growth and potential yield parameters. In recent years there have been a number of attempts to assess mutagen-induced genetic variability in quantitative traits of different crop plants such as *Oryza sativa,Cajanus cajan, Vigna unguiculata, Vigna mungo, Lathyrus sativus, Brassica napus, Lens culinaris, Cicer arietinum, Citrus sinensis,* and *Lepidium sativum.*

Mutation breeding in tree crops is not considered attractive because of lacunae in conventional breeding like time-consuming, unpredictable results, long juvenile phase, high heterozygosity and fear in loss of the unique genotype. However, studies on induced mutation in fruit crops have been performed particularly in apple, pear, peach, grapevine, etc. Mutation studies undertaken at the National Botanical Research Institute (NBRI), Lucknow, India have led to induction of cotyledonary variabilities in *J. curcas* (Pandey and Datta, 1995). Mutation breeding studies in *J. curcas* have been carried out in Thailand using fast neutrons and have isolated dwarf and early flowering mutants from the M3generation. The potential productivity of these variants under intensive cultivation conditions was not proved (Sakaguchi and Somabhi, 1987). Dwimahyani and Ishak (2004) used induced mutations in *J. curcas* for improvement of agronomic characters with an irradiation dose of 10 Gy and identified mutant plants with early maturity, 100 seeds weight (30% over control) and better branch growth.

2.3.1 Choice of Mutagen

Since the discovery of mutagenic effects of ionizing radiations and chemical compounds the list of mutagens has become so large that a beginner is often confused about how to make a choice of the mutagen. The difficulties are compounded by reports attributing extraordinary efficacy to a particular mutation in a particular system or a specificity of action to certain compounds. The first of these attributions is now born out by a large body of data while the validity of the second still needs to be established. A safe recommendation would be the use of chemical mutagens like ethyl methanesulphonate (EMS) and some nitrosoureas (nitrosomethyl urea) which have produced high mutation frequencies in a large number of plant systems. The inclusion of an ionizing radiation treatment, if the scale of the experiment permits, would also be useful on account of the total randomness of action of radiations on genetic material as also the facts that at optimal doses radiations produce effects both through gene mutations and chromosomal mutations.

2.3.2 Mutagenic Agents

Agents of artificial mutations are called mutagens. They may be grouped into two broad categories — physical and chemical mutagens. The specific agents vary in ease of use, safety issues, and effectiveness in inducing certain genetic alternations, suitable tissue, and cost, among other factors.

(a) **Physical mutagens:** The principal physical mutagens are ionizing radiations. X-rays were the first to be used to induce mutations. Since then, various subatomic particles (neutrons, protons, beta particles, alpha particles) have been generated using nuclear reactors. Gamma rays from radioactive cobalt (CO^{60}) is widely used. It is very penetrating and dangerous. Neutrons are hazardous and less penetrating, but they are known to seriously damage chromosomes. They are best used for materials such as dry seed, whereas the gamma radiation is suitable also for irradiating whole plants and delicate materials such as pollen grains. The relative biological effectiveness (RBE) of fast neutrons is higher than for gamma rays and X-rays. The breeder is interested in identifying and using treatments with high RBE for maximizing the number of mutants produced. Treatments with high RBE have high ionization density. Modifying the treatment environment (oxygen, moisture content of tissue) can increase the effectiveness of the treatment.

Ionizing radiations cause mutation by breaking chemical bonds in the DNA molecule, deleting a nucleotide, or substituting it with a new one. The radiation should be applied at the proper dose, a factor that depends on radiation intensity and duration of exposure. The dosage of radiation is commonly measured in roentgen (r or R) units. The exposure may be chronic (continuous low dose administered for a long period) or acute (high dose over a short period). The quality of mutagen (proportion of useful mutations) is not necessarily positively correlated with dose rate. A high dose does not necessarily yield the best results (Table 1).

(b) **Chemical mutagens:** Chemical mutagens are generally milder in their effect on plant material. They can be applied without complicated equipment or facilities. The ratio of mutational to undesirable modifications is generally higher for chemical mutagens than for physical mutagens. However, practical success with chemical mutagens lags behind achievements with physical mutagens. Usually, the material is soaked in a solution of the mutagen to induce mutations. Chemical mutagens are generally carcinogenic and must be used with great caution. One of the most effective chemical mutagenic groups is the group of alkylating agents (these react with the DNA by alkylating the phosphate group as well as the purines and pyrimidines). Another group is the base analogues (they are closely related to the DNA bases and can be wrongly incorporated during replication).

Table 1 Commonly used physical mutagens in plants. The concentration/doses of chemical and physical mutagens are given approximately by referring previous studies

Class	*Name*	*Characteristics*	*Recommended doses*	
			Sexually propagated crops	*Asexually propagated crops*
(A) Ionizing Radiations				
1. Particulate radiation	Alpha-rays	Derived from radioisotopes; a helium nucleus capable of heavy ionization; very shallowly penetrating	–	–
	Beta-rays	Produced in particle accelerators of from radioisotopes; are electrons; ionize; shallowly penetrating; sources include P^{32} and C^{14}.	–	–
	Neutrons	A variety exists (fast, slow, thermal); produced in nuclear reactors; uncharged particles; penetrate tissues to many centimetres; source is U^{235}.	600-800 k rad	600-800 k rad
2. Non-particulate radiations	X-rays	Electromagnetic radiation; penetrate tissues from a few millimetres to many centimetres	10-28 k rad	1-1200 k rad
	Gamma-rays	Electromagnetic radiation produced by radioisotopes and nuclear reactors; very penetrating into tissues; sources are Co^{60} and Ce^{137}.	0.6-60 k rad	2-60 k rad
(B) Non-ionizing radiation	UV-rays	Produced in nuclear reactors and accelerators; derived from hydrogen nucleus; penetrate tissues up to several centimetres.	NA (Not available)	(NA)

List of commonly used chemical mutagens are listed in the Table 2. The half-life period (time taken for degradation of the initial amount of alkylaing agent) for EMS in water is about 93 hours at 20°C but only 10 hours at 37°C. Consequently, chemical mutagens are best freshly prepared for each occasion.

Table 2 Commonly used chemical mutagens in plants

Name	*Formula*	*Nature*	*Concentration used generally*	*Source*
Ethyl methanesulphonate (EMS)	CH3SO2OC2H5	Colourless	0.1-0.3%	Kodak (USA), Kochi-light (England).
N-methyl-N-nitrosourea (MNH)	CH3 N — CONH2 ON	Yellow solid	0.01-0.03%	Fluka, Buchs (Switzerland); Aristo custom chemical Inc. (USA).
N-ethyl-N-nitrosourea (ENH)	C2H5 N — CONH2 ON	Yellow solid	0.01-0.03%	Fluka, Buchs (Switzerland); Aristo custom chemical Inc. (USA).
Sodium azide	NaN_3	White solid	0.001-0.004 M	E. Merck

2.3.3 Types of Mutations

In terms of origin, mutations may be spontaneous or induced. Spontaneous mutaiotns arise at the very low rate of about 10^{-5} or 10^{-6} per generation for most loci in most organisms. This translates to one in 1,00,000 or one in 10,00,000 gametes that may carry a newly mutated allele at any locus. They are caused by mistakes in molecular processes associated with the replication of DNA, recombination, and nuclear division. However, because mutagenic agents are common in the general environment induced mutations, as a result of these agents (natural radiations), are hard to distinguish from spontaneously induced mutation due to cellar processes. Mutations may be classified according to the type of structural change produced.

- Genomic mutation: changes in chromosome number due to gain or loss in complete sets of chromosomes or parts or a set.
- Structural mutation: changes in chromosome structure due to duplications and translocation of segments.
- Gene mutation: changes in the nucleotide constitution of DNA due to deletion or substitution.
- Mutation may occur in the nuclear DNA or chromosomes, or in extranuclear (cytoplasmic) genetic systems. In terms of gene action, a mutation may be recessive or dominant.
- Recessive mutation: change of a dominant allele to a recessive allele ($A \rightarrow a$).
- Dominant mutation: change of a recessive allele to a dominant allele ($a \rightarrow A$).

Mutations that convert the wild type (known as common phenotype) to the mutant form (known as the rare phenotype) are called forward mutations, while those that change a mutant phenotype to a wild phenotype are called reverse mutations. Forward mutations are more common than reverse mutations. Recessive mutations are the most common types of mutations. However, recessive alleles in a diploid are expressed only when they are in the homozygous sate. Consequently, an organism may accumulate a genetic load without any consequence because of heterozygous advantage.

2.3.4 Induced Mutation Versus Spontaneous Mutations

Spontaneous mutations produce novel alleles for the evolutionary process. Natural mutations have the benefit of being subjected to the evolutionary process whereby viable mutants become recombined with existing forms and become adapted under the guidance of natural selection. Mutagenesis can be used to create new alleles that can be incorporated into existing cultivars through recombination following hybridization and under the guidance of artificial selection. Modern crop production systems are capable of providing supplemental care to enable a mutant that would not have survived natural selection to become productive. Furthermore, the rate of spontaneous mutation is low (10^{-6} per locus). Artificial mutagenesis aims to increase mutation rates for desired traits. A significant number of commercial cultivars originated from mutation breeding techniques (Table 3).

Table 3 Some important varieties of selected field crops released in different countries through mutation breeding

Plant species and variety	*Country*	*Important characters*
Allium cepa (onion) - Brunette	Netherlands	Early, good quality and high yield
Arachis hypogaea - TG1	India	Late maturing (130)
Brassica juncea (raya)		
RLM 198	India	Higher yield and oil content
RLM 514	India	Early maturity, higher yield, low erucic acid
Capsicum annuum (pepper)		
Krichimsky ran	Bulgaria	Early maturing, high yield
Glycine max (Soybean)		
Raiden	Japan	Earliness
Raikô	Japan	Earliness
KEX-2	Korea	Earliness and high yield
Gossypium hirsutum (cotton)		
MCU7	India	Earliness and high yield
Rasmia	India	Day length tolerant, high yield, superior quality
Pusa Ageti	India	Short duration, high ginning, jassid tolerant
Oriza sativa (rice)		
KT 20-74	China	Early ripening, higher yield
SH 30-21	China	Early ripening, higher yield
Reimei	Japan	Low temperature tolerant

Contd...

Contd...

BPI-121-407	Philippines	Early ripening, short straw, high tillering.
IIT 48	India	Early maturity, day length tolerant
IIT 60	India	Flowers within 65 days, ripens one month earlier
Nucleoryza	Hungary	Earliness, better adapted to cold climate
Kashmir Basmati	Pakistan	Earliness, low temperature tolerant.
R.D. 15	Thailand	Earliness, better drought tolerance
Miyanishiki	Japan	Early heading, short culm, improved lodging resistance
CNM 6		
CNM 20	India	All early maturing by 10-25days
CNM 25		
CNM 31		
Zey mays (maize)		
CE 200 hybrid	CSSR	Early maturity, high yield, good stalk quality
Phaseolus vulgaris		
Pusa parvati	India	Early by 45 days to maturity, high yield
Pisum sativum		
Wasata	Poland	Early maturing, high yield potential, lodging resistant
Triticum aestivum(bread wheat)		
Lewis	USA	Early maturity, lodging resistance, high yield
Stadler	USA	Early maturity, lodging resistance, high yield
Zenkouzi-Komugi	Japan	Earlier, shorter culm, high yield
Sharbati Sonora	India	Better grain colour, quality

2.4 Polyploidy

Individuals having chromosome number more than two sets ($2x$) are referred to as polyploidy. In addition to mutation and hybridization, the change in number of chromosomes is an important source of genetic variation that resulted in the evolution of a number of crop species. The genes, which produce enzymes and finally a trait, are located on the chromosomes as a result of which the change in chromosome number leads to observable change in the trait of an individual. Each chromosome exists as a member of the pair and the number of such pair of chromosomes is specific to a particular species. For example, the human beings have 23 pairs, physic nut 11 pairs, maize 10 pairs and Arabidopsis 5 pairs only. A basic set of chromosomes constitute a genome containing one member of each pair of chromosomes and represents the minimum number of chromosomes that has survived in any species. The basic chromosome number of any species is designated as x which represents a complete set of non-homologous chromosomes of one genome. The individual having two sets of chromosomes ($2x$) is called a diploid while individuals with only single set of chromosome is called monoploid (x). The somatic chromosome number of an individual (e.g., sporophyte) is always designated as 2n (diploid condition) and chromosome number of the gamete as n (haploid condition) whereas x is basic chromosome number of the genome. The haploids (haplont) have gametic chromosome number which is half the number present in the somatic cell of an individual.

2.4.1 Types of Polyploids

Polyploidy has played an important role in the speciation within the plant kingdom and a number of our cultivated crops are polyploids. It can be broadly classified as euploids and aneuploids. An euploid contains a multiple of basic chromosome number where the same basic number may be multiplied which referred as autopolyploid or different genomes are combined which referred as allopolyploid each of which is present twice. In autopolyploids the same genome gets multiplied, i.e., by increasing an individual's own chromosome number while the allopolyploids arise by combining full genomes from two or more distinct but related species through hybridization (Table 4).

Table 4 Types of polyploidy

Ploidy term	*Type of change*	*Symbol*
1. Euploid	**Number of genomes different from two (a multiple of basic chromosome number)**	
Monoploid	Only one genome present	x
Haploid	Gametic chromosome number of the concerned species present	n
(i) Autopolypoid	More than two copies of the same genome present	
(a) Autotriploid	Three copies of the same genome	$3x$
(a) Autotetraploid	Four copies of the same genome	$4x$
(b) Autopentaploid	Five copies of the same genome	$5x$
(c) Autohexaploid	Six copies of the same genome	$6x$
(d) Autoseptaploid	Seven copies of the same genome	$7x$
(e) Autooctaploid	Eight copies of the same genome	$8x$
(ii) Allopolyploid	Two or more different genomes: generally each genome has two copies	
(a) Allotetraploid	Two different genomes: each has two copies	$(2x_1 + 2x_2)$
(b) Allohexaploid	Three different genomes: each has two copies	$(2x_1 + 2x_2 + 2x_3)$
(c) Alloctapoid	Four different genomes: each has two copies	$(2x_1 + 2x_2 + 2x_3 + 2x_4)$
2. Aneuploid	**One or few chromosomes extra or missing from 2*n***	**$2n$ ± few**
(a) Monosomic	One chromosome missing	$2n - 1$
(b) Double Monosomic	Two non-homologous chromosomes (each from a different pair) missing	$2n - 1 - 1$
(c) Nullisomic	One chromosome pair missing	$2n - 2$
(d) Trisomic	One chromosome extra	$2n + 1$
(e) Tetrasomic	One chromosome pair extra	$2n + 2$

genotypes within species to obtain compatible interaction, normal chromosomal pairing and seed set. Such a process should be carried though several cycles of hybridization and selection involving species with low ploidy level.

(iii) Aneuploids: The most commonly studied aneuploids in genetics and breeding are monosomics, nullisomics and trisomics. These may originate spontaneously through meiotic irregularity but their frequency is very low. The occurrence of any one type of polyploids like triploids or tetrasomic plants produces aneuploids directly or though crosses with normal diploid individuals.

The aneuploids arise by fusion of unbalanced (aneuploid) gametes, which result from meiotic irregularities. Either the unpaired chromosome fails to reach at the pole or non-disjunction leads to the unequal sets of chromosomes reaching to opposite poles. Thus, fusion of (n + 1) gamete with normal (n) gamete leads to trisomic plants. Monosomic results from the union of normal (n) gamete with aneuploid (n-1) gamete.

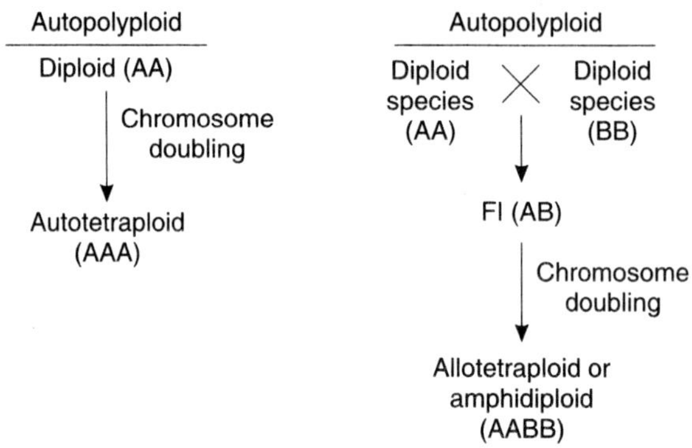

2.4.3 *Significance of Polyploids*

(a) Autopolyploid: The autotetraploids have been relatively more important than other types of autoploids in the evolution of crop plants. But in plant breeding, on the other hand, haploids, triploids and tetraploids have been used. Banana and some clones of mulberry, poplar, chrysanthemum and apple are natural triploids which are commercially grown in many countries. Autopolyploid ryegrass, sugarbeets, turnips, and fodder beets are commercially grown in many European countries. The sugarbeet triploids are preferred over the tetraploids. The seed for raising triploid crop is produced by pollinating diploid cytoplasmic male sterile lines with the pollen from tetraploid lines. Van Bogaert (1975) compared the field performance of colchicine induced tetraploid and diploid cultivars of ryegrass where tetraploids gave 18% higher fodder yield as compared to diploids. Moreover, tetraploids were more resistant to diseases and possessed better palatability and digestibility. Rye (*Secale cereale*) is the only grain producing crop in which induced tetraploids are successful. But tetraploid rye has 20% sterility which reduced its acceptability by growers. Pusa Giant Berseem tetraploid turnips, spinach, pierce grapes, Tetra petkus, steel rye and tetraploid *Brassica campestris* var. toria are some other examples of induced autopolyploidy. Sterility of triploids is being exploited for producing seedless watermelons. Diploids are used as pollinators and the triploid seed set on tetraploid females is harvested to grow the crop. The triploid fruits possess rudimentary seeds. The triploids may be possible in field crops, e.g., potato, cassava, *Allium* sp. that can be reproduced vegetatively. Autopolyploidy plays an important role in bridging ploidy levels in the interspecific crosses where hybridizing species differ in ploidy levels. For example, in potato it is difficult to hybridize the wild diploid species with cultivated tetraploid potato. Under such situations, the chromosome number of diploid species is doubled and the resulting tetraploids are crossed with the cultivated varieties. Though autotriploid and autotetraploids have been used for the development of successful crop varieties but monoploids have not been used to that extent. Haploids can be used to produce completely homozygous lines in a short period of time as compared to 8-10 years required in self-pollinated crops. They can also be used to screen for induced mutagenesis especially of the recessive genes which can express in haploid state. The breeding of some autotetraploids like potato is much easier at haploid stage.

(b) **Allopolyploids:** Natural allopolyploidy plays an important role in the evolution of our major field crops such as wheat, cotton, *Brassica* species and sugarcane. *Brassica* triangle provides an interesting example of the role of allopolyploidy in the evolution of different *Brassica* species. *Brassica nigra*, *B. campestris* and *B. oleracea* are diploid species, which got combined and led to the development of three allotetraploid species, viz., *B. juncea*; *B. napus* and *B. carinata*. It has been experimentally proved by resynthesis of these allotetraploid species which further created interest in the development of new crops through allopolyploidy. *Raphano brassica* (radish-cabbage allotetraploid) was developed but unfortunately it possessed radish-like leaves and cabbage-like roots. Triticale, a man-made cereal is an interesting example of artificially synthesized allopolyploid.

(c) **Aneuploids:** The aneuploids have not been of much direct use in plant breeding. But these provide a viable source to incorporate alien variation into the cultivated varieties through addition of complete chromosomes or preferably segments of chromosomes. A still more common use of aneuploids especially monosomics has been in gene-mapping. Effective use of aneuploids in plant breeding aiming at gene location depends upon the availability of complete set of a particular type of aneuploid.

Aneuploids are an important tool in the genetic studies to identify the chromosome or arm of the chromosome on which a particular gene is located. Determination of chromosome carrying a particular allele depends upon the distorted segregation that occurs for allelic differences when the chromosome with relevant genes is in monosomic condition. In a monosome the transmission of n-1gametes occurs only through female gamete. To determine the location of a dominant gene, each monosomic is crossed as female with homozygous recessive line. The F1s of all monosomics show dominant phenotype except for one monosomic for the chromosome on which gene is located. The location of gene on missing chromosome shall result in segregation for dominant and recessive type because the monosomic progeny of such a cross shall produce recessive phenotype. If the gene is on missing chromosome, the monosomic F_1 plants of such a cross shall have no recessive phenotype in F_2 progeny.

Monosomics are also very useful for intervarietal chromosome substitution where this technique results in the transfer of a single unchanged pair of chromosomes of a donor variety into recipient variety. Aneuploidy has made it possible to develop alien chromosome addition lines where a single pair of chromosomes from another species is added. In wheat, addition lines have been developed with the addition of *Secale*, *Aegilops* or *Agropyron* chromosomes to wheat genome.

2.5 Modification of Storage Products through Conventional Plant Breeding Approaches

(a) **Starch:** Advances in polymer science have permitted development of models that relate starch structure to functionality, thus permitting genetic selection for functional behaviour and such as food attributes. Much of the breeding efforts at starch modifications have been directed at altered amylase: amylopectin ratios through simple backcrossing utilizing several starch mutants available in maize, rice and wheat.

(b) **Protein:** Breeding attempts at enhancing protein content in cereals and pulses have largely been hampered by a poor response to selection pressure which is a consequence of complex inheritance, strong environment influence and biological penalty in terms of decreased proportions of crop yield and quality limiting amino acids. Induced mutagenesis and simple selection have helped in identifying several genetic sources with improved protein content, especially in wheat and barley (Johnson *et al.,* 1973). Cross specific genetic introgressions from wild species (e.g., *Triticum dicoccoides* to *Triticum aestivum*) have helped in raising the protein content to 16-19% in some genetic stocks without sacrificing the yield. High protein genotypes of rice (by using induced mutagenesis) and barley (by using simple selection) could be developed. Adequate mapping of genes affecting protein content in wheat is expected to increase the pace and efficiency of selection for increased protein content. High lysine mutants have been identified in maize barley and sorghum. The enhanced lysine content in these mutants is largely due to decreased prolamin protein fraction and is associated with drastically altered starch metabolism affecting several other seed attributes. Simple backcross method (barley and sorghum) and backcross in combination with recurrent selection (maize) have helped greatly in developing high lysine genotypes. Most of the breeding experiments were, however, limited by the adverse linkages between high lysine and low starch composition. High lysine rice genotypes are still elusive. Attempts with simple selection techniques have been successful in improving bread and pasta making quality in wheat. Backcross method has been used with limited success to improve seed protein concentration in pulses (Wehrman *et al.,* 1987).

(c) **Oil:** Because of a complex inheritance with additive type of gene action and strong environmental influence, only small incremental oil content improvements have been made in oilseed crops by using methods like simple selection and recurrent selection which exploit additive type of genetic variation. Classical case of Illinois selection experiments was the development of high oil /low oil variants of maize using a single base population having 4.68% oil content. Fifty years of selection resulted in the genotypes having oil content as high as 14.83% and as low as 0.77%. Increased oil content was, however, associated with strong yield depression. Pustovoit method of reserves, comparable to simple recurrent selection, is credited with raising oil content in sunflower from 30% to 54% without any negative impact on crop productivity. Due to additive genetic effects, strong environmental influence and inverse relationship with yield; hybrids have generally not been associated with heterosis for oil yield. Plant breeding methodologies like selection and mutagenesis have, however, helped greatly in modifying the fatty acid composition of oilseed crops like sunflower (high oleic acid), soybean (low linolenic acid, low palmitic acid), rapeseed-mustard (high oleic acid, low erucic acid and low linolenic acid) and safflower (high oleic acid). Due to a comparatively simple inheritance [major gene(s) with additive genetic effects], besides availability of rapid, nondestructive screening techniques (e.g., half seed method in *Brassica*), excellent analytical techniques and gene sources; fatty acid composition of elite commercial strains could be easily tailored using various modifications of standard backcross procedure.

2.6 Limitations of Conventional Breeding

The major limitations of conventional breeding derive from the limitations of the sexual process itself, and include constraints on the amount of genetic variation available within the crop and the fact that all traits differing between the parents are subject to segregation, and thus large populations and multiple generations of selection are required to identify rare individuals that combine the best qualities of both parents. In addition, sexual methods are useless for improving crops that are sexually sterile, such as banana, cassava, sugarcane, etc.

3. BIOTECHNOLOGY

It is said that the term 'biotechnology' was first coined around 1917 by the Hungarian engineer Karl Ereky to describe a process that he developed for large scale production of pigs using sugar beet as the source of feed. According to him biotechnology was all lines of work by which products are made using living systems, be it animals, plants or microorganisms.

Biotechnology may be defined as "any technique that uses living organisms, or substances from those organisms, to make or modify a product, to improve plants or animals, or to develop microorganisms for specific uses." In agriculture, biotechnology has found applications in cell and tissue culture for the rapid propagation of plant species; in diagnostics for detecting plant pests and diseases based on the use of monoclonal antibodies and nucleic acid probes; in genetic engineering of plant species to introduce new traits and in aiding conventional plant breeding programmes for crop improvement using molecular markers.

3.1 Genetic Engineering

In its current usage, the term "biotechnology" is defined either broadly or narrowly. It may be defined broadly as the use of techniques based on living systems to make products or improve other species. This would include the use of microbes to make products via fermentation, an age-old practice. In narrower definition, biotechnology refers to the genetic manipulation of organisms for specific purposes. The term genetic engineering is sometimes used to describe this practice. Some argue that classic plant breeding is genetic engineering, since the genetics (DNA) of plants are manipulated by breeders, albeit indirectly. Consequently, a much narrower definition of genetic engineering is used to describe the manipulation of organisms at the molecular level, directly involving the DNA. However, it is the revolutionary technology of recombinant DNA (rDNA), which enables researchers to transfer genes from any organism to another, that some accept as genetic engineering.

3.1.1 General Steps in rDNA Technology

Even though crossing of two different parent products new recombinant in the segregating population, the term recombinant DNA is restricted to the product of the union of DNA segments of different biological origins. A cultivar developed by the rDNA procedure is called a transgenic cultivar or a genetically modified (GM) cultivar. Generally, an organism developed by the rDNA procedure is called a genetically modified organism (GMO).

Certain basic steps are common to all rDNA projects:

1. The DNA of interest that is to be transferred (the transgene) is extracted from the source organism. The specific DNA sequence of interest is cut out using special enzymes.
2. The transgene is inserted into a special DNA molecule (a cloning vector) and joined to produce a new rDNA molecule.
3. The rDNA is transferred into and maintained in a host cell (bacterium) by the process of transformation. The vector replicates, producing identical copies (called clones) of the insert DNA.
4. The host cells with the cloned transgene are identified and isolated from untransformed cells.
5. The cloned transgene can be manipulated such that the protein product it encodes is expressed by a host cell.

Confirmation of transformation: Confirmation of putative transgenic plants is obtained in several ways depending upon the vector used in the transformation. Several independent tests carried out include: (i) phenotypic assay, (ii) enzymatic assay, (iii) PCR analysis, (iv) southern blot analysis (v) Western blot analysis and (vi) progeny analysis. Besides these, there are three more methods are available to confirm the transformation, viz: (i) transgene integration into the host genome, (ii) transgene expression in transgenic plants and (iii) stability of transgene expression.

3.1.2 Modification of Storage Products through Genetic Engineering

Genetic modification of storage products involves expression of alien genes in new host, site directed mutagenesis, reduction or elimination of deleterious components by antisense technology or regulating the activity of step limiting enzymes to enhance or reduce accumulation of specific storage products. Since DNA is direct genetic blueprint for proteins, manipulations are directly for gene products. For both starch and lipids, manipulations are possible only through altering the metabolic precursor levels or regulatory enzymes of biosynthetic pathways as both starches and lipids are not direct gene products. Recent attempts to genetically engineer the content and composition of various storage products are made by several workers which described here briefly.

(a) Starch: Enzymes like ADP-glucose pyrophosphorylase (AGPase), starch branching enzymes or starch synthase are currently the targets of genetic manipulation using mainly the "reverse genetic approach" with antisense plants. Transgenic potato tubers transformed with AGPase gene in reverse orientation showed 70-75% less starch levels than control tubers (Muller-Rober *et al.,* 1992). Conversely the elevated levels of AGPase correlate with increased starch yields in transgenic tubers. Transgenic tubers having expression of *E. coli* AGPase gene (*glg*16) under the control of tuber specific patatin promoter (Stark *et al.,* 1992) contained about 35% more starch than controls. Similarly, decline in activity of starch synthases at temperatures above 25°C could be associated with reduced starch deposition in endosperm tissue of wheat (Keeling *et al.,* 1993) and maize (Keeling *et al.,* 1994). Thus, over-expression of starch synthase can help in increasing starch synthesis. In experiments to modify starch composition using gene technology, success was achieved to modify degree of branching and reduce viscosity by introducing *E. coli* glucogen synthase (*glo*A) gene into potato tubers (Shewmaker *et al.,* 1994). Selective modification of branching enzymes is also expected to result in starch compositional changes as is evident from the fact that maize *ae* mutants lacking in SBEIIb produce with 66% amylase (Boyer and Preiss 1981).

(b) Protein: Transgenic approach for protein modification has several advantages over the conventional approaches. Most important advantages are the ability to overcome deleterious pleiotropic effects normally associated with mutated genes, simple inheritance and possibility to introduce different genes together for breeding as a single locus.

Proteins which are rich in S-amino acids have been identified in a number of species. Two low molecular albumins in pea contain about 7.5% and 16% cystine (Higgins *et al.,* 1986), a 2S albumin from Brazil nut contains 18% methionine and 8% cysteine (Altenbach *et al.,* 1987; Sun *et al.,*1987). A similar protein has been identified in sunflower (Lilley *et al.,* 1989). Transgenic genotypes of oilseed rape, tobacco and Arabidopsis carrying Brazil nut gene construct revealed variable but largely encouraging results for enhancing methionine content (Altenbach *et al.,* 1992; Conceicao *et al.,*1994). 2S albumin from sunflower is also being used to improve forage legumes (Tabe *et al.,* 1993). Use of antisense technology for 11S storage protein cruciferin helped in increase of about 32%, 10% and 8% in levels of cysteine, lysine and mehionine in transgenic rapeseed. For engineering improved nutritional quality of cereal proteins, two approaches are largely being used. These are adding lysine and tryptophan residues to prolamin polypeptides and supplementing or replacing them with other proteins of higher nutritional quality.

To develop Japonica rice suitable for breeding (Ajisaka *et al.,* 1994), primary storage protein was suppressed by introducing antisense construct for rice glutenin gene. Transformed rice showed lower glutenin content when expressed as ratio to globulin.

(c) Oil: Genetic engineering of fatty acid composition can improve nutritional value, enhance oil stability and create new industrial and food applications for speciality oils. A large number of lipid genes controlling various enzymes involved in lipid biosynthesis have been cloned. These include fatty acid elongase and fatty acid desaturase. For reducing polyunsaturates, antisense or co-suppression approaches using cloned homologs of FAD2 and FAD3 genes were thought to be ideal for effecting changes in oil composition for both soybean and rapeseed. Transgenic plants with an antisense oleate or linoleate desaturase cDNA under the control of seed specific conglycinin promoter showed a reduced (2%) level of 18:3 fatty acids as compared to 90% in conventional cultivars (Hitz *et al.,* 1994).

In rapeseed-mustard initial research has been carried out using conventional plant breeding techniques and was successful in developing genotypes having higher levels of palmitic acid (12%), oleic acid (80%) and linolenic acid (22%). However, the transformation technology has helped greatly in achieving necessary oil modifications through enhancing or suppressing expression of genes controlling a particular biosynthetic pathway using novel seed specific promoters. Too much talked about transgenic varieties, viz., stearate rapeseed (high stearic acid) and laurate rapeseed (high lauric acid) have entered commercial phase. Lauric acid (12:0) is an important industrial feed stock in detergent and surfactant industry. This fatty acid is absent from conventional varieties. Transgenic rapeseed with 30-40% lauric acid was developed by inserting an additional thioestrase gene from California bay (*Umbellularia californica*). To further increase content of lauric acid, attempts are now underway to introduce coconut LPA-AT gene into laurate canola lines (Davis *et al.,* 1994). Stearate rapeseed was developed by using antisense RNA of stearoyl-ACP desaturase, which reduced the activity of this enzyme in a transgenic rapeseed. This increased the stearic acid content to 40%. The stearic acid rich rapeseed oil will have low melting point and it will be solid at room temperature. This will avoid the need for hydrogenation of oils to make them solid.

Erucic acid rich rapeseed is prized for its own use in polymers, cosmetics, inks, pharmaceuticals and oleochemicals. Theoretically in conventional varieties erucic acid can be increased only upto 66% as it is excluded from second position of diglyceride. But the transgenic rapeseed with *Nasturtium* gene having erucic acid in all three positions will make it possible to develop very high erucic acid rapeseed. Molecular biologists are also attempting trierucin biosynthesis by exchanging the rapeseed LPA-AT for the corresponding enzyme from *Limananthes douglasii,* L. alba, which has been shown to possess preference for erucoyl-CoA as substrate (Hanke *et al.,* 1994a; 1994b; Walter *et al.,* 1995). Lassner *et al.* (1995) isolated from jojoba a cDNA clone which is involved in the formation of very long chain fatty acids. Low-erucic acid cultivars transformed with this gene had higher erucic acid levels. Transgenics carrying thio-esterase of *Cuphea lanceolate* were recently released in field. These had higher level of myristic acid without any decrease in vigour.

3.1.3 Tissue Culture in Selection of Transformation Events

Tissue culture is a critical part of a genetic engineering project. This is because transgenes are usually delivered into cells before they differentiate. Common explants used are protoplasts, cell suspension, immature embryos, shoot meristem, and immature inflorescences. Because a transgene may not be successfully delivered into all cells in a mass of callus, it is important to have a selection system to discriminate among cells to identify and isolate only genuine transformants.

Cloning vector is designed for specific purposes. Selectable marker systems built into vectors may be grouped as follows: antibiotic selection (i.e., kanamycin), herbicide selection (e.g., acetolactate synthase genes are used in herbicide selection (for glyphosate or sulfonylurea herbicides)), scorable gene-mediated selection, and positive selection. The selection agent is included in the tissue culture medium. Selection systems differ in their ease of use and efficiency.

3.2 Molecular Approaches in Crop Improvement

Molecular approaches are an active area of biotechnology, being applied in plant breeding towards the crop improvement in modern science especially for the identification, location, isolation, characterization, cloning and transfer of specific genes from one to another in crop plants. Plant breeding is a designer's activity. Conventionally, economically important agronomic traits dispersed in different parents lines are combined together for tailoring new genotypes solely through phenotype based assays. The performance of a genotype is determined by its own genetic constitution and the environment in which it grows. The effects of the environment in many cases mask those characteristics of the genotype. The phenotype, thus, provides an imperfect measure of a plant's genetic potential. In contrast, assays based on genotype would be more precise and reliable.

DNA contains an individual's genetic blueprint. The sequence of nucleotides in DNA of an individual is unique and thus determines the identity. The ultimate difference between individuals lies in the nucleotide sequence of their DNA. Detection of such differences employing different molecular approaches has led to development of molecular markers which follow simple mendelian pattern of inheritance. They are stable and not influenced by developmental and environmental factors. Thus, the use of molecular marker is suitable to assist plant breeding and has given birth to the subspecialization of molecular breeding. This specialization is two decades old. In 1980, a group of researchers working in human genetics at the Whitehead Institute of Biomedical Research in USA published a landmark research paper that gave a conceptual framework for constructing saturated genome maps by employing DNA based genetic markers in humans. This was followed by rapid developments in the field concerning both theoretical and practical aspects. In plants, molecular marker systems were first developed in 1985-1986 by two groups of researchers working independently at Native Plants Incorporated, USA and Cornell University, Ithaca, USA. Subsequently, through collaborative efforts, molecular marker systems were developed in almost all major field crops, many important vegetable crops and in some tree species including fruit-yielding. Molecular approaches consists of developing molecular marker systems, construction of high density linkage maps, mapping of genes, marker assisted selection and map-based cloning of genes which are the techniques using in the field of crop improvement.

3.2.1 Basic Techniques Involved in the Development of Molecular Markers

Molecular markers are based on the two basic techniques: (i) southern blot hybridization and (ii) polymerase chain reaction (PCR). The southern blot hybridization involves several steps which include isolation of high molecular weight DNA, restriction digestion of DNA using a special class of enzymes called restriction endonuclease, size separation of DNA fragments by agarose gel electrophoresis, transfer of DNA fragments from gel to nitrocellulose/nylon membranes after denaturation, hybridization of probe DNA with the membrane bound DNA and autoradiography. The probe DNA is prepared by incorporating radioisotopes through the use of labelled nucleotide, α^{32} PdCTP during fresh DNA synthesis. Specific DNA fragment on the membrane hybridizing with the probe DNA is detected by the presence of a dark coloured spot band at the corresponding position on the autoradiogram. The polymerase chain reaction is a much simpler and quicker process that allows automation.

(a) **Development of DNA molecular marker and its first generation:** After the invention of the PCR technique by Kary Mullis in 1986 while working as a chemist at the Cetus Corporation, a biotechnology firm in Emeryville, California, it has been modified in a number of ways to suit specific situation applications as are follows: (i) Asymmetric PCR, (ii) Nested (drop-in-drop-out) PCR, (iii) Inverse PCR, (iv) Anchored PCR, (v) RT (Reverse Transcriptase) PCR or Qualitative PCR, (vi) Real Time PCR or Quantitative PCR, (vii) Hot Start PCR, (viii) Touchdown PCR and (ix) Multiplex PCR. Later, the concept of using variations at DNA level as genetic markers started with the Restriction Fragment Length Polymorphism (RFLP). When the DNA of different individuals is digested with restriction enzymes, differences in size of the resulting fragments of DNA can be visualized *via* Southern hybridization with labelled probe (Southern, 1975). The differences are due to evolutionary changes in sequence of nucleotides in the DNA of different individuals. The first documentation of RFLP came from viruses (Grodzicker *et al.,* 1974) followed by a subsequent elegant demonstration made in the human-globin gene cluster (Jeffreys, 1979). Since then, most organisms have been explored for the presence of RFLP and application of technology has evolved in various fields. Subsequent to RFLP, several other methods such as Variable Number Tandem Repeats (VNTR), Allele Specific Oligonucleotide (ASO), Allele Specific Polymerase Chain Reaction (AS-PCR), Oligonucleotide Polymorphism(OP), Single Stranded Conformational Polymorphism(SSCP) and Sequence Tagged Sites (STS). The conventional hybridization based assay of detecting DNA level variations was replaced by the Polymerase Chain Reaction based assay and it has been evolved to detect variations at DNA level.

(b) **Second generation DNA markers:** The next generation of molecular markers responsible for various revolutions in the field of molecular genetics revolution are microsatellites — arrays of tandemly repeated di-, tri-, tetra and penta nucleotide DNA sequences which occur dispersed throughout the genomes of all eukaryotic organisms investigated to date. The microsatellites are otherwise called Sequence Tagged Microsatellite Sites (STMS) or Simple Sequence Repeats (SSR). SSRs are currently considered the molecular markers of choice within the genome mapping community and are rapidly being adopted by plant researchers as well. SSRs consist of around 10-50 copies of motifs from 1 to 5 basepairs that can occur in perfect tandem repetition, as imperfect (interrupted) repeats or together with another repeat type. These repeated motifs are flanked by unique or single copy sequences, which provide a foothold for specific amplification via PCR. Primers complimentary to the unique sequences in those flanking regions can be designed to amplify single copy products. The other marker systems developed during this period include Restriction Landmark Genome Scanning (RLGS), Cleaved Amplified Polymorphic Sequence (CAPS), Degenerate Oligonucleotide Primer PCR (DOP-PCR), Single Strand Conformation Polymorphism (SSCP), Multiple Arbitrary Amplicon Profiling (MAAP) and Sequence Characterized Amplified Region (SCAR) (Table 5).

(c) **New Generation DNA markers:** Recent developments in molecular biology have opened the possibility of employing various types of molecular tools to identify and use genomic

Table 5 List of first, second and new generations DNA markers used in plant breeding

Year	*Acronym*	*Nomenclature*	*Reference*
		First generation DNA markers	
1974	RFLP	Restriction Fragment Length Polymorphism	Grodzicker *et al.* (1974)
1985	VNTR	Variable Number Tandem Repeats	Jeffreys *et al.* (1985)
1986	ASO	Allele Specific Oligonucleotides	Saiki *et al.* (1986)
1988	AS-PCR	Allele Specific Polymerase Chain Reaction	Landegren *et al.* (1988)
1998	OP	Oligonucleotide Polymorphism	Beckmann (1988)
1989	SSCP	Single Stranded Conformational	Orita *et al.* (1989)
1989	STS	Polymorphism Sequence Tagged Site	Olsen *et al.* (1989)
Second Generation DNA markers			
1990	RAPD	Randomly Amplified Polymorphic DNA	Williams *et al.* (1990)
1990	AP-PCR	Arbitarily Primed Polymerase Chain Reaction	Welsh and McCleland (1990)
1990	STMS	Sequence Tagged Micro Satellite Sites	Beckmann and Soller (1990)
1991	RLGS	Restriction Landmark Genome Scanning	Hatada *et al.* (1991)
1992	CAPS	Cleaved Amplified Polymorphic Sequence	Akopyanz *et al.* (1992)
1992	DOP-PCR	Degenerate Oligonucleotide Primer - PCR	Akkaya *et al.* (1992)
1992	SSR	Simple Sequence Repeats	Telenius (1992)
1993	MAAP	Multiple Arbitrary Amplicon Profiling	Caetano-Anollés *et al.* (1993)
1993	SCAR	Sequence Characterized Amplified Region	Paran and Michelmore (1993)
New Generation DNA markers			
1994	ISSR	Inter Simple Sequence Repeats	Zietkiewicz *et al.* (1994)
1994	SNP	Single Nucleotide Polymorphisms	Jordan and Humphries (1994)
1995	AFLP (SRFA)	Amplified Fragment Length Polymorphism (selective Restriction Fragment Amplification)	Vos *et al.* (1995
1995	ASAP	Allele Specific Associated Primers	Gu *et al.* (1995)
1996	CFLP	Cleavase FragmentLength Polymorphism	Brow (1996)
1996	ISTR	Inverse Sequence-tagged Repeats	Rhode (1996)
1997	S-SAP	Sequence-specific Amplified Polymorphism	Waugh *et al.* (1997)
1998	RBIP	Retrotransposon Based Insertional Polymorphism	Flavell *et al.* (1998)
1999	IRAP	Inter-retrotransposon Amplified Polymorphism	Kalendar *et al.* (1999)
2000	MITE	Miniature Inverted-repeat Transposable Element	Casa *et al.* (2000)
2000	TE-AFLP	Three Endonuclease AFLP	van der Wurff *et al.* (2000)
2001	IMP	Inter-MITE Polymorphisms	Chang *et al.* (2001)
2001	SRAP	Sequence-related Amplified Polymorphism	Li and Quiros (2001)

variation improvement of various organisms. Information concerning the basis of these techniques and their applications are from the technology spillover of several genome projects. The last ten years have witnessed the birth of an array of molecular markers with high-throughput performance coupled with shift from manual mode of detection to complete automation. The list of recently developed markers with tremendous potential in understanding the variation at DNA level is given in the Table 5.

3.2.2 Marker-Assisted Breeding

Molecular markers may be used in several ways to make the plant breeding process more efficient. The adoption of a marker-assisted selection (MAS) or marker-aided selection (MAS) in a breeding programme hinges on the availability of useful molecular marker. Fortunately, this resource is becoming increasingly available to many species, thanks to the advances in biotechnology. This breeding approach is applicable to improving both simple and complex trait, as a means of evaluation of a trait that is difficult or expensive to evaluate by conventional methods. The basic requirement is to identify a marker that co-segregates with a major gene of the target trait. MAS is more beneficial to breeding quantitative traits with low heritability.

The key steps in the implementation of MAS in breeding quantitative trait loci (QTL) are summarized by D.W. Podlich and his colleagues as follows:

(a) Creation of a dense genetic map of molecular markers.
(b) Detection of QTLs based on statistical association between markers and phenotypic variability.
(c) Definition of a set of desirable marker alleles based on the results of the QTL analysis.
(d) The use of and/or extrapolation of this information to the current set of breeding germplasm to enable marker-based selection decisions to be made.

However, as these researchers noted, MAS for more complex traits is still challenging, partly because of the difficulty of effective detection, estimation, and utility of QTLs and their effects. The problem is more significant with more complex traits (such as grain yield) that are controlled by many genes under the influence of epistasis (gene-by-gene-interaction) and gene-by-environment (G×E) interaction effects. On the contrary, most researchers engaged in evaluations of mapping and MAS tend to assume that QTLs act independently (i.e. no interaction with other genes and/or environment). To overcome this problem, Podlich and colleagues proposed a new approach to MAS from the conventional one, which assumes that desirable QTL alleles, one identified, will remain relevant throughout many cycles of selection during plant breeding. In the conventional breeding, researchers tend to estimate QTL effects at the beginning of the project and continue to apply the estimates to new germplasm created during the breeding process. The assumption of fixed QTL values is appropriate if the traits are controlled by additive genes. This will allow MAS to be conducted by independently assembling or stacking desirable alleles. This assumption is not applicable to situations in which context dependencies occur. On such occasions, the value of QTL alleles can change depending on the genetic structure of the current set of germplasm in the breeding programme. These progressive changes in genetic structure may make the initial combinations of alleles no longer the best target or no longer significant in increasing the trait performance in future breeding cycles.

The development and use of molecular markers for the detection and exploitation of DNA diversity is one of the significant achievements in the field of molecular genetics which accelerate breeding by establishment of molecular fingerprints for distinct and most divergent genotypes using diversity analysis. DNA based markers (see Table 5) could use to assist breeding through Marker Assisted Selection (MAS) to select prospective genotypes in a variety of plant species such as *Jatropha curcas* (Pamidimarri *et al.,* 2009), *Citrus* (Biswas *et al.,* 2010), *Fritillaria imperialis* (Badfar-Chaleshtori *et al.,* 2012), *Ricinus communis* (Gajera *et al.,* 2010), *Hordeum vulgare* (Varshney *et al.,* 2007) and *Pongamia pinnata* (Kesari and Rangan, 2011). Besides these, the DNA based markers have also been widely used as a tool for assessing genetic diversity in number of insect and pest species (Behura, 2006).

3.2.3 Applications of Molecular Markers in Plant Breeding

Molecular markers may be used by breeders to increase the rate of genetic gain, verification or identification of parentage, characterization of germplasm, and quantification of genetic variability.

(a) **Screening single traits:** Molecular markers associated with many agronomic traits of importance (e.g., nematode resistance in sugar beet, blast resistance in rice) have discovered and used for screening for these traits in breeding. MAS is useful for simple traits, or those for which genetic gain per unit time (rate of genetic gain) is of high economic return.

(b) **Speeding up breeding programmes:** Markers add speed and precision to backcross breeding programmes. They help to identify the gene of interest to be transferred and facilitate the elimination of the undesirable genome in the donor parent. A fewer number of backcrosses are needed to recover the genotype of the adapted parent.

(c) **Germplasm evaluation:** Markers can distinguish homology and analogy. They can provide a common measure of assessing divergence in germplasm. Molecular markers have been used to assist in the discovery of heterotic patterns in order to guide the selection of parents for use in a hybrid breeding programme.

(d) **Genetic Dissimilarity/similarity between germplasm:** Marker analysis may be used to ascertain genetic dissimilarity/similarity among germplasm/lines of different heterotic groups to enable the breeder to predict the performance of hybrids to be developed from different intergroup crosses.

(e) **Cultivar identification and protection:** Molecular markers may be used to fingerprint genotypes for the protection of proprietary materials.

(f) **Mutation detection:** Although the most of common use of a molecular marker is related to **genetic mapping, taxonomic and phylogenetic studies**, the method is also used to detect the DNA alterations induced by chemical and physical mutagens.

4. CASE STUDY – BIODIESEL CROP (*JATROPHA CURCAS*)

The increase in petroleum prices has prompted the study and development of renewable fuels worldwide. Currently, the mass production of biodiesel from crop species (canola, soybean, sunflower) on commercial level is difficult; many of these crops used for human consumption, and altering their use to include biodiesel production would create a supply chain shortage

a sensitive plant to flooding and waterlodging conditions in winter season. Days to flowering and maturity in case of irradiated populations were consistently shifted towards earliness. It is valuable in obtaining varieties associated with escape from pests, drought and other stress injuries that occur in late growing season. Difference in effect of gamma rays might be due to seed metabolism and onset of DNA synthesis (Shah *et al.,* 2008).

In the present study, when markers obtained by the end of RAPD analyses done at the plant treated with gamma rays and EMS were examined, it was seen that there was a potential diversity between mutants and control (Table 8). In the dendrogram (Fig. 1) done according to RAPD markers, the genetic distance to control and mutants replying to gamma radiation and EMS in different doses in particular 25Kr and 4% EMS, and 5Kr and 1% EMS were rather different. Atak *et al.* (2004) found that there was potential polymorphism and improvement between mutants and control while they determined the differences among mutants applying to various doses of atrazine and radiation in soybean varieties with RAPD marker. The RAPD markers, which can quickly detect a large number of genetic polymorphism, have led to creation of genetic maps in a number of woody fruit crops (Lue *et al.,* 2002), small tree (Dhakshanamoorthy *et al.,* 2011) and sunflower (Erdem and Oldacay, 2004) including changes due to DNA damage and thus, RAPD marker can be used to detect mutation and DNA damage. RAPD being dominant, detect multiple loci distributed throughout the genome and hence strongly preferred for genetic variation studies in tree species. The results of the present study pave the way to marker assisted selection in the mutation breeding technique in *J. curcas* that leads to significant increment in seed yield production and early flowering genotypes within a short period of time which will be the key factor in reducing production costs and avoiding loss occurs in late growing season. This is the first report on molecular characterization of induced mutagenesis through gamma irradiation and EMS using RAPD marker analysis in *J. curcas*.

5. CONCLUSION

It may thus be concluded that conventionally that are a number of breeding strategies have been made aiming at incorporating the new gene combination in the genotype for crop improvement. Induced mutagenesis has been increased the wealth of germplasm resource and income of farmers by removing obvious defects of commercial varieties. It may not be successful where intensive cultivation is in progress. But mutation breeding can act as an important tool to conventional breeding approaches. Although conventional breeding has developed a number of varieties and cultivars worldwide, it has its own limitations because it is a life-time consuming process that producing unpredictable results and heterozygosity with long juvenile phase, and it may lose the unique genotype. Since the oil crisis of the 1970's, recognition of limitation of the world oil resources and the food crisis due to the overkilling of world populations, most of the developing countries including India have been motivated to develop an alternative approach of plant breeding to meet out their domestic demand for energy and food. In this context, we, plant breeders should have to work together to ease the process of developing a new genotype using currently available biotechnological approaches. Among these, molecular marker is one of the better approaches to analysis qualitative trait loci (QTL) and genetic diversity in the genotype. Major limitation in the use of molecular markers, at present, is the cost and time required.

Table 7 Effect of mutagens on seed germination, growth and yield parameters of *Jatropha curcas*

Treatment	*TG (%)*	*DFF*	*DTFF*	*DTFM*	*TNFI*	*NFB*	*NFBP*	*100-SW (g)*	*SYP (g)*
Control	58.33 ± 4.42	306 ± 8.02	8.00 ± 1.00	56.66 ± 4.72	220 ± 20.90	6.90 ± 0.70	8.96 ± 0.80	66.33 ± 5.86	123 ± 10.13
Gamma rays									
5Kr	80.66 ± 4.50	255 ± 21.54	6.00 ± 1.02	49.00 ± 1.00	255 ± 22.00	10.33 ±1.30	12.06 ± 1.00	86.66 ± 1.53	319 ± 30.11
10Kr	57.00 ± 3.60	289 ± 15.71	7.33 ± 1.52	57.33 ± 2.51	238 ± 12.66	8.96 ± 0.72	9.83 ± 1.01	77.00 ± 6.55	200 ± 17.51
15Kr	54.66 ± 4.03	299 ± 10.06	9.00 ± 1.00	59.66 ± 3.05	205 ± 18.66	7.70 ± 0.78	9.33 ± 0.81	56.00 ± 6.24	113 ± 4.04
20Kr	44.99 ± 2.00	294 ± 32.34	8.00 ± 1.20	59.66 ± 6.80	174 ± 16.82	5.76 ± 0.52	8.06 ± 0.91	54.00 ± 4.00	71 ± 7.20
25Kr	37.00 ± 3.35	279 ± 27.79	9.50 ± 0.50	58.66 ± 1.52	137 ± 16.37	4.20 ± 0.40	4.50 ± 0.70	62.33 ± 8.32	34 ± 3.60
EMS									
1%	52.56 ± 2.09	276 ± 18.52	8.00 ± 2.00	52.50 ± 1.50	232 ± 23.06	7.66 ± 0.53	9.33 ± 0.52	66.00 ± 5.29	139 ± 13.22
2%	39.56 ± 3.50	303 ± 8.02	9.00 ± 1.00	58.00 ± 1.52	159 ± 16.01	4.90 ± 0.10	6.73 ± 0.61	48.00 ± 3.00	46 ± 4.43
3%	29.66 ± 2.50	308 ± 16.07	10.16 ± 0.76	60.00 ± 2.51	135 ± 13.22	4.06 ± 0.84	6.56 ± 0.60	46.66 ± 3.21	36 ± 2.04
4%	24.66 ± 2.00	310 ± 8.08	10.16 ± 1.75	62.33 ± 2.00	112 ± 14.08	3.40 ± 0.35	4.36 ± 0.45	41.00 ± 2.00	17 ± 1.03
F-Value	38.525	3.021	3.396	4.681	10.650	7.465	6.357	24.457	42.545
Level of significant	0.05	0.05	0.05	0.05	0.05	0.05	0.05	0.05	0.05

TG-Total germination, **DFF**- Days to first flowering (DAS), **DTFF**- Days taken for flowering to fruiting, **DTFM**- Days taken for fruiting to maturity, **TNFI**- Total number of flowers/inflorescence, **NFB**-Number of fruits/bunch, **NFBP**-Number of fruit bunches/plant, **100-SW**:100 Seed weight, **SWP**: Seed Yield/plant, ± - Standard Deviation (SD); Three replicates.

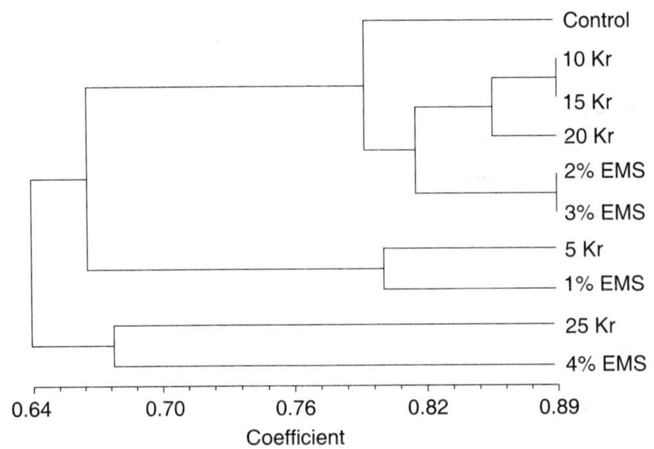

Fig. 1 Dendtrogram derived from a UPGMA cluster analysis using Jaccard's similarity coefficients of *J. curcas*, showing the clustering of 10 mutants (Control, 5Kr, 10Kr, 15Kr, 20Kr, 25Kr, 1% EMS, 2% EMS, 3% EMS and 4% EMS).

Besides, the lack of detectable polymorphism in certain crops also restricts the use of such marker systems in plant breeding programme. Even though the RAPD marker is included in the list of second generation DNA marker, it is fielding till now as new generation DNA marker due to its simplicity, reliability and reproducibility. A number of crop plants including maize, sorghum, rice, soybean and castor are improved through assessment of variation, selection of superior elite genotypes, alien gene transfer through inter-specific hybridization and biotechnological approaches to bring the change in the desired traits. Therefore, the same technologies may be applied in biodiesel and other food crops, to bring the change in the genotypes through crop improvement to meet their demand in the era of climate change.

Table 8 Jaccard's similarity matrix among the gamma rays and EMS treated plants of *J. curcas*

	Control	*5Kr*	*10Kr*	*15Kr*	*20Kr*	*25Kr*	*1% EMS*	*2% EMS*	*3% EMS*	*4% EMS*
Control	1.000									
5Kr	0.664	1.000								
10Kr	0.844	0.737	1.000							
15Kr	0.801	0.671	0.886	1.000						
20Kr	0.797	0.664	0.830	0.881	1.000					
25Kr	0.681	0.643	0.671	0.664	0.659	1.000				
1% EMS	0.640	0.805	0.702	0.647	0.664	0.592	1.000			
2% EMS	0.778	0.647	0.851	0.821	0.778	0.628	0.692	1.000		
3% EMS	0.758	0.618	0.805	0.840	0.823	0.644	0.640	0.886	1.000	
4% EMS	0.659	0.559	0.638	0.631	0.647	0.677	0.606	0.654	0.659	1.000

References

ADB, 1994. Climate Change in Asia: Sri Lanka Country Report. Asian Development Bank, Manila.

Ajisaka, H., Maruta, Y. Kumashiro T. and Jones D.D. (1994). Evaluation of transgenic rice carrying an antisense glutelin gene in an isolated field. In: the biosafety results of field tests of genetically modified plants and micro-organisms. Proceeding of the 3rd International Symposium, Monterey, California., USA, 13-16 November, 291-298.

Akkaya, M.S., Bhagwat, A.A. and Cregan, P.B. (1992). Length polymorphisms of simple sequence repeat DNA in soybean. Genetics 132: 1131-1139.

Akopyanz, N., Bukanov, N.O., Westblom, T.U. and Berg, D.E. (1992). PCR-based RFLP analysis of DNA sequence diversity in the gastric pathogen Helicobacter pylori. Nucleic Acid Res. 20: 6221- 6225.

Altenbach, S.B., Kuo, C.C., Staraci, L.C., Pearson, K.W., Wainwright, C. and Geogescu, A. (1992). Accumulationof a Brazil nut albumin in seeds of transgenic canola results in enhanced levels of seed protein methionine. Plant Mol. Biol. 18: 235-245.

Altenbach, S.B., Pearson K.W. Leung F.W. and Sun, S.S.M. (1987). Cloning and sequence analysis of a cDNA encoding a Brazil nut protein exceptionally rich in methionine. Plant Mol. Biol. 8: 239-250.

Atak, C., Alikamanoglu, S., Acik, L. and Canbolat, Y.(2004). Induced of plastid mutations in soybean plant (Glycine max L. Merrill) with gamma radiation and determination with RAPD, Mutat. Res. 566: 35-44.

Auerbach, C. (1943). *Drosophila.* inf. Serv. 17: 48-50.

Badfar-Chaleshtori, S., Shiran, B., Kohgard, M., Mommeni,H., Hafizi, A., Khodambashi, M., Mirakhorli, N. and Sorkheh, K. (2012). Assessment of genetic diversity and structure of Imperial Crown (*Fritillaria imperialis L.*) populations in the Zagros region of Iran using AFLP, ISSR and RAPD markers and implications for its conservations. Biochem. Systematics and Ecol. 42: 35-48.

Beckmann, J.S. (1988). Oligonucleotide polymorphisms: A new tool for genomic genetics. Bio/Technology 6: 161-164.

Beckmann, J.S. and Soller, M. (1990). Toward a unified approach to genetic mapping of eukaryotes based on sequence tagged micro satellite sites. Bio/Technology 8: 930-932.

Behura, S.K. (2006). Molecular marker systems in insects: current trends and future avenues. Mol. Ecol. 15: 3087-3113.

Biswas, M.K., Xu, Q. and Deng, X. (2010). Utility of RAPD, ISSR, IRAP and REMAP markers for the genetic analysis of *Citrus* spp. *Scientia Hort.* 124: 254-261.

Boyer, C.D. and Priess, J. (1981). Evidence for independent genetic control of the multiple forms of maize endosperm branching enzymes and starch synthases. Plant Physiol. 67: 1141-1145.

Brow, M.A., Oldenburg, M.C., Lyamichev, V., Heisler, L.M., Lyamicheva, N., Hall, J.G., Eagan, N.J., Olive, D.M., Smith, L.M., Fors, L. and Dahlberg, J.E. (1996). Differentiation of bacterial 16S rRNA genes and intergenic regions and Mycobacterium tuberculosis katG genes by structure-specific endonuclease cleavage. J. Clinical Microbiol. 34: 3129-3137.

Caetano-Anolles, G., Bassam, B.J. and Gresshoff, P.M. (1993). Enhanced detection of polymorphic DNA by multiple arbitrary amplicon profiling of endonuclease-digested DNA: identification of markers tightly linked to the super nodulation locus in soybean. Molec. Gen. Genet. 241: 57-64.

Casa, A.M., Brouwer, C., Nagel, A., Wang, L., Zhang, Q., Kresovich, S. and Wessler, S.R. (2000). The MITE family Heartbreaker (Hbr): Molecular markers in maize. Proc. Natl. Acad. Sci (USA) 97: 10083-10089.

Chang, R.Y., Donoughue, L.S. and Bureau, T.E. (2001). Inter-MITE polymorphisms (IMP): a high throughput transposon-based genome mapping and fingerprinting approach. Theor. Appl. Genet. 102: 773-781.

Conceicao, A. de S., Van-Vliet, A. and Krebbers, E. (1994). Unexpectedly higher expression levels of a chimeric 2S albumin seed protein transgene from a tandem array construct. Plant Mol. Biol. 26: 1001-1005.

Davis, H.M., Hawkins, D.J. Nelsen J.S. (1994). Utilization of laurate by the kennedy pathway in developing seeds of *Brassica napus* expressing a 12:0-ACP thioesterase gene. 11th International Symposium Metabolism Structure and Utilization of Plant Lipids. June 26-July1, 1994 Paris, France, L51(Abstr.).

Dhakshanamoorthy, D., Selvaraj, R. and Chidambaram, AL.A. (2011). Induced mutagenesis in Jatropha curcas L. using gamma rays and detection of DNA polymorphism through RAPD marker. C.R. Biologies 334: 24-30.

Dhakshanamoorthy, D.and Selvaraj, R. (2009). Extraction of genomic DNA from Jatropha sp. using modified CTAB method. Rom. J. Biol. - Plant Biol. 54(2): 117-125.

Doyle, J.J. and Doyle, J.L. (1990). Isolation of DNA from fresh plant tissue. Focus 12: 12-15.

Dwimahyani, I. Ishak (2004). Induced mutation on Jatropha (*Jatropha curcas* L.) for improvement of agronomic characters variability, http://www.digilib.batan.go.id/atom indonesia/fulltex/v 30-n2-7-2004/ Ita-Dwimahyani-Ishak.pdf (2004).

Erdem, G., and Oldacay, S. (2004). Employment of RAPD technique to assess the genetic stability of *Helianthus annuus* treated with different mutagenic agents, J. Appl. Sci. 4(2): 277-281.

FAO/IAEA. (1965). Use of induced mutations in Plant Breeding, Symposium Bub. Div., Pergamon Press. Oxford, New York.

Flavell, A.J., Knox, M.R., Pearce, S.R. and Ellis, T.H.N. (1998). Retro transposon-based insertion polymorphisms (RBIP) for high throughput marker analysis. Plant Journal 16: 643-650.

Gajera, B.B., Kumar, N., Singh, A.S., Punvar, B.S., Ravikiran, R., Subhash, N. and Jadeja, G.C. (2010). Assessment of genetic diversity in castor (Ricinus communis L.) using RAPD and ISSR markers. Industrial Crops and Products 32: 491-498.

Grodzicker T, Williams J, Sharp P and Sambrook J (1974) Physical mapping of temperature sensitive mutations. Cold Spring Harbor Symp. Quart. Biol. 39: 439-446.

Grodzicker, T.,Williams, J., Sharp, P. and Sambrook, J. (1974). Physical mapping of temperature sensitive mutations. Cold Spring Harbor Symp. Quart. Biol. 39: 439-446.

Gu, W.K., Weeden. N.F., Yu. J. and Wallace. D.H. (1995). Large-scale, cost-effective screening of PCR products in marker-assited selection applications. Theor. Appl. Genet. 91: 465-470.

Hanke, C., Peterek, G., Walter, F.P. and Frentzen, M. (1994a). Molecular approaches to the biosynthesis of trierucin in *Brassica napus* for technical use. Fat Science Technology 96: 418.

Hanke, C., Peterek, G., Walter, F.P. and Frentzen, M. (1994b). Cloning of cDNAs from *Limnanthes douglasii* showing similarity to *Escherichia coli* encoding 1-acylglycerol-3-phosphate acyltransferase. 11th International Symposium Metabolism Structure and Utilization of Plant Lipids. June 26-July1, 1994 Paris, France, F12 (Abstr.).

Hatada, I., Hayashizaki, Y., Hirotsune, S., Komatsubara, H. and Mukai, T. (1991). A genome scanning method for higher organism using restriction sites as landmarks. Proc. Natl. Acad. Sci. USA, 88: 397-400.

Higgins, T.J.V., Chandler, P.M., Randall, P.J., Spencer, D., Beach, L.R., Blagrove R.J., Kortt, A.A. and Inglis, A.S. (1986). Gene structure protein structure and regulation of the synthesis of a sulfur-rich protein in pea seeds. J. Biol. Chem. 261: 11124-11130.

Hitz, W.D., Carlson, T.J., Kinney A.J., Stecca, K.L. and Yadav, N.S. (1994). Cloning of a higher plant plastid ω-6 fatty acid desaturase cDNA and its expression in a cyanobacterium. Plant Physiology 105:635-641.

IPCC, 2001a. Climate Change 2001: Impacts, adaptation, and vulnerability. In: McCarthy, J.J., Canziani, O.F., Leary, N.A.

IPCC, 2001b. IPCC Third Assessment Report, Synthesis Report. Cambridge University Press, Cambridge.

Jeffreys, A.J., Wilson, V. and Thein, S.L. (1985). Hyper variable mini satellite regions in human DNA. Nature 314: 67-73.

Johnson, V.A. Mattern, P.J., Schmidt J.W. and Stroike J.E. (1973). Genetic advances in wheat protein quality and composition. In: proceedings International Wheat Genetics Symposium, University Missouri, Columbia pp 547-556.

Jordan, S.A. and Humphries, P. (1994). Single nucleotide polymorphism in exon 2 of the BCP gene on 7q31-q35. Human Molecular Genetics 3: 1915.

Kalendar, R., Grob, T., Regina, M., Suoniemi, A. and Schulman, A. (1999). IRAP and REMAP: two new retrotransposon-based DNA fingerprinting techniques. Theor. Appl. Genet. 98: 704-711.

Keeling, P.L., Bacon P.J. and Holt D.C. (1993). Elevated temperature reduces starch deposition in wheat endosperm by reducing the activity of soluble starch synthase. Planta 191: 342-345.

Keeling, P.L., Banisadr, R., Barone, L., Wasserman, B.P. and Singletary, G.W. (1994). Effect of temperature on enzymes in the pathway of starch biosynthesis in developing wheat and maize grain. Aus. J. Plant Physiol. 21: 807-810.

Kesari, V. and Rangan, L. (2011). Genetic diversity analysis by RAPD markers in candidate plus tree of *Pongamia pinnata*, a promising source of bioenergy. Biomass and Bioenergy 35: 3123-3128.

Landegren, U., Kaiser, R., Sanders, J. and Hood, L. (1988). DNA diagnostics. Molecular techniques and automation. Science 241: 1077-1080.

Lassner, M.W., Levering, C.K., Davies, H.M. and Knutzon D.S. (1995). Lysophosphatidic acid acyltransferase from meadowfoam mediates insertion of erucic acid at the SN-2 position of triacylglycerol in transgenic rapeseed oil. Plant Physiology 109: 1389-1394.

Li, G. and Quiros, (2001). Sequence-related amplified polymorphism (SRAP), a new marker system based on a simple PCR reaction: its application to mapping and gene tagging in *Brassica*. Theor. Appl. Genet. 103: 455-461.

Lilley, G.G., Caldwell, J.B., Kortt, A.A., Higgins, T.J. and Spencer D. (1989). In: Proceeding of the World Congress on Vegetable Protein Utilization in Human Foods and Animal Feedstuffs, Champaign, Illinois, American oil Chemist's Society, pp. 497-502.

Lue, S.P.H., Zheng, X. and Zhou, P. (2002). Inheritance of RAPD markers in an interspecific F1 hybrid of grape between *Vitis quinquangularis* and *V. Vinifera. Scientia Hort.* 93(1): 19-28.

Muller, H.J. (1927). Artificial transmutation of the gene. Science 66:84-87.

Muller-Rober, B.,Sonnewald, U., Willmitzer (1992). Inhibition of ADP-glucose pyrophosphorylase in transgenic potatoes leads to sugar-storing tubers and influences tuber formation and expression of tuber storage genes. EMBO Journal 11: 1229-1231.

OECD, 2001. Climate change. In: Sustainable Development: Critical Issues, Paris, France (Chapter 11).

Oehlkers, F. (1943). Dia Austrosung von. Chromosome mutation Elnwirkung von chemikalien. Zeitchr. F. inductive Abtammungs and Verebunghere, 81: 313-341.

Olsen, M., Hood, L., Cantor, C. and Botstein, D. (1989). A common language for physical mapping of the human genome. Science 245: 1434-1435.

Orita, M., Suzuki, Y., Sekiya, T. and Hayashi, K. (1989). Rapid and sensitive detection of point mutations and DNA polymorphisms using polymerase chain reaction. Genomics 5: 874-879.

Pamidimarri, D.V.N.S., Singh, S., Mastan, S.G., Patel, J. and Reddy, M.P. (2009). Molecular characterization and identification of markers for toxic and non-toxic varieties of *Jatropha curcas* L. using RAPD, AFLP and SSR markers. Mol. Biol. Rep. 36: 1357-1364.

Pandey, R.K. and Datta, S.K. (1995). Gamma ray induced cotyledonary variabilities in *Jatropha curcas* L., J. Nucl. Agric. Biol. 6 (1995) 24-62.

Paran, I. and Michelmore, R.W. (1993). Development of reliable PCR-based markers linked to downy mildew resistance genes in lettuce. Theor. Appl. Genet 85: 985-993.

Rahman, A.A., Ahmed, A.U., Alam, M., Mallick, D. Climate change, sustainable development and long term equity issues: perspectives of South Asia, Paper commissioned by OECD. OECD, Paris, in press.

Rapoport, J.A. (1947). Carboxyl compound and the chemical mechanism of mutation. C.R. Doklady. Agri. Sci. URSS, 54: 65-67.

Rhode, W. (1996). Inverse sequence-tagged repeat (ISTR) analysis, a novel and universal PCR-based technique for genome analysis in the plant and animal kingdom. J. Genetics Breed. 50: 249-261.

Saiki, R.K., Bugawan, T.L., Horn, G.T., Mullis, K.B. and Erlich, H.A. (1986). Analysis of enzymatically amplified beta-globin and HLA-DQ alpha DNA with allele-specific oligonucleotide probes. Nature 324: 163-166.

Sakaguchi, S., and Somabhi, M. (1987). Exploitation of Promising Crops of Northeast Thailand, Siriphan Offset, Thailand, Khon Kaen.

Shah, T.M., Mirza, J.I., Haq, M.A. and Atta, B.A. (2008). Induced genetic variability in Chickpea (*Cicer arietinum* L.) II. Comparative mutagenic effectiveness and efficiency of Physical and chemical mutagens. Pak. J. Bot. 40 (2): 605-613.

Shewmaker, C.K., Boyer C.D., Wiesenborn, D.P., Thompson, D.B. Boersig, M.R. Oakes J.V. and Stalker, D.M. (1994). Expression of Escherichia coli glycogen synthase in the tubers of transgenic potatoes (*Solanum tuberosum*) results in a highly branched starch. Plant Physiology 104: 1159-1162.

Stark, D.M., Timmerman, K.P., Barry, G.F., Preiss J. and Kishore G.M. (1992). Regulation of the amount of starch in plant tissues by ADP glucose phosphorylase. Science 258: 287-290.

Sun, S.S.M., Leung, F.W. and Tomic, J.C. (1987). Brazil nut (*Bertholletia excelsa* H.B.K.) proteins. Fractionation, composition, and identification of sulfur-rich protein. J. Agri. and Food Chem. 35: 232-235.

Tabe, L.M., Higgins, C.M., Mc Nabb, W.C. and Higgins T.J.V. (1993). Genetic engineering of grain and pasture legumes for improved nutritive value. Genetica 90: 181-200.

Telenius, H., Carter, N.P., Bebb, C.E., Nordenskjold, M., Ponder, B.J. and Tunnacliffe, A. (1992). Degenerate oligonucleotide-primed PCR: general amplification of target DNA by a single degenerate primer. Genomics 13: 718-725.

Tol, R.S.J., Fankhauser, S., Richels, R.G., Smith, J.B., 2000. How much damage will climate change do? Recent Estimates, Research Unit Sustainability and Global Change SCG-2. Centre for Marine and Climate Research, Hamburg University, Hamburg.

Tolleanaar, D. (1934). Untersuchungen uber mutation bei Tabak. I. Entstehungsweise und wesen kunstlich erzengter Gen-mutanten. Genetic. 16: 111-152.

Van Bogaert, G. (1975). A comparison between colchicine induced tetraploid and diploid cultivars of Lolium species. In Polyploidy in Fodder crops, B. Neusch (ed.). Eucarpia Report Zurich.

Van der Wurff, A.W.G., Chan, Y.L., van Straalen, N.M. and Schouten. J. (2000). TE-AFLP: combining rapidity and robustness in DNA fingerprinting. Nucleic Acids Res. 28: 105-109.

Varshney, R.K., Chabane, K., Hendre, P.S., Aggarwal, R.K. and Graner, A. (2007). Comparative assessment of EST-SSR, EST-SNP and AFLP markers for evaluation of genetic diversity and conservation of genetic resources using wild, cultivated and elite barleys. Plant Science 173: 638-649.

Vos, P., Hogers, R., Bleeker, M., Lee, T., Hornes, M., Frijters, A., Pot, J., Peleman, J., Kuiper, M. and Zabeau, M. (1995). AFLP: A new technique for finger printing. Nucleic Acids Res. 23: 4407-4414.

Walter, F.P., Hanke, C., Eickelkamp, A. and Frentzen, M. (1995). Trierucin biosynthesis in transgenic rapeseed: Cloning and expression of cDNAs encoding and erucoy 1-CoA specific acyltransferase. Proceedings of 9th International Rapeseed Congress, Cambridge, UK. D26.

Watson, R.T., 2001. IPCC Special Report: Climate Change 2001: Synthesis Report. Cambridge University Press, UK, pp. 398.

Waugh, R., McLean, K., Flavell, A.J., Pearce, S.R., Kumar, A., Thomas, B.T. and Powell, W. (1997). Genetic distribution of BARE-1 retrotransposable elements in the barley genome revealed by sequence-specific amplification polymorphisms (S-SAP). Mol. Gen. Genet. 253: 687-694.

Wehrman V.K., Fehr, W.R., Cianzio S.R., Cavins J.F. (1987). Transfer of high seed protein to high yielding soybean cultivars. Crop Science 27: 927-931.

Welsh, J. and McClelland, M. (1990). Fingerprinting genomes using PCR with arbitrary primers. Nucleic Acids Res. 18: 7213-7218.

Weubles, D.J., Jain, A.K., 2001. Concerns about climate change and the role of fossil fuel use. Fuel Processing Technology 71: 99-119.

Williams, J.G.K., Kubelik, A.R., Livak, K.J., Rafalsk, J.A. and Tingey, S.V. (1990). DNA polymorphisms amplified by arbitrary primers are useful as genetic markers. Nucleic Acid Research 18: 6531-6535.

Zietkiewicz, E., Rafalski, A. and Labuda, D. (1994). Genome fingerprinting by simple sequence repeat (SSR)-anchored polymerase chain reaction amplification. Genomics 20: 176-183.

19

Functional Genomics and Bioinformatics Approach to Understand Regulation of Abiotic Stress in Cereal Crops

Mousumi Sahu, Budheswar Dehury, Mahendra K Modi and Madhumita Barooah*

Department of Agricultural Biotechnology, Assam Agricultural University
Jorhat-785013, Assam, India

ABSTRACT

The environmental constraints such as drought, flood, cold, high temperature, increased soil salinity and acidity severely affect the crop productivity including the cereal crops like rice (*Oryza sativa* L.), sorghum (*Sorghum bicolor* L.), maize (*Zea mays* L.) barley (*Hordeum vulgare* L.) and wheat (*Triticum aestivum* L. and *T. turgidum* L.). Over the few decades several traditional and modern breeding methods have been used for the development of stress tolerance in crop plants to combat against the notorious abiotic stresses. The involvement of molecular breeding and genetic engineering technology has a major contribution towards solving the complexity of stress response in cereal crops. The complex cellular processes that regulate transduction, downstream regulatory factors and operate signal perception are the specific areas which are considered and analyzed for substantial information on abiotic stress response. In this chapter the application of functional genomics or genomic-assisted methods and bioinformatics approaches are briefly described to understand the complex regulatory networks of abiotic stress tolerance mechanism by gene expression profiling.

Keywords: Abiotic stress, cereal crop, gene expression profiling, genetic engineering, molecular breeding.

1. INTRODUCTION

Gramineae (also known as grass family or Poaceae) is one of the largest families of monocot plants where mostly the herbaceous grass-like plants includes several important staple cereals crops such as wheat, rice, maize, sorghum, barley, oats, and millet. These staple cereals are cultivated for the edible components of their grain or caryopsis, composed of the endosperm, germ, and bran. Cereal grains are grown in greater quantities and provide more food energy worldwide than any other type of crop; they are therefore staple crops. Cereals also represent the vital source of vitamins, minerals, carbohydrates, fats and protein in the form of whole grain. These plants have evolved to live in environments where they are often exposed to

*Corresponding author's E-mail: *m17barooah@yahoo.co.in*

Roychowdhury R. (Ed.), Crop Improvement in the Era of Climate Change.

various stress factors in combination such as high temperature, low water availability, salinity and mineral toxicity and deficiency throughout their life span.

The influence of non-living factors on living organisms in a specific environment is defined as abiotic stress. This type of stress condition is generated by high temperature and associated with low water availability, salinity and mineral toxicity and deficiency along with the geographic distribution pattern of plants. In general, abiotic stress factors such as heat, cold, drought, salinity, and nutrient stress are the major abiotic stresses which have made a huge impact on world agriculture, and it has been reported that they reduce the average yields by >50% for most major crop plants (Wang *et al.*, 2003). Furthermore, a wide range of pests and pathogens, including fungi, bacteria, viruses, nematodes, and herbivorous insects attack the plants. Each of these stresses (biotic or abiotic) elicits a complex cellular and molecular response system implemented by the plants in order to prevent the damage and survivability, but often at the detriment of growth and yield (Herms and Mattson, 1992). Under these adverse environmental conditions plant roots encounter high concentrations of salt and boron in the subsoil that constrains root growth. The abiotic stress condition affects both the yield and quality. More often, there are several secondary damages that develop associated with the low water content and photon irradiance in the subsoil which reduce the ability of the plant to reoxidize NADPH and causes an array of morphological, physiological, biochemical and molecular changes that unfavourably affect plant growth, development and productivity. Increasing industrialization with insufficient pollution control and growing greenhouse effect result into a serious abiotic stress condition like heavy metal contamination and elevating the levels of CO_2 respectively. The current global climatic change, change in variability of rainfall and temperature may adversely affect the yields and nutritional quality of crops (Porter and Semenov, 2005) and influence the habitat range of pests and pathogens, with increasing temperature facilitates spread of pathogens (Bale *et al.*, 2002; Luck *et al.*, 2011; Madgwick *et al.*, 2011; Nicol *et al.*, 2011). Crop plants are therefore likely to encounter a greater range and number of environmental stresses, which when occurring simultaneously can have severe consequences. The changing climatic conditions, combined with an increasing pressure on global food productivity due to population increase, result in a demand for stress-tolerant crop varieties that has never been greater (Takeda and Matsuoka, 2008; Newton *et al.*, 2011). Understanding the mechanisms of plant responses to multiple simultaneous stresses is therefore crucial in providing opportunities for the development of broad-spectrum stress-tolerant crops. So it is very important to assess and understand the abiotic stressors and in the mean time development novel and improved strategy to encounter their negative impact on agricultural system is needed. Modern biological approach of science such as System Biology, Metagenomics, Nano-Biotechnology, Bioinformatics, Phenomics, have aided in mining, isolating, characterizing and deploying novel genes for abiotic stress tolerance.

Identification and characterization of the genes and the specific genetic regions associated with both the quantitative and qualitative traits has been a major challenge in crop breeding technique. In order to improve the selection efficiency and accuracy, a high throughput marker assisted system is extensively being used in recent breeding programmes.

The key fundamental aspect of research with regard to abiotic stress tolerance encompasses the factor of abiotic stress responsive genes gene discovery in naturally occurring gene diversity.

This is the basic aim of the abiotic stress tolerance gene discovery and research in cereals to reveal novel way to adopt agriculture to climate change adaptation. For many functional genomics projects the initial strategy is to screen genetic diversity. For this propose, the grasses always makes an easy attraction as the content of extensive genetic diversity.

1.1 Abiotic Stresses on Cereal Crops

The abiotic stress condition is influenced by major players like atmosphere, soil, water and their associated factors. It is evident from Fig. 1 that several groups of abiotic stressors affect modern agricultural system.

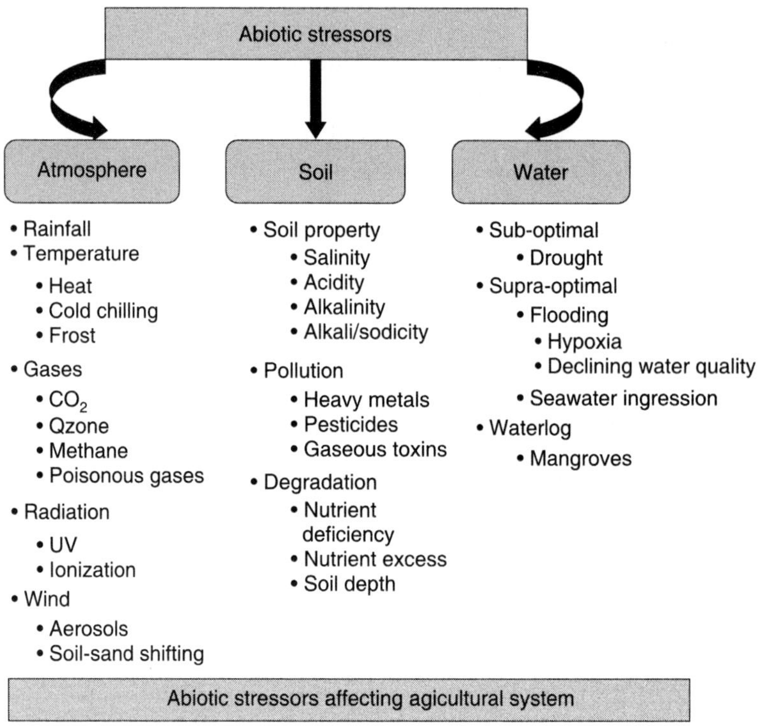

Fig. 1 A schematic representation of different group of abiotic stressors affecting the modern agricultural system.

1.1.1 Atmospheric Factor

1.1.1.a. Rainfall

Atmospheric pollution is another source of stress. Uncertainty or decline in rainfall adversely controls the soil salinity and acidic property. Sulfur dioxide and nitrogen oxides from fossil fuel burning result in acid rain. Acid rain is considered the foliage of sensitive plants. It also reduces soil pH, harming or ceasing plant-growth that is not adapted to acidic conditions.

1.1.1.b. Temperature

In subtropical condition the plant experiences extreme temperature that produces certain mechanical damages in plant tissue. Mechanical damage includes expansion-induced-lysis, phase transitions and fracture lesions in membranes, and physical damage. Freezing may induce the production of ROS that damages membrane components, and results protein denaturation. Different phenotypic symptoms develop in response to chilling stress include reduced leaf expansion, wilting, chlorosis and may lead to necrosis (death of tissue). In addition, chilling also affects the reproductive development of plants.

1.1.1.c. Gases

Increase in the CO_2 and other greenhouse gases in the atmosphere such as water vapour (H_2O), carbon dioxide (CO_2), methane (CH_4), nitrous oxide (N_2O) and ozone (O_3) result to the increase in atmospheric temperature and inversely affect the crop physiology and growth. The common damages are produced under this condition result into poor edaphic properties.

1.1.1.d. Radiation

The radiations such as ultraviolet and ionization radiations highly affect the growth and survival of the crop. These are the major water vapour (H_2O), carbon dioxide (CO_2), methane (CH_4), nitrous oxide (N_2O) and ozone (O_3) cause of photon irradiation that generates cellular damages in root and leaf tissue.

1.1.1.e. Wind

Plant tissue damage is very common during hot, humid, hazy weather with little wind. Air inversions, when warm air at the surface is trapped by even hotter air in the atmosphere above, lead to build up of air pollutant that harms plant yield.

1.1.2 Soil Factor

1.1.2.a. Soil Property

Salinity is another major abiotic stress found all over the globe. Hence, 800 million hectares of land and more than 6% of the total land area are affected by salt. The major toxic component, sodium ion (Na^+) has been detected to interfere in several metabolic processes such as protein translation, transcription and enzyme activity those in a result cause osmotic stress. These events gradually and significantly reduce the crop yield. Functional genomics provides new opportunities to understand these processes in the plant, enabling the identification of genes involved and providing opportunities to generate plants that are able to survive and produce viable yields on saline soils. Three types of major mechanisms produce tolerance to salinity stress such as osmotic stress tolerance, Na^+ exclusion and tissue tolerance. Osmotic stress tolerance is the ability to survive under osmotic stress causing stomatal closure and reduced cell expansion in root tips and leaves. Na^+ exclusion is the tolerance produced by the reduction of Na^+ accumulation in shoots by Na^+ exclusion in root. Tolerance can also be acquired exposure of tissues to the accumulated Na^+ and Cl^- at cellular and intracellular levels. Again, the soil pH property highly affects the crop growth and physiology.

1.1.2.b. Pollution

Rapid application of pesticide extensively produces soil pollution. Rapid industrialization near agricultural land and low pollution management control affect the crop growth and production. Other than the above stressors the boron content of the soil of an area also significantly affects the cereal survival and yielding. Boron is among the eight essential elements for plant growth. The high level of boron content drastically affects the cereal growth and can limit productivity. In several studies it has been studied that 30% of the soil in South Australia cultivated for cereal is affected by boron toxicity.

1.1.2.c. Degradation

Climate and soil interactions provide suitable environment for agricultural production, and also affect physical processes of land degradation. Regular land degradation results to sever root damage that governs nutrient irregularity and gaseous toxin. These conditions cause severe cellular damage through photon irradiation.

1.1.3 Water Factor

1.1.3.a. Suboptimal

The major trait of interest for cereal breeding is drought tolerance. The rapid growth of human population, declining rainfall, increasing temperature and water supply pattern extensively affect the growth of crop-plant area. To encounter this type of stress the scientific approaches of adapting germplasm has been developed. A wide range of characteristics such as multiple genetic, physiological, cellular and biochemical strategies are defined in the plant. However, the identification and characterization of a single gene or group of genes for dissecting drought tolerance is difficult.

1.1.3.b. Supraoptimal

A high salt content can be lethal for many plants, but some known as halophytes have adapted to salty conditions, which are found in both in coastal areas, such as salt marshes, and inland arid areas where high evaporation tends to concentrate salts originating in the mineral content of the soil. These plants may excrete salt from their leaves or store it within their cells in bodies known as vacuoles so that it is kept separate from the cell cytoplasm. Deficiencies in nutrients or the presence of toxic substances such as heavy metals in the soil can also result in abiotic stress.

1.1.3.c. Waterlog

In flooded soils, the oxygen concentration drops to near zero within 24 hours because water replaces most of air in the soil pore space. Oxygen diffuses much more slowly in water filled pores than in open pores. Roots need oxygen to respire and have normal cell activity. When any remaining oxygen is used up by the roots in flooded or waterlogged soils, they cease to function normally. Therefore, mineral nutrient uptake and water uptake are reduced or stopped in flooded conditions. There is also a buildup of ethylene in flooded soils, the plant hormone that in excess amounts can cause leaf drop and premature senescence.

2. FUNCTIONAL GENOMICS AND BIOINFORMATICS APPROACH TO ELUCIDATE ABIOTIC STRESS TOLERANCE MECHANISM

The plants go through a process of regulatory modification in several cellular, physiological and biochemical processes changed environmental condition by performing gene alteration mechanism. To analyze the process involved in this gene regulatory mechanism several functional genomics projects have been initiated across the globe. Moreover, several modern techniques have been employed for the gene expression study such as application of powerful microarray technology for transcriptome analysis of abiotic stress treated tissue. A large number of proteins such as transcription and signalling factor that are associated with salt, drought and temperature tolerance have been extensively studied to understand the mechanisms involved in plant growth and survival under these stress condition. There also exists a curiosity of cross-talk and interconnections that are involved in stress response, for example, Tran- and Cis-acting factors. Again the understanding of the post-transcriptional protein modification, protein degradation and non-coding microRNA activity allow the modulation of the target proteins. Some of the siRNAs play an important role as stress-inducible and affect protein synthesis including alternative splicing. Several functionally characterized genes have been identified, introduced into crop plants and used in building up tolerance against various abiotic stress conditions. A number of abiotic stress responsive genes have been functionally characterized to build up tolerance against various abiotic stress conditions. Many studies reported the increase stress tolerance in transgenic lines compared to the control plant. It is expected that the genetically engineered plants may have higher tolerance to water and cold stress as they contain higher accumulation of osmolytes and stress protein. Introduction of scientific promoters in order to achieve organ-specific and stress-responsive expression of the introduced gene is most desirable. In this regard, several stress-inducible genes have been identified along with their promoters and analyzed for specificity. Several genomics approaches have been used to study whole metabolic network involved in abiotic stress response and their relationships. However, to avoid unwanted side effects, intensive research and breeding efforts are needed.

2.1 Gene Discovery

The generation of ESTs from different cDNA libraries collected from abiotic stress treated tissues at different stages of development is considered a significant genomic approach to identify abiotic stress related gene. The dbEST (http://www.ncbi.nlm.nih.gov/dbEST/dbEST_summary.html.) of the National Center for Biotechnology Information (NCBI) (http://www.ncbi.nlm.nih.gov/) index detailed information on the type and number of different abiotic stress tolerant species. In order to enrich plant EST datasets with stress-responsive genes, specific sequencing programmes based on cDNA libraries from stress-treated plant tissues and organs of diverse species at many developmental time points are necessary. Moreover, the information of gene number, gene content and possible evolutionary background is studied by clustering the EST sequences resulted from abiotic stress treated cDNA libraries. Putative functional characterizations of the stress-treated-responsive genes are studied using BlastX (Altschul. *et al.*, 1990) by comparing the Swissprot dataset (http://www.uniprot.org/). These studies are very informative regarding gene index associated with stress-responsive genes among various species. Moreover, the clustering data provides consensus sequences which are more informative dataset than typical EST data.

species is extensively considered an important approach to identify the evolutionarily conserved mechanism (Sreenivasulu *et al.*, 2002). Several promoters of a group of stress-responsive gene from various species of rice, maize and Arabidopsis harbour DRE, ABF, GCC-box and w-box cis elements were reported (Chena *et al.*, 2002; Kizis *et al.*, 2002; Ito *et al.*, 2002; Dubouzet *et al.*, 2003).

2.3 Functional Aspects of Stress Tolerance Mechanisms Identified through Transgenic Approaches in Cereals

Based on the functional characters the abiotic stress responsive genes are classified in to 4 groups such as (i) osmolyte biosynthesis, (ii) antioxidant protectants, (iii) ion homeostasis (Sangam *et al.*, 2005; Valliyodan *et al.*, 2006). The abiotic stress responsive genes are classified into 4 classes based on their functional aspect. During the pre-genomics era the inadequate knowledge of genes and specific promoters (Zhu, J.K., 2001; Zhu, J.K., 2002; Zhu *et al.*, 1997; Zhu *et al.*, 1998; Ishitani *et al.*, 1997) was the main obstacle in the functional characterization of a large set of genomics analysis. In the modern stress biology now it is possible to analyze a large scale genes simultaneously on a genome wide scale considering their structure and function.

2.4 High-throughput Gene Inactivation Techniques to Characterize Putative Abiotic Stress Responsible Gene

Despite several genomic-assisted programmes, still now about 40% to 50% of the identified genes are yet to be characterized. To identify their putative functions in regard to abiotic stress tolerance, a number of methods have been developed. Two complementary approaches to identify mutations in target genes, namely, TILLING (Targeted Induced Local Lesions In Genomes) and T-DNA insertion mutant lines were employed. Three major groups of mutants such as los-low expression of osmotically responsive genes, cos-constitutive expression of osmotically responsive genes and hos-high expression of osmotically responsive genes were identified and studied for osmotic and cold stress signal transduction pathways regulated by ABA-dependent and ABA-independent pathway (Xiong *et al.*, 2001). Several other genes such as COR15, COR47, P5CS and RD22 expression were characterized for the tolerance to drought, salt stress and freezing in los5-1 and los5-2 mutants (Xiong *et al.*, 2002). Among the hos class of mutants hos5 showed osmotic stress specific response. Similarly, in fiery1 mutant, super-induction of RD29A-luciferase expression is studied in response to exogenous ABA, low temperature and osmotic stress in ABA and stress signalling.

2.5 Genomics-Assisted Breeding Approach

The traditional breeding programme shows significant successes in developing abiotic stress-tolerant lines, but it inversely relates to yield (Xiong *et al.*, 1999). In recent researches, two newly developed genomics approaches are used to exploit the gene pool responding abiotic stress tolerance. The first one is based on the identification of stress tolerant genes by functional genomics approaches and transformation technology is used for stress-tolerant genes into crops of interest. The second one is based on QTL/genes identification those confer tolerance to stress in germplasm collections and development of specific molecular markers and use in marker-assisted breeding programmes. It is important to develop genomics-assisted breeding programme to develop stable populations basing on the outcome of functional genomics. In this regard, variation between species and accessions for abiotic stress tolerance is needed to be explored (Lee *et al*., 2001). The unique resistance alleles, those are not found in the cultivated gene pool are considered special patent sources of abiotic stress tolerance traits of wild species and landraces (Merlot *et al*., 2002). Focus more on the cereal plants it was found that wild barley contains significant variation than the cultivated species and many alleles are associated with adaptation to specific environments. For example, some of the wild species such as *Hordeum jubatum, H. chilense* and *H. spontaneum* were identified as tolerant to salinity (Mustilli *et al*., 2002). In a report it was suggested that the D genome of hexaploid bread wheat provides a source of salt tolerance (Ceccarelli, S., 1987). Similarly, the international maize and wheat improvement center (CIMMYT) has significantly contributed during the past 25 years to improve drought tolerance in maize (Nevo, E., 1992). The genetic loci responding to the abiotic stresses in cultivated species are now analyzed by investigating in wild gene pool (Ellis *et al*., 2000). Keeping this aspect in mind the abiotic stress genes in wild species, progress was made in direction of producing salt-tolerant wheat (*Thinopyrum bessarbicum)* amphiploids by transferring salt tolerance gene of *T. bessarbicum* to wheat (Forster *et al*., 1997). Advanced Backcross QTL Analysis' is the term signifying the process of transfer of desirable QTLs from unadapted to elite germplasm was devised by Tanksley and Nelson (1996) (Gorham *et al*.,1990). In a very recent study, Tuberosa and Salvi (2006) reported QTL studies of drought in cereals by employing marker-assisted selection to introgress QTL alleles responding to drought tolerance (Bruce *et al*., 2002). In several studies the study of genes and regulatory networks in abiotic stress tolerance and the QTL analysis those are related to salt tolerance were reported by using genomics approaches (King *et al*., 1997). Considering the gene expression profiling and marker based fingerprinting of related lines or individuals in a segregating population, a new approach has been developed known as 'Genetical Genomics' (Tanksley *et al*., 1996). Instead of a segregating population, by using a set of introgression lines where an alien genome (in barley line, *Hordeum spontaneum* HS213 conferring higher abiotic stress tolerance) was introgressed in the genetic background of an elite line (Hordeum vulgare cv. Brenda) which resulted higher yield (Tuberosa *et al*., 2006). Jansen and Nap (2001) originally proposed the extensive expression profiling from introgression lines to understand the QTLs expression (eQTLs) (Jansen *et al*., 2001). This approach is useful to analyze cis- and trans-acting factors and to delineate a trait-related genetic network (Ren *et al*., 2005; Xu *et al*., 2006; Jansen, R.C., 2003). Such data can be used to develop trait-linked molecular markers and direct transgenic approaches, both important components of molecular breeding strategies.

2.6 Enhancing Yield Stability of Crops that are Challenged by Abiotic Stresses

The multidimensional impact of environmental stress was studied in Arabidopsis and other crop plants. These studies were generally carried out at the vegetative stage. In this regard, the alteration mechanism of ABA-dependent, ABA-independent signalling cascades, manipulation of osmo-protectant accumulation, chaperone production and cell integrity protection were studied (Li *et al*., 2005; Umezawa *et al*., 2006; Valliyodan *et al*., 2006).

3. CHARACTERIZATION OF GENES CONTROLLING IMPORTANT AGRONOMIC TRAITS

A case study on evaluation of seven function-known candidate genes for their effects on improving drought resistance of transgenic rice under field conditions (Adopted from Xiao *et al.*, 2009). Abiotic stresses such as drought, extreme high temperature, low water availability, mineral deficiencies or toxicities, high salt affects the productivity of cereal crops. With the declining in the availability of water resources, land degradation and community pressures these prominent stresses play an important role in agriculture prospective across the globe. Although extensive genetic studies and surveys of landrace and wild germplasm have indicated extensive variation for abiotic stress tolerance but it has failed to attain its maximum goal due to the relatively poor background knowledge of the molecular basis for stress cereal crop plants. As a network of signal transduction pathways has led to multiple responses to abiotic stresses, it has been quite difficult to study the abiotic stress tolerance mechanism using traditional approaches because of their complexity and the large number of genes and gene products involved in the various defensive and developmental responses of the plant. In this case functional genomics plays a vital role by providing tools for dissecting abiotic stress responses in cereals especially in rice, wheat and barley, through interrogation of genes that may be useful for improving resistance to abiotic stresses.

Xiao and co-workers (Xiao *et al.*, 2009) have selected seven well documented genes for drought resistance breeding, namely, CBF3, SOS2, NCED2, NPK1, LOS5, ZAT10, and NHX1 and transformed into rice cultivar Zhonghua 11 under the control of constitutive promoter Actin1 and stress-inducible promoter of a rice HVA22 homolog. The developed transgenic rice lines and subsequently tested for drought resistance under field conditions. Out of the 1598 independent transgenic T_0 plants, 36.7% showed single copy and 57.6% showed expression of the transgenes. For each gene construct, 30 T_1 families with expression of transgene were selected for drought resistance testing at the reproductive stage in field, and 10 of them were tested in PVC pipes with a defined stress. They considered two parameters such as relative yield (RY) and relative spikelet fertility to evaluate drought resistance performance because significantly decreased yield was observed in the T_1 generation. The transgenic families of eight constructs (HVA22P:CBF3, HVA22P:NPK1, Actin1:LOS5, HVA22P:LOS5, Actin1:ZAT10, HVA22P:ZAT10, Actin1:NHX1, and HVA22P:NHX1) showed significantly higher RY than wild-type (WT) under both drought stress field and PVC tube conditions whereas transgenic families of 9 constructs (HVA22P:SOS2 and CBF3, LOS5, ZAT10, and NHX1 by both promoters) showed significantly higher relative spikelet fertility than wild type in the field or PVC pipes. In the field drought resistance testing of T_2 families derived from the T_1 families with relatively lower yield decrease, transgenic families of seven constructs (HVA22P:CBF3, Actin1:NPK1, HVA22P:NPK1, Actin1:LOS5, HVA22P:LOS5, Actin1:ZAT10, and HVA22P:ZAT10) showed significantly higher yield per plant than wild type, and families of nine constructs (Actin1:CBF3, HVA22P:CBF3, HVA22P:SOS2, HVA22P:NPK1, Actin1:LOS5, HVA22P:LOS5, Actin1:ZAT10, HVA22P:ZAT10, and Actin1:NHX1) had higher spikelet fertility than wild type. It was seen that LOS5 and ZAT10 showed relatively better effect than the other five genes in improving drought resistance of transgenic rice under field conditions. The result from this study is an example which can be a useful reference for drought resistance engineering in rice. Genes controlling important agronomic traits can be effectively introduced into transgenic lines improve abiotic stress in selected cereals species.

4. CONCLUSION

Abiotic stresses in cereal crops such as boron tolerance, salinity tolerance, reproductive frost tolerance and drought tolerance a major genetic control of traits is well known and used for aid selection based on molecular method. In several cases, specific gene evolutions have been studied and adaptive-evolution mechanisms in plants have been identified. Gene identification is also helpful in generating new varieties in the future by applying genetic transformation with superior abiotic-stress tolerance. Functional genomics study is considered the core approach to provide molecular solution to the existing abiotic stress. Certainly regulatory proteins such as transcriptional and signalling factors related to salt, drought and temperature tolerance significantly contribute in the understanding of the corresponding molecular mechanisms and pathways associated with them as well as crop survival and yielding. However, a large amount of cross-talk and interconnections are found under these conditions. These are generally the trans- and cis- acting factors. Again understanding the post-translational modification of proteins, degradation of proteins and non-coding microRNA interactions will significantly aid to modulation of the target proteins. Similarly, it has been found that siRNAs play as certain stress-inducible and they affect transcriptional and translational processes including alternative splicing. Considering all these aspects, further investigations are needed for further development of products (varieties) that can overcome the environmental constraints and perform better. In this regard, identification and characterization of specific stress responsive genes and the specific promoters using gnomic and bioinformatics tool will be helpful in the stress responsive researches. Such promoters could be used to prevent gene silencing when gene pyramiding is sought as a feasible strategy to obtain higher tolerance levels.

References

Altschul, S.F., Gish, W., Miller, W., Myers, E.W., Lipman, D.J. (1990) "Basic local alignment search tool." J. Mol. Biol. 215(3): 403-410

Bruce, W.B., Edmeades, G.O., Barker, T.C., (2002). "Molecular and physiological approaches to maize improvement for drought tolerance." J. Exp. Bot. 53(366): 13-25.

Buitink J, Leger JJ, Guisle I, Vu BL, Wuillème S, Lamirault G, Le Bars A, Le Meur N, Becker A, Küster H, Leprince O, (2006) "Transcriptome profiling uncovers metabolic and regulatory processes occurring during the transition from desiccation sensitive to desiccation-tolerant stages in Medicago truncatula seeds." Plant J. 47(5): 735-750.

Ceccarelli, S., (1987). "Yield potential and drought tolerance of segregating populations of barley in contrasting environments." Euphytica 36(1987): 265-273.

Chen W, Provart NJ, Glazebrook J, Katagiri F, Chang HS, Eulgem T, Mauch F, Luan S, Zou G, Whitham SA, Budworth PR, Tao Y, Xie Z, Chen X, Lam S, Kreps JA, Harper JF, Si Ammour A, Mauch-Mani B, Heinlein M, Kobayashi K, Hohn T, Dangl JL, Wang X, Zhu T. (2002) "Expression profile matrix of Arabidopsis transcription factor genes suggests their putative functions in response to environmental stresses." Plant Cell 14(3): 559-574.

Chen WJ and Zhu T (2004). "Networks of transcription factors with roles in environmental stress response." Trends Plant Sci. 9(12): 591-596.

Dubouzet JG, Sakuma Y, Ito Y, Kasuga M, Dubouzet EG, Miura S, Seki M, Shinozaki K, Yamaguchi-Shinozaki K. (2003). "OsDREB genes in rice, Oryza sativa L., encode transcription activators that function in drought-, high-salt- and cold responsive gene expression." Plant J. 33(4): 751-763.

Ellis RP, Forster BP, Robinson D, Handley LL, Gordon DC, Russell JR, Powell W. (2000). Wild barley: a source of genes for crop improvement in the 21st century. J. Exp. Bot. 51(342): 9-17.

FORSTER, B. P., RZUSSELL, J. R., ELLIS, R. P., HANDLEY, L. L., ROBINSON, D., HACKETT, C. A., NEVO, E., WAUGH, R., GORDON, D. C., KEITH, R. and POWELL, W. (1997), Locating genotypes and genes for abiotic stress tolerance in barley: a strategy using maps, markers and the wild species. New Phytologist, 137: 141-147.

Fowler, S. and Thomashow, M.F. (2002). Arabidopsis transcriptome profiling indicates that multiple regulatory pathways are activated during cold acclimation in addition to the CBF cold response pathway. Plant Cell 14(8): 1675-1690.

Gorham, J., Wyn-Jones, R.G., Bristol, A., Jones, R.G.W. (1990). Partial characterization of the trait for enhanced K+–Na+ discrimination in the D genome of wheat. Planta 180(4): 590-597.

Hoth, S., Morgante, M., Sanchez, J.P., Hanafey, M.K., Tingey, S.V., Chua, N.H. (2002). Genome-wide gene expression profiling in *Arabidopsis thaliana* reveals new targets of abscisic acid and largely impaired gene regulation in the abi1-1 mutant. J. Cell Sci. 115(): 4891-4900.

http://www.ncbi.nlm.nih.gov/

http://www.ncbi.nlm.nih.gov/dbEST/dbEST_summary.html.

http://www.uniprot.org/

Ishitani, M., Xiong, L.M., Stevenson, B., Zhu, J.K., (1997). Genetic analysis of osmotic and cold stress signal transduction in Arabidopsis: interactions and convergence of abscisic acid dependent and abscisic acid-independent pathways. Plant Cell 9(11): 1935-1949.

Jansen, R.C., (2003). Studying complex biological systems using multifactorial perturbation. Nature Rev. Genet. 4(2): 145-151.

Jansen, R.C. and Nap, J.P., (2001). Genetical genomics: the added value from segregation. Trends Genet. 17(7): 88-391.

Kawasakia, S., Borcherta, C., Deyholosb, M., Wangb, H., Brazillea, S., Kawaia, K., Galbraitha, D., Bohnert, H.J. (2001). Gene expression profiles during the initial phase of salt stress in rice. Plant Cell 13(4): 889-905.

King, I. P., Forster, B. P., Law, C. C., Cant, K. A., Orford, S. E., Gorham, J., Reader, S. Miller, T. E. (1997), Introgression of salt-tolerance genes from Thinopyrum bessarabicum into wheat. New Phytologist, 137: 75-81.

Kizis, D. and Pages, M., (2002). Maize DRE-binding proteins DBF1 and DBF2 are involved in rab17 regulation through the drought-responsive element in an ABA-dependent pathway. Plant J. 30(6): 679-689.

Kreps, J.A., Wu, Y.J., Chang, H.S., Zhu, T., Wang, X., Harper, J.F., (2002). Transcriptome changes for Arabidopsis in response to salt, osmotic, and cold stress. Plant Physiol. 130(4): 2129-2141.

Lee, H.J., Xiong, L.M., Gong, Z.Z., Ishitani, M., Stevenson, B., Zhu, J.K., (2001). The *Arabidopsis* HOS1 gene negatively regulates cold signal transduction and encodes a RING finger protein that displays cold-regulated nucleo cytoplasmic partitioning. Genes Dev. 15(7): 912-924.

Li, J., Huang, X.Q., Heinrichs, F., Ganal, M.W., Roder, M.S. (2005). Analysis of QTLs for yield, yield components, and malting quality in a BC3-DH population of spring barley. Theor. Appl. Genet. 110(2): 356-363.

Merlot S, Mustilli AC, Genty B, North H, Lefebvre V, Sotta B, Vavasseur A, Giraudat J. (2002). Use of infrared thermal imaging to isolate *Arabidopsis* mutants defective in stomatal regulation. Plant J. 30(5): 601-609.

Mustilli, A.C., Merlot, S., Vavasseur, A., Fenzi, F., Giraudat, J. (2002). *Arabidopsis* OST1 protein kinase mediates the regulation of stomatal aperture by abscisic acid and acts upstream of reactive oxygen species production. Plant Cell 14(12): 3089-3099.

Nevo, E., (1992). Origin, evolution, population genetics and resources for breeding of wild barley Hordeum spontaneum in Fertile Crescent. In: Sherwy, P.R. (Ed.), Barley Genetics Biochemistry, Molecular Biology and Biotechnology. CAB International, Wallingford UK, pp. 19-43.

Ren Z.H., Gao J.P., Li L.G., Cai X.L., Huang W., Chao D.Y., Zhu M.Z., Wang Z.Y., Luan S., Lin. H.X., (2005). A rice quantitative trait locus for salt tolerance encodes a sodium transporter. Nat. Genet. 37: 1141-1146.

Sahi, C., Agarwal, M., Reddy, M.K., Sopory, S.K., Grover, A., (2003). Isolation and expression analysis of salt stress associated ESTs from contrasting rice cultivars using a PCR-based subtraction method. Theor. Appl. Genet. 106, 620-628.

Sangam, S., Jayasree, D., Reddy, K.J., Chari, P.V.B., Sreenivasulu, N., Kavi Kishor, P.B., (2005). Salt tolerance in plants—transgenic approaches. J. Plant Biotechnol. 7(1): 1-15.

Seki M, Narusaka M, Ishida J, Nanjo T, Fujita M, Oono Y, Kamiya A, Nakajima M, Enju A, Sakurai T, Satou M, Akiyama K, Taji T, Yamaguchi-Shinozaki K, Carninci P, Kawai J, Hayashizaki Y, Shinozaki K., (2002a). Monitoring the expression profiles of 7000 Arabidopsis genes under drought, cold and high-salinity stresses using a full-length cDNA microarray. Plant J. 31(3): 279-292.

Seki M, Ishida J, Narusaka M, Fujita M, Nanjo T, Umezawa T, Kamiya A, Nakajima M, Enju A, Sakurai T, Satou M, Akiyama K, Yamaguchi-Shinozaki K, Carninci P, Kawai J, Hayashizaki Y, Shinozaki K., (2002b). Monitoring the expression pattern of around 7000 *Arabidopsis* genes under ABA treatments using a full-length cDNA microarray. Funct. Integr. Genomics 2(6) 282-291.

Seki M, Narusaka M, Abe H, Kasuga M, Yamaguchi-Shinozaki K, Carninci P, Hayashizaki Y, Shinozaki K., (2001). Monitoring the expression pattern of 1300 *Arabidopsis* genes under drought and cold stresses by using a full-length cDNA microarray. Plant Cell 13(1): 61-72.

Shinozaki, K. and Yamaguchi-Shinozaki, K., (2000). Molecular responses to dehydration and low temperature: differences and cross-talk between two stress signaling pathways. Curr. Opin. Plant Biol. 3(3):217-223.

Sreenivasulu, N., Kavi Kishor, P.B., Varshney, R.K., Altschmied, L. (2002). Mining functional information from cereal genomes—the utility of expressed sequence tags (ESTs). Curr. Sci. 83(8): 1-9.

Sreenivasulu, N., Miranda, M., Prakash, H.S., Wobus, U., Weschke, W. (2004b). Transcriptome changes in foxtail millet genotypes at high salinity: identification and characterization of a PHGPX gene specifically upregulated by NaCl in a salt-tolerant line. J. Plant Physiol. 161(4): 467-477.

Sreenivasulu, N., Radchuk, V., Strickert, M., Miersch, O.,Weschke,W.,Wobus, U. (2006). Gene expression patterns reveal tissue-specific signaling networks controlling programmed cell death and ABA-regulated maturation in developing barley seeds. Plant J. 47(2): 310-327.

Sreenivasulu, N., Varshney, R.K., Kavi Kishor, P.B., Weschke, W. (2004a). Tolerance to abiotic stress in cereals: a functional genomics approach. In: Gupta, P.K., Varshney, R.K. (Eds.), Cereal Genomic, pp. 483-514.

Tanksley, S.D. and Nelson, J.C. (1996). Advanced backcross QTL analysis: a method for the simultaneous discovery and transfer of valuable QTLs from unadapted germplasm into elite breeding lines. Theor. Appl. Genet. 92: 191-203.

Tuberosa, R. and Salvi, S. (2006). Genomics-based approaches to improve drought tolerance of crops. Trends Plant Sci. 11(8): 405-412.

Umezawa T, Okamoto M, Kushiro T, Nambara E, Oono Y, Seki M, Kobayashi M, Koshiba T, Kamiya Y, Shinozaki K. (2006b). CYP707A3, a major ABA 8′-hydroxylase involved in dehydration and rehydration response in Arabidopsis thaliana. Plant J. 46(2): 171-182.

Umezawa, T., Fujita, M., Fujita, Y., Yamaguchi-Shinozaki, K., Shinozaki, K. (2006a). Engineering drought tolerance in plants: discovering and tailoring genes to unlock the future. Curr. Opin. Plant Biotech. 17(2): 113-122.

Valliyodan, B., Nguyen, H.T. (2006). Understanding regulatory networks and engineering for enhanced drought tolerance in plants. Curr. Opin. Plant Biotech. 9: 189-195.

Valliyodan, B. and Nguyen, H.T. (2006). Understanding regulatory networks and engineering for enhanced drought tolerance in plants. Curr. Opin. Plant Biotech. 9(2): 189-195.

Wanga, Z.l., Lia, P.H., Fredricksen, M., Gong, Z., Kim, C.S, Zhangd, C., Bohnertb, H.J., Zhud, J.K., Bressane, R.A., Hasegawae, P.M., Zhaoa, Y.X., Zhang H. (2004b). Expressed sequence tags from *Thellungiella halophila*, a new model to study plant salt-tolerance. Plant Sci. 166(3): 609-616.

Xiao, B.Z., Chen, X., Xiang, C.B., Tang, N., Zhang, Q.F., Xiong, L.Z. (2009). Evaluation of seven functionknown candidate genes for their effects on improving drought resistance of transgenic rice under field conditions. Mol Plant. 2(1): 73-83.

Xiong, L.M., Ishitani, M., Lee, H., Zhu, J.K. (2001a). The Arabidopsis LOS5/ ABA3 locus encodes a molybdenum cofactor sulfurase and modulates cold stress- and osmotic stress-responsive gene expression. Plant Cell 13(9): 2063-2083.

Xiong, L.M., Ishitani, M., Zhu, J.K. (1999). Interaction of osmotic stress, temperature, and abscisic acid in the regulation of gene expression in Arabidopsis. Plant Physiol. 119: 205-211.

Xiong, L.M., Lee, B.H., Ishitani, M., Lee, H., Zhang, C.Q., Zhu, J.K. (2001b). FIERY1 encoding an inositol polyphosphate 1-phosphatase is a negativeregulator of abscisic acid and stress signaling in Arabidopsis. Genes Dev. 15(15): 1971-1984.

Xu, K., Xu, X., Fukao, T., Canlas, P., Maghirang-Rodriguez, R., Heuer, S., Ismail, A.M., Bailey-Serres, J., Ronald, P.C., Mackill, D.J. (2006). Sub1A is an ethylene-response-factor-like gene that confers submergence tolerance to rice. Nature 442: 705-708.

Yoshida, R., Hobo, T., Ichimura, K., Mizoguchi, T., Takahashi, F., Aronso, J.,Ecker, J.R., Shinozaki, K., (2002). ABA-activated SnRK2 protein kinase is required for dehydration stress signaling in Arabidopsis. Plant Cell Physiol. 43(12): 1473-1483.

Yoshida, T., Nishimura, N., Asami, T., Shinozaki, K., Hirayama, T. (2005). Characterization of Arabidopsis ABA hypersensitive mutant, ahg3. Plant Cell Physiol. 46 (S51-S51, Suppl. S.).

Zhu, J.K.(2001). Cell signaling under salt, water and cold stresses. Curr. Opin. Plant Biol. 4(5): 401-406.

Zhu, J.K. (2002). Salt and drought stress signal transduction in plants. Ann. Rev. Plant Biol. 53: 247-273.

Zhu, J.K., Hasegawa, P.M., Bressan, R.A. (1997). Molecular aspects of osmotic stress in plants. Crit. Rev. Plant Sci. 16(3): 253-277.

Zhu, J.K., Liu, J.P., Xiong, L.M. (1998). Genetic analysis of salt tolerance in Arabidopsis: evidence for a critical role of potassium nutrition. Plant Cell 10(7): 1181-1191.

Index